MW00674908

Self-Organization and Green Applications in Cognitive Radio Networks

Anwer Al-Dulaimi
Brunel University, UK

John Cosmas
Brunel University, UK

Abbas Mohammed
Blekinge Institute of Technology, Sweden

Managing Director: Lindsay Johnston
Editorial Director: Joel Gamon
Book Production Manager: Jennifer Yoder
Publishing Systems Analyst: Adrienne Freeland
Development Editor: Monica Speca
Assistant Acquisitions Editor: Kayla Wolfe
Typesetter: Alyson Zerbe
Cover Design: Jason Mull

Published in the United States of America by
Information Science Reference (an imprint of IGI Global)
701 E. Chocolate Avenue
Hershey PA 17033
Tel: 717-533-8845
Fax: 717-533-8661
E-mail: cust@igi-global.com
Web site: http://www.igi-global.com

Library of Congress Cataloging-in-Publication Data

Self-organization and green applications in cognitive radio networks / Anwer Al-Dulaimi, John Cosmas, and Abbas Mohammed, editors.
p. cm.
Includes bibliographical references and index.
Summary: "This book provides recent research on the developments of efficient cognitive network topology, offering the most current procedures and results to demonstrate how developments in this area can reduce complications, confusion, and even costs"--Provided by publisher.
ISBN 978-1-4666-2812-0 (hardcover) -- ISBN 978-1-4666-2813-7 (ebook) -- ISBN 978-1-4666-2814-4 (print & perpetual access) 1. Cognitive radio networks. I. Al-Dulaimi, Anwer, II. Cosmas, John, 1956- III. Mohammed, Abbas.
TK5103.4815.S45 2013
621.39'8--dc23

2012032526

British Cataloguing in Publication Data
A Cataloguing in Publication record for this book is available from the British Library.

Editorial Advisory Board

Table of Contents

Detailed Table of Contents

This chapter studies the application of femtocells as part of the future cognitive 4G networks. It starts with a demonstration for the evolution of cellular and wireless networks. The developing technology that leads towards a converged LTE-Femtocell wireless environment is described in detail. The chapter presents the key challenges of deploying cognitive femtocell in the macrocell networks. As spectrum utilisation management is the main concern in the future network, the main models for spectrum allocation used to provide enough bandwidth to the femtocell in coexistence with the LTE systems are incorporated for further investigation. In addition, the Quality of Service (QoS) provisioning and the main approaches for measuring end user performance are given as function small range transmission domains. The requirement of an effective mobility management solution in such systems is analysed for future development. The chapter is concluded with a summary.

Cognitive radio has been proposed to have spectrum agility (or opportunistic spectrum access). In this chapter, the authors introduce the extended network architecture of cognitive radio network, which accesses not only spectrum resource but also wireless stations (networking nodes) and high-level application data opportunistically: the large-scale cognitive wireless networks. The developed network architecture is based upon a re-definition of wireless linkage: as functional abstraction of proximity communications among wireless stations. The operation spectrum and participating stations of such abstract wireless links are opportunistically decided based on their instantaneous availability. It is able to maximize wireless network resource utilization and achieve much higher performance in large-scale wireless networks, where the networking environment can change fast (usually in millisecond level) in terms of spectrum and wireless station availability. The authors further introduce opportunistic routing and opportunistic data aggregation under the developed network architecture, which results in an implementation of cognitive unicast and cognitive data-aggregation wireless-link modules. In both works, it is shown that network performance and energy efficiency can improve with network scale (such as including station density). The applications of large-scale cognitive wireless networks are further discussed in new (and smart) beyond-3G wireless infrastructures, including for example real-time wireless sensor networks, indoor/underground wireless tracking networks, broadband wireless networks, smart grid and utility networks, smart vehicular networks, and emergency networks. In all such applications, the cognitive wireless networks can provide the most cost-effective wireless bandwidth and the best energy efficiency.

In this chapter the authors propose a new approach for optimizing the sensing time and periodic sensing interval for energy detectors in cognitive radio networks. The optimization of the sensing time depends on maximizing the summation of the probability of right detection and transmission efficiency, while the optimization of periodic sensing interval is subject to maximizing the summation of transmission efficiency and captured opportunities. Since the optimum sensing time and periodic sensing interval are dependent on each other, an iterative approach to optimize them simultaneously is proposed and a convergence criterion is devised. In addition, the probability of detection, probability of false alarm, probability of right detection, transmission efficiency, and captured opportunities are taken as performance metrics for the detector and evaluated for various values of channel utilization factors and signal-to-noise ratios.

Distributing Radio Resource Management (RRM) in heterogeneous wireless networks is an important research and development axis that aims at reducing network complexity, signaling, and processing load in heterogeneous environments. Performing decision-making involves incorporating cognitive capabilities into the mobiles such as sensing the environment and learning capabilities. This falls within the larger framework of cognitive radio (Mitola, 2000) and self-organizing networks (3GPP, 2008). In this context, RRM decision making can be delegated to mobiles by incorporating cognitive capabilities into mobile handsets, resulting in the reduction of signaling and processing burden. This may however result in inefficiencies such as those known as the "Tragedy of commons" (Hardin, 1968) that are inherent to equilibria in non-cooperative games. Due to the concern for efficiency, centralized network architectures and protocols keep being considered and being compared to decentralized ones. From the point of view of the network architecture, this implies the co-existence of network-centric and terminal-centric RRM schemes. Instead of taking part within the debate among the supporters of each solution, the authors propose a hybrid scheme where the wireless users are assisted in their decisions by the network that broadcasts aggregated load information (Elayoubi, 2010). At some system's states, the network manager may impose his decisions on the network users. In other states, the mobiles may take autonomous actions

mon benefits. The primary terrestrial system is defined to have exclusive rights to access the spectrum, which is shared by the secondary HAP system upon request. The Markov chain is presented to model two spectrum-sharing scenarios and evaluate the performance of spectrum sharing between primary terrestrial and secondary HAP systems. Simulation results show that to reserve an amount of spectrum from a primary system could encourage spectrum sharing with a secondary system, which has a frequent demand on requesting spectrum resources.

Foreword

Cognitive Radio is considered as one of the main tools in combating the problem of spectrum availability in future mobile communications systems. It utilizes innovative technologies that increase the efficiency of spectrum usage, and therefore, solve the problem of bandwidth requirement for highly demanding applications. The shift towards digital TV and the latest developments in adaptive wireless technologies make it possible to fundamentally change the way radio spectrum is allocated and used. This is an outcome of a wide range of developments in the field of cognitive radio networks, including spectrum sensing, dynamic access of the possible transmission opportunities, increasing knowledge about the surrounding wireless environment through cognition systems, and managing network operations and charging systems. The momentum for application has also gained a further step forward with the latest LTE air interfaces and the application of different transmission power levels for the future green network. The innovations have not stopped the challenging changes in network design and the integration of femtocell stations into wider network deployments that employ cognitive communications.

The authors of *Self-Organization and Green Applications in Cognitive Radio Networks* have contributed a milestone to the evolution of cognitive radio networks with the publication of this book. Specifically, this book has marked the latest findings in emerging research topics. The participation of many experts from the mobile industry has marked the importance of this text with solid contributions towards the transition to a new era of applications in the field of wireless networks.

The assembly of state-of-the art contributions in this book gives a rich set of ideas that will potentially influence the development of efficient cognitive networks. However, the dynamic spectrum assignment has been achieved with flexible mobile terminals that can be reconfigured at all layers of the protocol stack. The key objective of this book is to deliver novel solutions to many issues that concern the deployment of cognitive radio communication networks. The book introduces the reader to new transparent concepts that involve designing/implementing cognitive radio systems and networks. Then, the book highlights how those concepts fit in real-world problems and applications. Hence, this book is an important reference for both professionals in industry and researchers in academia.

Besides addressing cognitive networks, the book also focuses on efficient resources allocation through the analysis of algorithms. These developments (at different layers) and newly proposed approaches consider cross-layer design and interference avoidance. The research scope of this book has been strengthened further through the investigation of the latest standardizations in self-organizing cognitive radio networks. This book expands the scope of the power issues and other emerging cognitive user technologies, such as high altitude platform-supported cognitive radio systems. To summarize, this textbook stands at the forefront of current literature in the field of cognitive radio networks.

Abbas Jamalipour
University of Sydney, Australia

Abbas Jamalipour *received the Ph.D. degree from Nagoya University, Nagoya, Japan. He is currently the Chair Professor of Ubiquitous Mobile Networking with the School of Electrical and Information Engineering, University of Sydney, Sydney, NSW, Australia. He is a Fellow of the Institute of Electrical and Electronics Engineers (IEEE), a Fellow of the Institute of Electrical, Information, and Communication Engineers (IEICE), and a Fellow of the Institution of Engineers Australia, an IEEE Distinguished Lecturer, and a Technical Editor of several scholarly journals. He has been an Organizer or the Chair of several international conferences, including the IEEE International Conference on Communications and the IEEE Global Communications Conference, and was the General Chair of the 2010 IEEE Wireless Communications and Networking Conference. He is the Vice President – Conferences and a Member of the Board of Governors of the IEEE Communications Society (ComSoc). He is the recipient of several prestigious awards, including the 2010 IEEE ComSoc Harold Sobol Award for Exemplary Service to Meetings and Conferences, the 2006 IEEE ComSoc Distinguished Contribution to Satellite Communications Award, and the 2006 IEEE ComSoc Best Tutorial Paper Award.*

Preface

Cognitive Radio (CR) is emerging as the dominant solution for the scarce spectrum in the near future. The cyclic recognition cycle of sensing, learning, and adaptation in the CR enables it to jump between channels according to temporal and spatial opportunities. This rapid behavior needs to be further analyzed in terms of accessibility for the information providers and tracking spectrum users. To illustrate this, CR transmission speed/reliability may face great obstacles due to the spectrum dynamic access models. Therefore, a cognitive base station may fail to transmit at irregular times. This will urge the CR network management entities to look for alternative routes to maintain online services. In addition, wireless broadcasting faces a drop in the delivery of high data rate requests, as there is no guarantee for the free channel availability in secondary systems. Here, designed solutions are required to secure high bandwidth and low power consumption for promising cognitive networking features. Low attenuation loss in short range transmissions, such as metropolitans (which are the ambitious environments for the future CR networks), enhance the emergence of smaller transmission domains such as cognitive femtocell networks. Different challenges are expected for such applications, starting from developed communication layers to improved corresponding technologies. Therefore, a fully steered "service delivery" network, which will revolutionize the traditional way of communications exchange, will soon be a reality.

This book is an advanced research work on cognitive networks that combines the network architectural design with the necessary technical modifications to deliver high quality services. It provides a state-of-the-art guide to cognitive networks and raises many discussions and solutions for the ongoing research challenges. The book covers all of the important aspects of cognitive networks, including chapters on concepts and fundamentals for beginners, advanced topics, and research-oriented chapters.

For the industry practitioner, this book is an instruction manual for immediate solutions that incorporate new concepts and techniques in this emerging area. There are a wide range of topics in this book that provide the readers with a clear vision of what they need to consider, while developing the technology and putting forward the challenges of standardizing and interfacing users. There are new approaches that employ practical solutions for realistic challenges and market demands that will be a source for designing/implementing cognitive radio systems and networks. The contributions of this book to the industry were certified by our editorial advisory board, which contains many specialists and experts from world-known mobile operators.

For academic researchers, there has been huge inertest in cognitive radio communications in the last few years. Although there are now a considerable number of publications in this field, there are still many open research problems remaining to be explored. This edited book presents the work of many internationally leading teams and experts with the latest solutions and algorithms. The book explores a wide range of challenges for cognitive radio networks, such as architecture, modeling, and coexistence.

This is not just a publication in the field of academia, but it is a step toward a fully applicable systematic modeling of what the cognitive network will be and what it needs for final deployment. This makes this book an interesting textbook for the courses offered by universities in cognitive radio networks and 4G wireless systems engineering.

The chapters in this book are arranged in an order that gradually builds upon the knowledge and information gained from previous chapters. However, we also put together similar topics and working aspects to provide the different solutions of certain layers together. This sequential management of chapters enables new learners in this field to build up their knowledge step-by-step, either in formal classrooms or via self-learning. The book is also an important source of knowledge for communication planning engineers seeking to develop their competence in new technologies, as it provides a vision of what the next generation networks may look like in the near future.

Anwer Al-Dulaimi
Brunel University, UK

John Cosmas
Brunel University, UK

Abbas Mohammed
Blekinge Institute of Technology, Sweden

Chapter 1
Technical Challenges in 4G Cognitive Femtocell Systems

Saba Al-Rubaye
Brunel University, UK

John Cosmas
Brunel University, UK

ABSTRACT

This chapter studies the application of femtocells as part of the future cognitive 4G networks. It starts with a demonstration for the evolution of cellular and wireless networks. The developing technology that leads towards a converged LTE-Femtocell wireless environment is described in detail. The chapter presents the key challenges of deploying cognitive femtocell in the macrocell networks. As spectrum utilisation management is the main concern in the future network, the main models for spectrum allocation used to provide enough bandwidth to the femtocell in coexistence with the LTE systems are incorporated for further investigation. In addition, the Quality of Service (QoS) provisioning and the main approaches for measuring end user performance are given as function small range transmission domains. The requirement of an effective mobility management solution in such systems is analysed for future development. The chapter is concluded with a summary.

INTRODUCTION

The evolving fourth generation (4G) mobile communication systems are expected to solve still-remaining problems of third generation (3G) systems and to provide a wide variety of new services, from high-quality voice to high-definition video to high-data-rate wireless channels. The term 4G is used broadly to include several types of broadband wireless access communication systems, not only cellular telephone systems. 4G is intended to provide high speed, high capacity, low cost per bit, IP based services. 4G is all about an integrated, global network that's based on an open system approach (Govil, 2007). With the successful deployment of 3G cellular networks worldwide, the attention of the cellular industry is now focused on the beyond-3G evolution of the wireless cellular network. The upcoming 4G mobile communications system is foreseeing

DOI: 10.4018/978-1-4666-2812-0.ch001

potentially a smooth merger of these technologies with a goal to support cost effective seamless communication at high data rate supported with global roaming and user customized personal services. The evolution of the cellular systems has come through three generations over the past 40 years (Raj & Gagneja, 2012).

There are three different potential approaches in developing the future 4G networks. Firstly is to increase the cell capacity using new techniques such as Long Term Evolution (LTE), which is replacing the Microwave Access (WiMAX) backbone stations. Secondly, is to improve the spectral efficiencies using reconfigurable technologies such as the Cognitive Radio (CR) and advanced antenna systems. The third approach is to develop new architectures for mobile networks that help to achieve an autonomous communications. A combination of these technologies and arrangements, if not all three, will lead to the new generation of efficient 4G networks that can be deployed to deal with huge traffic requirements and various corresponding technologies. Figure 1 illustrates the various given solutions for future standardization of 4G systems.

The future networks employ an IP-based environment for all traffic requests counting voice, video, broadcasting media and Internet that can access landline and wireless networks. This is integrated in the future 4G solutions and its applications. Together with intelligent terminals, 4G can provide easy access to the broadband services and understanding of the personal downloading profiles. This allows uninterrupted coverage for a user that change terminals or switch unnoticeably between the underlying fixed and mobile networks (UMTS, WLAN, etc.). This is very important for ad-hoc networking and for a mobile user that travels among different terminals of a single network or with the terminals of third parties. In short, a 4G network provide its individual users with full control over privacy and costs. This is a natural extension of the current technologies of broadband Internet and 3G mobile networks like UMTS.

This chapter addresses the key technical challenges encountering suitable femtocell system deployment management. There are so many requirements for keeping femtocell costs as low as possible for effectively competing against the ubiquitous Wi-Fi technology. The main challenge for the femtocell deployments is the interference with the macrocell base station. Other femtocell functions are addressed, including resource management, spectrum allocation management, providing QoS over an Internet backhaul and allowing access to femtocells (Chandrasekhar, Andrews, & Gatherer, 2008). Handover and mobility also very important aspect in femtocell networks, as there are different types in femtocell handover from/to macrocell. Furthermore, power consumption is very important consideration to be taken into account in the next generation wireless networks.

Figure 1. Future 4G cognitive systems with different corresponding technologies

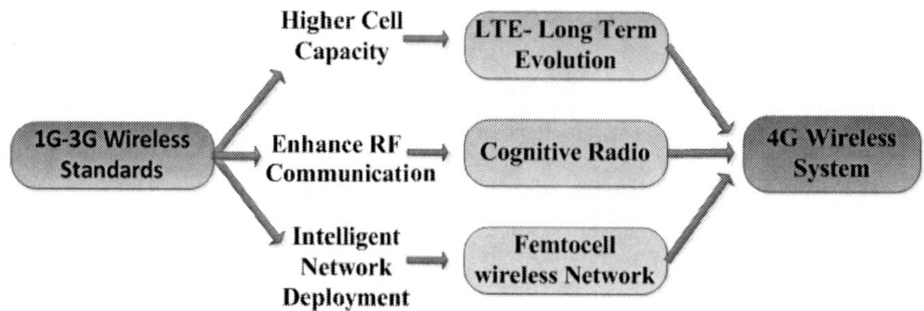

CURRENT STANDARDIZATION

Just as for any other new technology, industry standardization is a very important factor from both the market acceptance and economy of scale (i.e., ecosystem) perspectives. Femtocells are no exception (Knisely, Yoshizawa, & Favichia, 2009). In April 2009, the 3rd Generation Partnership Project (3GPP), in collaboration with the Femto-Forum and the Broadband Forum released the world's first femtocell standard. This standard covers different aspects of femtocell deployments including the network architecture, radio interference, femtocell management, provisioning and security. Concurrently, a number of operators-AT&T (USA), O2 and Motorola (Europe), and Softbank (Japan)—have started to conduct femtocell trials prior to any commercial deployments (Chandrasekhar, 2009).

The 3GPP is the primary international standards organization dealing with mobile networks using the GSM/UMTS and LTE family of air interface technologies. Therefore, 3GPP has been linking the standards of UMTS femtocells. In 3GPP, most mobile operators and vendors preferred the legacy architecture with a direct interface to the circuit and packet core network based on the Iu interface (including Iufcs for circuit services and Iu_ps for packet data services). Early in 2008, the UMTS femtocell standardization topic gained tremendous interest from many mobile operators, and efforts to find a way to have the capability incorporated into UMTS Release 8 began with great intensity. Areas of 3GPP femtocell standards included the following. Interface between the femtocell (Home Node B – HNB) and the femto network gateway (HNB Gateway, HNB-GW) Security protocols to authenticate femtocell (HNBs) and secure communications across the un-trusted Internet Management protocols for "touch free" Operations, Administration, and Management (OA&M) of femtocells (HNB devices) (Knisely, 2010).

Although the standardisation efforts are still ongoing process, the current applicable technologies are the inspiring guide to develop any models for the future.

LONG TERM EVOLUTION (LTE) TECHNOLOGY

LTE is intended to provide high data rate, low latency and packet optimized radio access technology supporting flexible bandwidth deployment and has considered a fourth generation technology. It was structured to meet the essential requirement for increased capacity and improved performance. The main differences to Third Generation 3G systems are a packet data optimized, cost more effective, IP architecture, and spectrally efficient air interface. Since there is no circuit switched domain in LTE, voice connectivity is based on Voice over IP (VoIP) on top of packet switched IP-protocol. LTE supports a wide range of bandwidth from 1.25 MHz to 20 MHz. The 20 MHz bandwidth gives peak data rate of 326 Mbps using 4x4 Multiple Input Multiple Output (MIMO). For uplink, MIMO currently is not realized, so the uplink data rate is limited to 86 Mbps (Xia, Chandrasekhar& Andrews, 2010).

LTE offers significant improvements over previous technologies such as Universal Mobile Telecommunications System (UMTS) and High-Speed Packet Access (HSPA) by introducing a novel physical layer and reforming the core network. The main reasons for these changes in the Radio Access Network (RAN) system design are the need to provide higher spectral efficiency, lower delay, and more multi-user flexibility than the currently deployed networks. In the development and standardization of LTE, as well as the implementation process of equipment manufacturers, simulations are necessary to test and optimize algorithms and procedures. This has to be performed on both, the physical layer (link-level) and in the network (system-level) context (Das & Ramaswamy, 2009).

LTE-OFDM TECHNIQUE

In LTE and LTE-Advanced networks, Orthogonal Frequency-Division Multiplexing (OFDM) has been chosen as the multiple access method since it can provide high data rates and spectrum efficiency. In OFDMA the whole bandwidth is divided into small subcarriers or parallel channels which is then used for transmission with reduced signalling rate. These sub-carriers are orthogonal which means that they are individual carrier and there is no correlations may happen. The subcarrier frequency is shown in the equation:

$$f_k = k\Delta f \quad (1)$$

where Δf is the subcarrier spacing.

Subcarrier is first modulated with a data symbol of either 1 or 0, the resulting OFDMA symbol is then formed by simply adding the modulated carrier signal. This OFDM symbol has larger magnitude than individual subcarrier and thus having high peak value which is the characteristics of OFDMA technique (AFRIDI, 2011). OFDM is a spectral efficient transmission structure in approach that it divides a high bit rate data stream into several parallel narrowband low-bit-rate data streams often called sub-carriers. This partition is made in the way that sub-carriers are orthogonal to each other which eliminates the need of non-overlapping sub-carriers to avoid inter carrier interference. The first carrier is selected so that its frequency contains integer number of cycles in a symbol period as shown in Figure 2.

In order to make sub-carriers orthogonal to each other, adjacent subcarriers are spaced by:

$$W_{SC} = \frac{W}{N_{SC}} \quad (2)$$

where W is the nominal bandwidth of high-bit-rate data stream, and N_{SC} refers to the number of sub-carriers. The OFDM downlink parameters are summarized in Table 1.

Another advantage of OFDM is the efficient use of the available frequency spectrum. In a conventional multi-carrier system, the frequency band is divided into non-overlapping subcarriers in order to eliminate the cross-talk between sub-carriers known as Inter-Carrier Interference (ICI). This non-overlapping design of the subcarriers leads to inefficient use of the available spectrum.

On the other hand, the OFDM offers high spectral efficiency by allowing the overlapping of the spectrum of the subcarriers. In order for this to work, the ICI between subcarriers must be mitigated. This is achieved by making the subcarriers mutually orthogonal. The orthogonality between subcarriers as shown in Figure 3 is maintained

Figure 2. Time-frequency representation of OFDM subcarrier

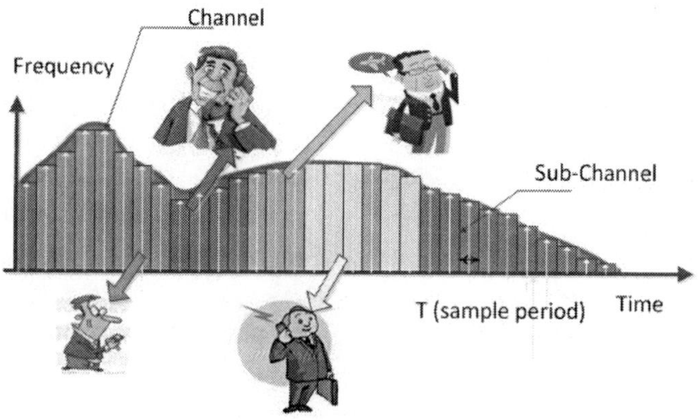

Table 1. OFDM downlink parameters

Parameter	Value					
Transmission BW	1.25 MHz	2.5 MHz	5 MHz	10 MHz	15 MHz	20 MHz
Sub-Carrier Duration T_{SC}	0.5ms					
Sub-Carrier Spacing	15kHz					

by carefully selecting the spacing between the subcarriers and the savings of bandwidth with OFDM compared to a conventional system with the same subcarrier bandwidth.

A saving of almost 50% of bandwidth can be achieved with OFDM due to the overlapping of subcarriers. The orthogonally of the subcarriers means that each subcarrier has an integral number of cycles over a symbol period consequently, there is a difference of an integral number of cycles between any two subcarriers over a symbol period. This ensures that the spectrum of each subcarrier has a null at the centre frequency of each of the other subcarriers in the system (Thanabalasingham, 2006). OFDMA is an excellent choice of multiplexing scheme for the 3GPP LTE downlink. Although it involves added complexity in terms of resource scheduling, it is vastly superior to packet-oriented approaches in terms of efficiency and latency. In OFDMA, users are allocated a specific number of subcarriers for a predetermined amount of time. These are referred to as Physical Resource Blocks (PRBs) in the LTE specifications. PRBs thus have both a time and frequency dimension. Allocation of PRBs is handled by a scheduling function at the 3GPP base station (eNodeB) (Zyren & McCoy, 2007).

LTE frames structure is 10 ms in duration as shown in Figure 4. They are divided into 10 sub-frames, each sub-frame being 1.0 ms long. Each sub-frame is further divided into two slots, each of 0.5 ms duration. Slots consist of either 6 or 7

Figure 3. Spectral efficiency of OFDM (Thanabalasingham, 2006): both systems have the same subcarrier bandwidth W_S. The conventional system uses a guard band W_g between adjacent subcarriers. On the other hand, the spectra of the OFDM subcarriers overlap, leading to higher spectral efficiency.

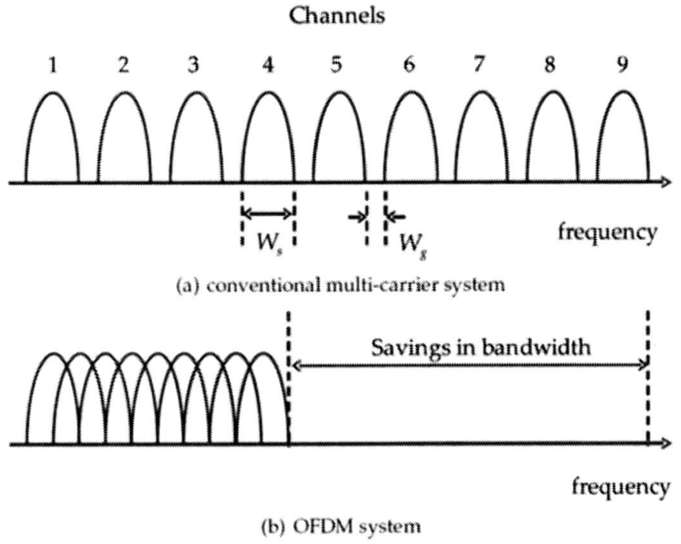

Figure 4. LTE frame structure

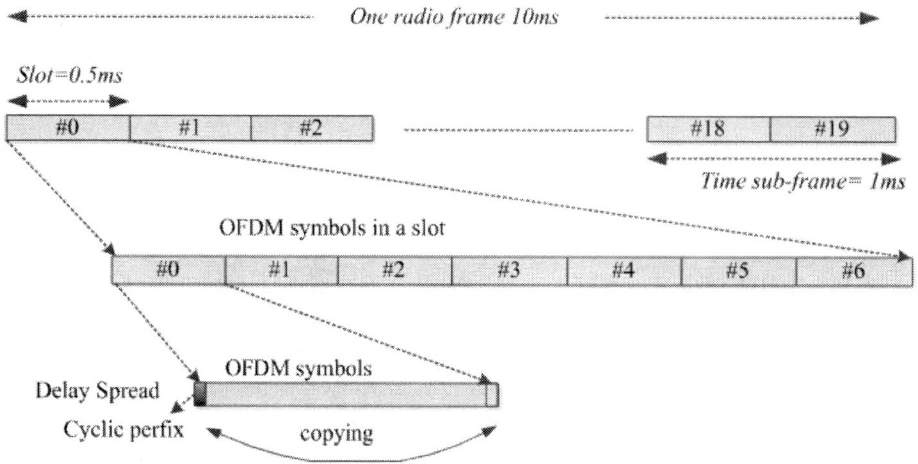

ODFM symbols, depending on whether the normal or extended cyclic prefix is employed.

The IEEE 802.11a specification offers support for a variety of modulation and coding alternatives in order to enhance OFDM system. For instance, the modification allows researchers to merge BPSK, QPSK, and 16-QAM modulations with difficult encrypting (L= 1/2 and constraint length seven) to generate data rates of 6, 12, and 24 Mbps. All other combinations of encoding rates, including L= 2/3 and L= 3/4 combined with 64- QAM, are used to produce rates up to 54 Mbps, which are possible in the new standardization as presented in Figure 5.

LTE uplink requirements differ from downlink requirements in several ways. Not surprisingly,

Figure 5. 64 QAM modulations

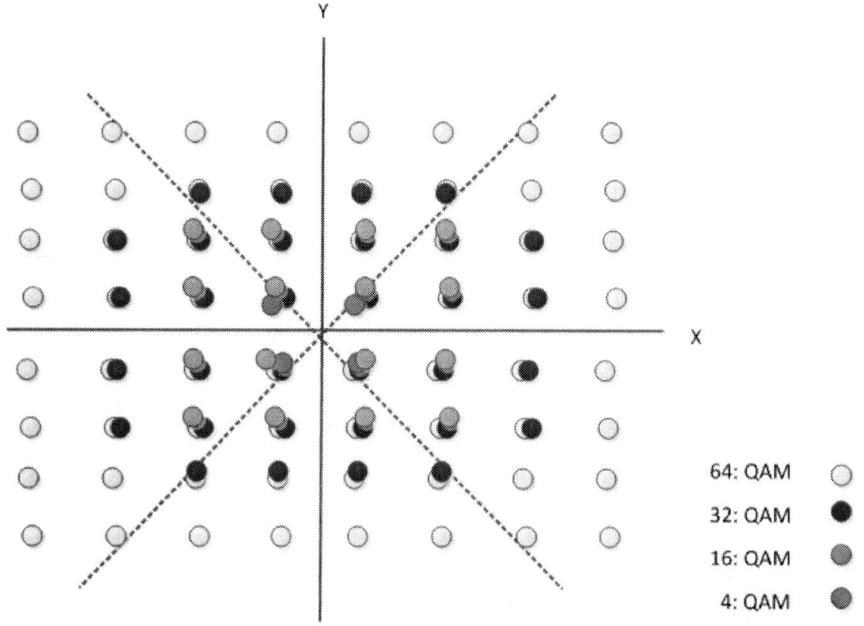

power consumption is a key consideration for UE terminals. The high PAPR and related loss of efficiency associated with OFDM signalling are major concerns. As a result, an alternative to OFDM was sought for use in the LTE uplink. Single Carrier – Frequency Domain Multiple Access (SC-FDMA) is well suited to the LTE uplink requirements. The basic transmitter and receiver architecture is very similar (nearly identical) to OFDMA, and it offers the same degree of multipath protection (Zyren & McCoy, 2007).

TRAFFIC IN 4G CELLULAR NETWORKS

The rapid growth of the Internet has placed new demands on communication networks and this has been the subject of periodic controversy. New technologies that generate large quantities of traffic include video-sharing, video conferencing, sites movie downloads, sites online gaming, sites remote medical imaging, and online storage of documents (ITU-T Report 10, 2009).

There is an increasing requirement to supply a communication medium appropriate for noisy building backgrounds, where there is a need for high data rate. Moreover, the number of mobile phone subscriptions is continuing to grow quickly, and are expected to reach 5.63 billion by 2013. Mobile phone data traffic is forecast to grow by between 10 and 30 times between 2008 and 2013, depending on the pricing and promotion of these services. In the same period, the Internet has also become a mass-market technology, growing to 1.6 billion users worldwide, nearly 25% of the world's population. Internet protocol traffic is forecast to grow by over 10 times in the period from 2006 to 2012 (Saunders, et al., 2009).

Projections from Cisco are somehow impressive with the global mobile data traffic doubling every year through 2014, increasing 39 times between 2009 and 2014, with a disproportionate trend toward video (by 2014, 2/3 of the world's mobile data traffic will be video and consumer mobile Internet traffic will account for 73% of all mobile data traffic). As a result, we are witnessing surging growth in global mobile data traffic, which is projected to rise by sixty-six times by 2013 as shown in Figure 6.

Therefore, it is necessary to investigate new technologies such as femtocells and the advan-

Figure 6. Global mobile data growth (source: Cisco)

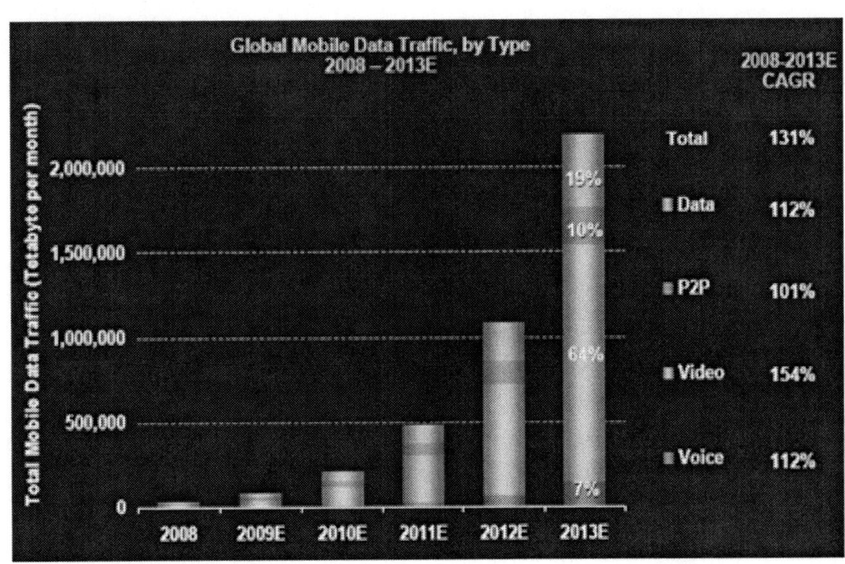

tages they can offer to overcome the challenges of huge traffic across indoor and outdoor applications.

FEMTOCELL NETWORKS

Femtocell is an economical solution to provide high speed indoor communications instead of the conventional macro-cellular networks. Especially, cognitive femtocell is considered in the next generation cellular network such as 3GPP LTE and WiMAX802.16e systems. Although the femtocell has great advantages to improve coverage for outdoor/indoor users, the interference and mobility management problems are a critical issue in the operation of femtocell networks.

Recent surveys have predicted that 50 percent of phone calls and 70 percent of data services will happen indoors. In the next generation systems, more intelligent devices will appear, and the contents of their services will require more network capacity than the services that exist today (Lopez-Perez, et al., 2009). Since it is expensive to serve indoor customers with increased service demands from macrocell base stations, new solutions for improving the indoor coverage/capacity are required. According to the femtocell working procedures, the femtocell unit adapts its coverage area to avoid harming other transmitting services in the area by serving around 4-8 customers. However, this definition does not include limits for the number of expected femtocells in a certain area. In other words, the growth of the number of Femtocell Base Stations (FBSs) may impact network performance by consuming all locally available resources without centralized mapping to the spectrum access function.

In the dynamic spectrum access function, the cognitive radio accesses the spectrum only when the primary user is off. Therefore, this is the most efficient way to utilise the scarce spectrum. With the current femtocells working schemes, the possibility of femtocells interfering with the macrocell communications becomes higher with the rapid growth of femtocells. FBSs are low-power base stations operating in the licensed spectrum that can integrate mobile and Internet technologies within the home using optical fibre connection or Digital Subscriber Line (DSL). FBSs are installed indoors in a small office or house by end users to provide exclusive access to the subscriber. For instance, instead of an operator owning all base stations in a city area, femtocells base stations could be acquired by individuals or small organizations.

Femtocell deployment network architecture is illustrated in Figure 7. Thus, a tendency can be identified for moving away from central control and fixed base stations to a more flexible distributed network approach. However, in the future there is a lot of attention paid to FBS in terms of infrastructure cost saving and enhanced user practice in covered environments. A femtocell unit generates a signal to the personal mobile equipments at home and connects this to the operator's network through the Internet. This allows improving coverage and capacity for each user within their household. Applications and services should be delivered at the appropriate security levels and finally, several "green" aspects like necessary power or energy levels needed, activation or deactivation of power consuming devices etc., should be considered in order to achieve future environment friendly networks and infrastructures.

THE DEMAND FOR SMALL-CELL BASE STATIONS

The two major limitations for wireless communication are range and capacity. If a service provider wants to improve coverage, either they install a macrocell and provide high power, or they can use a smaller base station that provides coverage only up to a few hundred meters and provide high data rates at lower power. Even from an economics point of view, femtocells have been

Figure 7. Femtocell deployments in 4G wireless systems

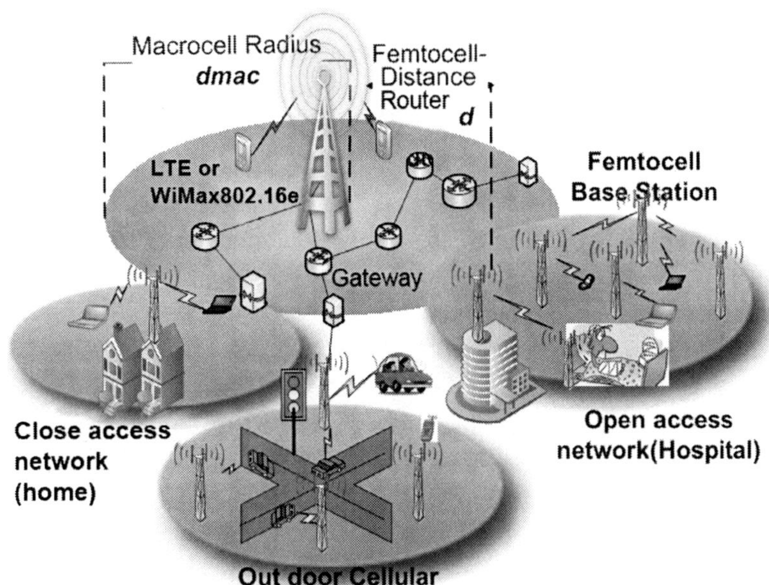

found to provide a low cost solution compared to installing higher power macrocells to provide the same quality of service. The goal of femtocells is to provide reliable communication using existing broadband Internet connection.

There are three main reasons why we need femtocells in the current cellular services:

1. **Coverage:** As macrocells cannot deliver good indoor coverage, femtocells can provide strong signal strength indoors. The coverage of femtocells is limited to a few hundred meters or less, which is within the setting of a home or an office. As the distance between the transmitter and the user is now smaller, it leads to higher received signal strength.
2. **Capacity:** Since the coverage of femtocells is small as compared to that of the macro base stations. Therefore, there is less number of users using the same spectral resources which saves more resources for other users. In other words, the user almost gets a private bandwidth from a shared bandwidth.
3. **Power:** The increasing number of mobile users leads to a huge load on the macrocellular base stations. The deployment of femtocells will divert many users from macrocell to the femtocell stations in order to reduce the load on macrocells. In addition, a balance can be maintained with the end users effectively using shared macrocells-to-femtocells infrastructure either outdoors or at indoors.

FEMTOCELL DEPLOYMENT ASPECTS

Fourth Generation broadband wireless mobile networks, specifically Long Term Evolution (LTE) and beyond are considering issues related to the deployment of femtocells (Das, et al., 2011). The feasibility of femtocells deployment in cellular network is studied by the standard development organizations like 3rd Generation Partnership Project (3GPP) (3GPP Tech. Report, 2010). Deployments of femtocells are proposed as a solution in

Long Term Evolution (LTE) networks as detailed in (Knisely, Yoshizawa, & Favichia, 2009). The deployment of femtocell, low power base stations, together with predictable sites is often believed to greatly lower energy consumption of cellular radio networks since the communication distance is decreased, less power per bit is needed and available spectrum resources are shared between fewer users.

The impact of access control for femtocells is particularly substantial in the integrate macrocell-femtocell deployment scenario. For example, when the macrocell and femtocell networks share the same carrier, this heterogeneous deployment scenario is expected to be the standard rather than the exception, since macrocell network operators possibly will not have any other choice to obtain as much user volume as they can from their costly spectrum by deploying co-channel femtocell systems.

In applied deployment of femtocell systems, the location of FBSs in a random and uncoordinated fashion is unavoidable and may generate high interference scenarios and dead spots particularly in an indoor environment (Sung, Haas, & McLaughlin, 2010). For dedicated channel assignments, femtocells are assigned a separate spectrum of bandwidth W_{fem}. Even though this mostly eliminates potential interference from the macrocell, frequency resources are not efficiently utilized. The capacity of number of mobile user N_{MS} with dedicated channel assignment can be written as:

$$C^{ch}_{N,i} = \frac{W_{mac} - W_{fem}}{N} \log\left(1 + \frac{NP_{m,i}}{\left(W_{mac} - W_{fem}\right)N_O}\right) \quad (3)$$

where ch refers to a dedicated channel deployment scenario, i is the index of N_{MS}, W_{fem} denotes the bandwidth dedicated to the femtocell networks, W_{mac} denoted the macro bandwidth, P_m is the receiving signal power from macrocell base station and N_O denotes the noise power.

From the above formula it is observed that with the introduction of femtocells there is less available spectrum for the macrocell network. Hence, in general, the capacity of macrocell users may improve for smaller values of W_{fem} and N (Mahmoud & G¨uven, 2010).

ACCESS MECHANISMS

Femtocell systems are expected to be deployed using one of the following three methods (Al-Rubaye, Al-Dulaimi, & Cosmas, 2011).

Close Access Method

A femtocell can be deployed in close access network areas, such as homes or offices, which mean that the FBS is providing services to fewer clients, and only registered mobiles can have access to such a femtocell and the macro-user has no access to the FBS as illustrated by the cross line in Figure 8.

From the description of closed access, only users who are listed in the allowed access list of the FBS are granted access to the base station. The main reason for this type of deployment is to guarantee user knowledge when they are within coverage of the FBS they learned. However, one critical issue arises with the deployment of this method when an unsigned user enters the femtocell coverage and that user is not on the allowed access list.

Generally, the visiting macro-user to the femtocell coverage area will still attempt to access with the FBS due to the fact that close FBS pilot power signal is usually much higher than macrocell BS pilot signal within the FBS coverage. Nevertheless, this effort will not be successful due to the visiting macro-users is not on the allowed access list of the FBS. This method has the advantage of decreasing the number of handovers in this par-

Figure 8. Femtocell close access method

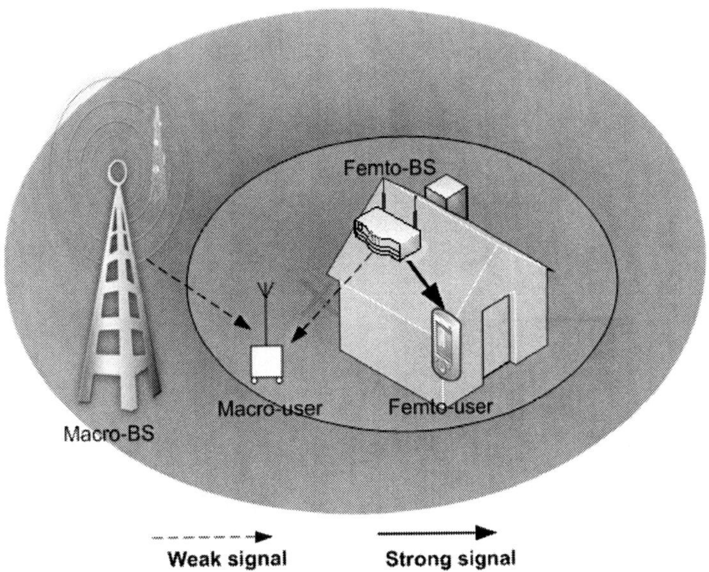

ticular network and that each user can get a high data rate for being close to the femtocell station because of the limited number of users.

Open Access Method

In this case, femto base stations are deployed in open areas like airports or hospitals. Any user has the right to access these FBSs and there is no need for registration in this case. Complete open access is when any user, as long as it is in the coverage of the femtocell is allowed to use the femtocell, if of course there is capacity available on it. While appropriate for an operator-owned femtocell deployment, a completely open access scheme may prevent, in the residential femtocell case, an entitled user to connect to its FBS due to capacity used up by macrocell users. These two extremes, and any scheme in between that allows partial open access, have different ramifications in terms of user experience, user capacity, and ultimately user churn and operator revenue (Das & Ramaswamy, 2009). The scenario for femtocell deployed as open access bases station is shown in Figure 9.

Apparently, open access is useful to network operators as an inexpensive way to expand their network capacities by leveraging third-party backhaul for free. Open access also reduces macro-to-femto interference by letting strong interferers simply use the femtocell and coordinate with the existing users through it. However, unwanted handover may be increased for many users entering and leaving such femtocells, causing a noticeable decline in the quality of service (Xia, Chandrasekhar, & Andrews, 2010).

Hybrid Access Method

The approach of open and closed access methods for femtocells applications are likely to be combined together in some cases. In such a model, the unregistered subscribers are allowed to access the femtocell base station, but only for limited usage of resources. A limited amount of the femtocell resources are available to all users as shown in Figure 10, while the rest are operated in a closed subscriber's group manner. When the access method blocks the use of femtocell resources to a subset of the users within its coverage area, a

Figure 9. Femtocell open access method

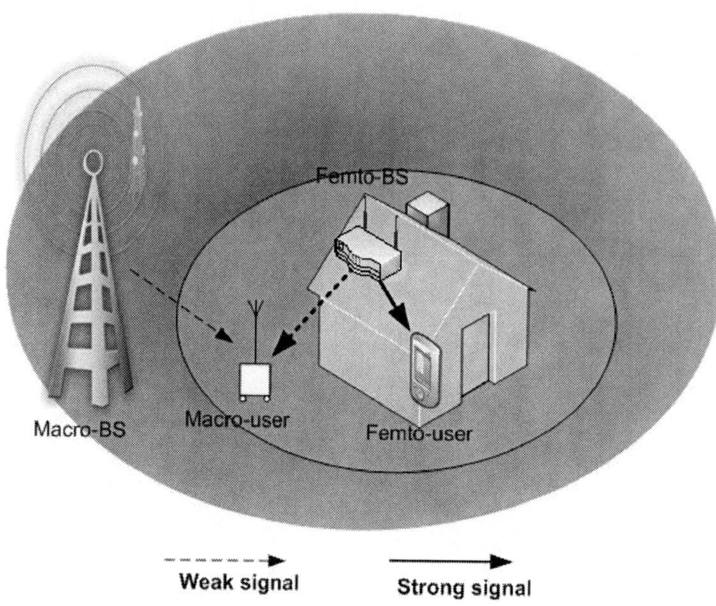

Figure 10. Femtocell hybrid access method

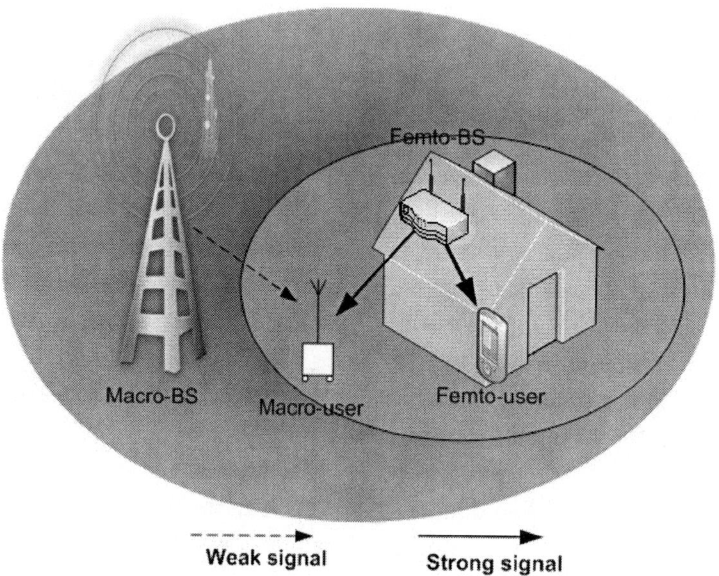

new set of interfering signals is implicitly defined in such area. Hence, the deployment of closed subscriber's group femtocells makes the problem of interference mitigation even more complex.

Hybrid access method reach a compromise between the impact on the performance of subscribers and the level of access granted to nonsubscribers. Therefore, the sharing of femtocell resources between subscribers and nonsubscribers needs to be finely tuned. Otherwise, subscribers might feel that they are paying for a service that is to be exploited by others. The impact on subscribers must thus be minimized in terms of performance or via economic advantages such as reduced costs (De la Roche, et al., 2010).

The hybrid access method in OFDMA femtocell networks consists of managing the sharing of the OFDMA resources (frequency and time) between subscribers and nonsubscribers. Therefore, these resources have to be defined. In OFDMA systems, sub-channels contain a series of subcarriers, which can be adjacent or pseudo randomly distributed across the spectrum in order to exploit either multi-user or frequency diversity and the choice of either one or the other channelization mode depends on the accuracy of the channel state information (FP7-ICT-248891, 2010).

TECHNICAL CHALLENGES

These sections indicate the key technical challenges facing functional deployments of femtocell network systems.

Spectrum Allocation Efficiency

Channel management in the femtocell networks is one of the main features that represent a radical departure from the existing mobile networks. The spectrum technology deployment is growing faster than the determination in spectrum availability. Therefore, as frequency bands or free white holes become more scarce and unavailable for the next generation wireless technologies then a wide range of services will require radio spectrum as a resource to operate properly. That leads to the important investigation of the spectrum efficient utilization.

The average cell spectral efficiency is defined:

$$S_{eff} = \frac{TP_{th}}{W_{eff}} \tag{4}$$

where TP_{th} is the average cell throughput, W_{eff} is the effective channel bandwidth.

The effective channel bandwidth is defined:

$$W_{eff} = W * R \tag{5}$$

where W is the used channel bandwidth and R is time ratio of the link.

Example: In case of downlink, R=(32/48), soW_{eff}= 10MHz*(32/48)= 6.67MHz. However, the dedicated channel deployment may not be applicable when femtocells are densely deployed. As a consequence, it may be practical to make macrocell and femtocells share the available spectrum together. Co-channel operation of macrocell and femtocells was considered by dynamically adjusting the transmit power (Oh, Lee, & Lee, 2010).

Assigning orthogonal spectrum resources between the central macrocell and femtocell BSs eliminates cross-tier interference. The orthogonal access spectrum allocation strategy is illustrated in Figure 11 (Chandrasekhar & Andrews, 2008).

To avoid persistent collisions with neighbouring femtocell in accessing the spectrum, each femtocell can access a random subset of the candidate frequency sub-channels, where each subchannel is accessed with equal probability. Mobile operators can deploy femtocell base stations that operate on a cross channel or a co-channel basis with existing macrocell. While operating on a dedicated channel is a pragmatic strategy, co-channel operating also has benefits due to the potentially increased spectral efficiency using spatial frequency re-use. On the other hand, co-channel operating may cause more interference between BSs.

Shared spectrum operation may be desirable to operators because of the scarce availability of spectrum and flexibility during deployment (Ho & Claussen, 2007) with shared spectrum between tiers. However, radio interference between cellular users to femtocell hotspots, and between

Figure 11. Spectrum sub-channels division

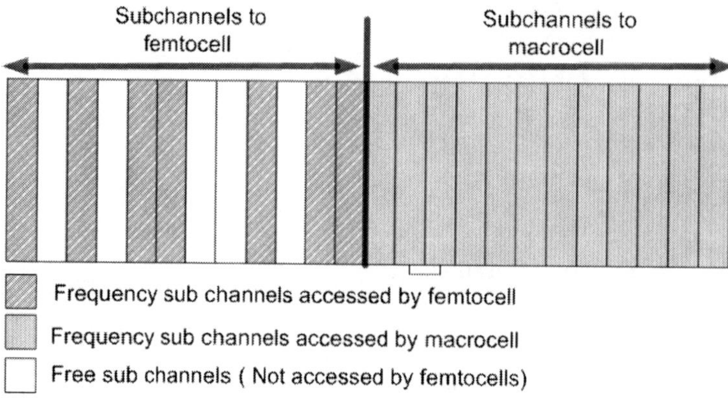

hotspot users to the macrocell BS, is likely to be the capacity-limiting factor.

Providing QoS over an Internet Backhaul

Allocated and distributed nature of FBSs raises many technical challenges for the increasing requirements of multimedia applications, and this requires more management over the systems accessible resources. Femtocells can interface with wireless communication system without the any requirements for installing additional equipments in wireless communication networks because of using a Radio Network Controller (RNC). Nevertheless, a femtocell use voice communication may still have a problem in data communication. There are also two methods to connection with mobile communication IP network, Unlicensed Mobile Access (UMA) and Internet Protocol Multimedia Subsystem (IMS) systems. Mobile Internet Protocol Television (IPTV) technology for mobile telecommunications has developed which is a representative service of broadcasting and telecommunications convergence. ′Mobile IPTV′ will be a killer service of International Mobile Telecommunications-Advanced (IMT-advanced) system. Therefore, the work should be performed towards developing core technologies for cellular network based ′Mobile IPTV′ and personal IP wireless broadcasting systems (see Figure 12).

IP backhaul needs QoS for delay-sensitive traffic and providing service parity with macrocells. Additionally, it should provide sufficient capacity to avoid creating a traffic bottleneck. While existing macrocell networks provide latency guarantees within 15 ms, current backhaul networks are not equipped to provide delay resiliency. Lack of net neutrality poses a serious concern, except in cases where the wire line backhaul provider is the same company or in a tight strategic relationship with the cellular operator (Chandrasekhar, Andrews, & Gatherer, 2008).

The literature lacks a comprehensive study for the femtocells performance from QoS perspective, because the LTE and femtocell coexistence is still a developing integration of technologies. The Internet connection over packet switched Internet Protocol (IP) is Best Effort (BE) type. Any connection in the telecom operator's network could be overfilled or even damaged. There are no guaranteed maximum levels for delay and jitter. This might require a more flexible interface between the femtocell and macrocell base station.

Figure 12. Mobile (IPTV) technologies for next-generation systems

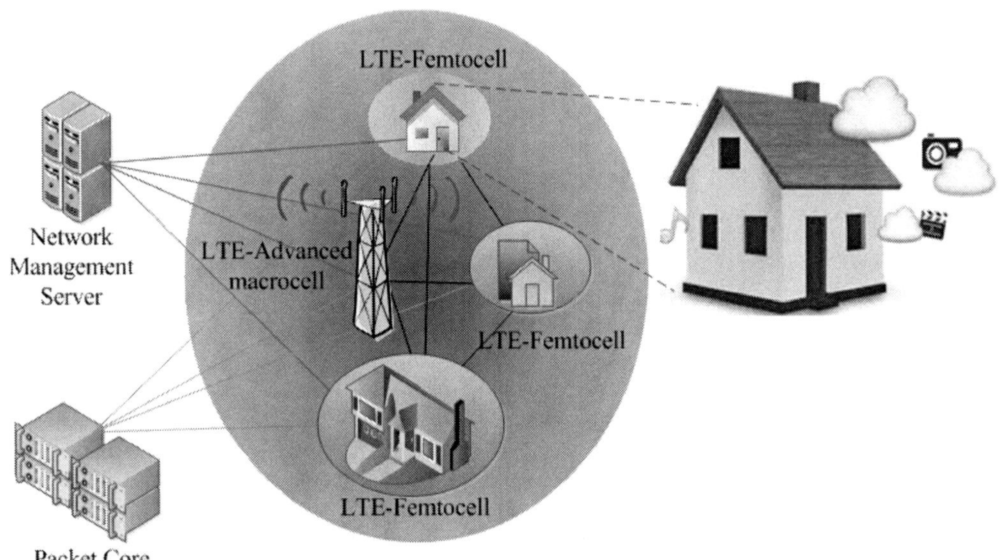

Interference Optimization Problem

In general, there are two types of interferences in two-tier network architecture as follows:

Co-Tier Interference

Co-tier interference occurs among network elements that belong to the same tier in the network. This takes place between elements of the same tier, e.g. between neighbouring femtocells. Variables in lower case represent magnitudes in natural units, while upper case indicates logarithmic scale, i.e. dB (Perez, et al., 2010). However, in OFDMA femtocell systems, the co-tier uplink or downlink interference occurs only when the aggressor (or the source of interference) and the victim use the same sub-channels. Therefore, efficient allocation of sub-channels is required in OFDMA-based femtocell networks to mitigate co-tier interference.

Cross-Tier Interference

This type of interference occurs among network elements that belong to the different tiers of the network, i.e., interference between femtocells and macrocells, as shown in Figure 13. For example, femtocell UEs and macrocell UEs (also referred to as MUEs) act as a source of uplink cross-tier interference to the serving macrocell base station and the nearby femtocells, respectively. On the other hand, the serving macrocell base station and femtocells cause downlink cross-tier interference to the femtocell UEs and nearby macrocell UEs, respectively. Again, in OFDMA-based femtocell networks, cross-tier uplink or downlink interference occurs only when the same sub-channels are used by the aggressor and the victim. However, this approach requires a higher spectral efficiency because both tiers access all resources. Nevertheless, in such configuration, cross-tier interference occurs, which could degrade the overall network performance unless interference is efficiently handled.

Figure 13. Macrocell downlink interference to the femtocell user

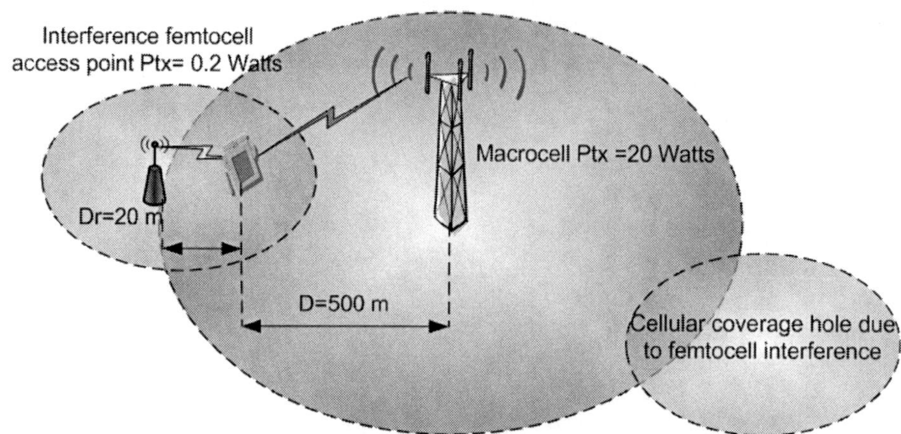

The Downlink Dedicated Physical Channel (DL-DPCH) performance for the femtocell operating co-channel to macrocell can be modelled as the ratio of the received DL-DPCH power to interference taking into account interference from the femtocell loading and interference from the co-channel microcell base station as formulated in (TR 2008 & Kim, et al., 2009):

$$\frac{DCH_{Ec}}{Io_{voice}} = \frac{\dfrac{P_{femto\text{-}call}}{L_{femto}}}{\left(\dfrac{0.5 * (1-\alpha) * P_{femto\text{-}total}}{L_{femto}}\right) + P_{micro\text{-}int}} \tag{6}$$

The interference from the co-channel micro-BS is modelled as the received power from the macro-BS as following:

$$P_{macro\_int} = 10^{\wedge}\left[\left(P_{macro\_total\_db} - 3 - L_{macro\_femto\_dB}\right)\right] \tag{7}$$

where $P_{femto\text{-}call}$ power allocated to the femto voice call, $P_{macro\text{-}int}$ microcell-bs interference to the femto user, $l_{macro\_femto\_db}$ path loss from femtocell user to the macro–bs, l_{femto}, dch_ec/io_voice dch performance requirement for 12k2 bits/second voice service, α femtocell downlink signal orthogonally.

The strong interference in the downlink (DL) from the femtocells should be well organized to satisfy the QoS requirements of both femtocell users and macrocell users. Interference problems could occur by femtocell base station Down Link (DL) connected to a far macro mobile user which may subsequently be jammed due to the presence of a closer DL femtocell user who is using the same frequency/time. On the other hand, Up Link (UL) mobile user connected to a femtocell could be jammed due to the attendance of a close UL user connected to a macrocell using the same frequency/time as shown in Figure 14. Therefore, the interference management techniques are essential to reduce the impact of femtocell on the macrocell.

Example: Consider an OFDM reverse link with parameters (Chandrasekhar, Andrews, & Gatherer, 2008):

Distance of macrocell user to macrocell
= 500 meters

Distance of macrocell user to femtocell
= 30 meters

Femtocell radius = 40 meters

Figure 14. Downlink interference in LTE-femtocell OFDM co-channel

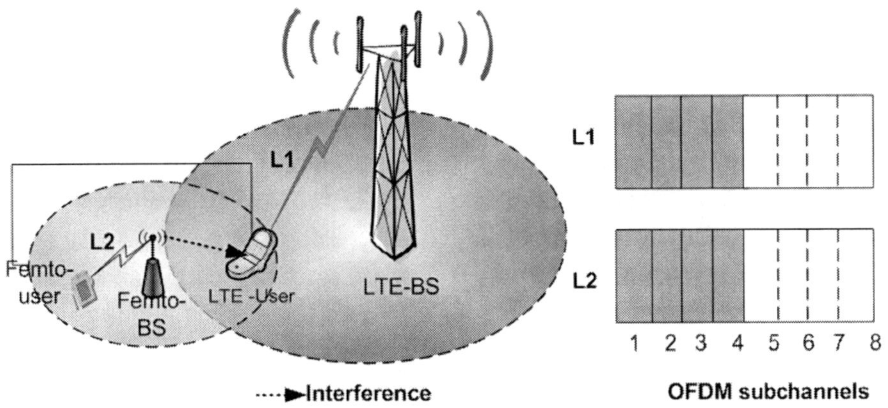

Processing gain =128

Path-loss exponent = 4

Desired receive power = 0 dBm (1 mW)

Interference Power from user at femtocell

$$= 10 * \log_{10}\left(\frac{500^4}{30^4}\right) = 4887 \text{ dB} \quad (8)$$

OFDM Interference suppression

$$= 10 * \log_{10}\left(128\right) = 21.07 \text{ dB} \quad (9)$$

Signal-to-Interference Ratio at femtocell
= -27.8 dB

In this scenario, the macrocell user is connected to the macro base station at the cell edge. This user is located in the same FBS zone where no access is allowed. Thus, this macrocell user is interfered in the downlink by the fully loaded FBS. The best solution for this case is to develop a procedure for adaptive power control to maintain capacity.

ACCESS AND MOBILITY OF FEMTOCELL USER

Handover from femtocell to macrocell is more complicated than the vice-versa case. Therefore, seamless handover between macrocell BSs and femtocell systems as well as between FBS should be supported, but at ordinary speeds. Furthermore, for the abovementioned system requirements, there is an expectation that the 802.16m air interface will provide further optimization for femtocell operation. For example:

Signal measurement report to support advanced interference management, radio resource management, and femtocells location; Optimization of UEs scanning, selection, network entry, and handover to the desired BSs in multilayer networks consisting of macrocell BSs and large numbers of femto BSs.

These features will further optimise femtocell operations and facilitate their usage (TR 36.942, 2009). Mobile users should be able to select a proper operating network so that corresponding assess can be employed. Handover issues among the cognitive femtocell networks can also be considered in the scope of cell selection.

MAC CONTROL PROCEDURES

The control procedures for radio resource management in networks with femtocells can manage radio resources despite the limited backbone capacity. To achieve an effective utilisation of backbone capacity, new scenarios will are required for data routings among any users served by the same FBS (Vivier, et al., 2010). Besides routing, spectrum-efficient techniques for power control, scheduling and broadcast services transmission are demanded. Additionally, novel user admission policies and FBSs identification techniques are necessary in a system with escalating numbers of users.

SELF-ORGANIZATION AND ENERGY EFFICIENCY CHALLENGES

Efforts to increase the energy efficiency of information and communication systems in mobile radio networks have recently gained a lot of attention. Besides reducing the carbon footprint of the industry, there is a strong economic incentive for network operators to reduce the energy consumption of their systems. Currently over 80% of the power in mobile telecommunications is consumed in the radio access network, more specifically the base stations (Fettweis & Zimmermann, 2008). Improvements can, in principle, be achieved in three ways: Firstly, reducing the power consumption of base station (either by using more power proficient hardware or by using new advanced software to adjust power consumption to the traffic situation based on the mobile distance). Secondly, by optimisation of individual sites, through the use of more efficient and load adaptive hardware components as well as software modules. Thirdly, by using intelligent network deployment strategies, effectively lowering the number of sites requisite in the network to achieve certain performance metrics such as spectral efficiency coverage. Femtocell is proposed as a low power base station that is supposed to decrease the power consumption compared to high power macro BSs.

To allow adaptation to changes in the network environment (i.e., configuration and properties of neighbouring macrocells/femtocells) an ongoing measurement and self-optimization process is performed to adapt parameters such as scrambling code, pilot- and maximum data transmit power, sleep mode technique and power saving. This ensures minimal impact on the macro-cellular network and ensures that femtocell performance is maximized under the given constraints.

To avoid high operating costs and redundant power consumption, femtocell should support plug-and-play and implementable self-configuration and self-optimization algorithms. In addition, the numbers of sites per square act with respect to inverse of the range. That may generate a lot of load on the energy consumption for a small coverage area. Throughput formula of Shannon capacity should be considered to evaluate system power saving and in order to achieve efficient network energy, Shannon's theory can be applied in a different. Shannon's theory provides guidance for learning and developing novel methodologies to reduce energy per data bit "Improved energy efficiency" while approaching the Shannon limit of maximized network capacity.

The Shannon capacity cannot be reached in practice due to several application issues. The following equations approximate the data rate "throughput" over a channel and to represent these mechanisms precisely we use adapted Shannon capacity formula which has suitable parameters B_{eff} and $\in_{eff}$:

$$\text{Data Rate, } R = \left\{ \begin{array}{ll} R = 0 & \text{for } \in < \in_{min} \\ R = B_{eff}.B.\log_2\left(1 + \frac{\in}{\in_{eff}}\right) & \text{for } \in_{min} \leq \in < \in_{max} \\ R = R_{max} & \text{for } \in < \in_{max} \end{array} \right\} \quad (10)$$

where R is the maximum throughput or data rate (Symbol Rate), B is the operation bandwidth (Samples/Sec= 1/Ts), $\in$ is the signal to interference and noise ratio (SINR), $\in_{min}$ is the minimum value for SINR such that connection is blocked if $\in < \in_{min}$ and $\in_{max}$ is the SINR needed to reach the maximum throughput R_{max}. The coefficients B_{eff} and $\in_{eff}$ are the bandwidth and SINR efficiency factors respectively (Koutsopoulos & Tassiulas, 2008).

COGNITIVE RADIO TECHNIQUE

Cognitive radio network has recently emerged as a promising technique to improve the utilization of the existing radio spectrum. Cognitive radio is an intelligent wireless communication system that is aware of its surrounding environment (i.e., outside world), and uses the methodology of understanding-by-building to learn from the environment and adapt its internal states to statistical variations in the incoming radio frequency by making corresponding changes in certain operating parameters (e.g., transmit power; carrier frequency, and modulation strategy) in real-time, with two primary objectives in mind: (1) highly reliable communication whenever and wherever needed and (2) efficient utilisation of the radio spectrum (Haykin & Soliman, 2005), as shown in Figure 15.

A radio is adaptive when it can autonomously modify its operating parameters in response to the characteristics of the environment in which it finds itself, for instance, a radio that modifies Intermediate Frequency (IF) filter characteristics in response to the characteristics of the channel it is using may be considered adaptive. In other words, an adaptive radio must be able to make changes to its operating parameters such as power level, modulation, frequency, etc. An 802.11a radio exhibits a level of adaptively as it is able to sense the Bit Error Rate (BER) of its link and adapts the modulation to a data rate and a corresponding Forward Error Correction (FEC) that sets the BER to an acceptable level for data applications (Fette, 2006). To perform the required CR characteristics, in addition to the Soft Defined Radio (SDR) based Radio frequency front to end in the CR physical layer, the different protocols in medium access control MAC, networks, transport and application layers should be adaptive to the variation in the CR environment like the licensed user activity, cognitive radio system requirements and the channel qualities. On the other side, a CR module is used to establish the interfaces among the different layers and control

Figure 15. Spectrum opportunities for cognitive radio

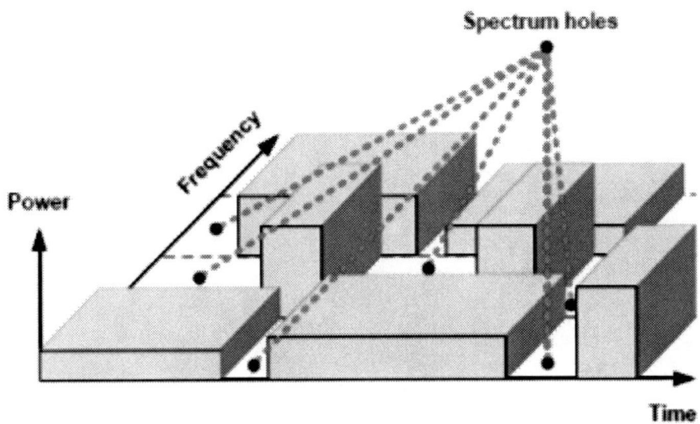

the protocol parameters based on intelligent algorithms (Hossain, Niyato, & Han, 2009).

The cognitive radio offers a novel way of solving spectrum underutilization problems. It does so by sensing the radio environment with a two-fold objective: identifying those sub-bands of the radio spectrum that are underutilized by the primary (i.e. legacy) users. However, we must note that in order to achieve these goals in an autonomous manner, multiuser cognitive radio networks will have to be self-organising and able to respond to the surrounding wireless changes including the power adaptability requirements.

COGNITIVE NETWORK FUNCTIONS

Although there are many scenarios for the coexistence between primary and secondary networks, it is mandatory to achieve a certain level of cooperation to prevent any interference resulting from the contention between the cognitive and primary networks or between the cognitive networks. This situation becomes even more complicated in a heterogonous wireless environment composed of many types of networks: primary and secondary. Thus, cooperative schemes are necessary to guarantee seamless communications and to achieve optimal spectrum access. The IEEE 802.11 networks perform the listen to talk operations in transmissions. Therefore, they are the best available standards to simulate the future cognitive network with zero inferences. However, there are several studies about the differences between the traditional networks and the cognitive capabilities as they have been reviewed in Al-Rubaye et al. (2010) and Chen et al. (2008).

Spectrum Sensing

A cognitive radio can sense spectrum and detect "spectrum holes" which are those frequency bands that are not used by the licensed users or having limited interference with them. Spectrum sensing can be performed in either centralized or distributed ways. In centralized spectrum sensing, a central unit, also called sensing controller, is in charge of the sensing process. The sensing information is shared with the different secondary users using a control channel, as shown in Figure 16.

Figure 16. Cognitive radio operations

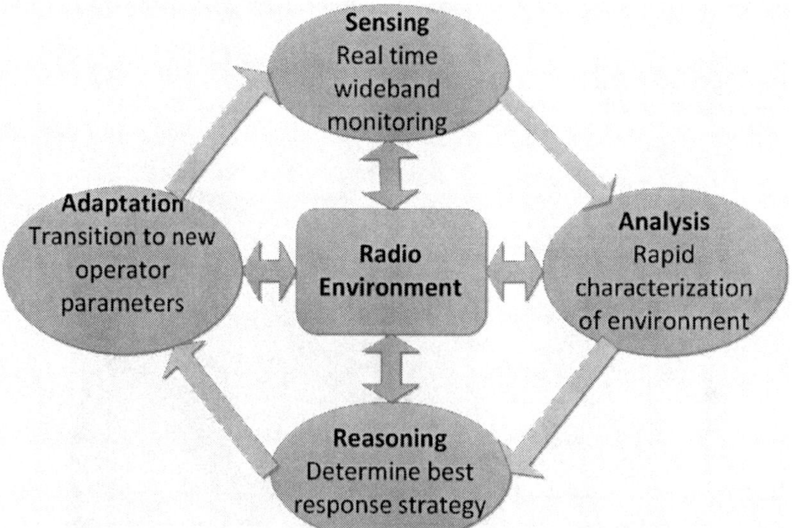

The cooperative spectrum sensing is more accurate and can reduce the primary signal detection time (Ganesan & Li, 2007).

Spectrum Sharing

A cognitive radio could incorporate a mechanism that would enable sharing of spectrum under the terms of an agreement between a licensee and a third party. This function picks the appropriate MAC protocol to access the spectrum holes. At the MAC layer protocol, the reasonable spectrum sharing between the different users can be guaranteed. Additionally, management between base stations can be achieved in order to avoid the collision with Primary User (PUs) as well as with other SUs pairs. This can be done by negotiating the spectrum opportunities available to share them on an ad-hoc or real-time basis, without the need for prior spectrum allocation between all users.

Spectrum Decision

This function able to analyse the information comes from the sensing the spectrum users. Before making the spectrum access decision, the characteristics of the identified spectrum holes, the probability of the PU appearance, and the possible sensing errors should be measured. Once the suitable band is selected, the CR has to optimize the available system resources in order to achieve the required objective. A novel method for decision making assumes the sharing of knowledge between CRs before allowing each CR to make its own decision of transmitting (Al-Dulaimi & Al-Saeed, 2010).

Location Identification

The ability to determine the spectrum users' locations in order to identify the appropriate operating parameters such as the power and frequency allowed at surrounding location is highly required. In bands such as those used for satellite downlinks that are receive-only and do not transmit a signal, location technology may be an appropriate method of avoiding interference because sensing technology would not be able to identify the locations of nearby receivers.

Network Discovery

For a cognitive radio terminal to determine the best way to communicate, it should first discover available access networks around it. These networks are reachable either via directed one hop communication or via multi-hop relay nodes. The ability to discover one hop or multi-hop located access networks is critical to secure reliable connections. This becomes very challenging in a dynamic spectrum access model, which requires developing more effective algorithms that can ease the task of interfacing mobile users to the access network terminals.

COGNITIVE FEMTOCELL IN LTE SYSTEMS

Cognitive radio interconnection of femtocell networks could be part of future Internet wireless networks. This new coexistence paradigm between cognitive radio and femtocells is based on predefined converged areas of services and the challenge is to combine the capacity of different resources to provide broadband access to both stationary and mobile cognitive users. However, highly throughput demands of mobile end users in the future may need more resources than what are currently available for the exiting macrocell networks. Hence, femtocell networks will play a vital role in supporting indoor environments, such as airports, hospitals, or houses. Enhancing data delivery for Internet services could be creating a novel cross layer framework in the gateway router. This improves the cognitive femtocells' flexibility and efficiency in accessing the spectrum. On the other hand, the new design concepts increase system's reconfigurability to respond to real-time changes in the wireless environment (Al-Rubaye, Al-Dulaimi, & Cosmas, 2011).

Future networks of 4G LTE (Long Term Evolution) networks are projected to have two-tier domains that consist of overlaying macrocell and cognitive femtocell systems. Within the overlapped coverage areas in such systems, an UE has the option to access either macrocell base station or femtocell base station, and it can switch the access tier by performing vertical handover.

After accessing the unlicensed spectrum, cognitive operator can adopt different approaches to utilize the spectrum in a wireless overlay network. The principal approach is that two tier networks share the unlicensed spectrum such that macrocell and femtocell operate in co-channel frequency reuse. This approach is called spectrum efficiency usage, as shown in Figure 17. Other aspects are focusing on resource management to avoid two-tier interference where each radio tier has a more available aggregate of spectrum but suffers higher cross-tier interference. Therefore, any potential increase in energy efficiency while deploying a certain number of low powered relay femtocell nodes can provide lower interference to both primary and secondary users at the same time. The investigation in the possible energy savings has to consider the daily deviations in network traffic as this may change the coverage areas and the times of activities of the femtocells. This can be achieved by dynamically sending the low loaded cells into sleep mode, during which radio circuits are efficiently powered down. This results in decreasing the overall networks energy consumption.

The main goal of the above proposal is to maximize the throughput of the system while minimizing the need of cognitive femtocell for the power. Therefore, a further work is necessary to investigate the effects of deploying cognitive femtocell network within the LTE system transmission from power side.

SUMMARY

The evolution of the cellular and wireless networks is presented which leads towards multi-technology heterogeneous wireless networks that employ 4G wireless macrocell and femtocell systems. The basic architecture of the main cellular and wireless technologies is described in details. The chapter starts by surveying the standardization efforts in the field of femtocells and also introduces a technical background of femtocell technologies and deployment issues. The emergence of the LTE and cognitive, as corresponding technologies, are identified for femtocells applications. There are numerous challenges that will require to be ad-

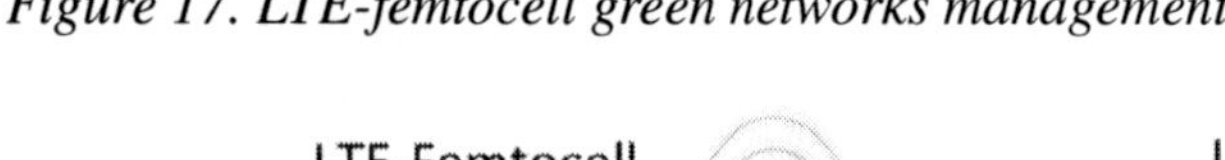

Figure 17. LTE-femtocell green networks management

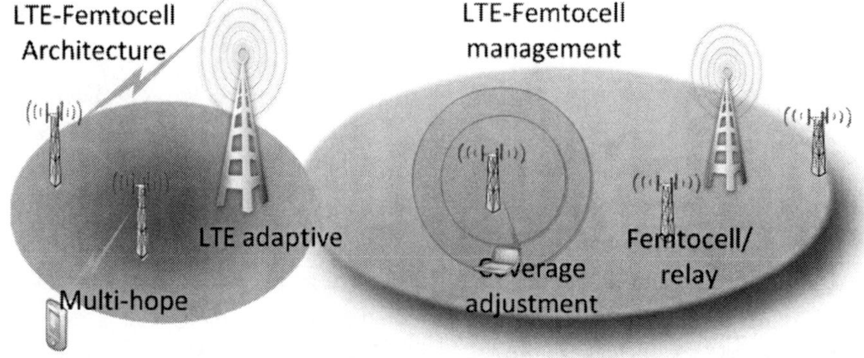

dressed by the service provider in order to make the femtocell able to impact on subscribers in terms of performance or via economic advantages (e.g., optimized power). The main challenges were described to enable a clear view for the main requirements of multi-tier model of cognitive macro and femtocells in the next 4G networks.

REFERENCES

Afridi, A. (2011). *Macro and femto network aspects for realistic LTE usage scenarios.* (Masters Dissertation). Royal Institute of Technology (KTH). Stockholm, Sweden.

Al-Dulaimi, A., & Al-Saeed, L. (2010). An intelligent scheme for first run cognitive radios. In *Proceedings of the IEEE International Conference and Exhibition on Next Generation Mobile Applications, Services, and Technologies (NGMAST 2010)*. Amman, Jordan: IEEE Press.

Al-Rubaye, S., Al-Dulaimi, A., Al-Saeed, L., Al-Raweshidy, H. S., Kadhum, E., & Ismail, W. (2010). Development of heterogeneous cognitive radio and wireless access network. In *Proceedings of the 24th Wireless World Research Forum (WWRF)*. Penang Island. Malaysia: WWRF.

Al-Rubaye, S., Al-Dulaimi, A., & Cosmas, J. (2011). Cognitive femtocells: Future wireless networks for indoor applications. *IEEE Vehicular Technology Magazine*, *6*(1), 44–51. doi:10.1109/MVT.2010.939902

Chandrasekhar, V. (2009). *Coexistence in femtocell-aided cellular architectures.* (PhD Dissertation). University of Texas at Austin. Austin, TX.

Chandrasekhar, V., Andrews, J., & Gatherer, A. (2008). Femtocell networks: A survey. *IEEE Communications Magazine*, *46*(9), 59–67. doi:10.1109/MCOM.2008.4623708

Chandrasekhar, V., & Andrews, J. G. (2008). Spectrum allocation in two-tier femtocell networks. In *Proceedings of the IEEE Conference Signals, Systems and Computers*. Pacific Grove, CA: IEEE Press.

Chen, K., Peng, Y., Prasad, N., Liang, Y., & Sun, S. (2008). Cognitive radio network architecture: Part I - General structure. In *Proceedings of the IEEE 2nd International Conference on Ubiquitous Information Management and Communication*. Suwon, Korea: IEEE Press.

Das, D., & Ramaswamy, V. (2009). Co-channel femtocell-macrocell deployments-access control. In *Proceedings of the IEEE 70th Vehicular Technology Conference (VTC-2009 Fall)*. Chelmsford, MA: IEEE Press.

Das, S., Chandhar, P., Mitra, S., & Ghosh, P. (2011). *Issues in femtocell deployment in broadband OFDMA networks: 3GPP–LTE a case study.* Paper presented at IEEE Vehicular Technology Conference (VTC-2011 Fall). San Francisco, CA.

De la Roche, G., Valcarce, A., Lopez-Perez, D., & Zhang, J. (2010). Access control mechanisms for femtocells. *IEEE Communications Magazine*, *48*(1), 33–38. doi:10.1109/MCOM.2010.5394027

FP7-ICT-248891. (2010). *Femtocell-based network enhancement by interference management and coordination of information for seamless connectivity*. STP FREEDOM.

Fette, B. (2006). *Cognitive radio technology*. New York, NY: Newnes.

Fettweis, G. P., & Zimmermann, E. (2008). ICT energy consumption – Trends and challenges. In *Proceedings of the 11th International Symposium on Wireless Personal Multimedia Communications*. Lapland, Finland: IEEE.

Ganesan, G., & Li, Y. (2007). Cooperative spectrum sensing in cognitive radio, part I: Two user networks. *IEEE Transactions on Wireless Communications*, *6*(6), 2204–2213. doi:10.1109/TWC.2007.05775

Govil, J., & Govil, J. (2007). *4G mobile communication systems: Turns, trends and transition.* Paper presented at IEEE International Conference on Convergence Information Technology (ICCIT). New York, NY.

Haykin, S., & Soliman, S. (2005). Cognitive radio: Brain-empowered wireless communications. *IEEE Journal on Selected Areas in Communications*, *23*(2), 201–220. doi:10.1109/JSAC.2004.839380

Ho, L. T. W., & Claussen, H. (2007). Effects of user-deployed, co-channel femtocells on the call drop probability in a residential scenario. In *Proceedings of the IEEE 18th International Symposium on Personal, Indoor and Mobile Radio Communications (PIMRC 2007).* Athens, Greece: IEEE Press.

Hossain, E., Niyato, D., & Han, Z. (2009). *Dynamic spectrum access and management in cognitive radio networks*. Cambridge, UK: Cambridge University Press. doi:10.1017/CBO9780511609909

Kim, R., Kwak, J., & Etemad, K. (2009). WiMAX femtocell: Requirements, challenges, and solutions. *IEEE Communications Magazine*, *47*(9), 84–91. doi:10.1109/MCOM.2009.5277460

Knisely, D. (2010). *Femtocell standardization.* White Paper. Chelmsford, MA: Airvana.

Knisely, D., Yoshizawa, T., & Favichia, F. (2009). Standardization of Femtocells in 3GPP2. *IEEE Communications Magazine*, *47*(9), 76–82. doi:10.1109/MCOM.2009.5277459

Koutsopoulos, I., & Tassiulas, L. (2008). The impact of space division multiplexing on resource allocation: A unified treatment of TDMA, OFDMA and CDMA. *IEEE Transactions on Communications*, *56*(2), 1–10. doi:10.1109/TCOMM.2008.050102

Lopez-Perez, D., Valcarce, A., de la Roche, G., & Zhang, J. (2009). OFDMA femtocells: A roadmap on interference avoidance. *IEEE Communications Magazine*, *47*(9), 41–48. doi:10.1109/MCOM.2009.5277454

Mahmoud, H., & G¨uven, I. (2010). *A comparative study of different deployment modes for femtocell networks.* Paper presented at IEEE 20th Symposium on Personal, Indoor and Mobile Radio Communications. Istanbul, Turkey.

Oh, D.-C., Lee, H.-C., & Lee, Y.-H. (2010). Cognitive radio based femtocell resource allocation. In *Proceedings of the 2010 International Conference on Information and Communication Technology Convergence (ICTC 2010),* (pp. 274-279). Jeju Island, Korea: ICTC.

Perez, D., Valcarce, A., Ladanyi, A., Roche, G., & Zhang, J. (2010). Intracell handover for interference and handover mitigation in OFDMA two-tier macrocell-femtocell networks. *EURASIP Journal on Wireless Communications and Networking*, *1*, 1–16.

Raj, R., & Gagneja, A. (2012). 4G wireless technology. *International Journal of Computer Science and its Applications*, 263-270.

Saunders, S., Carlaw, S., Giustina, A., Bhat, R., Rao, V., & Siegberg, V. (2009). *Femtocells book: Femtocells opportunities and challenges for business and technology*. New York, NY: John Wiley & Sons Ltd.

Sung, K., Haas, H., & McLaughlin, S. (2010). A semianalytical PDF of downlink SINR for femtocell networks. *EURASIP Journal on Wireless Communications and Networking*, 9.

3Tech, G. P. P. Report. (2010). *Further advancements for e-utra physical layer aspects*. Retrieved from http://www.3gpp.org

Technology Watch Report, I. T. U.-T. 10. (2009). *The future internet*. Retrieved from http://www.itu.int

Thanabalasingham, T. (2006). *Resource allocation in OFDM cellular networks*. (PhD Dissertation). University of Melbourne. Melbourne, Australia.

TR. (2008). Interference management in UMTS femtocells. *Femto Forum- Technical Report*. Retrieved from http://www.femtoforum.org

TR 36.942. (2009). *Radio frequency (RF) system scenarios (release 8)*. 3GPP Technical Report, V. 8.2.0. Retrieved from http://www.3gpp.org

Vivier, G., Kamoun, M., Becvar, Z., de Marinis, E., Lostanlen, Y., & Widiawan, A. (2010). Femtocells for next-G wireless systems: The FREEDOM approach. In *Proceedings of Future Network and Mobile Summit*. Paris, France: La Défense.

Xia, P., Chandrasekhar, V., & Andrews, J. (2010). Open vs. closed access femtocells in the uplink. *IEEE Transactions on Wireless Communications*, *9*(12), 798–809. doi:10.1109/TWC.2010.101310.100231

Zyren, J., & McCoy, W. (2007). *Overview of the 3GPP long term evolution physical layer.* White Paper. LTD Free scale Semiconductor. Retrieved from http://www.3gpp.org

Chapter 2
Large Scale Cognitive Wireless Networks: Architecture and Application

Liang Song
University of Toronto, Canada & OMESH Networks Inc., Canada

Petros Spachos
University of Toronto, Canada

Dimitrios Hatzinakos
University of Toronto, Canada

ABSTRACT

Cognitive radio has been proposed to have spectrum agility (or opportunistic spectrum access). In this chapter, the authors introduce the extended network architecture of cognitive radio network, which accesses not only spectrum resource but also wireless stations (networking nodes) and high-level application data opportunistically: the large-scale cognitive wireless networks. The developed network architecture is based upon a re-definition of wireless linkage: as functional abstraction of proximity communications among wireless stations. The operation spectrum and participating stations of such abstract wireless links are opportunistically decided based on their instantaneous availability. It is able to maximize wireless network resource utilization and achieve much higher performance in large-scale wireless networks, where the networking environment can change fast (usually in millisecond level) in terms of spectrum and wireless station availability. The authors further introduce opportunistic routing and opportunistic data aggregation under the developed network architecture, which results in an implementation of cognitive unicast and cognitive data-aggregation wireless-link modules. In both works, it is shown that network performance and energy efficiency can improve with network scale (such as including station density). The applications of large-scale cognitive wireless networks are further discussed in new (and smart) beyond-3G wireless infrastructures, including for example real-time wireless sensor networks, indoor/underground wireless tracking networks, broadband wireless networks, smart grid and utility networks, smart vehicular networks, and emergency networks. In all such applications, the cognitive wireless networks can provide the most cost-effective wireless bandwidth and the best energy efficiency.

DOI: 10.4018/978-1-4666-2812-0.ch002

INTRODUCTION

Modular design and well-defined architecture have played an important role in many engineering success. In the world of communications and networking, the fundamentals of Open System Interconnection (OSI) architecture have defined multiple hierarchy layers in the communication protocol stack, which provide abstracts of network functionalities and hide implementation complexity. Among these hierarchy layers, the physical layer defines the waveform being transmitted in the communication medium and the conversion of information (modulation/demodulation). The data-link layer, including Medium Access Control (MAC) sub-layer, provides the abstraction of communication channel where information is transmitted. The network layer routes the communication across the network, and the transport layer defines an end-to-end tunnel to hide the network complexity from higher layers. In cable and computer networks, the definition of these layers has been quite appropriate, since 1) it converts complicated system into simplified modules (layers); 2) methods developed for particular module will benefit overall system as well; 3) modifications on a single module will not need a system re-design. The related success stories include the telephone networks and the Internet.

When wireless communications and networking are being considered, engineering efforts have been trying to adapt the hierarchy layers, for the communication protocol stack to work appropriately in wireless medium. For example, in the physical layer, digital communications have defined mechanisms to modulate (demodulate) digital sequence onto (from) a carrier frequency as mapped in the radio spectrum. The MAC layer is developed to set up point-to-point wireless linkage over wireless medium, with predetermined spectrum (or wireless channel) allocation. A network topology can then be determined, where the network layer implements routing protocols to further set up end-to-end communications.

This adaptation can be appropriate only when the network resource availability can be predetermined, including such as radio spectrum and wireless station. In addition, a "virtual wired" network topology shall be set up, with wireless links as "virtual lines" and wireless stations as "virtual dots." In engineering practices, it has limited most real-world wireless networks to a star-topology, or to the very last hop. In a star-topology, the predetermined spectrum allocation is feasible to avoid intra-network interference as introduced by the broadcasting nature of wireless medium; and predetermined wireless station availability can be realized by having the base-station (or point coordinator) to coordinate access terminals. The related success stories in wireless communications include such as cellular networks and Wireless Local Area Networks (WLAN).

However, the methodology above has also made large-scale wireless infrastructure very expensive to build and maintain; and radio spectrum becomes a scarce resource. For example, in cellular networks, operators need to acquire spectrum license to prevent other networks using the same spectrum and offer the required Quality of Service (QoS). The cellular base-stations shall also be installed on towers to provide necessary coverage; and network cable/fiber is needed for every base-station to provide connection to backbone networks. In WLAN, although unlicensed spectrum is utilized, setting up multiple access points to provide sufficient coverage and bandwidth is usually a challenging and costly task in municipal and enterprise applications, due to spectrum planning and device installation costs.

In beyond 3G smart infrastructures, most applications require cost-effective wireless bandwidth and energy efficiency (green communications) that current wireless networks cannot support. In this chapter, we introduce the architecture and application of large-scale cognitive wireless networks that differentiate from traditional wireless networks. Instead of depending on predetermined spectrum allocation and wireless stations, the principal of cognitive wireless networking (Song, CCNC, 2008) opportunistically utilizes these network resources, so as to realize reliable communications. In such typical scenario, the cognitive wireless networks operate in unlicensed spectrum, or in licensed spectrum as a secondary network (using the spectrum white space of primary licensed network). The spectrum availability of a communication channel can typically change in millisecond interval, which introduces random spectrum availability. At the same time, the network can also take a mesh and mobile topology by multiple wireless hops instead of a star topology. Traffic congestion and station power saving (duty-cycled sleep mechanism), as well as mobility, introduce random station (wireless nodes) availability that can also change in millisecond scale. By opportunistically utilizing radio spectrum and wireless stations, the cognitive wireless networking can realize much higher network bandwidth and energy efficiency.

The network architecture of cognitive wireless networking necessarily takes a cross-layer approach as compared to traditional networking, at the same time it is also designed as compatible to traditional networks. The developed network architecture (Song & Hatzinakos, 2009, 2011), Embedded Wireless Interconnection (EWI), is based upon a redefinition of wireless linkage as functional abstraction of proximity communications among wireless stations. The operation spectrum and participating nodes of such abstract wireless links are opportunistically decided based on their instantaneous availability (see Figure 1).

We then present the investigation in multi-hop wireless communications under the developed network architecture. It is essentially an integration of opportunistic routing and opportunistic spectrum access which results in the implementation of a unicast wireless-link module (Song & Hatzinakos, 2008; Spachos, et al., 2011). Compared to traditional networks based on predetermined link spectrum allocation and routing tables, the cognitive unicast is demonstrated to achieve much higher end-to-end communication throughput (network bandwidth), and to support real-time communications over multiple wireless hops. By network simulation under realistic channel model, the performance metrics of throughput, delay, and energy efficiency are evaluated, where magnitudes of improvement are shown.

Figure 1. Cognitive networking concept (International Journal of Communication Networks and Distributed Systems, Volume 2/2009) (URL: http://www.inderscience.com/search/index.php?action=record&rec_id=26558) © Inderscience Enterprises Ltd.

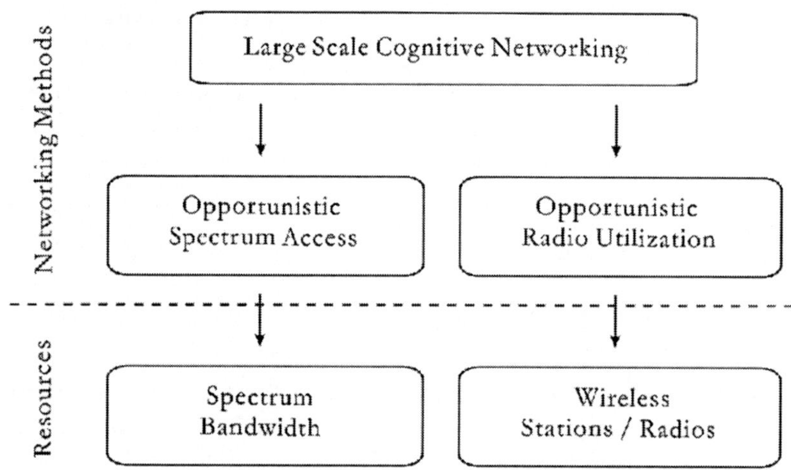

We further present the investigation in opportunistic data aggregation under the developed network architecture, which results in the implementation of a data-aggregation wireless link module (Song & Hatzinakos, 2007). The cognitive data aggregation opportunistically collect sensory and management data in the network and optimize the trade-off between application QoS and energy/ resource consumptions. In principle, the best available application data are collected in real-time in self-organized networks; the aggregated data can then be delivered to remote server by multi-hop cognitive unicast. An example application of target tracking wireless sensor networks is studied by network simulation, where the tracking error (application QoS) and energy consumption are compared.

In both of the above investigations, that network and application performance can improve with network scale (such as including station density) under the architecture of large-scale cognitive wireless networks. The chapter then introduces the prototyping and production of cognitive radios that can support the cognitive wireless networking architecture, with a further discussion on the applications in B3G smart infrastructures, where application features of large-scale cognitive wireless networks are summarized.

HISTORIC REVIEW

The concept of cognitive radio was proposed by Mitola (1999), in which the radio is envisioned with the intelligence to exploit ambient environment for user centric communications. Since regulation authorities, e.g., Federal Communications Commission (FCC) in the United States, have recognized the inefficiency of legacy spectrum allocation, researches in cognitive radio have been focused on opportunistic spectrum access (Zhao & Sadler, 2008).

One of the earlier attempts to define the concept of cognitive network was made by Thomas et al. (2005), where cognitive network is described as a network with a cognitive process that can perceive current network conditions, plan, decide, act on those conditions, learn from consequences of its actions, and follow end-to-end goals. This cognition loop senses the environment, plans actions according to input from sensors and network policies, decides which scenario fits best its end-to-end purpose using a reasoning engine, and finally acts on the chosen scenario as discussed in the previous section. The system learns from the past (situations, plans, decisions, actions) and uses this knowledge to improve the decisions in the future.

This early definition of cognitive network however did not explicitly describe what the knowledge of the network is. It only described the cognitive loop and adds end-to-end goals that would distinguish it from cognitive radio.

As a potential integration of cognitive radio in wireless mesh networks, Chen et al. (2007) proposed the cognitive mesh networks, which investigates cognitive spectrum allocation schemes in wireless ad hoc and mesh networks. Chowdbury and Akyildiz (2008) also aimed to have the spectrum agility of cognitive radio without changing mesh networking protocols, and formulate optimal channel assignment problems to wireless mesh routers. The researches in cognitive radio networks have been considered as a bridge toward the future cognitive networks.

Fortuna and Mohorcic (2009) reviewed cognitive network as a communication network augmented by a knowledge plane that can span vertically over layers and/or horizontally across technologies. The knowledge plan is composed of at least two elements: 1) a representation of relevant knowledge about the scope; 2) a cognition loop, which has the intelligence inside its states.

In this chapter, the full potential of cognitive-networking concept (Song, 2008) is further best detailed, and is interpreted as a network that can utilize both radio spectrum and wireless station resources opportunistically, based upon the knowledge of such resource availability. Since cognitive radio has been developed as a radio transceiver that can utilize spectrum channels opportunistically (or opportunistic spectrum access), the cognitive network is therefore a network that can opportunistically organize cognitive radios.

NETWORK ARCHITECTURE

Principles

The key architectural differentiation of Embedded Wireless Interconnection is based on the new definition of wireless linkage. The new *abstract wireless links* are redefined as arbitrary mutual cooperations among a set of neighboring (proximity) wireless nodes. In comparison, traditional wireless networking relies on point-to-point "virtual wired-links" with a predetermined pair of wireless nodes and allotted spectrum. The architectural diagram of EWI is illustrated in Figure 2.

The EWI network architecture has the following three primary principles:

- **Functional Linkage Abstraction:** Based on the definition of abstract wireless linkage, wireless link modules are implemented in individual wireless nodes, which can set up different types of abstract wireless links. According to the functional abstractions, categories of wireless link modules can include: broadcast, unicast, multicast, and data aggregation, etc. Therefore, network functionalities can be integrated in the design of wireless link modules. This also results in two hierarchical layers as the architectural basics, including the system layer and the wireless link layer, respectively. The bottom wireless link layer supplies a library of wireless link modules to the upper system layer; the system layer organizes the wireless link modules to achieve effective application programming.
- **Opportunistic Wireless Links:** In realizing the cognitive wireless networking concept, both the occupied spectrum and the participating nodes of an abstract wireless link are opportunistically determined by their instantaneous availabilities. This principle decides the design of wireless link modules in the wireless link layer. The system performance can improve with

Figure 2. Layered network architecture (International Journal of Communication Networks and Distributed Systems, Volume 2/2009) (URL: http://www.inderscience.com/search/index.php?action=record&rec_id=26558) © Inderscience Enterprises Ltd.

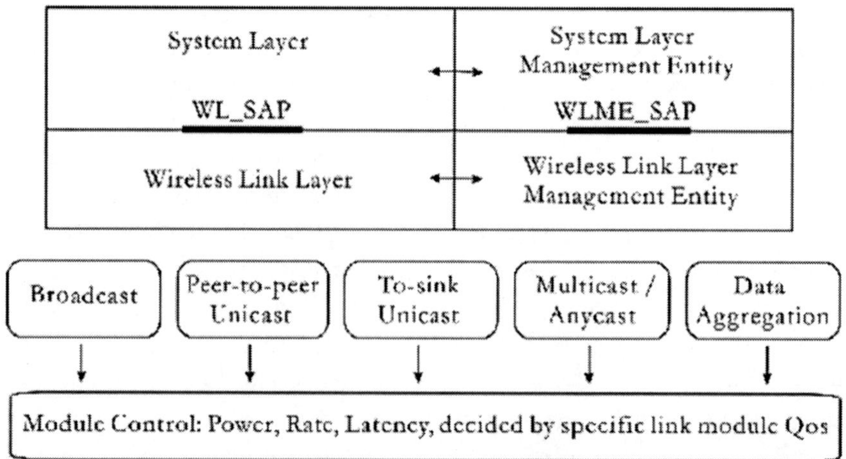

larger network scale, since higher network density introduces extra diversity in the opportunistic formation of any abstract wireless links. This is also provided by the radio implementation that will be elaborated later.

- **Global QoS Decoupling:** Global application or network QoS is decoupled into local requirements of co-operations in neighboring wireless nodes, i.e., wireless link QoS. More specifically, by decoupling global application-level QoS, it allows the system layer to better organize the wireless link modules that are provided by the wireless link layer. For example, by decoupling global network-level QoS, such as throughput, end-to-end delay, and delay jitter, the wireless link module design can achieve the global QoS requirements. Based on the provided wireless link modules, the complexity at individual nodes can be independent of the network scale.

Architecture Interface

Wireless link modules provide system designers with reusable and open network abstractions, where the modules can be individually updated, or new modules may be added into the wireless link layer. The high modularity and flexibility could be essential for middleware or application developments.

EWI is also an organizing-style architecture, where the system layer organizes the wireless link modules (at the wireless link layer); and peer wireless link modules can exchange module management information by padding packet headers to the system-layer information units.

Five types of wireless link modules are illustrated in Figure 2, including broadcast, peer-to-peer unicast, multicast, to-sink unicast, and data aggregation, respectively. Other arbitrary types of modules may be added, establishing other types of abstract wireless links without limitation. For example, the broadcast module simply disseminates data packets to neighboring nodes. The peer-to-peer unicast module can deliver data packets from source to destination over multiple wireless hops. The multicast module sends data packets to multiple destinations, as compared to peer-to-peer unicast. The to-sink unicast module can be especially used in wireless sensor networks, which utilizes higher capabilities of data collectors (or sinks), so as to achieve better data delivery. The data-aggregation module opportunistically collects and aggregates the context related data from a set of proximity wireless nodes.

Shown in Figure 2, two Service Access Points (SAPs) are defined on the interface between the system layer and the wireless link layer, which are WL_SAP (Wireless Link SAP) and WLME_SAP (Wireless Link Management Entity SAP), respectively. WL_SAP is used for the data plane, whereas WLME_SAP is used for the management plane. The SAPs are utilized by the system layer in controlling the QoS of wireless link modules.

Radio Implementation

Previous researches in cognitive radio have been focused on opportunistic spectrum access (Zhao & Sadler, 2008), which locates the "white-space" in spectrum for communications. In order to implement the principles of abstract wireless linkage, there are two basic propositions for the radio in cognitive wireless networks, which are the two processes of "sensing" and "polling," respectively:

- The radio can opportunistically sense available spectrum resource, so that the selected spectrum usage will not be interfering with other co-existing wireless communications.
- The radio can opportunistically poll one or more other neighboring radios (nodes) on the selected spectrum, so as to realize certain types of local co-operations.

Under the EWI architecture, a wireless node with the proposed radio can initiate an abstract wireless link, i.e., certain types of local co-operations among proximity wireless nodes. Both the wireless nodes and the occupied spectrum are therefore opportunistically decided by their instantaneous availability. The initiated abstract wireless link would not be interfering with other wireless communications, as protected by the first cognitive proposition.

Based on the above two propositions, we further describe a prototype implementation of the radio for large-scale cognitive wireless networks. In the example implementation, the radio can access a group of predetermined data channels, where every data channel is also associated with two distinctive frequency tones, i.e., one sensing tone and one polling tone. The radio hardware is therefore comprised of two transceivers, which are the tone transceiver and the data transceiver, respectively. When the radio initiates an abstract wireless link, it senses for an available channel, with the vacant data channel and the vacant sensing/polling tones. The radio then broadcasts the polling tone associated to the selected channel, to poll neighboring nodes. The neighboring nodes can decide autonomously whether to join in the initiated abstract wireless link on the data channel, when they detect the rising edge of the associated polling tone. On joining in an abstract wireless link, the radio of the neighboring nodes also broadcasts the associated sensing tone. As such, both sensing and polling tones protect abstract wireless links from spectrum interferences. The demonstration and experiment results with this prototype appeared in Song (2008).

Now assume that there are totally N data channels, as differentiated by frequency, time, or spreading codes. Let S_n and P_n denote respectively the sensing and polling tones associated to one data channel $n(0<n<N+1)$. A state diagram of the described radio implementation is shown in Figure 3.

Network Addressing

In traditional computer networks, network addresses are based on symbols, e.g., Internet Protocol (IP) or MAC address. There have also been one-to-one mapping between network addresses

Figure 3. Prototype radio state diagram (International Journal of Communication Networks and Distributed Systems, Volume 2/2009) (URL: http://www.inderscience.com/search/index.php?action=record&rec_id=26558) © Inderscience Enterprises Ltd.

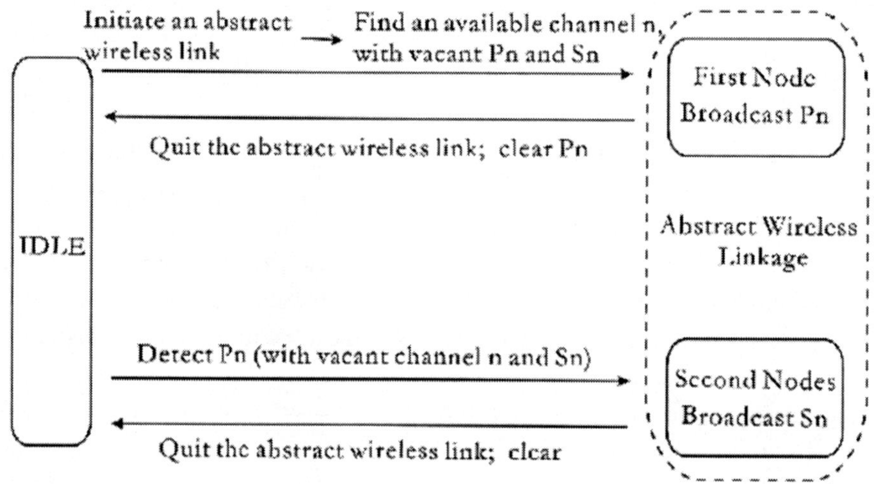

and networking devices, where an address can be globally unique as assigned to one networking device. In the large-scale cognitive wireless networks, the network addresses can be based on context, e.g., of location coordinates, application-specific address, or logic address, etc. In respect of different contexts, a wireless node (networking device) can acquire more than one address. Meanwhile, more than one node, e.g., in proximity, can also share the same context address, where these nodes can be considered as identical in the network for the corresponding context.

In the design of wireless link modules, a *cost of delivery criterion* is defined, being applied to the context-based network addresses. Given two arbitrary nodes n and m, C_n and C_m denote their addresses related to certain context, respectively. According to the criterion, an estimated cost of delivery $C_{n,m}$ can be locally calculated from, C_n and C_m, which indicates the expected or average cost of sending one packet (information unit) from n to m, or vice versa, independent of dynamic changes in the network. The cost of delivery criterion is necessary so that effective packet routing can be set up based on network address (as comparable to IP address) without networking loops.

For example, in location-centric networks, where wireless nodes are aware of their own locations, e.g., by Global Position Systems (GPS), or by triangulation estimations, the cost of delivery can be location distance, and the network address can be derived from the location coordinates. In data-centric networks (e.g., Intanagonwiwat, et al., 2002), the network address is decided by application-specific context. The cost of delivery can be the application data gradient, and can be assumed as a monotonically increasing function of the distance between peer wireless nodes, since the correlation of application data decreases with spatial separation. In data-collecting or fusion networks, e.g., wireless sensor networks, the sink (or data collector) can broadcast a number of identity advertisement packets, which is then flooded in the network, by broadcasting. Every node can count the average smallest number of hops to the sink, on receiving the advertisement packets. The count number can be used as the cost of delivery for one node.

New nodes joining in the network may also dynamically acquire its own (context) network address by querying such addresses of neighboring nodes. If a node leaves the network, it will not take part in any abstract wireless links, and not create bottlenecks to the network.

Wireless Link Modules

Figure 4 illustrates a state diagram of the wireless link layer, where a library of wireless link modules is provided. Other types of modules may also be added in the library. Table 1 further provides a list of primitive functions related to these modules, at the service access points (i.e., WL_SAP and WLME_SAP in Figure 2). System layer can control wireless link modules by calling their defined primitive functions.

The wireless link layer remains in the IDLE state, when no wireless link module is invoked. The pair of primitives, *Module.start()* and *Module.sleep()*, are utilized to switch between the IDLE state and the SLEEP state, i.e., a further power-saving mode. While in the IDLE state, the wireless link layer monitors if there are new abstract wireless links being initiated. On detecting an initiation from any one of the neighboring nodes, the wireless link layer transfers from the IDLE state to the Module Request state, where the primitive *Module.request()* is invoked. It provides the system layer with the control of wireless link activities. When the primitive *Module.response()* is received, the wireless link layer either transfers back to the IDLE state, or joins in the abstract wireless link, as decided by the response result. On the other hand, when receiving a command to initiate an abstract wireless link, e.g., by the primitives *Broadcast/UnicastP. send(...)* or *Aggregate.*

Figure 4. Wireless link layer state diagram (International Journal of Communication Networks and Distributed Systems, Volume 2/2009) (URL: http://www.inderscience.com/search/index.php?action=record&rec_id=26558) © Inderscience Enterprises Ltd.

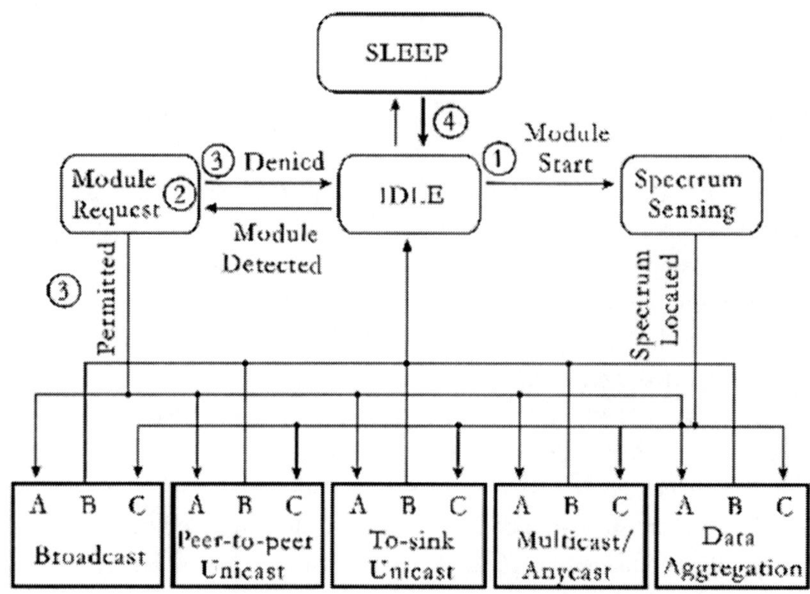

start(...), the wireless link layer transfers from the IDLE state to the Spectrum Sensing state. A corresponding wireless link module can be started, after available spectrum resource has been located.

The implementations of "spectrum sensing" and "module detecting" were specified previously. By the "spectrum sensing," available spectrum resource is located opportunistically for the abstract wireless link operation, without interfering with other wireless communications. By the "module detecting," an abstract wireless link is set up among a group of proximity wireless nodes, where the wireless nodes are opportunistically determined by their autonomous availability.

COGNITIVE UNICAST

Background

Cognitive unicast presents the design of peer-to-peer unicast wireless link module, which can transmit packets from source to destination over multiple wireless hops. The design by itself is an integration of opportunistic routing and opportunistic spectrum access.

During the last decade, a number of opportunistic protocols have been developed. The first opportunistic routing method is introduced in Biswas and Morris (2004). Extremely Opportunistic Routing (ExOR) selects the next relay node by a slotted ACK (acknowledge) mechanism. Having successfully received a data packet, the node calculates a priority level, which is inversely proportional to the expected transmission count metric (ETX) (De

Table 1. Primitive functions (International Journal of Communication Networks and Distributed Systems, Volume 2/2009) (URL: http://www.inderscience.com/search/index.php?action=record&rec_id=26558) © Inderscience Enterprises Ltd.

Primitive Function	Description	SAP	Caller	Responder	Module
Broadcast.send (*DATA*, *Priority*)	Command of broadcasting the *DATA* (information unit);	WL_SAP	System layer	Wireless link layer	Broadcast
Broadcast.indicate (*SA*, *DATA*)	Indicate the received broadcast *DATA* (information unit) from the source address *SA*;	WL_SAP	Wireless link layer	System layer	Broadcast
UnicastP.send (*SA*, *DA*, *DATA*, *Priority*)	Command of sending the *DATA* (information unit) from the source address *SA* to the destination address *DA*;	WL_SAP	System or Wireless link layer	Wireless Link layer	Peer-to-Peer unicast
UnicastP.indicate (*SA*, *DATA*)	Indicate the received unicast *DATA* (information unit) from the source address *SA*;	WL_SAP	Wireless link layer	System layer	Peer-to-Peer unicast
Aggregation.start (*Context*)	Command of starting data aggregation related to the *Context*;	WL_SAP	System layer	Wireless link layer	Data aggregation
Aggregate.indicate (*SA*, *DATA*)	Indicate the received *DATA* (context-related information unit) from the source address *SA*;	WL_SAP	Wireless link layer	System layer	Data aggregation
Aggregate.stop (*Context*)	Command of stopping data aggregation related to the *Context*;	WL_SAP	System layer	Wireless link layer	Data aggregation
Aggregate.request (*AA*, *Context*)	Indicate the data aggregation request related to the *Context*, from the aggregation address *AA*;	WL_SAP	Wireless link layer	System layer	Data aggregation
Aggregate.send (*AA*, *DATA*, *Priority*)	Command of sending the *DATA* (context related information unit) to the aggregation address *AA*;	WL_SAP	System layer	Wireless link layer	Data aggregation
Module.status (*&Status*, *Module*)	Indicate the status of the module initiations: Broadcast/ UnicastP.send(...), or Aggregation.start(...), e.g., Success or Fail;	WLME_SAP	Wireless link layer	System layer	N/A
Module.request ( )	Request to start a wireless link module, from the wireless link layer;	WLME_SAP	Wireless link layer	System layer	N/A
Module.response ( )	Response to Module.request(), i.e., Permitted or Denied, from the system layer;	WLME_SAP	System layer	Wireless link layer	N/A
WLME.reset ( )	Reset the wireless link layer to the IDLE state, i.e., to terminate running wireless link modules;	WLME_SAP	System layer	Wireless link layer	N/A
WLME.getState (*&State*)	Get the current state of the wireless link layer;	WLME_SAP	System layer	Wireless link layer	N/A
WLME.start ( )	Turn on the wireless link layer operations;	WLME_SAP	System layer	Wireless link layer	N/A
WLME.sleep ( )	Turn off the wireless link layer operations;	WLME_SAP	System layer	Wireless link layer	N/A
WLME.getAddress (*&Address*, *Context*)	Obtain the *Context* related network *Address*;	WLME_SAP	Wireless link layer	System layer	N/A

Couto, et al., 2005), defined based on the distance between the node and the destination. The shorter the distance is, the higher the priority is. The node with the highest priority will then be selected as the next relay node. The main drawback of ExOR is that it prevents spatial reuse because it needs global coordination among the candidate nodes. Candidate nodes transmit in order, only one node is allowed to transmit at any given time while all the other candidate nodes attempt to overhear the transmission in order to learn which node will be the next relay node. Moreover, the simple priority criteria that it uses, ETX distance, may lead packets toward the destination through low-quality routes. To overcome this problem, Opportunistic Any-Path Forwarding (OAPF) (Zhong, et al., 2006) introduces an expected any-path count (EAX) metric. This approach can calculate the near-optimal candidate set at each potential relay node to reach the destination. However, it needs more state information about the network and it has high computational complexity. ExOR ties the MAC with routing, imposing a strict schedule on routers access to the medium. The scheduler goes in rounds. MAC-Independent Opportunistic Routing and Encoding Protocol (MORE) (Chachulski, et al., 2007) tries to enhance ExOR. MORE uses the concept of innovative packets in order to avoid duplicate packets, which might occur in ExOR.

In Zorzi and Rao (2003a, 2003n), a Geographic Random Forwarding (GeRaF) technique was proposed. In GeRaF each packet carries the location of the sender and the destination, so that the prioritization of the candidate nodes is based on location information. This technique is simple to be implemented, but requires location information for all the nodes in the network. Hybrid ARQ Based Inter-cluster Geographic Relaying (HARBINGER) (Zhao, et al., 2004) is a combination of GeRaF with hybrid Automatic Repeat Request (ARQ). In GeRaF, when there is no forwarder within the range of the sender node, everything must start over again, while in HARBINGER hybrid ARQ is used for a receiver to combine the information accumulated over multiple transmissions from the same sender.

In other works, Coding-Aware Opportunistic Routing Mechanism (CORE) (Yan, et al., 2008) is an integration of localized interflow network coding and opportunistic routing. By integrating localized network coding and opportunistic forwarding, CORE enables a packet holder to forward a packet to the next hop that leads to the most coding changes among its forwarder set. Opportunistic Routing in Dynamic Ad Hoc Networks (OPRAH) (Westphal, 2006) builds a braid multipath set between source and destination via on-demand routing to support opportunistic forwarding. For this purpose, OPRAH allows intermediate nodes to record more sub-paths back to the source and also those sub-paths downstream to the destination via received Route Request and Route Replies.

Module Design

A state diagram of the cognitive unicast is illustrated in Figure 5, whereas the analysis, simulations, and experiments were reported in Song and Hatzinakos (2008). The peer-to-peer unicast module can send unicast data packets from source to destination; and by the cognitive wireless networking, data packets can travel along opportunistically available paths with opportunistically available spectrum on every hop.

The source and relays can use the primitive *UnicastP.send(...)* to initiate an unicast wireless link. And the destination uses *UnicastP.indicate(...)* to forward received data to the system layer. Shown in Figure 5, unicast control information includes the source address, the destination address, the data-sender (current relay) address, and the QoS control. On receiving those control information, the wireless nodes joining in the unicast wireless link, i.e., relay candidates, locally calculate a time-delay parameter T_a, which can be decided by the wireless channel status and the cost of delivery to the destination address, from the *cost of delivery criterion*. In principle, the relay candidate with satisfying wireless channel status and the smallest cost, should obtain the smallest T_a. As such, this preferred node broadcasts a relay announcement packet, and serves as the next-hop relay by receiving the data. Other relay candidates can exit the unicast module on receiving the relay announcement. After having received the data packet, the next-hop relay can invoke the primitive *UnicastP.send(...)* for further relaying. Otherwise, if the destination has been reached, the primitive function *UnicastP.indicate(...)* is invoked, which forwards the received data to the system layer. Moreover, if the next-hop relay is not found, the current relay (data-sender) can invoke the primitive *UnicastP.send(...)* again to re-send the data. A resolution process may be further employed, if more than one relay candidates obtain the same smallest T_a.

Figure 5. Cognitive unicast state diagram (International Journal of Communication Networks and Distributed Systems, Volume 2/2009) (URL: http://www.inderscience.com/search/index.php?action=record&rec_id=26558) © Inderscience Enterprises Ltd.

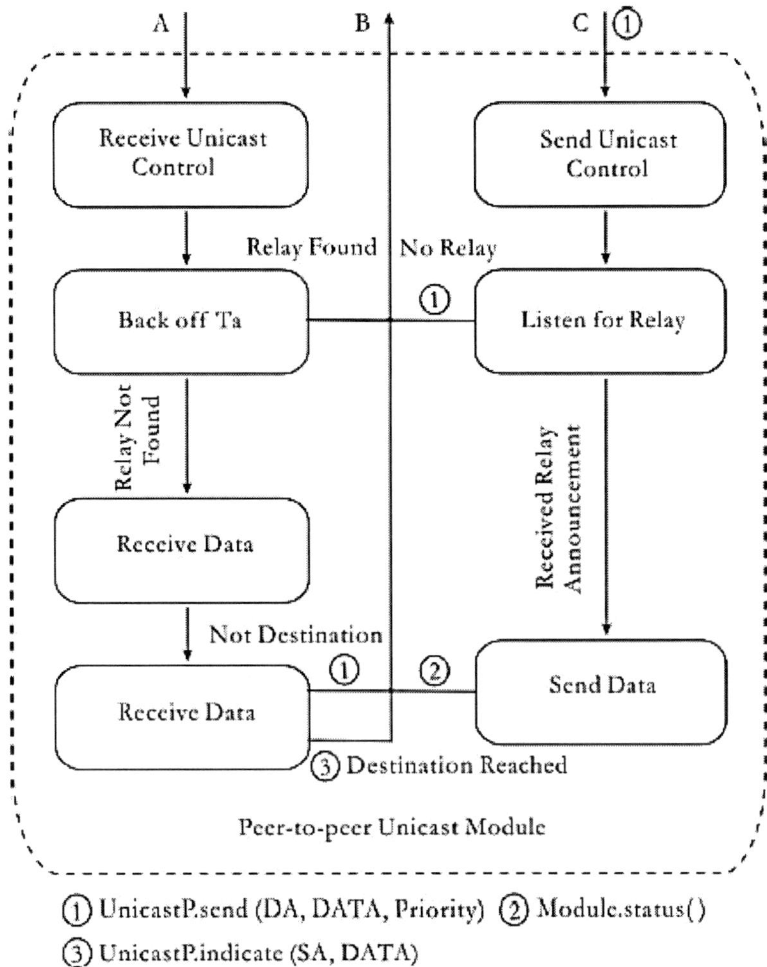

The QoS class of the peer-to-peer unicast module, as denoted by the parameter *Priority* in the primitive *UnicastP.send(...)*, can be decided by the data flow QoS levels of throughput, end-to-end delay, and delay variance (jitter). These global network QoS can be supported in the unicast module design, by transmitting power control (or joint power and rate control).

Performance Analysis

Song and Hatzinakos (2008) analyzed the performance of cognitive unicast under a 2-dimentional node Poisson distribution model. These analytical results indicate that the investigated QoS metrics, including multi-hop throughput, end-to-end delay, delay variance (jitter), and energy consumption, can improve with larger network scale. In particular, the maximal multi-hop throughput increases with the network density to a predetermined bound, while the expected end-to-end delay decreases with network density, approaching a predetermined bound, as decided by other related parameters. On the other hand, the delay variance reduces to zero with higher network density, which indicates that the network performance can be made arbitrarily stable, simply by putting more nodes in the network.

For supporting long-range communications over a relatively large number of wireless hops, the analytical results show that the maximal throughput can be independent of the source-to-destination distance, while both the expected delay and the delay variance increase linearly with the source-to-destination distance. Furthermore, it is identified that the radio transmitting power can be an ideal control knob for the tradeoff between the QoS requirements and the resource consumptions, for example, network energy consumption or network capacity.

We further study the performance difference between cognitive unicasting and traditional multi-hop wireless networking (Spachos, et al., 2011) in a realistic indoor setup (Spachos, et al., 2011, 2012), as shown in Figure 6.

In order to evaluate the wireless network in a realistic environment, our approach consists of modeling the indoor wireless channels based on an accurate ray-based simulator (Volcano Lab, 2012). The network simulator is based on the discrete event simulation system OMNeT++ (2012) with 30 nodes as shown in Figure 6. The physical communication parameters are chosen based on IEEE 802.15.4. The simulation is conducted in two steps. In the first step, we calculate the power of received signal between all the nodes in the network. In the second step, the channel simulation results are used in the network simulator to calculate packet error rate in the transmission among different nodes. The routing strategies of cognitive unicast and traditional networking are then applied.

The traditional networking is simulated based on predetermined shortest-path routes over high quality links; and ideal scheduling is assumed for

Figure 6. Indoor network setup © 2011 IEEE. Reprinted with permission

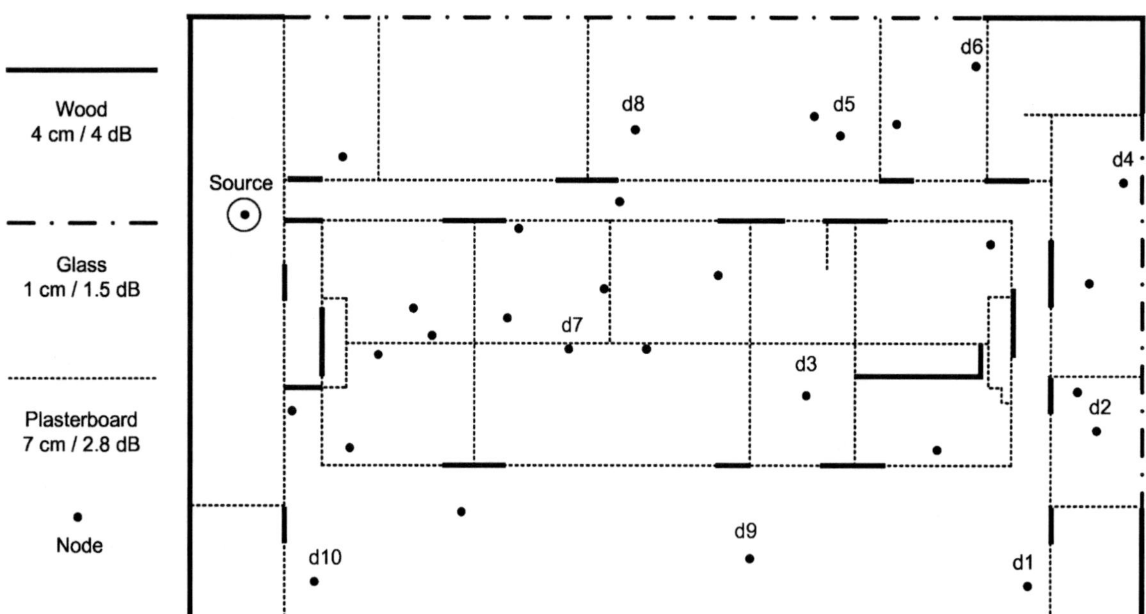

packet transmission and channel allocation. No external interference is assumed. Therefore, the obtained results can be virtually viewed as an upper bound of any traditional networking performance.

Throughput. Throughput is the number of bits divided by the time needed to transport the bits. From the source node 1000 packets were transmitted toward each of the 10 different destinations, *d1–d10*, in Figure 6. The packet size is 200 bytes and the bit rate is 250kbps, hence, the data packet transmission time is 6.4 ms. The results can be seen in Figure 7.

Traditional routing follows the path that was discovered during the initialization phase for all the packet transmission. For the indoor environment of the simulation, traditional routing is following paths around the different rooms, avoiding the plasterboard.

Cognitive routing tends to find the best available and shorter path in each time slot toward the destination, leading to better throughput compared with the traditional approach. The path changes dynamically in each packet transmission, and it uses nodes that are not used from traditional routing. As a result, the path for each packet might be different and shorter than that of the traditional routing. In this manner, it can achieve better performance in terms of throughput.

Delay. Delay of a packet in the network is the time it takes the packet to reach the destination after leaving the source. The source node sends 1000 packets toward each destination, *d1–d10*. The results can be seen in Figure 8.

In traditional routing every packet transmission needs exactly the data transmission time to be transmitted between any two nodes. In cognitive routing, extra handshake time (i.e., for unicast control and relay announcement information) is included any one transmission of the unicast module.

Energy consumption is the total energy consumed (in the unit of Joule) in all the nodes of the network for the source to send 1000 packets to each destination, d1–d10. The results can be seen in Figure 9.

Figure 7. Throughput in different destinations from source © (2011) IEEE. Reprinted with permission

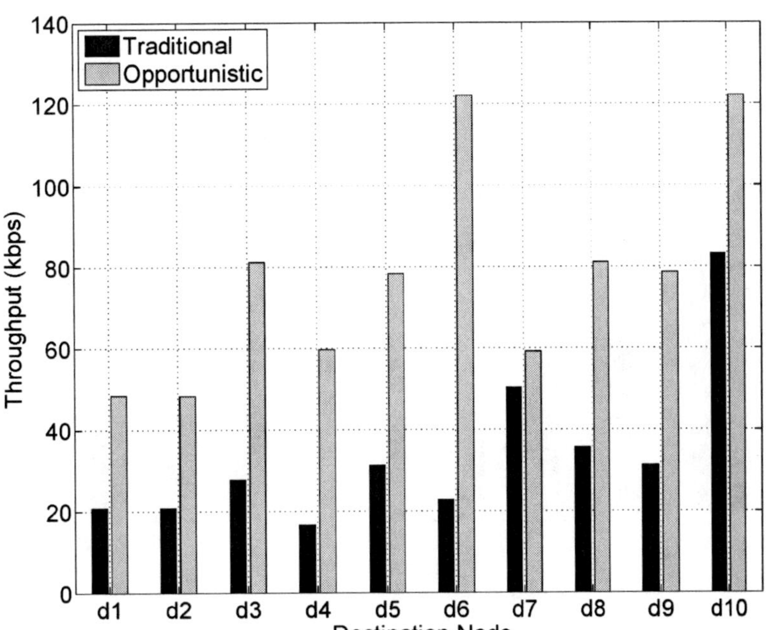

Figure 8. Delay in different destinations from source © (2011) IEEE. Reprinted with permission

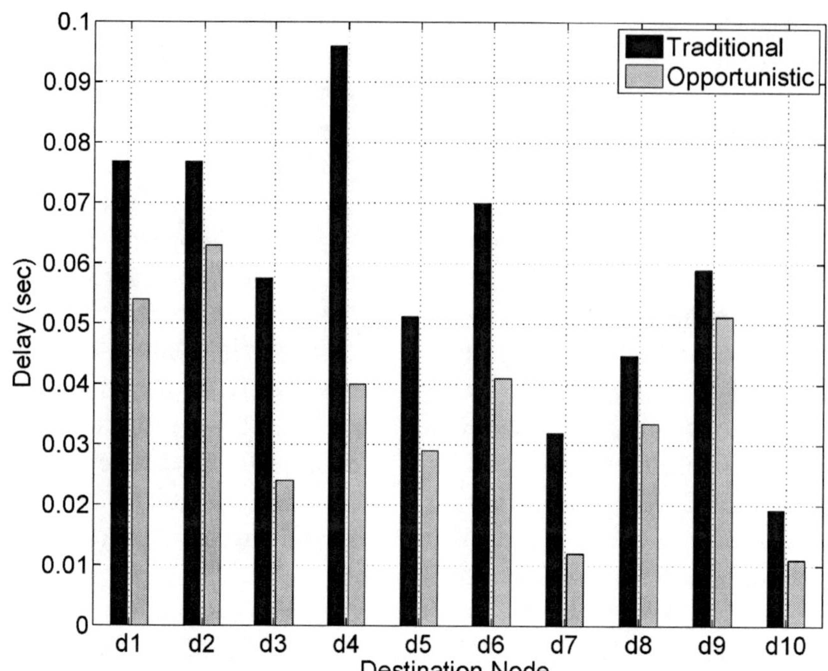

Figure 9. Network energy consumption in different destinations from source

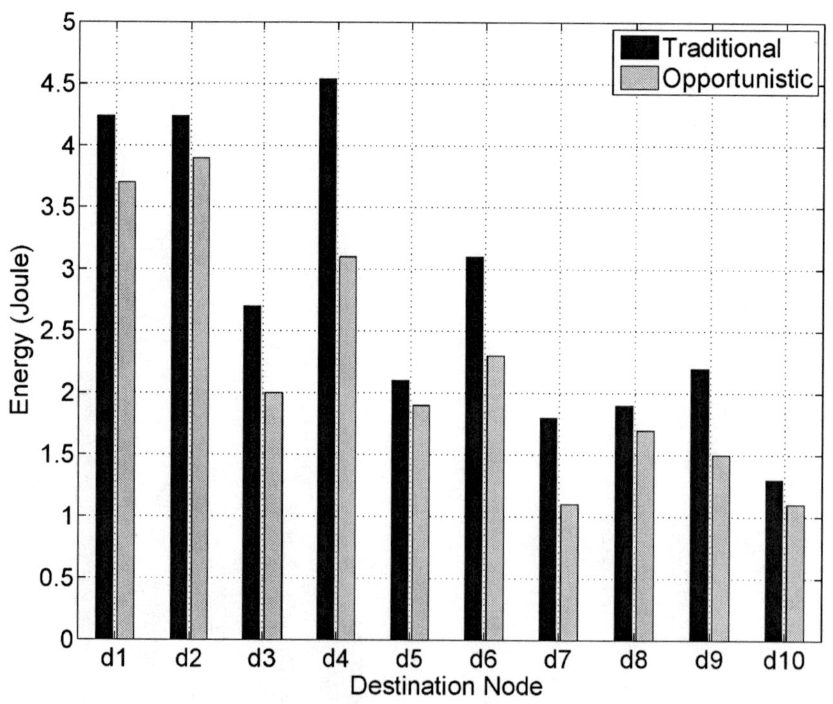

From the above results, it can be observed that cognitive unicast performs better than the upper bound of traditional networking in all circumstances. In addition, the performance gain can grow with the size of the network, the number of hops from the source to destination, and the complexity of environment. For example, in the setup of Figure 6, the destinations *d4* and *d6* show the largest improvements.

COGNITIVE DATA AGGREGATION

Background

Cognitive data aggregation presents the design of data-aggregation wireless link module, which can perform real-time and distributed data aggregation (signal processing) in the network. It is realized by the integration of opportunistic spectrum access and opportunistic data aggregations that can maximize the (energy) resource efficiency in meeting the application QoS.

In-network signal processing and distributed data aggregation have been studied in the context of wireless sensor networks for a long time. For target tracking sensor networks, research efforts have been focused on the handover of target tracking duty among leader nodes (or cluster heads). Zhao et al. (2002) and Chu et al. (2002) proposed the IDSQ (Information Driven Sensor Querying) where a leader sensor node is intelligently selecting the best neighbor node to perform sensing and serve as the next leader. A cost function was employed by jointly considering the energy expenditure and information gain. Based on a similar idea, Guo and Wang (2004) applied the Bayesian SMC (Sequential Monte Carlo) methods to the problem of optimal sensor selection and fusion in target tracking. These approaches require that individual sensor nodes process detailed information about all nodes in neighborhood, such as the location and residual energy level, which limits the protocol scalability. Moreover, the complexity of node selection algorithms might impose high constraints on sensor node processing capability.

Brooks et al. (2003) and Moore et al. (2003) proposed location centric CSP (Collaborative Signal Processing) approaches for target tracking sensor networks, where a selected region instead of an individual sensor node is activated. However, since they are focused on upper layers (application and network) design, it is unclear how the CSP methods can be efficient implemented in wireless sensor networks. Moreover, energy efficiency was not considered in the work of CSP. Zhang and Cao (2004) proposed optimized tree reconfiguration for target tracking networks, which is concentrated on the Network layer domain, and shaped by the tracking application requirements. Potential optimization in lower layers, however, was also not considered.

Module Design

A state diagram of the cognitive data aggregation is illustrated in Figure 10. The data-aggregation module collects context-related data packets from proximity nodes, which may be used in the applications of context-aware search, mobile computing, and wireless sensor networks. The aggregation node can use the primitive *Aggregate. start(...)* to initiate a data-aggregation wireless link, and collect context-related data packets. It can also terminate the data-aggregation wireless link with the primitive *Aggregate.stop(...)*, when enough data has been collected. The primitive *Aggregate. indicate(...)* is further used by the aggregation node to forward the received data packets to the system layer. For the wireless nodes joining in a data-aggregation wireless link, they can use the primitive *Aggregate.send(...)* to transmit their context-related data to the aggregation node, and use the primitive *Aggregate.request(...)* to supply the system layer with the requested context.

Shown in Figure 10, the aggregation control information can include the aggregation address (i.e., of the aggregation node), and the context

Figure 10. Cognitive data aggregation diagram (International Journal of Communication Networks and Distributed Systems, Volume 2/2009) (URL: http://www.inderscience.com/search/index.php?action=record&rec_id=26558) © Inderscience Enterprises Ltd.

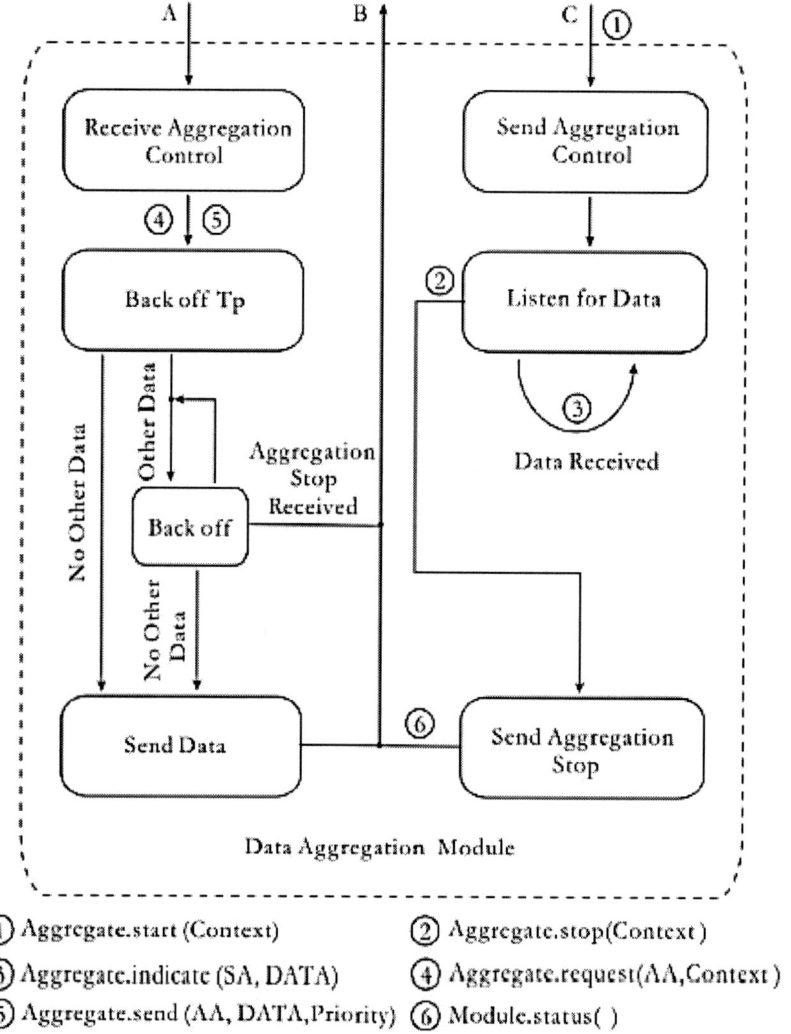

request, e.g., temperature, humidity, or other measurement metrics related to certain events. The QoS class parameter *Priority*, being used in the primitive function *Aggregate.send(...)*, indicates the specific context-related data quality. Intuitively, data with higher quality should obtain higher priority for being sent to the aggregation node. Therefore, based on *Priority*, a time-delay parameter T_p can be calculated at the nodes joining in the aggregation, which is a backoff period inversely proportional to the data quality. If the radio channel is busy after the backoff period T_p, i.e., another context-related data packet with higher priority is being transmitted by another node, the current data at the local node backoffs for another fixed time-period corresponding to the transmission of one data packet. As such, the context-related data packets will be broadcasted to the aggregation node sequentially, according to the data quality. Once enough data has been collected, the aggregation node broadcasts an "aggregation stop" control packet to terminate the data-aggregation wireless link.

Therefore, by data-aggregation wireless links, the best-quality data is collected, where the quality is defined according to the requested context. The optimization between application QoS and network energy consumption can be resolved at the system layer by a control knob.

Performance Analysis

We further do performance analysis of the cognitive data aggregation based on an application of real-time event location (tracking) in wireless sensor networks (Song & Hatzinakos, 2007).

Shown in Figure 11, we now consider that a large number of wireless sensor nodes are deployed in a surveillance area for certain event detection and tracking. More specifically, a set of leader nodes are elected along the moving track of the target event, which is shown as an intruding vehicle in Figure 11. A leader node collects sensing data from surrounding sensor nodes, and generates context-related tracking records. The programming of wireless sensor nodes is done with the primitive functions of wireless link modules shown in Table 1. Figure 12 shows the protocol stack of a wireless sensor node. And some additional system primitive functions are defined in Table 2.

Derived from the sensor reading, the parameter *SenPriority* can be obtained, indicating the

Figure 11. Sensor networks for event detection and tracking (International Journal of Communication Networks and Distributed Systems, Volume 2/2009) (URL: http://www.inderscience.com/search/ index.php?action=record&rec_id=26558) © Inderscience Enterprises Ltd.

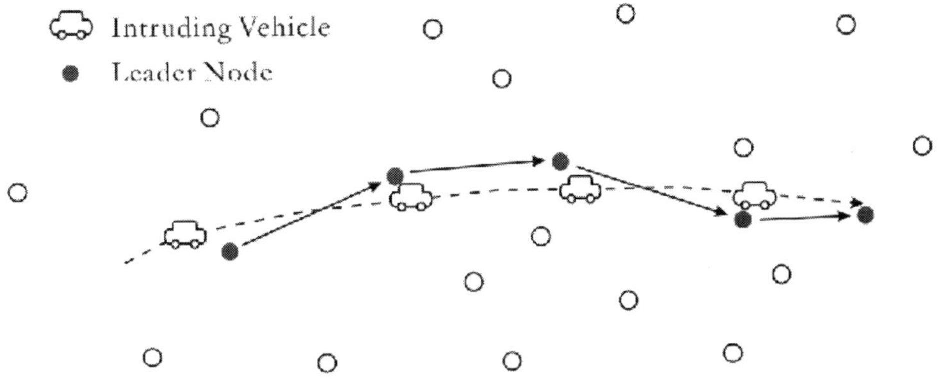

Figure 12. Protocol stack of wireless sensor nodes (International Journal of Communication Networks and Distributed Systems, Volume 2/2009) (URL: http://www.inderscience.com/search/index.php?action=record&rec_id=26558) © Inderscience Enterprises Ltd.

System Layer

Sensor

Wireless Link Layer

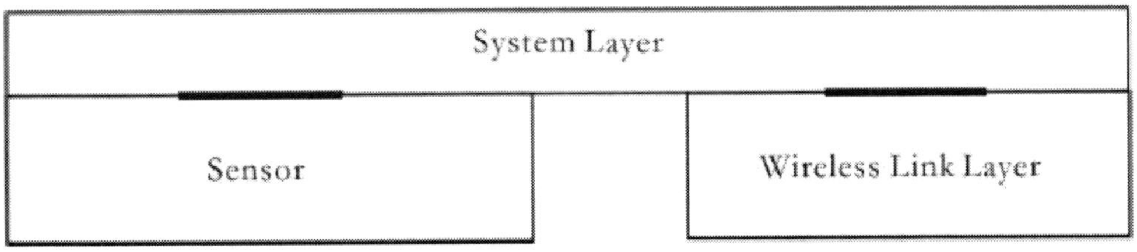

Table 2. System primitive functions (International Journal of Communication Networks and Distributed Systems, Volume 2/2009) (URL: http://www.inderscience.com/search/index.php?action=record&rec_id=26558) © Inderscience Enterprises Ltd.

Function	Description	Caller	Responder
SensorReading (*Context*, *&DATA*)	Read the *DATA* (sensing data), related to the *Context*;	System layer	Sensor
CalcAddress (*DATA*, *Context*, *&SenPriority*, *&Candidacy*)	Calculate the network address, i.e., composed of *SenPriority* and *Candidacy* from the *DATA* (sensing data), based on the *Context*.	System layer	N/A

sensing data quality as related to the context. Nominally, *SenPriority* can be proportional to the physical proximity between the wireless sensor node and the event under surveillance. Based on the parameter *SenPriority*, the system layer also decides whether the node can serve as a leader node, i.e., a leader candidate, which is denoted by the boolean parameter *Candidacy*. For example, *Candidacy* may be decided by whether *SenPriority* is above a predetermined threshold, and other related metrics such as the node computation capability. The combination of the parameters *SenPriority* and *Candidacy* decides the network address, which can be acquired by using the primitive function CalcAddress(...) in Table 2.

Some information variables are further defined in Table 3. With the provided primitive functions in Table 1 and 2, the primitive-programming pseudo-code of the wireless sensor nodes is shown in Table 4. In principle, the programming utilizes data-aggregation modules to opportunistically collect context-related sensing data at leader nodes. The cognitive unicast is used for forwarding the generate tracking-record packet to the next leader node along the event track. Since these two abstract wireless links are used consecutively along the event moving trajectory, the related sensing data along the track is collected by the leader nodes, generating the event profile. The use of unicast modules and the group address of leader node *LeaderAddr* can pick up one leader candidate, if more than one such candidate exists. Not shown in the primitive programming, the wireless sensor nodes also periodically read their sensors for the initial event detection.

The application QoS is determined by the event location estimation error. The optimization between location error and network energy consumption can be controlled by a knob parameter at the system layer, which determines how much data to collect and aggregate at a leader node. When

Table 3. Sensor node information variables (International Journal of Communication Networks and Distributed Systems, Volume 2/2009) (URL: http://www.inderscience.com/search/index.php?action=record&rec_id=26558) © Inderscience Enterprises Ltd.

Variable	Description
Addr	Network address of the local sensor node;
EventContext	Context information about the interested event under surveillance;
LeaderAddr	Group network address of the leader nodes, i.e., *Candidacy =1*;
TrackPriority	Traffic priority of the event tracking.

Table 4. Programming of wireless sensor nodes (International Journal of Communication Networks and Distributed Systems, Volume 2/2009) (URL: http://www.inderscience.com/search/index.php?action=record&rec_id=26558) © Inderscience Enterprises Ltd.

```
Switch (message)
{
Case UnicastP. indicate (SA, DATA):
  if ((SA is of LeaderAddr)&&(DATA is of tracking record packets))
    Aggregate.start(EventContext);
  break;
Case Aggregate. indicate (SA,DATA):
  if (DATA is related to EventContext)
    if (Enough sensing data has been collected)
    {
      Aggregate.stop (EventContext);
      Generate the tracking record packet TraDATA;
      UnicastP.send (Addr, LeaderAddr, TraDATA, TrackPriority);
    }
    else
      Save the DATA for further generating the tracking record;
  break;
Case Aggregate. request (AA, Context):
  EventContext = Context;   SensorReading(EventContext, &DATA);
  CalcAddress(DATA, EventContext, &SenPriority, &Candidacy);
  Addr=[SenPriority, Candidacy, EventContext];
  Aggregate.send(AA, DATA, SenPriority);
  break;
}
```

sufficient data has been collected, *Aggregate.stop(...)* can be issued to terminate the local data aggregation and initiate the next iteration of processing. It is intuitive that collecting more date would contribute to higher location accuracy, but consume more energy at the same time.

Song and Hatzinakos (2007) presents network simulation of such an event detection and tracking wireless sensor networks, via the discrete event simulation system OMNeT++ (2012). Let 500 wireless nodes be randomly deployed in a 50X50 meter square region. The network simulation time duration is set to be 120 *sec*, and the target appears in the surveillance square at the time 30 *sec*, and disappears at the time 90 *sec*. Without loss of generality, the target mobility is configured as follows: the target velocity is fixed at 10 m/s, and the direction is a random variable uniformly distributed in $\left[0, 2\pi\right)$. The mobility direction is independently updated every 0.5 *sec*, while the generator guarantees that the target would not move out of the surveillance region in the next time period of 0.5 *sec*. The physical communication parameters are chosen based on IEEE 802.15.4. Acoustic sensors are assumed where the sensor detection *SenPriority* is the energy detection over multiple sensing samples. Two configurable parameters as shown in the simulation results of Figure 12, 13, and 14 are: 1) the number of sensor samples N per detection; 2) the QoS knob $\rho\,(0<\rho<1)$ deciding how much data to collect in an iteration of data aggregation.

Figure 13 shows that the network detects the target at the 30 *sec* and the network energy consumption increases linearly with the time until

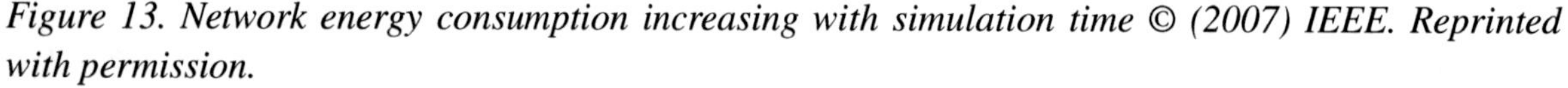

Figure 13. Network energy consumption increasing with simulation time © (2007) IEEE. Reprinted with permission.

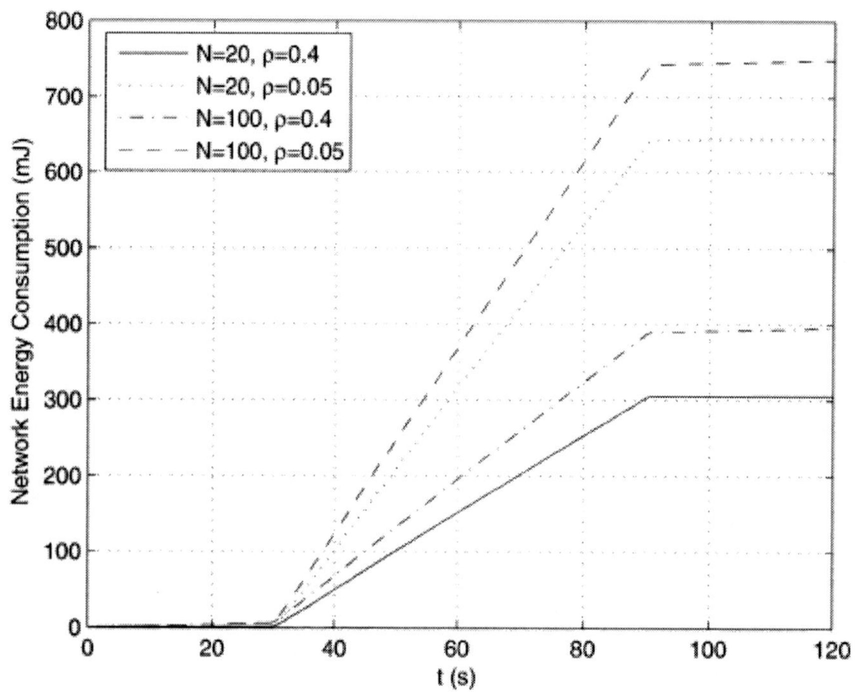

Figure 14. Network energy consumption at 120 sec time © (2007) IEEE. Reprinted with permission.

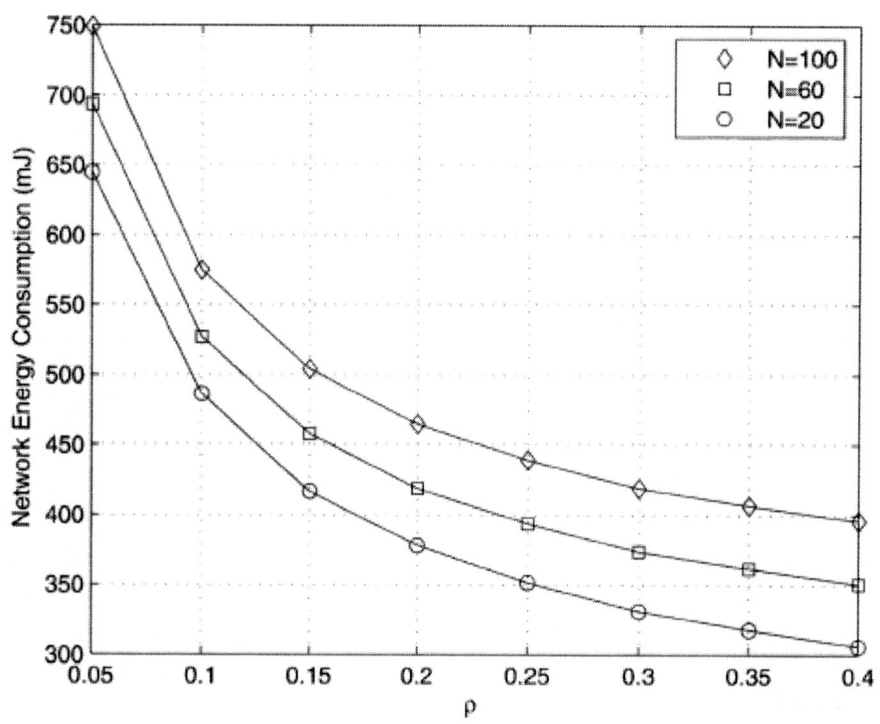

the 90 *sec*, when the target leaves the sensor deployed area. It suggests the network can well track the target when it is moving in the area under surveillance.

Figure 14 and Figure 15 further compares the network energy consumption and location error respectively, when the QoS knob ρ varies. With higher ρ, less the network energy consumption can be achieved, but higher location error is also introduced. The location accuracy also improves with sensor sampling rate *N*, at the cost of higher network energy consumption. It is apparent that ρ is used for controlling the application QoS in the cognitive data aggregation, so as to determine how much application data is collected at the leader (aggregation) node in the target location estimation. The design of cognitive data aggregation ensures that the data with the highest quality is collected.

SMART INFRASTRUCTURE

Applications

Beyond 3G smart infrastructures usually requires cost-effective wireless bandwidth that current network infrastructures cannot support. In the business world of wireless telecommunications, a killer application comparable to wireless voice has not been identified in wireless data services. Killer application such as wireless voice consumes very little bandwidth but can bring great profit to service providers, which is a primary reason why cellular infrastructures can be deployed across the globe. Due to the lack of such killer applications in wireless data services, significant business barriers exist for the world's telecom operators to deploy more wireless bandwidth with the current technologies.

Cognitive networking can provide the most cost-effective bandwidth in large-scale wireless

Figure 15. Target localization error comparison © (2007) IEEE. Reprinted with permission.

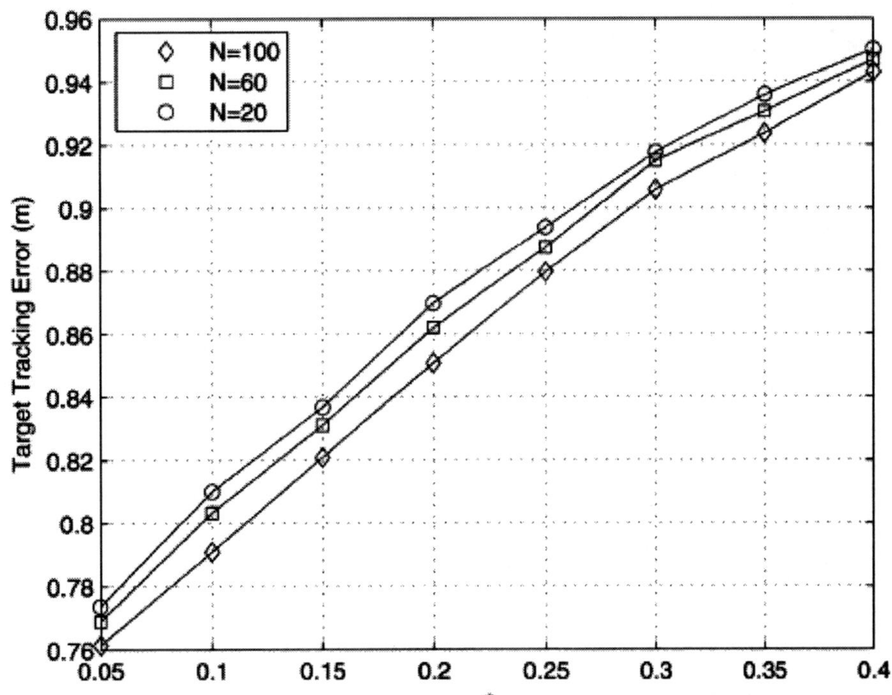

infrastructures. Compared to traditional wireless networks, it has been identified to achieve 5-10 times higher throughput (bandwidth), with a fraction of the cost in terms of materials, installation, and maintenance. The full potential can be a game-changer that brings the most cost-effective wireless connectivity to "everybody" and "everything." For "everybody," it can address the pain of insufficient bandwidth and coverage of wireless Internet used by smart phones; and for "everything" it can bring smart wireless connectivity to sensor-devices in many traditional industries, including smart utility networks, indoor location/tracking (context computing) networks, mining, healthcare, surveillance and emergency communications, agriculture, home/building automation, and retailers.

Thus such a new network platform as cognitive networking can have great potentials in numerous vertical applications. Once it becomes ubiquitous, its full impact can be comparable to how packet switching technology (Internet) has differentiated from circuit switching technology (telephone networks) in wire-line networks. It creates a new world where high-bandwidth and low-cost communications may become possible, for wireless and mobile devices (see Table 5).

More detailed application features of cognitive networks are further elaborated in what follows:

- **Dynamic Network Planning and Deployment Model:** Since the radio resource is opportunistically utilized, no deterministic network topology needs to be maintained. The wireless stations can be "drop-and-play" in the network deployment. Putting more nodes can improve the radio resource to be opportunistically exploited, and therefore increase the network capacity. Likewise removing any individual nodes shall not create bottlenecks in the network. This fluid "drop-and-play" nature offers the potential of great cost-saving in the infrastructure planning and deployments. The setup also does not need expensive planning and calibration, as multi-tier new deployments (for example introduced by service providers or subscribers) guarantee improved network capacity. High mobility can be supported.
- **Better Network Resource Utilization:** The network resource in large-scale wireless networks includes: the amount of spectrum bandwidth and the number of wireless stations. Theoretical network capacity is decided by the network resources, and the multiplication of these two factors (Gupta & Kumar, 2003). Traditional wireless networking depends on a deterministic mesh-network topology. It is therefore difficult to efficiently utilize the network resources, subject to a dynamic wireless networking environment where both spectrum bandwidth and mesh radio availability cannot be predetermined. Cognitive networking offers a means of better network-resource utilization, approaching the information-theoretical limit on wireless-network capacity.
- **Supporting High-Quality and Real-Time Services:** Due to the opportunistic network-resource utilization, reliable wireless communications with specified dataflow throughput, end-to-end delay, and delay variance can be supported over multiple wireless hops. Therefore, real-time services, including high-quality multimedia, can be supported in smart mesh infrastructures by the cognitive networking. Dataflow throughput can be independent of the num-

Table 5. Evolution of networking technology

Networking	Medium	Traffic
Circuit Switch	Reliable	Predetermined
Packet Switch	Reliable	Random
Cognitive Wireless Networking	Unreliable	Random

ber of wireless hops; end-to-end delay and delay variance only increase linearly with the number of wireless hops; and delay variance can also diminish to zero with higher network density. Therefore, network operators only need to assure that sufficient network resources are deployed to support the applications, where the resources, e.g., gateway capacity and wireless stations, can be deployed with low cost.

- **Robust to Wireless Interferences:** Provided by the opportunistic spectrum access, the network can be robust to interferences, which can be substantial in unlicensed spectrum bands (e.g., ISM bands). For example, effective operation within unlicensed bands brings large free bandwidth to smart infrastructures, which results in large network capacity with virtually zero cost.
- **Supporting Scalable Radio Complexity (Low Power):** The complexity of individual wireless nodes (with cognitive-networking capabilities) is low and independent of network scale. The low radio complexity can result in low power consumption, lower cost, and long battery life. When needed, it is also possible to power the cognitive radio by cost-effective solar panel, which will further reduce installation cost by removing any cable attachment.
- **Better Economics and Business Case (Low Cost):** Cognitive networking can offer excellent economics in large-scale wireless systems, by which 1) the costs of deploying network resources could be greatly reduced by the utilization of unlicensed spectrum bands and drop-and-play (mobile and mesh) stations; 2) much higher efficiency in network-resource utilization results in excellent performance with all the available resources being used to their instantaneous maximum.

Standards

Standards have played important roles in telecommunications, which make equipment from different vendors interpretable. For example, in cellular networks and WLAN, different brands of cellular phones or WLAN cards can connect to the base stations or access points according to the respective standards. Standardization in wireless mesh (multi-hop) networks, such as IEEE 802.11S (broadband) WiFi mesh and Zigbee (low power) mesh, however, hasn't been comparably successful, since most current wireless infrastructures takes a star topology, and self-organizing multi-hop wireless networks are still very limited. Once the cognitive networking devices can become ubiquitous, standards are necessarily required especially for the potential of massive consumer applications.

The principles of cognitive networks can be compatible with all established wireless radio standards, so that the implementation can be independent of physical radios. Therefore, radio

Figure 16. RapidMesh™ OPM15 development board (IEEE 802.15.4 compatible) © OMESH Networks Inc., with permission

modules (with cognitive-networking capabilities) can use off-the-shelf RF technology. The implementation can also be seamlessly integrated with existing network-layer protocols, including for example Internet Protocols. Given the standard compatibility, it can be relative straightforward to extend the cognitive-networking capabilities to respective standards, so that inter-vendor interpretability can be realized for massive applications with such interoperability requirements. Towards this perspective, OPM Radios (2012) are having radio modules with the cognitive-networking capabilities, being compatible to standards including IEEE 802.11 and IEEE 802.15.4 (see Figure 16).

To the best of our knowledge, such cognitive-networking radios have been deployed in fields for the applications of Real-Time Location Systems (RTLS), and Wireless Sensor Networks (WSN). Upcoming deployments in emergency applications, and broadband networks are well expected. Massive consumer applications in integrated smart infrastructures and services can be a hot topic with greater impact for the technologies on the road.

CONCLUSION

Cognitive radio has received great research attentions, especially since Federal Communications Commission (FCC) decided to open white spaces of TV band for secondary radios to access. Cognitive network is still in a comparatively earlier research stage, although recent development and pilots have shown a lot of promising applications. In this chapter, we have elaborated the major advantage of cognitive wireless networks in providing cost-effective wireless bandwidth of B3G smart infrastructures. We have also introduced the current research results on cognitive wireless networks, including the principles, network architecture, as well as the design and analysis of two wireless link modules. The contribution may be viewed as a fundamental work based on which we hope many interesting networking and application researches can be performed for cognitive wireless networks.

REFERENCES

Biswas, S., & Morris, R. (2004). Opportunistic routing in multi-hop wireless networks. *SIGCOMM Computer Communication Review*, *34*(1), 69–74. doi:10.1145/972374.972387

Brooks, R., Ramanathan, P., & Sayeed, A. (2003). Distributed target classification and tracking in sensor networks. *Proceedings of the IEEE*, *91*(8), 1163–1171. doi:10.1109/JPROC.2003.814923

Chachulski, S., Jennings, M., Katti, S., & Katabi, D. (2007). Trading structure for randomness in wireless opportunistic routing. In *Proceedings of the 2007 Conference on Applications, Technologies, Architectures, and Protocols for Computer Communications,* (pp. 169–180). New York, NY: ACM Press.

Chen, T., Zhang, H., Maggio, G. M., & Chlamtac, I. (2007). CogMesh: A cluster-based cognitive radio network. In *Proceedings of the 2007 IEEE Symposium on New Frontiers in Dynamic Spectrum Access Networks (IEEE DySPAN 2007)*. Dublin, Ireland: IEEE Press.

Chowdbury, K. R., & Akyildiz, I. F. (2008). Cognitive wireless mesh networks with dynamic spectrum access. *IEEE Journal on Selected Areas in Communications*, *26*(1), 168–181. doi:10.1109/JSAC.2008.080115

Chu, M., Haussecker, H., & Zhao, F. (2002). Scalable information-driven sensor querying and routing for ad hoc heterogenous sensor networks. *International Journal of High Performance Computing Applications*, *16*(3), 293–313. doi:10.1177/10943420020160030901

De Couto, D., Aguayo, D., Bicket, J., & Morris, R. (2005). A high-throughput path metric for multi-hop wireless routing. *Wireless Networking*, *11*(4), 419–434. doi:10.1007/s11276-005-1766-z

Fortuna, C., & Mohorcic, M. (2009). Trends in the development of communication networks: Cognitive networks. *Computer Networks*, *53*(9), 1355–1376. doi:10.1016/j.comnet.2009.01.002

Guo, D., & Wang, X. (2004). Dynamic sensor collaboration via sequential Monte Carlo. *IEEE Journal on Selected Areas in Communications*, *22*(6), 1037–1047. doi:10.1109/JSAC.2004.830897

Gupta, P., & Kumar, P. R. (2003). Towards an information theory of large net-works: An achievable rate region. *IEEE Transactions on Information Theory*, *49*(8), 1877–1894. doi:10.1109/TIT.2003.814480

Intanagonwiwat, C., Govindan, R., Estrin, D., Heidemann, J., & Silva, F. (2002). Directed diffusion for wireless sensor networking. *IEEE/ACM Transactions on Networking*, *11*(1), 2–16. doi:10.1109/TNET.2002.808417

Mitola, J. (1999). Cognitive radio: Making software radios more personal. *IEEE Personal Communications*, *6*(4), 13–18. doi:10.1109/98.788210

Moore, J., Keiser, T., Brooks, R., Phoha, S., Friedlander, D., Koch, J., et al. (2003). Tracking targets with self-organizing distributed ground sensors. In *Proceedings of the IEEE Areospace Conference*. IEEE Press.

OMNeT++. (2012). *Website.* Retrieved from http://www.omnetpp.org

Radios, O. P. M. (2012). *Website.* Retrieved from http://www.omeshnet.com

Song, L. (2008). Cognitive networks: Standardizing the large scale wireless systems. In *Proceedings of IEEE Consumer Communications and Networking Conference*, (pp. 988 – 992). Las Vegas NV: IEEE Press.

Song, L. (2008). Mesh infrastructure supporting broadband Internet with multimedia services. In *Proceedings of the IEEE International Conference on Circuits and Systems for Communications*. Shanghai, China: IEEE Press.

Song, L., & Hatzinakos, D. (2007). A cross-layer architecture of wireless sensor networks for target tracking. *IEEE/ACM Transactions on Networking*, *15*(1), 145–158. doi:10.1109/TNET.2006.890084

Song, L., & Hatzinakos, D. (2008). Real-time communications in large scale wireless networks. *International Journal of Multimedia Broadcasting*. Retrieved from http://www.hindawi.com/journals/ijdmb/2008/586067/

Song, L., & Hatzinakos, D. (2009). Cognitive networking of large scale wireless systems. *International Journal of Communication Networks and Distributed Systems*, *2*(4), 452–475. doi:10.1504/IJCNDS.2009.026558

Song, L., & Hatzinakos, D. (2011). Wireless sensor networks: From application specific to modular design. In Foerster, A. (Ed.), *Emerging Communications for Wireless Sensor Networks*. New York, NY: IN-TECH.

Spachos, P., Bui, F., Song, L., Lostanlen, Y., & Hatzinakos, D. (2011). Performance evaluation of wireless multihop communications for an indoor environment. In *Proceedings of IEEE International Symposium on Personal, Indoor and Mobile Radio Communications*. Toronto, Canada: IEEE Press.

Spachos, P., Song, L., & Hatzinakos, D. (2011). Performance comparison of opportunistic routing schemes in wireless sensor networks. In *Proceedings of the Ninth Annual Communication Networks and Services Research Conference*, (pp. 271-277). IEEE.

Spachos, P., Song, L., & Hatzinakos, D. (2012). Opportunistic multihop wireless communications with calibrated channel model. In *Proceedings of IEEE International Conference on Communications (ICC)*. Ottawa, Canada: IEEE Press.

Thomas, R. W., DaSilva, L. A., & MacKenzie, A. B. (2005). Cognitive networks. In *Proceedings of the First IEEE International Symposium on New Frontiers in Dynamic Spectrum Access Networks*. Baltimore, MD: IEEE Press.

Volcano Lab. (2012). *Website.* Retrieved from http://www.siradel.com

Westphal, C. (2006). Opportunistic routing in dynamic ad hoc networks: The oprah protocol. In *Proceedings of the IEEE International Conference on Mobile Adhoc and Sensor Systems,* (pp. 570 –573). IEEE Press.

Yan, Y., Zhang, B., Mouftah, H., & Ma, J. (2008). Practical coding aware mechanism for opportunistic routing in wireless mesh networks. In *Proceedings of the IEEE International Conference on Communications,* (pp. 2871 –2876). IEEE Press.

Zhang, W., & Cao, G. (2004). Optimizing tree reconfiguration for mobile target tracking in sensor networks. In *Proceedings of IEEE INFOCOM* (pp. 2434–2445). IEEE Press.

Zhao, B., Seshadri, R., & Valenti, M. (2004). Geographic random forwarding with hybrid-arq for ad hoc networks with rapid sleep cycles. In *Proceedings of IEEE Global Telecommunications Conference,* (pp. 3047 – 3052). IEEE Press.

Zhao, F., Shin, J., & Reich, J. (2002). Information-driven dynamic sensor collaboration. *IEEE Signal Processing Magazine*, *19*(2), 61–72. doi:10.1109/79.985685

Zhao, Q., & Sadler, B. (2007). A survey of dynamic spectrum access. *IEEE Signal Processing Magazine*, *24*(3).

Zhong, Z., Wang, J., Nelakuditi, S., & Lu, G. (2006). On selection of candidates for opportunistic anypath forwarding. *SIGMOBILE Mobile Computer Communication Review*, *10*(4), 1–2. doi:10.1145/1215976.1215978

Zorzi, M., & Rao, R. (2003a). Geographic random forwarding (geraf) for ad hoc and sensor networks: Energy and latency performance. *IEEE Transactions on Mobile Computing*, *2*(4), 349–365. doi:10.1109/TMC.2003.1255650

Zorzi, M., & Rao, R. (2003b). Geographic random forwarding (geraf) for ad hoc and sensor networks: Multihop performance. *IEEE Transactions on Mobile Computing*, *2*(4), 337–348. doi:10.1109/TMC.2003.1255648

Chapter 3
Iterative Optimization of Energy Detector Sensing Time and Periodic Sensing Interval in Cognitive Radio Networks

Mohamed Hamid
University of Gävle, The Royal Institute of Technology (KTH), Sweden

Niclas Björsell
University of Gävle, Sweden

Abbas Mohammed
Blekinge Institute of Technology, Sweden

ABSTRACT

In this chapter the authors propose a new approach for optimizing the sensing time and periodic sensing interval for energy detectors in cognitive radio networks. The optimization of the sensing time depends on maximizing the summation of the probability of right detection and transmission efficiency, while the optimization of periodic sensing interval is subject to maximizing the summation of transmission efficiency and captured opportunities. Since the optimum sensing time and periodic sensing interval are dependent on each other, an iterative approach to optimize them simultaneously is proposed and a convergence criterion is devised. In addition, the probability of detection, probability of false alarm, probability of right detection, transmission efficiency, and captured opportunities are taken as performance metrics for the detector and evaluated for various values of channel utilization factors and signal-to-noise ratios.

DOI: 10.4018/978-1-4666-2812-0.ch003

INTRODUCTION

Demands on high data rate applications are increasing and consequently demands on spectral resources are increasing as well. Several studies, initiated recently by the US regulator Federal Communications Commission (FCC), have shown that the frequency spectrum is underutilized and inefficiently exploited: some bands are highly crowded, at some day hours or in dense urban areas, while others remain poorly used. Regulators worldwide are beginning to recognize that the traditional way of managing the electromagnetic spectrum, called Fixed Spectrum Access (FSA), in which the licensing method of assigning fixed portions of spectrum, for very long periods, is inefficient (Wellen, Wu, & Mahonen, 2007).

Among the efforts taken, by regulators worldwide, in order to achieve better usage of spectrum is the introduction (promotion) of secondary markets. In a secondary usage context, the spectrum owned by the license owner (also called primary user) can be shared by a non-licensee referred to as a secondary user. Besides the promotion for secondary markets, we are currently experiencing rapid evolutions of Software Defined Radio (SDR) techniques. Such techniques allow reconfigurable wireless transceivers to change their transmission/reception parameters, such as the operating frequency that can be modified over a very wide band, according to the network or users' demands. The efforts taken by regulators in order to make better usage of spectrum, in particular the promotion for secondary market, together with the rapid evolution of the SDR techniques, have led to the development of opportunistic Cognitive Radio (CR) systems. The term Cognitive Radio was first introduced by Mitola in 1999. CR generally refers to a radio device that has the ability to sense its Radio Frequency (RF) environment and modify its spectrum usage based on what it detects (Mitola, 1995).

In cognitive radio environments the primary users are allocated licensed frequency bands while secondary cognitive users can be dynamically allocated the empty frequencies within the licensed frequency band, according to their requested Quality of Service (QoS) specifications. Spectrum sensing is commonly recognized as the most fundamental task in cognitive radio based on dynamic spectrum access due to its important role in discovering the spectrum holes (Haykin, 2005). To achieve this goal the unlicensed user should monitor the licensed channels to identify the spectrum holes and to properly utilize them. The concept of spectrum hole or spectrum opportunity imposes a multi dimensional spectrum awareness (Yucek & Arslan, 2009) since a spectrum hole is a function of frequency, time and geo-location. Besides, considering noise existence all over the radio spectrum, then spectrum hole is a function of the received power as well since a received power equal to the noise floor means spectrum hole. The concept of spectrum holes is as demonstrated in Figure 1.

The detection of the existence of the primary user is a fundamental task to be performed by the Secondary User (SU); the following three approaches have been proposed for achieving this task (Ghasemi & Sousa, 2008):

- **Database Registry:** The PU activity information is obtained through an accessible database by the SUs.
- **Beacon Signals:** A beacon signal is sent via standardized channel to tell which channels are occupied by the PUs.
- **Spectrum Sensing:** The SU is responsible for checking the PU signal existence and accordingly finding the spectrum holes.

The database registry and beacon signals approaches require supporting positioning in SUs and implies cost burden on the PUs since they are responsible for providing information about spectrum availability. Moreover, Internet connection is necessary to adopt database registry and standardized channel is needed for beacon

Figure 1. Spectrum holes concept

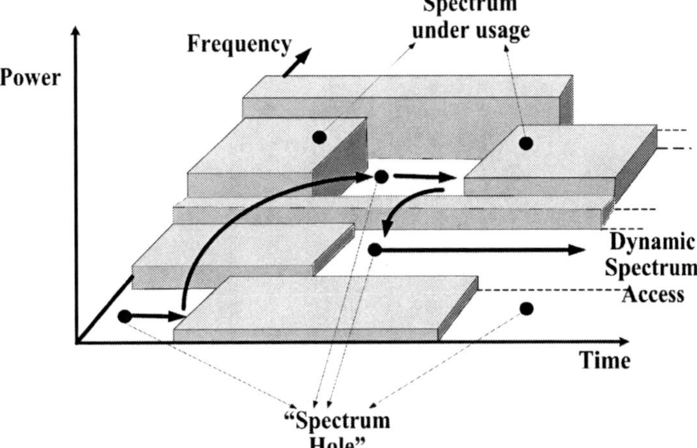

signals. On the other hand, spectrum sensing does not charge PU; hence, it attracts considerable attention as the most likely approach to be used. Furthermore, spectrum sensing can be used as a priori step for the database registry and beacon signals approaches where the sensing is done by a centralized device, which populates a database or spreads out the sensing information via a beacon signal (Weiss, Hillenbrand, & Jondral, 2003). Various spectrum sensing techniques have been proposed, such as:

- **Energy Detection (Kyungtae *et al.*, 2010; Urkowitz, 1967):** Energy detector captures the signal and calculates its energy over a specific period of time. This energy is compared with a predefined threshold, the existence of a PU signal is confirmed when the signal energy exceeds the threshold and vice versa. Energy detection based spectrum sensing imposes some challenges including threshold selection, inefficiency of detecting spread spectrum signals and inability to distinguish between PU signals and interference (Cabric, Mishra, & Brodersen, 2004).
- **Waveform Sensing (Mishra *et al.*, 2007):** Here a captured signal is tested to verify if it is a PU signal or not. The test is done by checking the signal pattern in terms of preambles, pilot patterns, spreading sequences, etc. The measurements findings reported in Cabric, Tkachenko, and Brodersen (2006) show that waveform sensing requires short sensing time; however, it is very sensitive to synchronization errors.
- **Matched Filter Detection (Kapoor, Rao, & Singh, 2011):** The detector is a matched filter with a PU signal as a transfer function. Matched filtering outperforms other sensing techniques since it requires shorter sensing time (Tandra & Sahai, 2005). The drawback of matched filtering method is that the SU needs to demodulate the received signal, thus perfect knowledge about PU signal features is required such as the modulation scheme, pulse shaping and bandwidth.
- **Cyclostationarity Feature Detection (Liu & Zhai, 2008):** The PU signal existence is detected by checking the cyclic correlation function of the received signal. The concept behind the cyclostationarity detection is that noise is uncorrelated Wide-Sense Stationary (WSS) process while modulated signals are cyclostationarity due to the

signal's periodicities. Compared to other sensing techniques cyclostationarity feature detection has medium accuracy and complexity.

- **Eigenvalues-Based Detection (Yonghong & Ying-Chang, 2007):** The ratio between the eigenvalues of the correlation matrix of the received signal can be used to declare the existence or absence of the PU signal.

In spite of the many detection methods, energy detection seems to be the simplest method (Won-Yeol & Akylidiz, 2008; Tang, 2005). Optimization of the sensing parameters for energy detector is of a great importance in order to optimize the sensing time and the periodic sensing interval as in Won-Yeol and Akylidiz (2008) where the sensing time is considered to be fixed and the periodic sensing interval which minimizes the mutual interference between SU and PU is computed. In Hai Ngoc et al. (2010) the sensing time which gives a specific value of the probability of false alarm is calculated and then the periodic sensing intervals which minimize the energy consumption in a sensing network are obtained. In Hamid and Mohammed (2012) the periodic sensing intervals for multi channels system are optimized considering a preset value for the sensing time. This chapter proposes a new method for optimizing the sensing time and periodic sensing interval simultaneously with the aim at maximizing the detector performance metrics without the need for setting these parameters in advance. The organization of this chapter is as follows. First, the theoretical background, the system model, performance metric and optimization method are introduced. Then we present the simulation assumptions, approaches and considerations, convergence criterion, simulation parameters and results. Finally, we provide conclusions and directions for future work.

THEORETICAL ANALYSIS

This section presents the theoretical aspects and it considers two parts: the system model and the optimization procedure of the sensing time and the periodic sensing interval.

System Model

The maximum *a posteriori* probability (MAP) detector is an optimal signal detector when it comes to the detection precision (Won-Yeol & Akylidiz, 2008) where the received signal by the Secondary User (SU) can be expressed as (Digham, 2007):

$$r(t) = \begin{cases} n(t), & H_0 \\ s(t) + n(t), & H_1 \end{cases} \tag{1}$$

where $s(t)$ is the primary user signal $n(t)$ is the noise. In Equation 1, the hypothesis H_1 corresponds to 'the existence of PU signal' and H_0 corresponds to 'the absence of PU signal.' For a detector which performs energy detection over a bandwidth W, the detector collects N samples of $r(t)$, squares and sums them up to test H_0 and H_1 as shown in Figure 2.

The output of the MAP energy detector has a Chi-square distribution (Hai Ngoc, et al., 2010); however, if we have a large number of samples and apply the central limit theorem, the Chi-square distribution can be approximated by Gaussian distribution. In this case, the output of the MAP detector Y, which is to be compared with a threshold level λ, can be expressed as:

$$Y \sim \begin{cases} \mathcal{N}\left(N\sigma_n^2, N\sigma_n^4\right), & H_0 \\ \mathcal{N}\left(\left(N\sigma_n^2 + \sigma_s^2\right), 2N\left(\sigma_n^2 + \sigma_s^2\right)^2\right), & H_1 \end{cases} \tag{2}$$

where $\mathcal{N}\left(\mu, \sigma^2\right)$ is a Gaussian distributed random variable with a mean of μ and a standard deviation

Figure 2. Block diagram of MAP energy detector

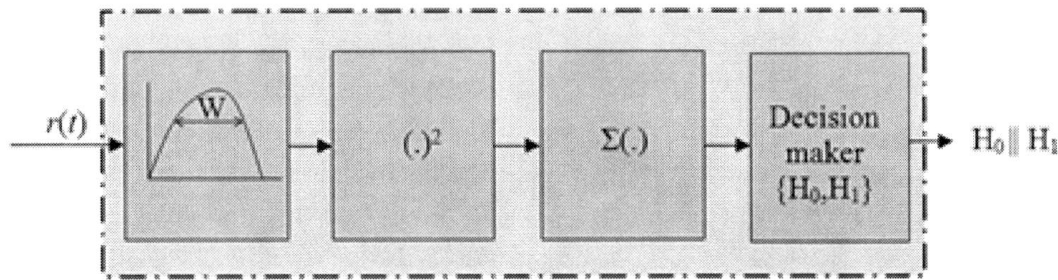

of σ, σ_n^2 is the noise variance and σ_s^2 is the signal variance.

In Hamid and Mohammed (2012), the primary channel is modeled as an alternating ON/OFF Markov process corresponding to the primary system activity, where the ON state corresponds to H_1 and the OFF state corresponds to H_0. The ON and OFF periods are assumed to be exponentially distributed with a mean of α and β seconds, respectively. Consequently, the channel utilization factor u, which is can defined as the fraction of time during which the channel is being utilized by its primary user, can be expressed as:

$$u = \frac{\alpha}{\alpha + \beta} \tag{3}$$

Optimization of the Sensing Time and Periodic Sensing Interval

Let us define the probability of false alarm p_{fa} to represent the probability of claiming a primary channel as an occupied channel while it is actually free, or simply declaring H_1 wrongly; this can be written as (Hai Ngoc, et al., 2010):

$$p_{fa} = \frac{\beta}{\alpha + \beta}\mathbb{Q}\left(\frac{\lambda - N\sigma_n^2}{\sqrt{2N}\ \sigma_n^2}\right) \tag{4}$$

where $\mathbb{Q}(.)$ is the Q function representing the complementary Cumulative Distribution Function (CDF) of the Gaussian distributed random variable. By using Nyquist criterion for sampling, W and N can be related by the sensing time t_s which is defined as the time during which the N samples are gathered. Thus, the relation between N and W can be expressed by:

$$N = 2Wt_s \tag{5}$$

Substituting Equation 5 in Equation 4 yields:

$$p_{fa} = \frac{\beta}{\alpha + \beta}\mathbb{Q}\left(\frac{\lambda - 2t_s W\sigma_n^2}{2\sqrt{t_s W}\ \sigma_n^2}\right) \tag{6}$$

From Equation 6, and by knowing p_{fa}, λ can be calculated as:

$$\lambda = 2\sqrt{t_s W}\ \sigma_n^2\mathbb{Q}^{-1}\left(\frac{\left(\alpha + \beta\right)p_{fa}}{\beta}\right) + 2t_s W\sigma_n^2 \tag{7}$$

where $\mathbb{Q}^{-1}(.)$ represents the inverse Q function.

Now, let us define the probability of detection p_d to be the probability of correct detection of a primary signal existence, which can be obtained by:

$$p_d = \frac{\alpha}{\alpha + \beta}\mathbb{Q}\left(\frac{\lambda - 2t_s W\left(\sigma_n^2 + \sigma_s^2\right)}{2\sqrt{t_s W}\left(\sigma_n^2 + \sigma_s^2\right)}\right) \tag{8}$$

Equation 8 can be rewritten in terms of the signal-to-noise ratio (SNR), γ, which is equivalent to $\left(\sigma_s^2 / \sigma_n^2\right)$, as:

$$p_d = \frac{\alpha}{\alpha+\beta}\mathbb{Q}\left(\frac{\lambda - 2Wt_s\left(\lambda+1\right)\sigma_n^2}{2\sqrt{Wt_s}\left(\lambda+1\right)\sigma_n^2}\right) \tag{9}$$

According to the best of our knowledge, the extensive work done on energy detection assumes a fixed value of p_{fa} and subsequently the calculations of other parameters are done. However, here we aim to minimize p_{fa} and maximize p_d simultaneously; it is important to note that these two parameters are contradictory processes according to Equations 6 and 9 with respect to t_s. Therefore, an optimization solution to evaluate the optimum value of t_s is needed. This optimization is accomplished by introducing a new performance metric: the probability of right detection p_{rd} which represents the probability of having no false alarm and correct detection and is defined as in Box 1 (see also Figure 3).

Box 1.

$$p_{rd} = p_d * \left(1 - p_{fa}\right) = \frac{\alpha}{\alpha+\beta}\mathbb{Q}\left(\frac{\lambda - 2t_sW\left(\lambda+1\right)\sigma_n^2}{2\sqrt{t_sW}\left(\lambda+1\right)\sigma_n^2}\right)\left(1 - \frac{\beta}{\alpha+\beta}\mathbb{Q}\left(\frac{\lambda - 2t_sW\sigma_n^2}{2\sqrt{t_sW}\sigma_n^2}\right)\right) \tag{10}$$

Figure 3. The impact of t_s on p_{fa}, p_d and p_{rd} for α =2 sec, β =3sec, γ =–10 dB, σ_n^2 =–174 dBm/Hz, and W= 1MHz

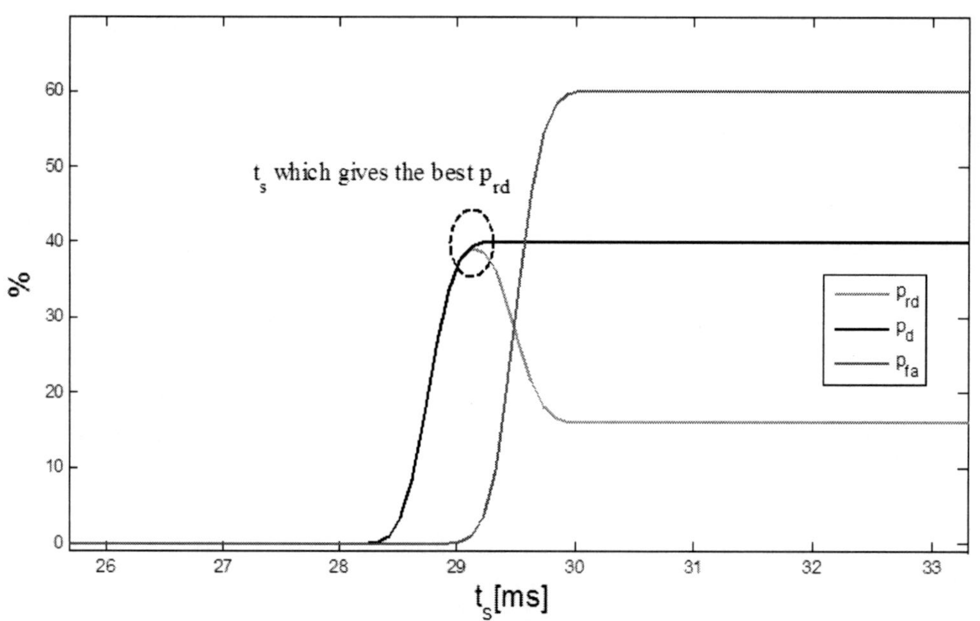

In addition to sensing when communications is needed, the secondary user required to check for the existence of primary user signal periodically:

- When proactive sensing is adopted. In this case, the channels are sensed periodically and the available channels are ranked according to their channel utilization factor in order to minimize the idle channel search delay (Hamid & Mohammed, 2012).
- If the primary system resumes transmission without broadcasting any information. That is, there is no beacon signal to notify the secondary system about the primary system start point of being active. Thus, the primary system reappearance check is done just via periodic sensing.

Regardless of the reason behind periodic sensing, the periodic time interval T_p affects both the transmission efficiency η, and the Unexplored Opportunities (UOP); these parameters are described below.

The transmission efficiency η is a measure of the free spectrum utilization which is defined as the ratio between time the secondary system spends on transmission and the time spent on transmission and sensing. Thus, η can be written as:

$$\eta = \frac{T_p}{T_p + t_s} \tag{11}$$

Higher η means better spectrum utilization. Consequently, from Equations 10 and 11, the optimum t_s which is referred to as t_s^* is defined as:

$$t_s^* = arg\ max_{ts} \left\{ \left(\eta + p_{rd} \right) \right\} \tag{12}$$

Figure 4 illustrates the change of the detection probability p_{rd}, transmission efficiency η, and

Figure 4. The impact of t_s on p_{rd}, η and $(p_{rd}+\eta)$ for $\alpha=2$ sec, $\beta=3$sec, $\gamma=-10$ dB, $\sigma_n^2=-174$ dBm/Hz, W=1 MHz, and $T_p=350$ ms

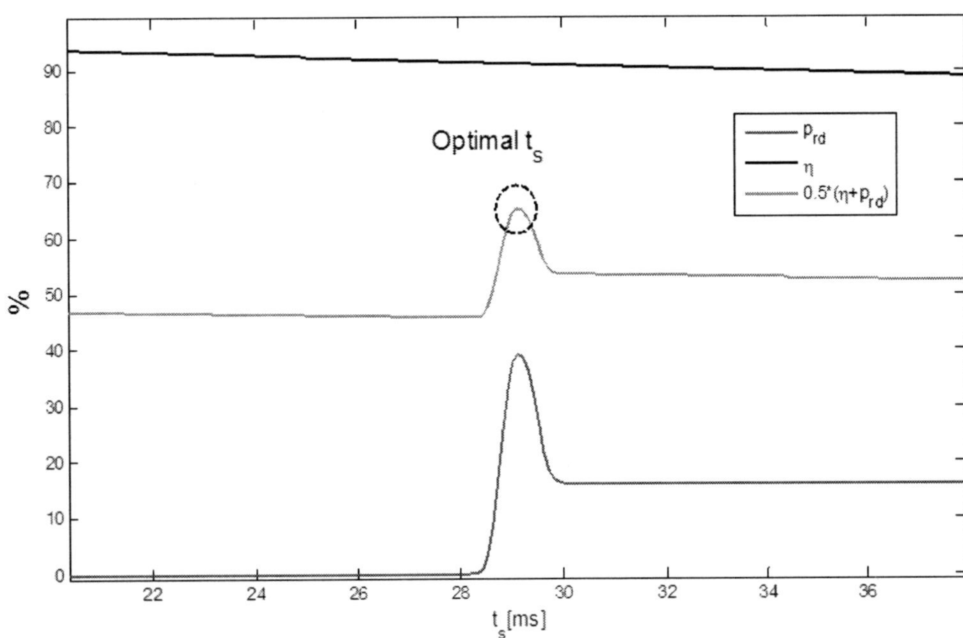

their normalized average $0.5^*(p_{rd}+\eta)$ for different values of the channel parameters (α, β, γ, σ_n, W and T_p).

An additional performance metric affected by T_p is the Unexplored Opportunities (UOP) (Hamid & Mohammed, 2012) which is the portion of time during which the secondary system misses to utilize the free spectrum due to detection process point's allocation in time. This can be written as:

$$UOP = \frac{\beta}{\alpha+\beta}\left(1+\frac{1}{\alpha T_p}\left(e^{-\alpha Tp}-1\right)\right) \tag{13}$$

Using the above UOP definition, we can find the captured opportunities (COP) defined as the portion of time during which the free spectrum is utilized by the secondary system and can be expressed as:

$$COP = 1 - UOP = 1-\frac{\beta}{\alpha+\beta}\left(1+\frac{1}{\alpha T_p}\left(e^{-\alpha Tp}-1\right)\right) \tag{14}$$

The concept of UOP is demonstrated in Figure 5.

From Equations 11 and 14, it is clear that increasing T_p would increase η and decrease COP. Therefore, the optimization of T_p can be accomplished by merging Equations 11 and 14 and finding T_p^* as the optimum T_p defined by:

$$T_p^* = arg\ max_{T_p}\left\{\left(\eta + COP\right)\right\} \tag{15}$$

The relationships between T_p and COP and between T_p and η are illustrated in Figure 5. Figure 6 shows the impact of T_p on COP, η and (η+COP) for different values of the channel parameters (α, β, γ, σ_n, W and t_s). In this figure, $0.5^*(\eta+COP)$ versus T_p are shown with the multiplication factor of 0.5 only to rescale the values to be between 0 and 1.

Equations 11, 12, and 15 show that t_s^* and T_p^* are both influenced by η which itself depends on t_s and T_p, respectively. Thus, in order to solve this optimization problem we propose an *iterative calculation of* t_s^* and T_p^* by starting from an arbitrary value for one of these parameters and stop when they converge; the procedure of these iterative calculations is explained in Algorithm 1.

SIMULATION RESULTS

In the following section, we present the simulation assumptions and approaches, convergence criterion, simulation parameters, and results.

Simulation Assumptions

In order to carry out the simulations, the following assumptions are taken into account:

Figure 5. The UOP concept: the sensing is done during the t_s intervals and the UOP intervals in the figure denote the missed opportunities

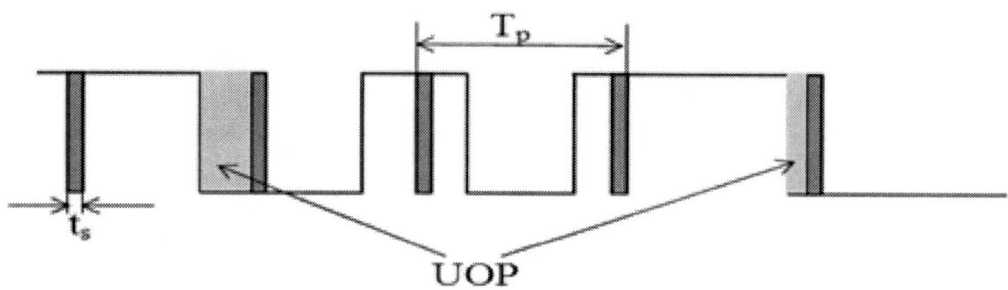

Figure 6. Impact of T_p on COP, η and (COP+ η) for α =2 sec, β=3sec, γ =–10 dB, σ_n^2 =–174 dBm/Hz, W=1 MHz, and t_s=30 ms

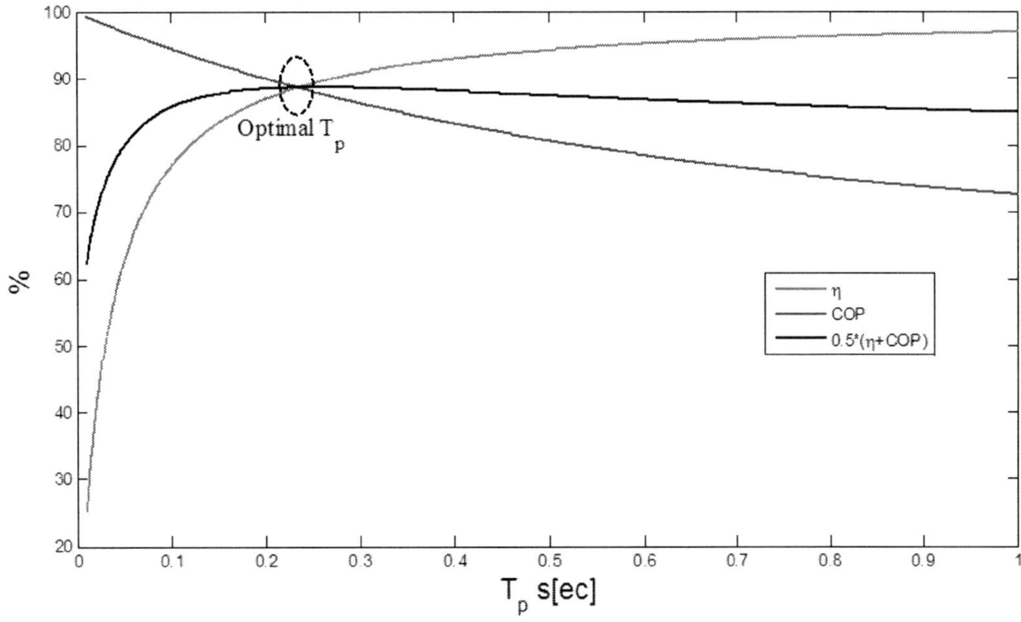

Algorithm 1. Iterative calculation of t_s^* and T_p^*

1. Note the values of α, β, γ, σ_n and W
2. Generate the vector of p_{fa} values
3. Set i to 0 and start with arbitrary values for $t_s(0)$ and $T_p(0)$
4. Calculate the λ vector according to Equation 7
5. Define the convergence criterion
6. Calculate p_d and p_{rd} vectors using Equations 9 and 10, respectively
7. Calculate η using Equation 11
8. Calculate COP using Equation 14
9. Calculate t_s^* (i)= t_s(i+1) using Equation 12
10. Calculate T_p^* (i)= T_p(i+1) using Equation 15
11. Iteratively repeat steps 6-10 until t_s^* and T_p^* are converged according to the defined convergence criterion used in step 5.

- The primary system activity is stationary and the channel parameters α and β are perfectly estimated. A suitable method for estimating α and β is the maximum likelihood estimator presented in Hamid and Mohammed (2012). Furthermore, σ_n and γ are assumed to be known.
- The time needed to perform the calculations for detection is neglected compared to the capturing of the N samples time, t_s. Therefore, the sensing and decision-making concerning the availability of the channel is made within t_s.

Simulation Approaches

Two simulations approaches will be considered in order to assess the performance of the proposed sensing time and periodic sensing interval optimization procedure as follows:

1. Optimization procedure implementation as described by Algorithm 1. This optimization is investigated for different pairs of α and β at a fixed value of γ of –10 dB. The main aim in this procedure is to assess the impact of α and β on the converged values of t_s^* and T_p^*, respectively.
2. Performance metric evaluation for various values of γ. In this perspective, the parameters p_{fa}, p_d, p_{rd}, η, COP, $(p_{rd} + \eta)$ and $(COP+ \eta)$ are considered as performance metrics and evaluated for different pairs of α and β.

Convergence Criterion

As in Algorithm 1, a convergence criterion is needed to stop iterating the calculation of t_s^* and T_p^*. This convergence criterion should highlight the optimal values of t_s^* and T_p^* which means further iterations would not provide significant improvement on these values and associated performance. The convergence criterion is defined as:

$$C = \left\| t_s^*\left(i\right) - t_s^*\left(i-1\right) \right\| < \varepsilon_{t_s} t_s^*\left(i-1\right) \quad \& \quad \left\| T_p^*\left(i\right) - T_p^*\left(i-1\right) \right\| < \varepsilon_{T_p} t_p^*\left(i-1\right) \tag{16}$$

where the logical value C allows continuation of iteration when it is 0 and stops it when it is 1, ε_{t_s} and ε_{T_p} are small values accounting for the tradeoff between convergence speed and accuracy, and & is the logical ANDING operation.

Simulation Parameters

For both simulation approaches 1 and 2 above, the simulation parameters listed in Table 1 is used. These simulation parameters are for a generalized system to investigate and assess the proposed optimization procedure characterized by Algorithm 1.

Simulation Results

The simulation results will be presented in two categories as follows.

The Sensing Time and Periodic Sensing Interval Optimization

Figure 7 shows the values of t_s for different values of mean ON and OFF periods (α and β) for an SNR, γ, of *–10 dB*, and starting from an arbitrary value of *10 ms* for t_s in each case. It is clear from Figure 7 that when the channel utilization factor, *u*, increases, the sensing time t_s increases. The simulation results show that for a channel with α=5 and β=1, t_s converges at *99 ms*, while it converges at *90 ms* when α=*3* and β=*2*, *20 ms* when α=2 and β=3 and finally t_s has the value of *6 ms* when α=1 and β=*5*.

In contrast to Figure 7, Figure 8 shows the values of T_p when γ is *–10 dB* which converges at

Table 1. Simulation parameters

Parameter	Value
(α, β) pairs	(2,3), (3,2), (1,5), (5,1)
W	1 MHz
σ_n^2	–174 dBm/Hz
γ	–20:1:0 dB[1]
p_{fa} vector	0.01%:0.01:10%[1]
t_s vector	1:1:100 ms[1]
T_p vector	0.01:0.1:50.01 s[1]
ε_{t_s}, ε_{T_p}	0.01

Figure 7. The optimal sensing time t_s^* with the iteration

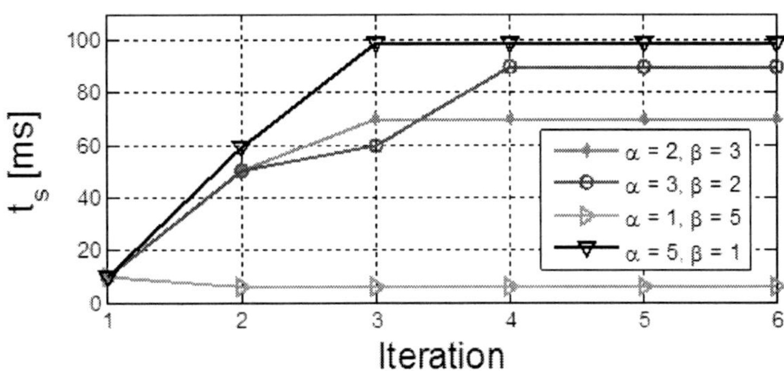

Figure 8. The optimal sensing time T_p^* with the iteration

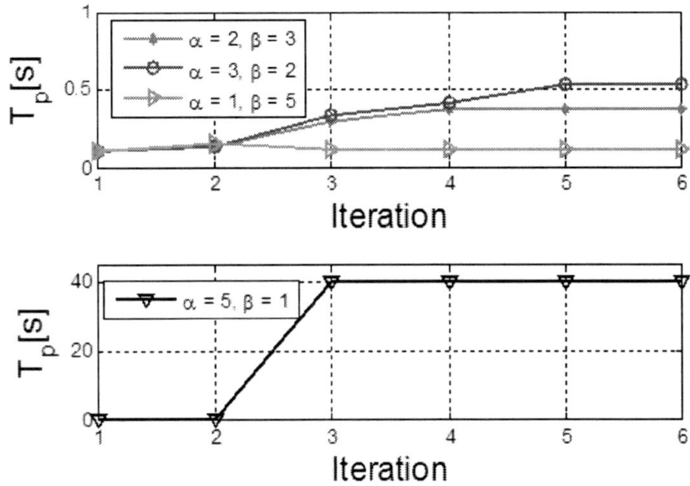

40 s when α=5 and β=*1, 0.53 sec* when α=3 and β=2, *170 ms* when α=2 and β=3, and finally *118 ms* when α=1 and β=5. Accordingly the value of η for the different values of α and β can be calculated and presented in Table 2.

Performance Metrics vs. SNR

As discussed before, p_{fa}, p_d, p_{rd}, η, COP, $(p_{rd}+\eta)$, and $(COP+\eta)$ are to be carefully considered in order to evaluate the performance of the detector for different values of γ at the optimal sensing time t_s and periodic sensing interval T_p. Figure 9 shows the achievable probability of false alarm p_{fa} after convergence for different values of α and β when γ varies from *–20 to 0 dB*. It can be clearly seen from this figure that the probability

Table 2. η versus α and β

α	β	η
1	5	95.16%
2	3	89.47%
3	2	85.48%
5	1	99.75%

Figure 9. The probability of false alarm p_{fa} versus the signal-to-noise ratio γ

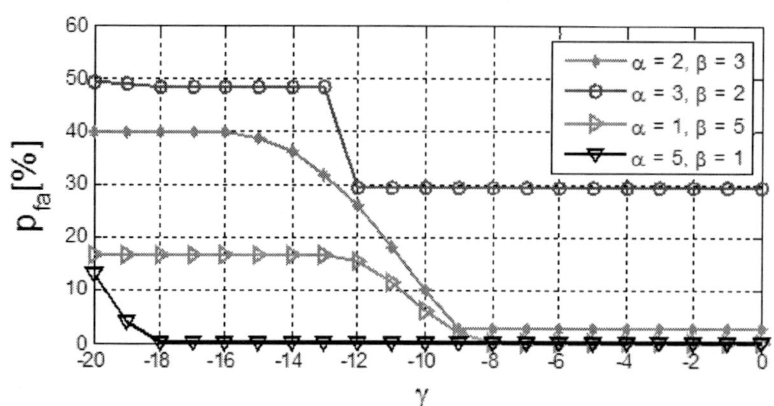

of false alarm p_{fa} decreases with increasing the signal-to-noise ratio γ.

Figure 10 shows the probability of detection p_d performance for different values of α, β, γ and u, respectively. As can be clearly seen from the figure p_d decreases when the value of u increases and it is not affected so much by γ. As explained in Equation 10, the probability of right detection p_{rd} is obtained by combining p_{fa} and p_d, and consequently, combining Figures 9 and 10 would generate the plots for p_{rd} as shown in Figure 11. This figure shows the increase of p_{rd} with increasing γ, and the decrease of p_{rd} with increasing u. Accordingly, it can be seen from Figure 11 that the best p_{rd} of *83.14%* is achieved when u is *0.17* and γ is greater than *–8 dB*, and the worst p_{rd} value of *12.84%* is realized when u is *0.83* and γ is *–20 dB*.

Figure 10. The probability of detection p_d versus signal-to-noise ratio γ

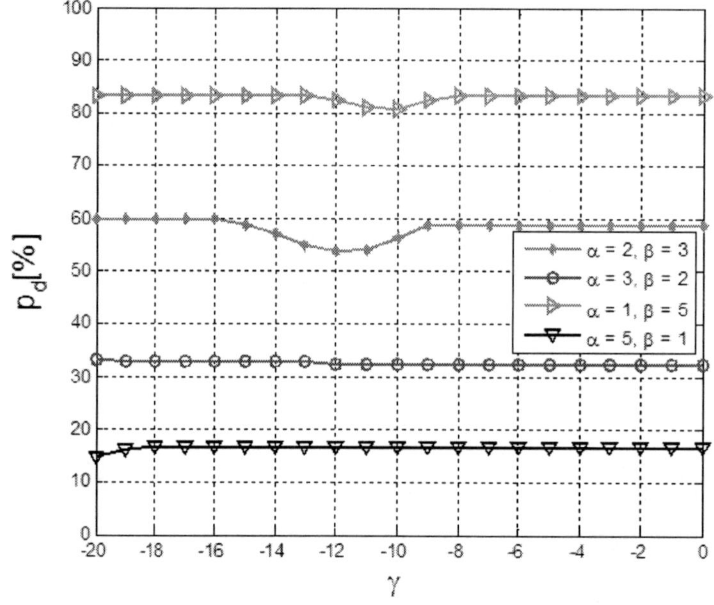

Figure 11. The probability of right detection p_{rd} versus signal-to-noise ratio γ

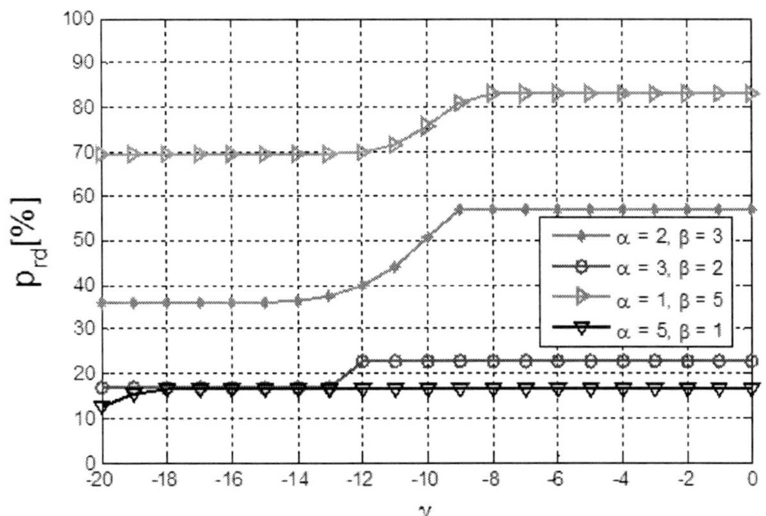

Figure 12 shows that the Captured Opportunities (COP) is proportionally varying with γ and inversely varying with u. The results show the best *COP* of *95.58%* is achieved when u=*0.17* and γ is *–9 dB*. On the other hand, the worst *COP* of *83.67%* is realized when u is *0.83* for all values of γ.

Figure 13 illustrates the transmission efficiency η for four values of u versus γ. The value of η when u is *0.83* is approximately *99.5%* for all values of γ. For u = *0.6*, η fluctuates between *95.05%* and *95.14%* for γ between *–20 dB* and *–13 dB*, and it drops down to *87.44%* when γ goes higher. For a primary channel with u = *0.4*, η

Figure 12. Captured opportunities COP versus signal-to-noise ratio γ

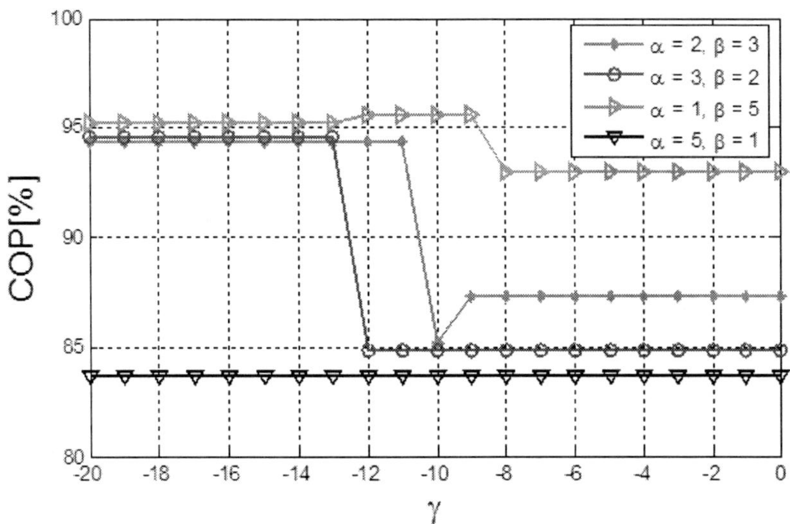

Figure 13. Transmission efficiency η versus signal- to-noise ratio γ

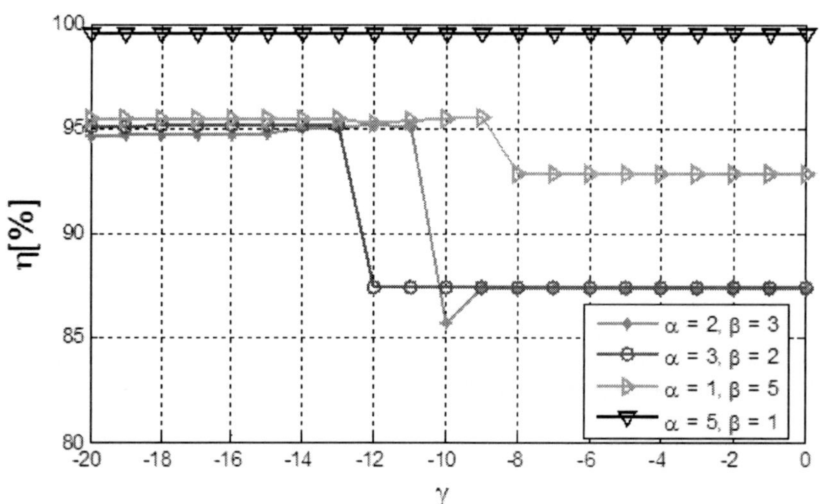

changes between *94.6%* and *95.14%* with the change of γ between *–20 dB* and *–11 dB* and then it drops down to *85.7%* for γ*=–10 dB*, and stabilize at *87.34%* for higher values of γ. Finally, for u*=0.17*, η changes between *95.56%* and *95.23%* when γ varies between *–20 dB* and *–9 dB* and drops down to *92.8%* for higher values of γ.

Figure 14 shows the average of $(p_{rd}+\eta)$ versus the signal-to-noise ratio. It can be clearly seen from this figure that the best average value of

Figure 14. The average of probability of right detection p_{rd} and transmission efficiency η versus signal-to-noise ratio γ

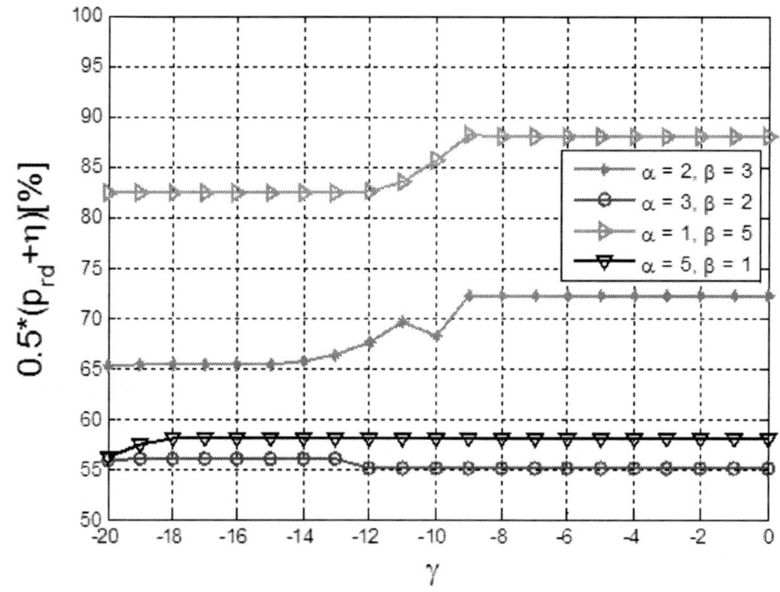

approximately *88.0%* is achieved for *u=0.17* and γ greater than *–9 dB*, and the worst average occurs at *u*=0.6 where it is about *55.0%*.

The curves in Figure 15 show the resulting average of the plots shown in Figures 12 and 13, respectively. From this figure we can clearly see that when *u* is 0.17 the average of *(COP+η)* achieves its highest value of *95.57%* at *γ=–9 dB* and its worst value of *92.88%* when *γ* is greater than *–8 dB*. For other simulated values of *u*, for low values of *γ*, it can be ordered descendingly according to the average value of *(COP+η)* as *u=0.6*, *0.4* and *0.83*, and for higher values of *γ* the order is reversed.

CONCLUSION AND FUTURE RESEARCH DIRECTIONS

In this chapter, we proposed a new optimization approach for the sensing time and periodic sensing interval built upon maximizing the probability of right detection, transmission efficiency and captured opportunities. The probability of right detection is defined as the probability of having correct detection and no false alarm. Since sensing time and the periodic sensing interval are dependent on each other, an iterative approach and a convergence criterion are needed is needed to solve this optimization problem. The simulation results show that the optimal sensing time and optimal periodic sensing interval both increase with increasing the channel utilization factor. In order to evaluate the detector performance the probability of detection, probability of false alarm, probability of right detection, transmission efficiency and the captured opportunities have all been considered as performance metrics. All of these metrics were found to be dependent on the channel utilization factor and SNR with different tendencies. For future work more complicated systems and non stationary channels should be investigated. It is also interesting to investigate the optimization of sensing time and periodic sensing interval for multiple channels systems where spectrum scanning is to be adopted.

Figure 15. The average of the captured opportunities COP and transmission efficiency η versus signal-to-noise ratio γ

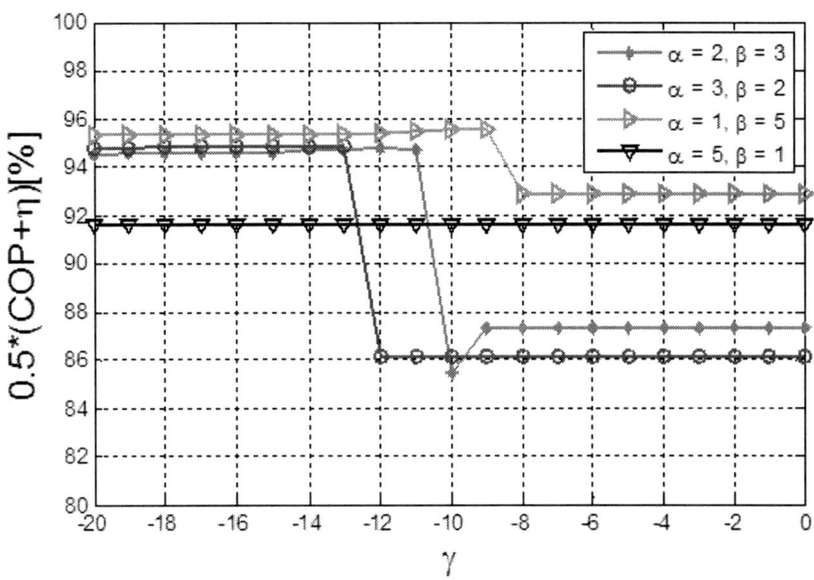

REFERENCES

Cabric, D., Mishra, S., & Brodersen, R. (2004). Implementation issues in spectrum sensing for cognitive radios. In Proceedings of the Asilomar Conference on Signals, Systems and Computers, (vol. 1, pp. 772–776). Asilomar.

Cabric, D., Tkachenko, A., & Brodersen, R. (2006). Spectrum sensing measurements of pilot, energy, and collaborative detection. In *Proceedings of the IEEE Military Communications Conference*, (pp. 1-7). Washington, DC: IEEE Press.

Cox, D. (1967). *Renewal theory*. London, UK: Butler & Tanner Ltd.

Digham, F., et al. (2007). On the energy detection of unknown signals over fading channels. *IEEE Transactions on Communications*, *55*, 21–24. doi:10.1109/TCOMM.2006.887483doi:10.1109/TCOMM.2006.887483

Ghasemi, A., & Sousa, E. (2008). Spectrum sensing in cognitive radio networks: Requirements, challenges and design trade-offs. *IEEE Communications Magazine*, *46*(4), 32–39. doi:10.1109/MCOM.2008.4481338doi:10.1109/MCOM.2008.4481338

Hai Ngoc, P., et al. (2010). Energy minimization approach for optimal cooperative spectrum sensing in sensor-aided cognitive radio networks. In *Proceedings of the 5th International Conference on Wireless Internet (WICON 2010)*. Piscataway, NJ: IEEE.

Hamid, M., & Mohammed, A. (2012). MAC layer spectrum sensing in cognitive radio networks. In *Self-Organization and Green Applications in Cognitive Radio Networks*. Hershey, PA: IGI Global. doi:10.1109/CMC.2010.342doi:10.1109/CMC.2010.342

Haykin, S. (2005). Cognitive radio: Brain-empowered wireless communications. *IEEE Journal on Selected Areas in Communications*, *23*(2), 201–220. doi:10.1109/JSAC.2004.839380doi:10.1109/JSAC.2004.839380

Hyoil, K., & Shin, K. (2008). Efficient discovery of spectrum opportunities with MAC-layer sensing in cognitive radio networks. *IEEE Transactions on Mobile Computing*, *7*, 533–545. doi:10.1109/TMC.2007.70751doi:10.1109/TMC.2007.70751

Kapoor, S., Rao, S., & Singh, G. (2011). Opportunistic spectrum sensing by employing matched filter in cognitive radio network. In *Proceedings of the International Conference on Communication Systems and Network Technologies (CSNT)*. CSNT.

Kyungtae, K., et al. (2010). Energy detection based spectrum sensing for cognitive radio: An experimental study. In *Proceedings of the 2010 IEEE Global Communications Conference (GLOBECOM 2010)*. Piscataway, NJ: IEEE Press.

Liu, X., & Zhai, X. (2008). Feature detection based on multiple cyclic frequencies in cognitive radios. In *Proceedings of the 2008 China-Japan Joint Microwave Conference (CJMW 2008)*, (pp. 290-393). Piscataway, NJ: CJMW.

Mishra, S., Ten, S., Mahadevappa, R., & Brodersen, R. (2007). Cognitive technology for ultra-wideband/wimax coexistence. *Proceedings of IEEE DySPAN*, *2007*. IEEE Press, 179–186.

Mitola, J. (1995). The software radio architecture. *IEEE Communications Magazine*, *33*(5), 26–38. doi:10.1109/35.393001doi:10.1109/35.393001

Tandra, R., & Sahai, A. (2005). Fundamental limits on detection in low SNR under noise uncertainty. In *Proceedings of the IEEE International Conference on Wireless Networks, Communication, and Mobile Computing*, (vol. 1, pp. 464-469). Maui, HI: IEEE Press.

Tang, H. (2005). Some physical layer issues of wide-band cognitive radio systems. In *Proceedings of the 2005 1st IEEE International Symposium on New Frontiers in Dynamic Spectrum Access Networks*, (pp. 151-159). Piscataway, NJ: IEEE Press.

Urkowitz, H. (1967). Energy detection of unknown deterministic signals. *Proceedings of the IEEE*, *55*(4), 523–531. doi:10.1109/PROC.1967.5573doi:10.1109/PROC.1967.5573

Weiss, T., Hillenbrand, J., & Jondral, F. (2003). A diversity approach for the detection of idle spectral resources in spectrum pooling systems. In *Proceedings of the 48th International Scientific Colloquium*. Ilmenau, Germany: IEEE.

Wellens, M., Wu, J., & Mahonen, P. (2007). Evaluation of spectrum occupancy in indoor and outdoor scenario in the context of cognitive radio. In *Proceedings of the 2nd International ICST Conference on Cognitive Radio Oriented Wireless Networks and Communications (CROWNCOM 2007)*, (pp. 420-427). Orlando, FL: CROWNCOM.

Won-Yeol, L., & Akyildiz, I. F. (2008). Optimal spectrum sensing framework for cognitive radio networks. *IEEE Transactions on Wireless Communications*, *7*, 3845–3857. doi:10.1109/T-WC.2008.070391doi:10.1109/T-WC.2008.070391

Yonghong, Z., & Ying-Chang, L. (2007). Maximum-minimum eigenvalue detection for cognitive radio. In *Proceedings of the 2007 IEEE 18th International Symposium on Personal, Indoor and Mobile Radio Communications*, (pp. 1165-1169). Piscataway, NJ: IEEE Press.

Yucek, T., & Arslan, H. (2009). A survey of spectrum sensing algorithms for cognitive radio applications. *IEEE Communications Surveys Tutorials*, *11*(1), 116–130. doi:10.1109/SURV.2009.090109doi:10.1109/SURV.2009.090109

KEY TERMS AND DEFINITIONS

Energy Detector: A signal detector which decide existence or absence of a signal upon it is energy.

Periodic Sensing Interval: The time spent between performing two periodic detection processes.

Probability of Detection: The probability of correctly detecting a signal.

Probability of False Alarm: The probability of declaring signal existence while only noise is received.

Sensing Time: The time needed to collect a number of samples used for the signal detection.

Transmission Efficiency: The fraction of time used for communication with respect to the one used for both communication and sensing.

Unexplored Opportunity (UOP): The fraction of time during which spectrum opportunities are missed.

ENDNOTES

1. *start:step:stop* notes a vector start with *start* and increase the next value by *step* up to *stop*.

Chapter 4
Self-Organization in IEEE Standard 1900.4-Based Cognitive Radio Networks

Majed Haddad
INRIA Sophia-Antipolis, France

Eitan Altman
INRIA Sophia-Antipolis, France

Sana ben Jemaa
Orange Labs, France

Salah Eddine Elayoubi
Orange Labs, France

Zwi Altman
Orange Labs, France

ABSTRACT

Distributing Radio Resource Management (RRM) in heterogeneous wireless networks is an important research and development axis that aims at reducing network complexity, signaling, and processing load in heterogeneous environments. Performing decision-making involves incorporating cognitive capabilities into the mobiles such as sensing the environment and learning capabilities. This falls within the larger framework of cognitive radio (Mitola, 2000) and self-organizing networks (3GPP, 2008). In this context, RRM decision making can be delegated to mobiles by incorporating cognitive capabilities into mobile handsets, resulting in the reduction of signaling and processing burden. This may however result in inefficiencies such as those known as the "Tragedy of commons" (Hardin, 1968) that are inherent to equilibria in non-cooperative games. Due to the concern for efficiency, centralized network architectures and protocols keep being considered and being compared to decentralized ones. From the point of view of the network architecture, this implies the co-existence of network-centric and terminal-centric RRM schemes. Instead of taking part within the debate among the supporters of each solution, the authors propose a hybrid scheme where the wireless users are assisted in their decisions by the network that broadcasts aggregated load information (Elayoubi, 2010). At some system's states, the network manager may impose his decisions on the network users. In other states, the mobiles may take autonomous actions in reaction to information sent by the network. Specifically, the authors derive analytically the utilities related to the Quality of Service (QoS) perceived by mobile users and develop a Bayesian framework to

DOI: 10.4018/978-1-4666-2812-0.ch004

obtain the equilibria. They then analyze the performance of the proposed scheme in terms of achievable throughput (for both mobile terminals and the network) and evaluate the price of anarchy which measures how good the system performance is when users play selfishly instead of playing to achieve the social optimum (Johari, 2004). Numerical results illustrate the advantages of using the hybrid game framework in a network composed of HSDPA and 3G LTE system that serve streaming and elastic flows. Finally, this chapter addresses current questions regarding the integration of the proposed hybrid Stackelberg scheme in practical wireless systems, leading to a better understanding of actual cognitive radio gains.

1. INTRODUCTION

In tomorrow's Beyond 3G context, numerous heterogeneous Radio Access Technologies (RAT) will have to coexist including of course the new 3GPP LTE, but also legacy 3GPP technologies (like GSM/GPRS/EDGE and UMTS/HSDPA) and also non-3GPP technologies like WiFi. Since mobile devices are also becoming multi-mode and may hence access several of these technologies, the need for a coordinated resource management encompassing all these technologies arises: an efficient traffic balancing is namely required for an optimal usage of the deployed network equipment and an improved user experience. In the medium term, new Software Defined Radio (SDR) equipment will enable flexible spectrum management like Dynamic Spectrum Assignment (DSA), and dedicated resource management mechanisms will be required (3GPP, 2008). Self-Organizing Network (SON) is currently considered as a key lever to minimize operational costs and optimize delays of deploying and running a network by reducing and eliminating manual configuration and maintenance of network operational parameters at the time of network planning, network deployment, network operations, and network optimization. Different standardization bodies have picked up this topic, and SON functionalities are expected to become widely commercially available with the introduction of 4G networks (3GPP, 2008).

This chapter presents some background on the research leading to the future deployment of the IEEE standard 1900.4. We highlight different classes of techniques and algorithms which attempt to realize the various benefits of introducing game theory as an efficient technique for the analysis of such cognitive radio networks. Based on tools from non-cooperative and Bayesian games, we develop a Bayesian framework to analyze the performance of the proposed IEEE 1900.4 cognitive radio network. In order to improve the performance of the non-cooperative scenario, we investigate the properties of an alternative solution concept named Stackelberg game, in which the network tries to control the users' behavior by broadcasting appropriate information, expected to maximize its utility, while individual users maximize their own utility.

The IEEE Standard 1900.4 offers a solution specifically tailored to answer these needs. It proposes a distributed approach for optimized inter-system resource management: a Terminal Reconfiguration Manager (TRM) entity located in each mobile takes autonomous decisions to access the RAT that will best fit with user-specific expectations (e.g. maximize throughput, or increase battery life, minimize cost, etc.). To guarantee that the global system efficiency will be preserved in spite of individual (and potentially selfish) decisions from the mobiles, a Network Reconfiguration Manager (NRM) entity located in the network gives global recommendations that help the mobiles to take appropriate decisions. These recommendations are conveyed by so-called "policies" broadcast to the mobiles, and constrain the mobiles to take acceptable actions from network point of view. The proposal to introduce a logical communication channel between NRM and the TRM, e.g., the radio enabler, into heterogeneous wireless systems is one of the main outcomes of the IEEE standard 1900.4 (see Figure 1). The objective is to support an efficient discovery of the available radio accesses and reconfiguration management in heterogeneous wireless environment between the NRM and the TRM. Note that the approach proposed in this chapter, while profiting from the IEEE standard 1900.4 capabilities, presents a key to understand the actual benefits brought by this standard. In fact, although IEEE standard 1900.4 have spurred great interest and excitement in the community, many of the fundamental theoretical questions on the limits of such a standard remain unanswered.

In this chapter, we investigate the association problem in the context of distributed decision making in a heterogeneous cognitive network. We wish to avoid completely decentralized solutions

Figure 1. The proposed hybrid 1900.4 system description

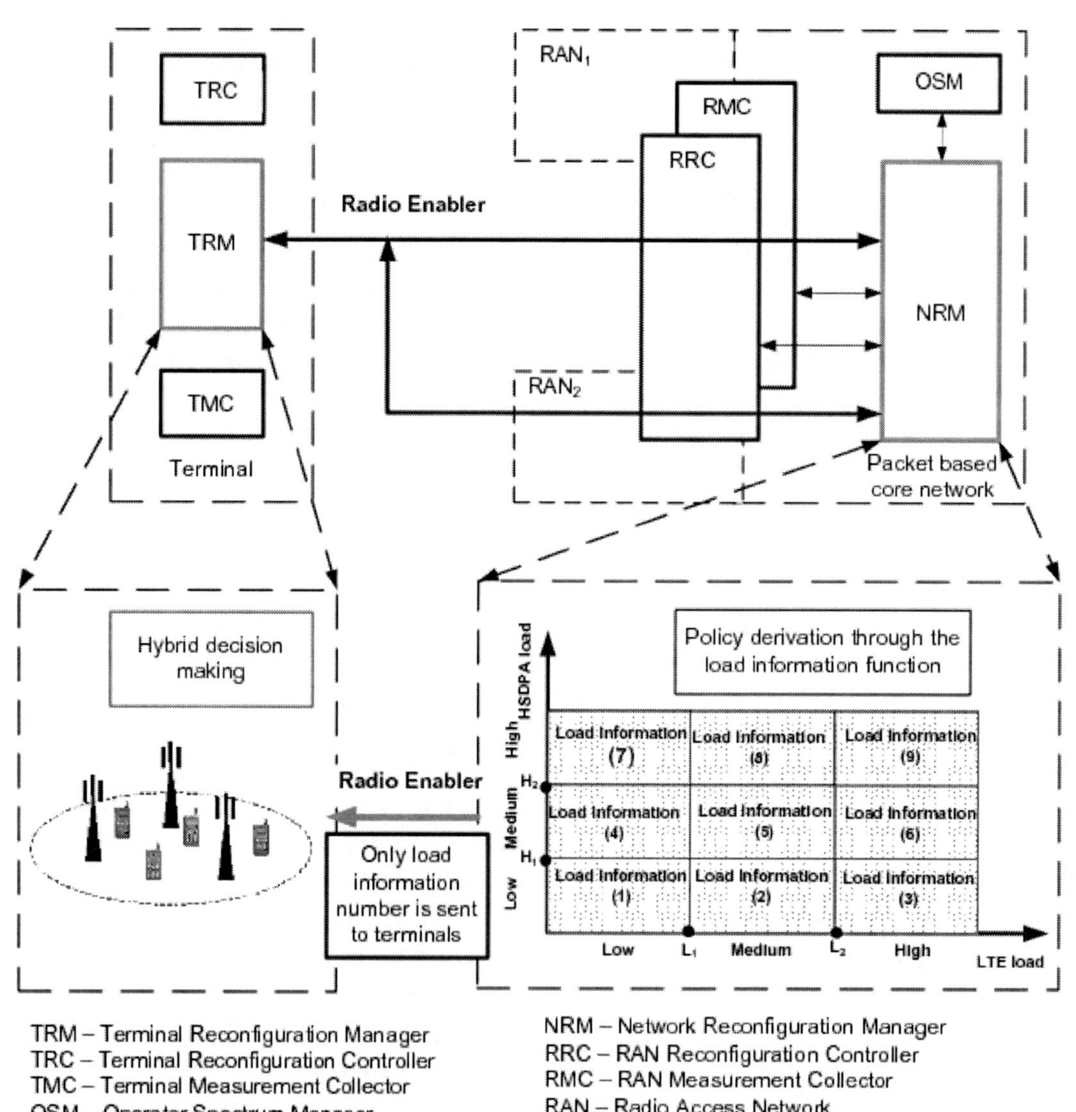

of the association problem in which all decisions are taken by the mobiles, due to well known inefficiency problems that may arise when each mobile is allowed to optimize its own utility. This inefficiency is inherent to the non-cooperative nature of the decision-making. Nevertheless, we wish to delegate to the mobiles a large part in the decision making in order to alleviate the burden from the base stations. We then propose association methods that combine benefits from both decentralized and centralized design. Central intervention is needed during severe congestion periods. At those instants, we assume that the mobiles follow the instructions of the base stations. Otherwise, the association decision is left to the mobiles, which make the decision based on aggregated load information. The decision-making is thus based on partial information that is signaled to the mobiles by the base station. A central design aspect is then for the base stations to decide how to aggregate information, which then determines what to signal to the users. This decision-making at the BS can be viewed as a mechanism design problem, or as a Bayesian game (in the case we wish to view the base station as a player on his own). We will exemplify our general analysis by investigating the possibility of offering real time or non-real time services.

2. BACKGROUND

When we deal with heterogeneous cognitive networks, interactions among selfish users sharing a common transmission channel can be modeled as a non-cooperative game using the game theory framework (Fudenberg, 1991). Game theory provides a formal framework for studying the interactions of strategic agents. Recently, there has been a surge in research activities that employ game theory to model and analyze a wide range of application scenarios in modern communication networks (Altman, 2006; Felegyhazi, 2006).

Moreover, radio access equipment is becoming more and more multi-standard, offering the possibility of connecting through two or more technologies concurrently. Switching between networks using different technologies is referred to as vertical handover. The association schemes actually implemented by network operators are fully centralized: the operator tries to maximize his utility (revenue) by assigning the users to the different systems (Stevens, 2008; Kumar, 2007; Horrich, 2008; Jiang, 2008). However, distributed Joint Radio Resource Management (JRRM) mechanisms are gaining in importance: Users may be allowed to make autonomous decisions in a distributed way. The association problem is related in nature to the channel selection problem. We note that when a single technology is used or, when the decision concerns the choice of channels of a given access point rather than the choice of an access point, one can often exploit simpler structure of the decision problem and obtain efficient decentralized solutions. Some examples of work in that direction are Kauffmann (2007) and Bejerano (2004). This has lead game theoretic approaches to the association problems in wireless networks, as can be found in Ozgur (2008), Shakkottai (2007), Shakkottai (2008), Kumar (2007), and Coucheney (2009). The potential inefficiency of such approaches have been known for a long time. In fact, this could likely lead to congestion and overload conditions in the RAT in question (which offers the best peak rate) and all users would lose. The term "The Tragedy of the Commons" has been frequently used for this inefficiency (Hardin, 1968); it describes a dilemma in which multiple individuals acting independently in their own self-interest can ultimately destroy a shared limited resource even when it is clear that it is not in anyone's long term interest for this to happen. To overcome this hurdle, we introduce a game theoretic framework with partial information to maximize the throughput while taking into account the system overload. This study requires particular attention when all users wish to maximize their individual throughput but each has a different approach (e.g. users may have different tolerance for delay, or may have a certain target throughput to guarantee).

The basic idea of the hybrid decision approach has been first presented in Elayoubi (2010), where the association problem was formulated as a non-cooperative game. In the present contribution, we prove formally that the Nash equilibrium exists in the case of mixed strategies. We further extend the model by allowing the network to control the users' behavior by choosing the information he broadcasts and formulate it as a Stackelberg game. The network model is extended to include intra- and inter-system mobility of users. We also present a detailed calculation of the individual utilities in the most general case, for streaming services.

This chapter is structured as follows. The association problem in heterogeneous cognitive networks is exposed in Section 3. In Section 4, we calculate the utility of the wireless users by a Markov analysis. In Section 5, we present the non-cooperative game framework adapted for the considered hybrid model and show how the base station can control the equilibrium of its users by means of a Stackelberg formulation. In Section 6, we provide numerical results to illustrate the theoretical solutions derived in the previous sections for streaming flows. Section 7 eventually concludes the chapter.

3. HYBRID DECISION FRAMEWORK

3.1. Network Resources

Consider a wireless network composed of S systems managed by the same operator. Clearly, the peak throughput that can be obtained by a user connected to system s, if served alone by a cell, differs depending on its position in the cell. This is illustrated in Figure 2, showing the peak throughputs for a cell served by HSDPA and 3G LTE as an example. In order to have realistic expressions of the effective throughput on each system and to include the impact of mobility on the performance, we decompose the cell into N location areas corresponding to concentric circles of radius d_n for $n=1,...,N$ with homogeneous radio characteristics.

Figure 2. LTE and HSDPA peak throughputs as function of user positions in the cell

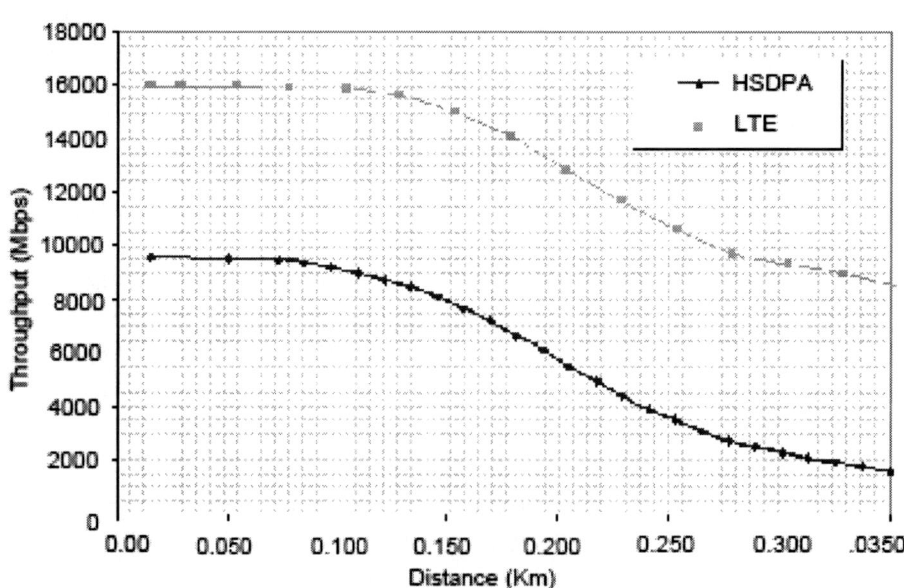

Users with radio condition n have a peak rate D_n^s if connected to system s. Let M_n^s represents the number of users with radio condition n connected to system s. Depending on the required service, some of these network states (referred to hereafter and interchangeably, as micro-states) are not admissible in the sense that the time-frequency resources are not sufficient to meet the service requirements of all mobiles given their location and radio conditions. We denote by $\mathcal{A}$ the corresponding state space composed of all admissible network states.

3.2. Cognitive Channel Signaling Information

In order to be able to achieve these goals, we make use of the IEEE SCC41 standard framework (IEEE SCC 41) for cognitive radio networks and particularly the IEEE standard 1900.4 (Buljore, 2008) that proposes scenarios and solutions to allow information exchange between the network and the end-user terminals; the aim is to allow devices to distributively and optimally choose among the available radio resources so that the overall efficiency and capacity of the resulting composite network is improved.

The network is fully characterized by its state $\boldsymbol{M}$. However, when distributing the RRM decisions, this complete information is not available to the users. In this setting, we assume that, using the *radio enabler* proposed by the IEEE standard 1900.4, the NRM broadcasts to the TRM an aggregated load information that takes values in some finite set $\{1,...,L\}$ indicating in which load state mobile terminals are (low, medium, or high) (see Figure 2). This reduces signalling overhead while staying inline with the IEEE standard 1900.4 requirements, which specifies that "the network manager side shall periodically update the terminal side with context information" (IEEE SCC 41).

More formally, an assignment f specifies for each network micro-state $\boldsymbol{M}$ the corresponding macro-state $f(\boldsymbol{M})$. We will call $f(.)$ the load information function. As an example, Figure 2 shows how the load information is aggregated by the NRM to the TRM for a network composed of two systems (HSDPA and LTE), indicating for each system if it is in a low, medium, or high load state. This figure illustrates the relationship between the loads of the systems $\boldsymbol{M}$ and the corresponding load information $l=f(\boldsymbol{M})$. In particular, a function f is constructed based on four thresholds: L_1, L_2, H_1 and H_2. The load of system X is considered low if it is less than threshold X_1, medium if it is between X_1 and X_2 and high if it is larger than X_2 where $X=L$ for LTE and $X=H$ for HSDPA. The load information l, an integer between 1 and 9, is then obtained and broadcasted to mobile users.

3.3. RRM Policies

Mobiles arrive in the network at random, attempt a call and leave the network immediately if blocked or persist until the end of the call if admitted. Within the space of admissible states $\mathcal{A}$, each mobile will decide individually to which of the available systems it is best to connect according to its radio condition and the load information l broadcast by the network. Their policies (or strategies) are then based on this information. Let $P^l = [P_{1,l},...,P_{N,l}]^T$ be the user's decision vector, knowing the aggregated load information l, whose element $P_{n,l}$ is equal to s if class-n users connect to system s.

The set of all possible choices is P. We then denote by $\boldsymbol{P}$ the strategy profile matrix defined as the actions taken by mobiles in the different load conditions. Equivalently, when the network load information is equal to l and the strategy profile is $\boldsymbol{P}$, we can determine the system to which users of class n will connect by the value $P_{n,l}$. As an example, knowing the function $f(.)$ and the strategy profile $\boldsymbol{P}$, if the network is in state $\boldsymbol{M}$, a class n user will connect to system $P_{n,l}$ where $l=f(\boldsymbol{M})$. Note that, if the function f is modified, the groups of states that are aggregated within the same macro-state are changed, leading to a different JRRM decision taken by a user that finds the system in micro-state ***M*** for the same policy ***P***.

Both admission control and vertical handovers can be involved: upon arrival, each user decides its serving system following the actual policy ***P*** and the load information function *f*. The association problem is then generalized to allow the mobile users to change their new serving system according to the aggregated load information while taking into account the reduction of unnecessary handovers, namely the ping-pong effect. Vertical handovers are possible only following changes in the radio conditions of a given mobile in an event-driven manner. The migrating mobiles check, once their radio conditions change, whether the serving Radio Access Technology (RAT) is still the best choice according to its policy. Otherwise, it can perform an inter-system handover to the other RAN after checking that it could be admitted on it.

4. UTILITIES

The first step before analyzing the hybrid decision scheme is to define the utilities of users. These latter are often related to throughput, whose variations are mainly due to network load, radio network conditions, and mobility such as handovers.

4.1. Steady State Analysis

We consider a Real-Time Transport Protocol (RTP) streaming service. As we consider cellular networks where Adaptive Modulation and Coding (AMC) ensure that the Block Error Rate (BLER) is lower than a certain target, the video quality is guaranteed when the throughput required by the codec is obtained. The goal of a streaming user is thus to achieve the best throughput, knowing that the different codecs allow a throughput between an upper (best) T_{max} and a lower (minimal) T_{min} bounds. His utility is expressed by the quality of the streaming flow he receives, which is in turn closely related to his throughput. Indeed, a streaming call with a higher throughput will use a better codec offering a better video quality.

This throughput depends not only on the peak throughput, but also on the evolution of the number of calls in the system where the user decides to connect. Note that a user that cannot be offered this minimal throughput in neither of the available systems is blocked in order to preserve the overall network performance. However, once connected, we suppose that a call will not be dropped even if its radio conditions degrade because of mobility. For the ease of comprehension, we will begin by considering static users.

4.2. Instantaneous Throughput

The instantaneous throughput obtained by a user in a system depends on both his own decisions and the decisions taken by the other users. Assuming proportional fair scheduling among different users, the throughput of a user with radio condition class *n* connected to system *s* is given by:

$$t_n^s(\mathbf{M}) = \min\left[D_n^s \frac{G(\mathbf{M})}{\sum_{n=1}^{N} M_n^s}, T_{max} \right] \tag{1}$$

where *G(M)* is the opportunistic scheduler gain. Note here that the admission control will ensure that $t_n^s(M) \geq T_{\min}$ by blocking new arrivals. The space of admissible states $\mathcal{A}$ is thus the set of all states ***M*** where the minimal constraint on the throughput is ensured:

$$\frac{G(\mathbf{M})}{\sum_{n=1}^{N} M_n^s} \geq \frac{T_{min}}{D_n^s}, \forall s \mid M_n^s > 0 \tag{2}$$

Note that $T_{min}<T_{max}$, so that the states where $t_n^s(M)=T_{max}$ verify the latter condition.

4.3. Steady State Probabilities

The throughput achieved by a user depends on the number of ongoing calls. This latter is a random variable whose evolution is described by a Markov chain governed by the arrival and departure processes. We assume that the arrival process of new connections with radio condition *n* is Poisson with rate λ_n uniformly distributed over the cell. Each arriving user makes a streaming connection whose duration is exponentially distributed with parameter $1/\mu$. Within the space of feasible states $\mathcal{A}$, transitions between states are caused by the departures, the arrivals and the subsequent decisions determined by the policy:

- Arrivals of new connections of radio condition *n*. Let $G_n^s(\mathbf{M})$ denote the state of the system if we add one mobile of radio condition *n* to system *s*. The transition from state **M** to $G_n^s(\mathbf{M})$ happens if the policy implies that system *s* is to be chosen for the load information corresponding to state **M**, and if the state $G_n^s(\mathbf{M})$ is an admissible state. The corresponding transition rate is thus equal to:

$$q(\mathbf{M}, G_n^s(\mathbf{M}) \mid \mathbf{P}, f) = \lambda_n \cdot 1\!\mathrm{I}_{P_{n,f(\mathbf{M})=s}} \cdot 1\!\mathrm{I}_{G_n^s(\mathbf{M})\in\mathcal{A}} \quad (3)$$

  where $1\!\mathrm{I}_C$ is the indicator function equal to 1 if condition C is satisfied and to 0 otherwise,
- End of a communication of class *(n,s)*. Let $D_n^s(\mathbf{M})$ denote the state with one less mobile of class *(n,s)*. The transition from state ***M*** to $D_n^s(\boldsymbol{M})$ is equal to:

$$q(\mathbf{M}, D_n^s(\mathbf{M}) \mid \mathbf{P}, f) = M_n^s \cdot \mu \cdot 1\!\mathrm{I}_{M_n^s>0} \quad (4)$$

Figure 3. Transitions of the Markov chain. We only illustrate arrivals and departures corresponding to matrix state **M***: Arrival transitions occur only if the policy* **P** *implies that users with radio conditions n connect to system s.*

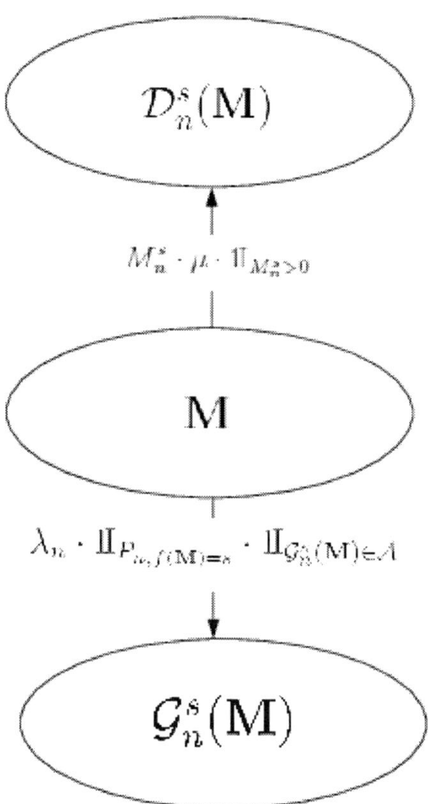

These transitions are illustrated in Figure 3. The transition matrix **Q (P,*f*)** of the Markov chain can then be easily written for each policy **P** knowing that its diagonal element is equal to:

$$q(\mathbf{M},\mathbf{M} \mid \mathbf{P}, f) = -\sum_{n=1}^{N}\sum_{s=1}^{S}\left[\begin{array}{l} q(\mathbf{M}, D_n^s(\mathbf{M}) \mid \mathbf{P}, f) \\ +q(\mathbf{M}, G_n^s(\mathbf{M}) \mid \mathbf{P}, f) \end{array}\right] \quad (5)$$

The steady-state distribution is then obtained by solving the following system of equation:

$$\begin{cases} \Pi(\mathbf{P},f)\cdot\mathbf{Q}(\mathbf{P},f) = 0 \\ \Pi(\mathbf{P},f)\cdot\mathbf{e} = 1; \end{cases} \quad (6)$$

where $\Pi(\mathbf{P},f)$ is the vector of the steady-state probabilities $\pi(\mathbf{M}|\mathbf{P},f)$ under policy **P** and load information function *f*, and **e** is a vector of ones of appropriate dimension.

Numerical resolution of this problem is possible, and once the vector π is obtained, the global performance indicators can be calculated. Among these performance metrics, we can cite the blocking rate of class-*n* calls knowing that the load information is equal to *l*:

$$b_n(l \mid \mathbf{P},f) = \frac{\sum_{\mathbf{M}\in\mathcal{A};G_n^s(\mathbf{M})\notin\mathcal{A},\forall s\in\mathcal{S}} \pi(\mathbf{M}\mid\mathbf{P},f)}{\sum_{\mathbf{M}\in\mathcal{A};f(\mathbf{M})=l} \pi(\mathbf{M}\mid\mathbf{P},f)} \quad (7)$$

In this equation, we consider as blocked all calls that arrive in states where each system are saturated, i.e., where $t_n^s(\mathbf{M}) < T_{min}$. We can also obtain the overall blocking rate:

$$b(\mathbf{P},f) = \sum_{n=1}^{N} \frac{\lambda_n}{\sum_{m=1}^{N}\lambda_m} b_n(l \mid \mathbf{P},f) \quad (8)$$

4.4. Transient Analysis

The steady-state analysis described above is not sufficient to describe the utility of the users as the throughput obtained by a user at his arrival is not a sufficient indication about the quality of his communication because of the dynamics of arrivals/departures in the system. In order to obtain the utility, we modify the Markov chain in order to allow tracking mobiles during their connection time. For users of radio condition *n* connected to system *s*, only states where there is at least one user (n,s) are considered. The calculation is as follows:

1. Introduce absorbing states A_n^s corresponding to the departure of mobiles that have terminated their connections. Additional transitions are thus added between **M** and A_n^s with rate equal to:

$$\tilde{q}_n^s(\mathbf{M}, A_n^s) = \mu\cdot 1\!\!1_{M_n^s>0} \quad (9)$$

The transitions to the neighboring states with one less user are then modified accordingly by subtracting μ from the original transition rates defined in Equation 10:

$$\tilde{q}_n^s(\mathbf{M}, D_n^s(\mathbf{M}) \mid \mathbf{P},f) = (M_n^s - 1)\cdot\mu\cdot 1\!\!1_{M_n^s>0} \quad (10)$$

The remaining transition rates remain equal to the original transitions:

$$\tilde{q}_n^s(\mathbf{M}, G_{n'}^{s'}(\mathbf{M}) \mid \mathbf{P},f) = q(\mathbf{M}, G_{n'}^{s'}(\mathbf{M}) \mid \mathbf{P},f), \quad \forall n', s' \quad (11)$$

and

$$\tilde{q}_n^s(\mathbf{M}, D_{n'}^{s'}(\mathbf{M}) \mid \mathbf{P},f) = q(\mathbf{M}, D_{n'}^{s'}(\mathbf{M}) \mid \mathbf{P},f), \quad \forall(n', s') \neq (n,s) \quad (12)$$

The transition rates for the modified chain are illustrated in Figure 4.

2. Define matrix $\tilde{\mathbf{Q}}_n^s$ of elements $\tilde{q}_n^s(\mathbf{M},\mathbf{M}')$ defined above and with diagonal elements as in Equation 13:

$$\tilde{q}_n^s(\mathbf{M},\mathbf{M} \mid \mathbf{P},f) = q(\mathbf{M},\mathbf{M} \mid \mathbf{P},f) \quad (13)$$

Figure 4. Transitions of the Markov chain, modified to calculate utilities of calls (n,s). Notice that departures to the absorbing state are added with respect to the Markov chain in Figure 3.

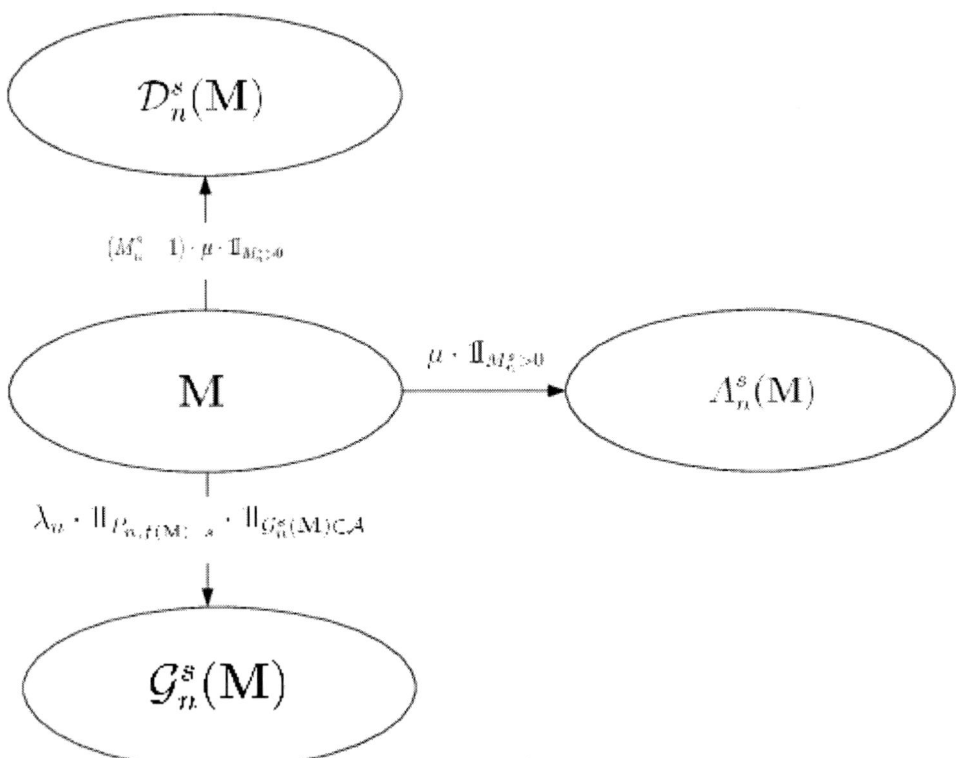

Knowing the load information function f and under policy **P**, the volume of information $I_n^s(\mathbf{M} \mid \mathbf{P}, f)$ sent by system s users subject to radio conditions n starting from state **M** is then equal to the volume of information sent between state **M** and the absorbing state A_n^s. Knowing that $I_n^s(A_n^s) = 0$, these values can be calculated by solving the set of linear equations for all states **M**:

$$\sum \tilde{q}_n^s(\mathbf{M}, \mathbf{M}' \mid \mathbf{P}, f) \cdot I_n^s(\mathbf{M}' \mid \mathbf{P}, f) = -t_n^s(\mathbf{M}) \quad (14)$$

where $t_n^s(\mathbf{M})$ stands for the throughput and $\tilde{q}(\mathbf{M}, \mathbf{M}' \mid \mathbf{P}, f)$ is the transition rate between state **M** and its neighboring state **M'** obtained as described above (due to arrivals and departures).

This calculation is based on the idea that the volume of information sent within state **M** before any transition occurs is equal to the throughput multiplied by the time spent in **M** (i.e., $-1 / \tilde{q}_n^s(\mathbf{M}, \mathbf{M} \mid \mathbf{P}, f)$).

Adding the information volume sent starting from **M'** (i.e., $I_n^s(\mathbf{M}' \mid \mathbf{P}, f)$) and the probability to move first to **M'** (i.e., $-\tilde{q}_n^s(\mathbf{M}, \mathbf{M}' \mid \mathbf{P}, f) / \tilde{q}_n^s(\mathbf{M}, \mathbf{M} \mid \mathbf{P}, f)$), we obtain the equation:

$$I_n^s(\mathbf{M} \mid \mathbf{P}, f) = \frac{-t_n^s(\mathbf{M})}{\tilde{q}_n^s(\mathbf{M}, \mathbf{M} \mid \mathbf{P}, f)} + \sum_{\mathbf{M}' \neq \mathbf{M}} \frac{-\tilde{q}_n^s(\mathbf{M}, \mathbf{M}' \mid \mathbf{P}, f)}{\tilde{q}_n^s(\mathbf{M}, \mathbf{M} \mid \mathbf{P}, f)} I_n^s(\mathbf{M}' \mid \mathbf{P}, f) \quad (15)$$

which is exactly Equation 14 after simple manipulations.

3. The utility of a class-n user that has found the network in state $\mathbf{M}$ and chosen to connect to system s is the volume of information sent starting from state $G_n^s(\mathbf{M})$. Recall that $G_n^s(\mathbf{M})$ is defined as the state with one more class-n call connected to system s:

$$u_n^s(\mathbf{M} \mid \mathbf{P}, f) = I_n^s(G_n^s(\mathbf{M}) \mid \mathbf{P}, f) \quad (16)$$

5. GAME THEORETIC FRAMEWORK

In this section, we use the users' utilities we derived above to derive the association policies. We first search for the global optimum policy, i.e. the policy that maximizes the global utility of the network. Nevertheless, as it is not realistic to consider that the users will seek the global optimum, we show how to find the policy that corresponds to the Nash equilibrium, knowing that users will try to maximize their individual utilities. We will next show, by means of a Stackelberg formulation, how the operator can control the equilibrium of its wireless users to maximize its own utility by sending appropriate load information.

5.1. Global Optimum Strategy

Cooperative approaches in communication theory usually focus on studying how users can jointly improve their performance when they cooperate. For example, the users may optimize a common objective function, which represents the optimal social welfare allocation rule based on which the system-wide resource allocation is performed. A profile of actions $\mathbf{P}$ is said to be global optimum if no other policy profile gives every agent as much utility while giving at least one agent a higher utility. For our specific problem, the global utility function can be written as in Box 1.

Note that, in this utility, we consider not only the QoS of accepted users, but also the blocking rate as the aim is also to maximize the number of accepted users. We also weight the users of different radio conditions with their relative arrival rates. The global optimum policy is the one among all possible policy profiles that maximizes this utility:

$$\mathbf{P}^{GO} = \underset{P}{argmax}\ U(\mathbf{P}, f) \quad (18)$$

It is worth mentioning that information exchanges among users are generally required to enable users to coordinate in order to achieve and sustain global efficient outcomes. In order to alleviate this hurdle, we turn to non-cooperative games.

5.2. Nash Equilibrium Strategy

There exist many systems where multiple independent users, or players, may strive to optimize

Box 1.

$$U(\mathbf{P}, f) = \sum_{n=1}^{N} \frac{\lambda_n + \lambda_{HO}}{\sum_{i=1}^{N} \lambda_i + \lambda_{HO}} \sum_{l} [(1 - b_n(l \mid \mathbf{P}, f)) \cdot \sum_{\mathbf{M} \mid f(\mathbf{M}) = l} u_n^{P_{n,l}}(\mathbf{M} \mid \mathbf{P}, f) \pi(\mathbf{M} \mid \mathbf{P}, f)] \quad (17)$$

their own utility or cost *unilaterally*, which can be regarded as non-cooperative games. In this context, users of different radio conditions are interested by maximizing their individual QoS given the load information broadcast by the network. The utility that a class n user might obtain if it chooses system s when the load information is l, while all other users follow policy profile **P** is then:

$$U^{s}_{n,l}(\mathbf{P}, f) = \frac{\sum_{\mathbf{M}|f(\mathbf{M})=l} u^{s}_{n}(\mathbf{M} \mid \mathbf{P}, f)\pi(\mathbf{M} \mid \mathbf{P}, f)}{\sum_{\mathbf{M}|f(\mathbf{M})=l} \pi(\mathbf{M} \mid \mathbf{P}, f)} \tag{19}$$

A strategy profile $\mathbf{P}^{NE}$, $\forall\, n, l$ corresponds to a Nash Equilibrium (NE) if, for all radio conditions and all load informations, any unilateral switching to a different strategy can improve user's payoff. Mathematically, this can be expressed by the following inequality, given the load information function f, for all radio condition $n \in \mathcal{N}$ and all load information $l \in \mathcal{L}$, $\forall \sigma_{n,l} \neq P^{NE}_{n,l}$:

$$U^{P^{NE}_{n,l}}_{n,l}(\mathbf{P}^{NE}, f) \geq U^{\sigma_{n,l}}_{n,l}(\mathbf{P}^{NE}, f) \tag{20}$$

Two points are noteworthy here. First, Nash equilibria may be globally inefficient. Second, Nash equilibria need not exist in general within pure policies (the reason is that this set is in non-convex).

5.3. Stackelberg Equilibrium Strategy

In the previous section, we derived the policy that corresponds to the Nash equilibrium for a game where players are the wireless users that aim at maximizing their payoff. However, there is another dimension of the problem related to the information sent by the network and corresponding to the different load information.

Motivated by the fact that when selfish users choose their policies independently without any coordination mechanism, Nash equilibria may result in a network collapse, we propose a methodology that transforms the non-cooperative game into a Stackelberg game. Stackelberg equilibria of the Stackelberg game can overcome the deficiency of the Nash equilibria of the original game. Concretely, the network may guide users to an equilibrium that optimizes its own utility if it chooses the adequate information to send. At the core lies the idea that introducing a certain degree of hierarchy in non-cooperative games not only improves the individual efficiency of all the users but can also be a way of reaching a desired trade-off between the global network performance at the equilibrium and the requested amount of signaling. The proposed approach can be seen as intermediate scheme between the totally centralized policy and the non-cooperative policy. It is also quite relevant for flexible networks where the trend is to split the intelligence between the network infrastructure and the (generally mobile) users' equipments.

More formally, let $\mathcal{C}$ be the finite set of all possible choices of the aggregating loads function f_j. The way of aggregating the loads in the broadcast information (expressed hereafter by the load information function f_j) is inherent to the previous analysis. In particular, the utilities of individual users, calculated in Equation 19, is function of $f_j(.)$:

$$U^{s}_{n,l}(\mathbf{P}, f_j) = \frac{\sum_{\mathbf{M}|f_j(\mathbf{M})=l} u^{s}_{n}(\mathbf{M} \mid \mathbf{P}, f_j)\pi(\mathbf{M} \mid \mathbf{P}, f_j)}{\sum_{\mathbf{M}|f_j(\mathbf{M})=l} \pi(\mathbf{M} \mid \mathbf{P}, f_j)} \tag{21}$$

We call the transformed game the Stackelberg game because the network manager chooses his strategy (by means of the load information function f_j) before the users make their decisions' policies. In this sense, the network manager can be thought of as a Stackelberg leader and the users as followers. The Stackelberg problem is thus defined as the maximization of the utility of the network by tuning the load information function $f_j(.)$. Suppose that the aim of the operator is to maximize its revenues by maximizing the acceptance ratio, the Stackelberg equilibrium verifies:

$$f^{SE} = argmax_f \min_j b(\mathbf{P}^{NE}, f_j) \quad (22)$$

where $\mathbf{P}^{NE}$ is the NE policy verifying Equation 20 and the blocking $b(.)$ is defined as in Equation 18. This leads the wireless users to a Stackelberg equilibrium that depends on the way the network aggregates the load information. Notice that the users still behave non-cooperatively and maximize their payoffs, and the intervention of the manager affects their selfish behavior even though the manager does neither directly control their behavior nor continuously communicate with the users to convey coordination. As a result, this tends to substantially reduce signalling overhead.

6. PERFORMANCE RESULTS

For the sake of simplicity, we suppose that users are classified between users with good radio conditions (or cell center users) and users with bad radio conditions (or cell edge users). We emphasize that the model we develop in this work is generic and applicable to different kinds of situations. We will focus on the more realistic and cost effective case where the operator uses the same cell sites to deploy the new system (e.g. 3G LTE), while keeping the old ones (e.g. HSDPA). We consider joint admission control and vertical (inter-system) handover between HSDPA and LTE. The network sends aggregated load information as shown in Figure 2 with the following thresholds: [L_1=0.3, L_2=0.7, H_1=0.3, H_2=0.7], meaning that a system is considered as highly loaded if its load exceeds 0.7 and as low-loaded if its load is below 0.3. For comparison purposes, we study four different association approaches:

- **Global Optimum Approach:** Obtained through an exhaustive search considering all possible strategy combinations in the vector profile $\mathbf{P} \in \mathcal{P}$. This will thus serve as the optimal social welfare solution for problem (1.18) in order to demonstrate just how much gain may theoretically be exploited through considering such a global optimal solution with respect to the other schemes.
- **Hybrid Decision Approach:** The proposed hybrid scheme where users receive aggregated load information and maximize their individual utility. We illustrate the global utility corresponding to the Nash equilibrium strategy.
- **Peak Rate Maximization Approach:** This is a simple association scheme where users do not have any information about the loads of the systems. They connect to the system s^{PR} offering them the best peak rate:

$$s^{PR} = \underset{s}{argmax}\, D_n^s \quad (23)$$

Note that this peak rate can be known by measuring the quality of the receiving signal.

- **Instantaneous Rate Maximization Approach:** The network broadcasts **M**, the exact numbers of connected users of different radio conditions. Based on this information and on the measured signal strength, the wireless users estimate the throughput they will obtain in each system. Any new user with radio condition n will then connect to the system s^{IR} offering him the best throughput:

$$s^{IR} = \underset{s}{argmax} \frac{D_n^s}{1 + \sum_{m=1}^{N} \sum_{r=1}^{S} M_m^r} \quad (24)$$

> Note that this scheme is not realistic as the network operator will not divulge the exact number of connected users in each system and each position of the cell.

We consider a streaming service where users require a minimal throughput of 1 Mbps and can profit from throughputs up to 2 Mbps in order to enhance video quality (T_{min}=1 Mbps and T_{max}=2 Mbps). The call duration is 1/μ= 120 seconds. We consider an offered traffic that varies such that the blocking rate remains below 1% and obtain numerically the Global-optimum and Nash equilibrium points. Figure 5 depicts the optimal policies for low and high traffic conditions. An important observation is that the policy chosen by cell edge users is different from that of cell center ones, as the throughputs they obtain are different: Cell edge users have a larger preference for LTE as their throughput in HSDPA is too low (see Figure 1). It is also shown that the optimal policy depends on the offered traffic: In a system with low traffic, it can be useful to connect to HSDPA even if it offers low peak throughputs as a throughput between 1 and 2 Mbps is sufficient. However, when the traffic increases, the number of simultaneous users sharing HSDPA capacity increases, and it is better for more users to connect to LTE.

We illustrate in Figure 6 and 7 the correspondence between the utility and the load information for the hybrid decision approach. These curves give, for each load information *l* (*l*= 1,...,9), the utilities of cell edge and cell center users that are connected to LTE and HSDPA. The utility is expressed in Mbits as users are interested in maximizing the information they send during their transfer time. The results are in concordance with those presented in Figure 5. For instance, when the load information index increases from 1 to 3, corresponding to a low HSDPA load and an increasing LTE load (see the load information function correspondence in Figure 2), users begin connecting to HSDPA instead of LTE. This is exactly what is illustrated in Figure 5. In particular, notice that in the edge cell, utility of HSDPA-edge users tends to become negligible as the load information increases from *l*=4 to *l*=9. As a consequence, cell edge users have a larger preference for LTE since their throughput in HSDPA is too low as shown in Figure 5.

Figure 5. Optimal policy as function of the offered traffic. This table indicates the optimal choice for cell center and cell edge users for different aggregated load conditions. center=j and edge=i indicate that cell center users choose to connect to system j while cell edge users connect to system i.

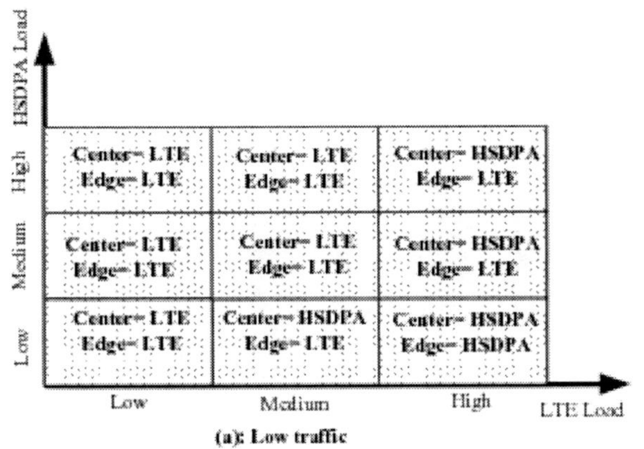

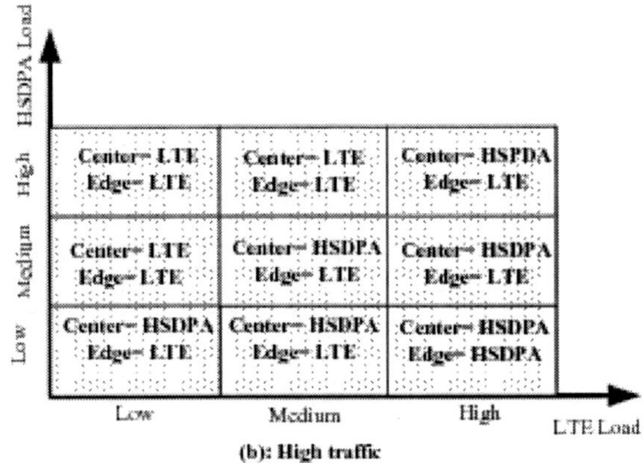

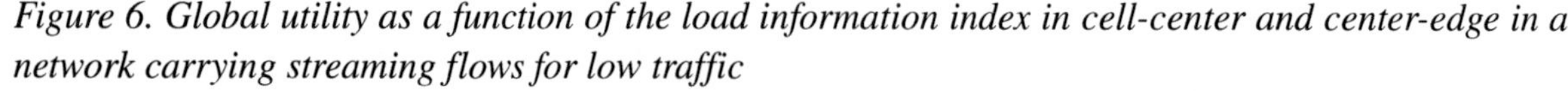

Figure 6. Global utility as a function of the load information index in cell-center and center-edge in a network carrying streaming flows for low traffic

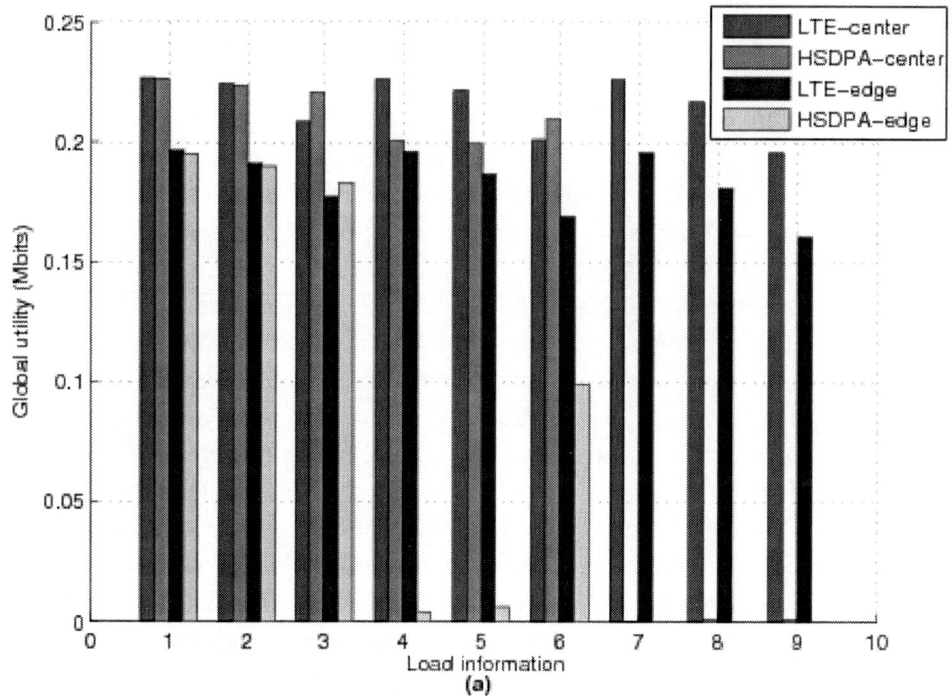

Figure 7. Global utility as a function of the load information index in cell-center and center-edge in a network carrying streaming flows for high traffic

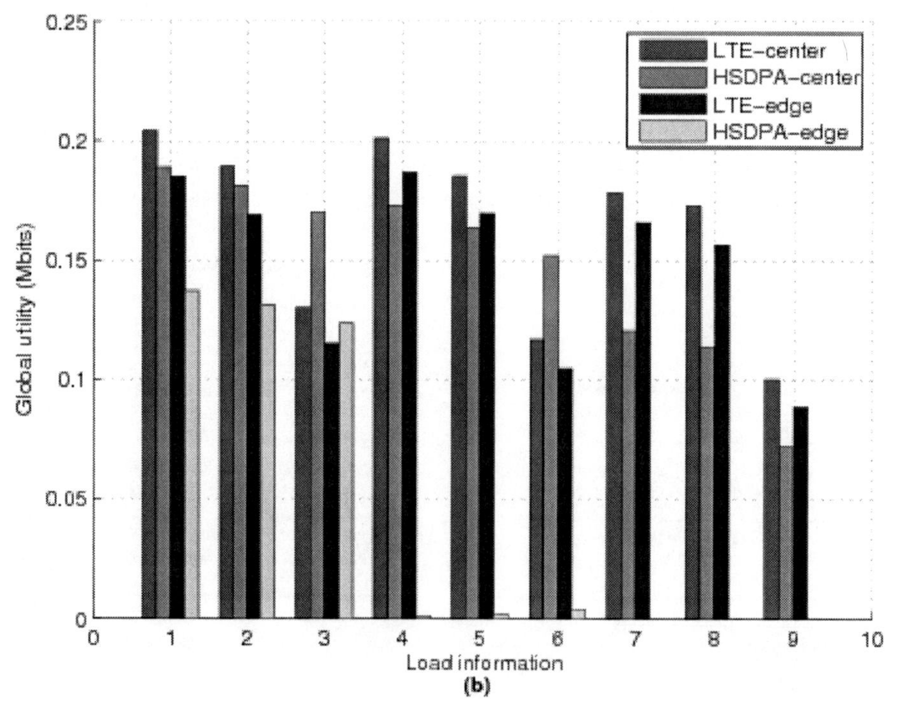

Figure 8. Global utility in a network carrying streaming flows

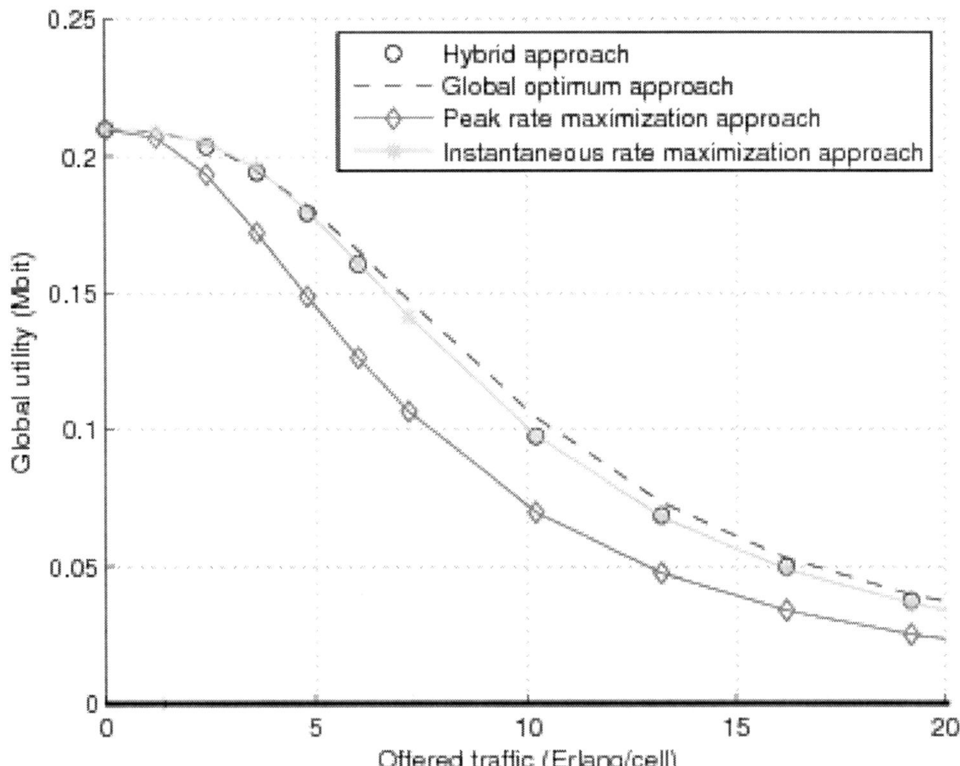

We plot in Figure 8 the global utility for the representative association approaches. As intuition would expect, the results show that the peak rate maximization approach presents the worst performance as the system that offers the largest peak throughput may be highly loaded, resulting in a bad QoS. However, a surprising result is that the hybrid scheme, based on partial aggregated information, is comparable to the instantaneous rate maximization approach when traffic increases. This is due to the fact that streaming users will have relatively long sessions, visiting thus a large number of network states; knowing the instantaneous throughput at arrival will not bring complete information about the QoS during the rest of the connection (i.e., short-term reward). In an opposite way, the proposed hybrid approach aims at maximizing the QoS during the whole connection time (i.e., long-term reward). Notice that NE points do not exist for some values of offered traffic. In fact, although a non-cooperative game always has a mixed-strategy equilibrium, it may in general not have a pure-strategy equilibrium. However, we focus on pure-strategy Nash equilibria since they are arguably more natural and, when they exist, they may better predict game play. In particular, we see that hybrid approach results exhibit approximately 20%, respectively 40%, of global utility gain beyond peak rate maximization approach at 5 Erlang per cell, respectively at 10 Erlang per cell.

We now turn to the Stackelberg formulation of our problem, where the network tries to control the users' behavior by broadcasting appropriate information, expected to maximize its utility while individual users maximize their own utilities. We

Figure 9. Blocking rate for different broadcast load information; a vector of thresholds [L_1, L_2, H_1, H_2] means that system s will be considered as highly loaded if its load exceeds s_2 and as low-loaded if its load is below s_1 (s=L for LTE and H for HSDPA)

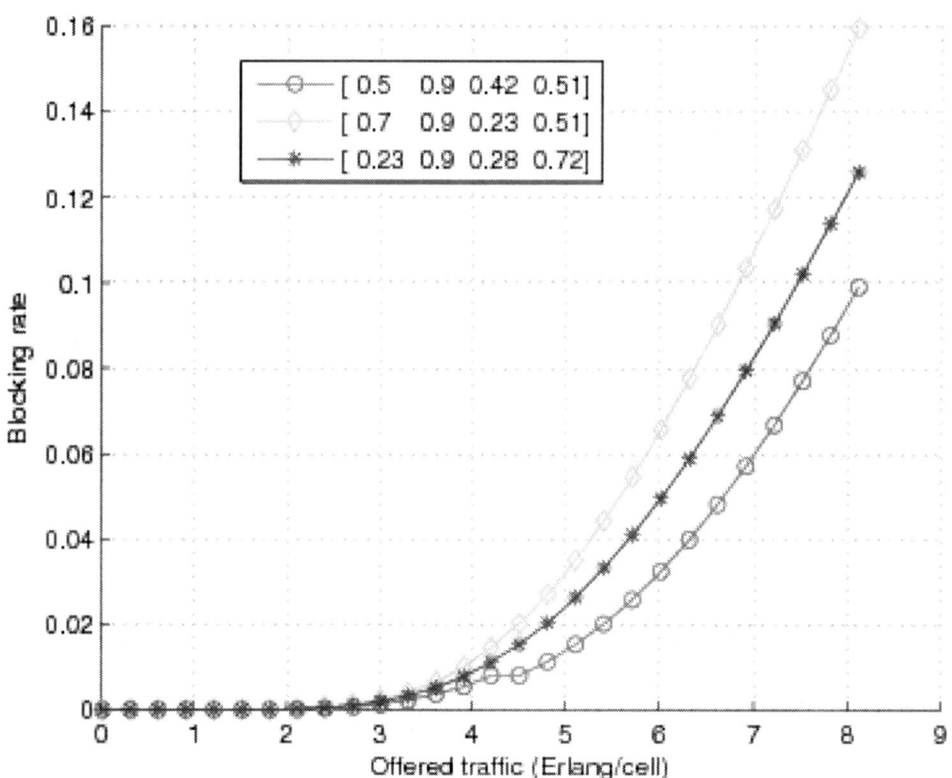

plot in Figure 9 the blocking rate for different ways of aggregating load information, obtained when users follow the policy corresponding to Nash equilibrium. In this figure, we plot the results for three cases: the optimal thresholds (in red circles) and two other sets of thresholds. We can observe that the utility of the network (expressed in the acceptance rate) can be substantially enhanced depending on the load information that is broadcasted. Such an accurate modeling of the Stackelberg problem is a key to understand the actual benefits brought by the proposed hybrid decision approach.

7. CONCLUSION

In this chapter, we studied a hybrid approach for radio resource management in heterogeneous cognitive networks in the presence of mobility. By hybrid we mean distributed decision making assisted by the network, where the wireless users aim at maximizing their own utility, guided by information broadcast by the network about the load of each system. We first showed how to derive the utilities of streaming and elastic flows that are related to the QoS they receive under the different association policies. We then derived the policy that corresponds to the Nash equilibrium and global optimum. Finally, we showed by means of a Stackelberg formulation, how the operator, by sending appropriate information about the state of the network, can optimize its global utility while users maximize their individual utilities. The proposed hybrid decision approach for cognitive radio networks can reach a good trade-off between the global network performance at the equilibrium and the requested amount of signaling.

REFERENCES

Altman, E., Boulogne, T., Azouzi, R., & Jimenez, T. (2006). A survey on networking games in telecommunications. *Computers & Operations Research, 33*(2). doi:10.1016/j.cor.2004.06.005

Bejerano, Y., Han, S.-J., & Li, L. E. (2004). Fairness and load balancing in wireless lans using association control. In *Proceedings of ACM MobiCom*. ACM Press.

Buljore, S., Muck, M., Martigne, P., Houze, P., Harada, H., Ishizu, K., … Stametalos, M. (2008). Introduction to IEEE p1900.4 activities. *IEICE Transactions on Communications, E91-B*(1).

Coucheney, P., Touati, C., & Gaujal, B. (2009). Fair and efficient user-network association algorithm for multi-technology wireless networks. In *Proceedings of the 28th Conference on Computer Communications Miniconference (INFOCOM)*. IEEE Press.

Elayoubi, S. E., Altman, E., Haddad, M., & Altman, Z. (2010). A hybrid decision approach for the association problem in heterogeneous networks. In *Proceedings of IEEE Infocom*. San Diego, CA: IEEE Press.

Felegyhazi, M., & Hubaux, J. (2006). *Game theory in wireless networks: A tutorial. EPFL Technical Report, LCA-REPORT-2006-002*. New York, NY: EPFL.

Fudenberg, D., & Tirole, J. (1991). *Game theory*. Cambridge, MA: MIT Press.

3GPP TS 32.500. (2008). *Telecommunication management: Self-organizing networks (SON): Concepts and requirements (rel. 8)*. Retrieved from http://www.3gpp.org

Haddad, M., Hayar, A., & Debbah, M. (2008). Spectral efficiency for spectrum pooling systems. *IET, 2*(6), 733–741. doi:10.1049/iet-com:20070469

Hardin, G. (1968). The tragedy of the commons. *Science, 162*(3859), 1243–1248. doi:10.1126/science.162.3859.1243

Horrich, S., Elayoubi, S.-E., & Jemaa, S. B. (2008). On the impact of mobility and joint rrm policies on a cooperative wimax/hsdpa network. In *Proceedings of IEEE WCNC*. Las Vegas, NV: IEEE Press.

IEEE Standards Coordinating Committee 41. (2012). *Dynamic spectrum access networks*. Retrieved from http://grouper.ieee.org/groups/scc41

Jiang, L., Parekh, S., & Walrand, J. C. (2008). Base station association game in multi-cell wireless networks. In *Proceedings of IEEE WCNC*. IEEE Press.

Kauffmann, B., Baccelli, F., Chaintreau, F., Mhatre, V., Papagiannaki, K., & Diot, C. (2007). Measurement-based self organization of interfering 802.11 wireless access networks. In *Proceedings of IEEE INFOCOM*. IEEE Press.

Kumar, D., Altman, E., & Kelif, J.-M. (2007). Globally optimal user-network association in an 802.11 wlan and 3G umts hybrid cell. In *Proceedings of 20th International Teletraffic Congress (ITC 20)*. Ottawa, Canada: ITC.

Mitola, J. (2000). *Cognitive radio: An integrated agent architecture for software defined radio*. (Doctoral Dissertation). Royal Institute of Technology. Stockholm, Sweden.

Ozgur, E. (2008). Association games in IEEE 802.11 wireless local area networks. *IEEE Transactions on Wireless Communications*, *7*(12), 5136–5143. doi:10.1109/T-WC.2008.071418

Shakkottai, S., Altman, E., & Kumar, A. (2006). The case for non-cooperative multihoming of users to access points in IEEE 802.11 wlans. In *Proceedings of IEEE INFOCOM*. IEEE Press.

Shakkottai, S., Altman, E., & Kumar, A. (2007). Multihoming of users to access points in wlans: A population game perspective. *IEEE Journal on Selected Areas in Communications*, *25*(6). doi:10.1109/JSAC.2007.070814

Stevens-Navarro, E., Lin, Y., & Wong, V. W. S. (2008). An MDP-based vertical handoff decision algorithm for heterogeneous wireless networks. *IEEE Transactions on Vehicular Technology*, *57*(2), 1243–1254. doi:10.1109/TVT.2007.907072

Chapter 5
Cross-Layer Design in Cognitive Radio Systems

Krishna Nehra
King's College London, UK

Mohammad Shikh-Bahaei
King's College London, UK

ABSTRACT

The main functionalities of a cognitive radio system, to ensure efficient operation of the primary users without harmful intervention from the secondary users and to simultaneously satisfy the requirements of the secondary users, are spectrum sensing, spectrum management, spectrum mobility, and spectrum management. These functions involve more than one layer of protocol stack rather than being performed at a single layer. This chapter briefly revisits these functions from the perspective of classification of the roles of different communication network layers in carrying out these functions. An exhaustive study is then presented of the key properties of cross-layer design applications in cognitive radio systems by taking examples from the existing literature and highlighting some open challenges and new opportunities. A cross-layer design example for interference-limited spectrum sharing systems is discussed in detail, which considers the parameters from the Physical Layer (PHY) and the Data Link Layer (DLL) in order to maximize the overall spectral efficiency of the Secondary User (SU). The numerical results show that the secondary link of spectrum sharing systems combining ARQ with adaptive modulation and coding achieves significant gain in throughput depending on the maximum number of retransmissions.

INTRODUCTION

The long familiar and widely referred Open Systems Interconnection (OSI) model is based on the hierarchical abstraction of various features of data communication networks. It organizes the network in certain number of layers (7 layers in the original OSI model), each layer performing a well-defined function to offer services to the higher layers without revealing the details of how the service was implemented. Although the traditional layered approach enjoys the benefits

DOI: 10.4018/978-1-4666-2812-0.ch005

of modularity, standardization, and expansibility, its rigid and strict architecture makes the layered structure inefficient to solve the problems related to wireless networks (the protocol stack was defined for wired networks). The boundary between different layers of network is blurring day by day because of evolution of the wireless networks, and rising demand of QoS satisfaction.

Cross-layer design, when it comes to Cognitive Radio (CR) systems, becomes even more trickier, because of the inherent characteristics (or required features) of observing, learning, reasoning, and adaptation (Rashid, 2009). A CR user needs to take into account several input sources at the same time including its own past observations as a result of learning property. Furthermore, a CR needs to consider a number of factors simultaneously such as application preferences of the Secondary Users (SU)s, several constraints such as interference limit to the Primary Users (PU)s and sensing capabilities, and its own capabilities to exploit the available primary spectrum and the channel conditions. Reaching to an optimal solution by merging all the requirements, constraints, and limitations into a single problem needs an adaptation and compromise covering multiple layers (Qing Zhao, 2007; Le & Hossain, 2008; Zhao & Sadler, 2007; Foukalas, et al., 2008; Foukalas, Gazis, & Alonistioti).

The main functionalities of a CR to ensure the efficient operation of the primary users without harmful intervention from the Secondary Users (SU) and simultaneously satisfying the requirements of the SUs are spectrum sensing, spectrum management, spectrum mobility, and spectrum sharing. These functions are inter-dependent and involve more than one layer of protocol stack rather than being performed at a single layer. The capacity of AWGN channels under received power constraints at the primary receiver for different scenarios including relay networks, multiple access channels with dependent sources and feedback, and collaborative communication, was analyzed in Gastpar (2004). The authors in Musavian and Aissa (2009a) derived capacity and optimum power-allocation schemes for different capacity metrics, e.g. ergodic, outage, and minimum-rate in Rayleigh fading channels under average and peak received-power constraints at the primary's receiver. Spectrum sharing systems with an additional statistical delay QoS constraint as well as interference-power constraint at the primary receiver were studied in Musavian and Aissa (2009b). The authors determined the maximal possible arrival rate supported by the secondary user's link under aforementioned constraints. Where majority of the available CR cross-layer literature focuses on joint optimization of the Physical Layer (PHY) and link (or MAC) layer (for example Digham, 2008; Vu et al., 2007; Bansal, 2008), interactions with higher layers have recently been subject of many works. Advantages of coordination between PHY and Network layer were analyzed in Xin et al. (2005), Wu and Tsang (2009), Yang and Wang (2008), and Shi et al. (2010). Whereas joint design of PHY- Data Link layer-TCP was presented in Luo et al. (2010a), in order to maximize TCP throughput. However, from user's viewpoint, QoS at the application layer is more important than that at the other layers. The improvement in performance using cross-layer design which considers application layer QoS, has been witnessed in Khan et al. (2006), Bobarshad et al. (2010), and references therein for non-CR systems. Recently application-layer QoS has captured attention in CR cross-layer research and few works including Luo et al. (2010b), Ali and Yu (2009), Luo et al. (2009), and Hu et al. (2010) have dealt with this.

This chapter describes the key properties of cross-layer design applications in CR systems. Furthermore, a new cross-layer design strategy for the systems employing spectrum underlay access technique is explained in detail. It is organized as follows. Section 2 revisits the main functionalities of CR: spectrum sensing, spectrum management, spectrum mobility, and spectrum sharing, by outlining the role played by different network layers in performing these functions. State-of-the-art and inspiration behind different cross-layer designs for CR systems is presented in Section 3. In particular integrated design of PHY-Data Link layer, PHY-network layer, and interactions with higher layers are discussed in subsections 3.1, 3.2, and 3.3, respectively. A model for cross-layer design for interference-limited spectrum sharing systems is explained in Section 4. Section 5 concludes the chapter by providing a brief summary.

COGNITIVE RADIO FUNCTIONALITIES AND ROLES PLAYED BY DIFFERENT LAYERS

Figure 1 shows different parameters from network layers, which determine/affect the CR functionalities. The outcomes of CR functionalities having an impact on the operation of network layers are also indicated in Figure 1. The color of CR functions indicates that those functions are performed at the layers with corresponding colors. CR functions take factors as input from different OSI layers, and the outputs of these functions are connected to those CR functions and OSI layers which get affected by the corresponding function. For example, spectrum sensing receives inputs from PHY (strength of current occupants) and DLL (sensing policies), and its output is connected to all other CR functions and all OSI layers (because every layer and CR functionality requires sensing information to perform their operation). Following subsections provide a brief overview of CR functionalities and responsibilities of different network layers involved.

Spectrum Sensing

Spectrum sensing is one of the main components of CR concept. It can be conceptualized as the task of gaining awareness about the spectrum usage and existence of PUs in the surrounding of a CR- enabled transceiver, across multiple dimensions such as time, space, frequency, and code. A great deal of literature including Yucek (2009), Zhu Han (2009), and Ying-Chang Liang (2008) addresses various techniques of spectrum sensing and their pros and cons. Data Link layer acquires information about channel occupancy with the help of different sensing policies using MAC protocols. Spectrum sensing also involves determining the strength of current occupants of the spectrum by having a priori knowledge of their modulation, waveform, bandwidth, carrier frequency etc., and PHY is responsible to obtain this information. Physical parameter detection and sensing algorithms, or spectrum sensing as a whole, require both PHY and data link layer to interact with each other. In particular, design of sensing algorithm has direct impact on accuracy of the parameters estimated at the PHY. On the other hand, information about parameters such as operating frequency and strength of existing occupants detected at PHY plays a critical role at data link layer in determining which channel to access.

Spectrum Management

The task of spectrum management consists of two parts, namely spectrum allocation and spectrum access. The process of spectrum allocation involves allocation of the underutilized spectrum bands to CR users based on a certain policy. As mentioned in the previous subsection information about channel occupancy is a result of sensing which is the responsibility of the PHY and the data link layers. After detecting the spectrum holes, the CR users analyze the available channels to fairly choose the appropriate channel/s over which they can transmit their signals. Selection of channels depends on parameters such as availability of CSI for those channels at certain CRs, interference inflicted on those if the channels are occupied by other SUs. PHY is also engaged in spectrum access by adapting transmit power and modulation scheme constrained with power and

Figure 1. Various protocol layers' parameters affecting CR functionalities and vice-versa

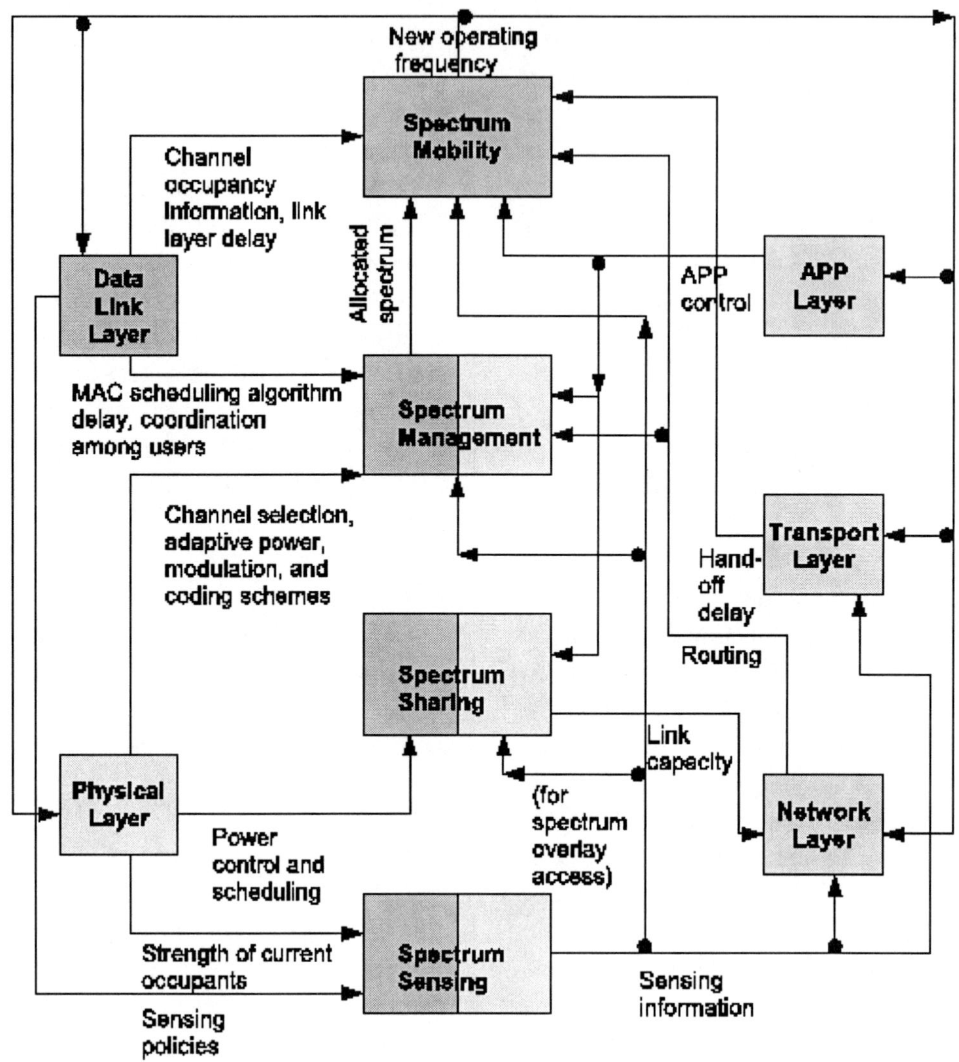

BER requirements. The outcome of spectrum management process also depends on data link layer parameters such as delay of MAC scheduling algorithms, and retransmissions.

Apart from PHY and data link layer, requirements at higher layers such as routing and application-specific requirements also demand for considerable attention in spectrum management operation. For example, in some applications data rate is more important than the packet loss performance, and therefore they require modulation and coding schemes to provide high spectral efficiency. Whereas in other applications error performance can be more critical, which demands modulation and coding schemes with low Bit Error Rate (BER). In addition, there is a possibility of multiple users competing to use the same portion of spectrum, which will result in collision among them. Spectrum access techniques take care of coordination among multiple users in order to allow the fair use of spectrum, which is the functionality of the data link layer.

Spectrum Mobility

The CR environment is dynamic and the frequency bands it is operating on may be re-occupied by the PUs at any instant. Therefore, a CR user also needs to be capable of changing its operating frequency when required, or randomly before the channel becomes unavailable. The process of SUs changing their operating frequencies can be defined as spectrum handoff, in which a SU needs to pause its current communication, vacate the channel, and determine a new available channel to continue its operation. Mobility process is again accomplished by data link layer, where MAC algorithms acquire the information about channel occupancy and use it to determine when to switch to a new channel. Along with sensing and data link layer delay, routing information, hand-off delay, and application-specific parameters are important factors affecting the mobility operation.

Spectrum Sharing

Spectrum sharing is concerned with allocation of bandwidth and power, which is again accomplished at the PHY. Strength of a signal at any SU receiver is determined by the power allocated to its transmitter. The power allocated to one SU will be interference for other SUs, and therefore power control policy for an SU has direct impact on interference levels to other SUs. Apart from PHY, power control and scheduling have impact on functioning of higher layers too. For example routing at transport layer is affected by the link capacity which itself is dependent on power control and scheduling. Constraints at APP layer should also be considered while designing a power allocation scheme. Because some applications such as video transmission are power hungry whereas another group of applications, e.g., those involving sensor networks are power savvy. There is rich literature available providing good insight on spectrum sharing fundamentals such as capacity and potential of spectrum sharing systems, optimal limits of spectrum sharing etc. (Akyildiz, et al., 2006; Ghasemi & Sousa, 2006; Srinivasa, 2008; Etkin, 2007).

CROSS-LAYER DESIGN

PHY: Data Link Layer

The functionalities defined for PHY and data link layer in CR networks are intertwined and designing them independently does not result in optimal performance, be it in terms of throughput, delay, or packet error rate. This calls for more frequent interactions between PHY and data link layer than that between other layers. Cross-layer design of PHY and data link layer enables the data link layer to adapt to the available communication resources to be in line with the time-varying channel and improve the QoS of SUs. One of the methods to achieve PHY-data link layer collaboration is to combine AMC at PHY and ARQ at data link layer. In a system implementing joint coding and scheduling, PUs' activity can be modeled as a two state first-order Markov process defined by channel transition matrix *P*, as:

$$P = \begin{bmatrix} 1 - p_{0\to 1} & p_{0\to 1} \\ p_{1\to 0} & 1 - p_{1\to 0} \end{bmatrix}$$

where $p_{0\to 1}$ refers to the probability of the occupied channel at time *t* to become free at time slot *t + 1*, and $p_{1\to 0}$ is the probability of a free channel to be occupied in the next time slot. The variations in quality of fading channel for SUs can be modeled as finite state Markov chain with N states, and the transmission modes of AMC can be adjusted according to the channel states. Waiting for idle time slots causes delays and increases chance of packet loss at SUs, and thus makes the SUs compromise QoS. One way to improve the QoS of SUs is to design joint channel aware and queue aware MAC scheduling protocols, which reduce the overall delay and packet loss probability.

In addition to joint implementation of AMC at PHY and data link layer protocols, another perspective of interactive PHY-data link layer design is combining spectrum sensing at PHY with packet scheduling at the data link layer. Some of the studies assume two transceivers available for SU, one to dedicatedly operate over control channel and the other to sense the vacant subbands in the spectrum. However adding one more transceiver would increase the cost of hardware extensively, therefore majority of practical systems have only one transceiver at every SU node. Simultaneous sensing and transmission in two transceivers case have the advantage that SUs do not need to interrupt their operation to periodically sense the channel availability, and this reduces the delay at SU. Delay sensitive (real time) applications as well as rate critical ones (non-real time) may coexist based on the users' requirements. Keeping this is mind, another point to be addressed by the CR systems is to satisfy the QoS of mixture of Real Time (RT) services and Non-Real Time (NRT) applications by combining packet scheduling with subchannel, bit, and power allocation for both the services.

The above-mentioned cross-layer design methods are based on physical-layer channel models which cannot completely capture the random behavior at various stages e.g., incursion of PUs on the licensed sub-channels, random fluctuations in received power levels and data rates, and sensing error in terms of false alarms. Furthermore, the link-layer QoS constraints such as delay bounds requiring the analysis of queuing behavior cannot be effectively addressed by the physical-layer channel models, because these models only provide the estimate of physical layer performance of the cognitive radio systems, for example, in terms of bit error performance as a function of SNR. The link-layer channel model, Effective Capacity (EC), developed in Wu and Negi (2003) can very well support the statistical QoS requirements and play a significant role in evaluating the QoS performance of cognitive radio systems. EC characterizes the wireless link by the functions of probability of non-empty buffer and the QoS exponent of the connection, which can be easily translated into link-level QoS metrics such as delay bound probability. EC has recently drawn attention of CR research community and effects of various random phenomena occurring in CR systems on their performance have been investigated in terms of effective capacity (Akin & Gursoy, 2010).

PHY: TCP

TCP layer is responsible for timely and reliable delivery of the transmitted data without error. Therefore packet loss probability is an important performance measure of TCP, which depends on access technology as well as on the frequency of operation, interference level, and the available bandwidth. Another performance metric of TCP, Round Trip Time (RTT), also depends indirectly on the frequency used. Packet loss may occur due to congestion, channel error, or node mobility. In CR systems employing Opportunistic Spectrum Access (OSA), the PUs can occupy a channel at any time irrespective of presence of any SU, forcing the current SU on that channel (if any) to vacate the channel. Therefore, the probability of availability of licensed channels to the SUs can have a significant impact on the packet loss performance of secondary systems. In layered approach, TCP considers only congestion as the cause of packet loss, even if it was due to instantaneous state of the wireless channel or PU activity. Upon detection of packet loss, it initiates the retransmission of TCP segments and reduces the congestion window without identifying the actual cause of the losses. Therefore, TCP performance optimization needs a model of PER combining parameters from different network layers, and cross-layer design coupling PHY-data link layer-

TCP is a deserving candidate in this direction. Information about the channel state at TCP can be used to differentiate the packet losses due to bad link-quality or congestion. Moreover, knowledge about the spectrum sensing and access decisions at TCP can be utilized to meet the QoS requirements. Cross-layer design involving TCP in CR networks is at its infancy and very little literature is available on this.

TCP throughput can be significantly improved by joint optimization of spectrum sensing, access decision, AMC, and data link layer frame size and by judiciously selecting and designing lower layer parameters. TCP throughput model as a function of packet loss Probability (*p*), Round Trip Time (*RTT*), maximum congestion window (*cwnd*), initial time out $\left(T_0\right)$, number of packets acknowledged by a receive ACK(*b*) can be defined as in Box 1 (Luo, et al., 2010a).

The design parameters used to maximize TCP throughput have significant impact on it. It was shown in Luo et al. (2010a) that lower order modulation AMC schemes outperform those with higher order in terms of TCP throughput for bad channels, however opposite performance results are obtained for good channel conditions. Similarly, an optimal frame size is required to gain optimal TCP performance.

PHY: Application Layer

End users have direct connection with Application Layer (AL) and are generally unaware of the functions, which are performed at other network layers to efficiently and effectively deal with their demands. User satisfaction is by far at the core of all the technologies, which is achieved if the user- perceived quality is met. Packet loss and delay jitter are among main factors attributing to reduction in the quality experienced by SUs and AL has the knowledge of the diminution in QoS caused by these parameters. Similar to non-CR systems, user-perceived video quality is the most crucial performance metric (AL QoS) in cognitive wireless multimedia networks. Most of the works focusing on throughput maximization do not necessarily guarantee best QoS at AL for some multimedia applications such as video.

Another important parameter at AL is intra refreshing rate, adaptive adjustment of which for on-line video encoding applications can improve error resilience of SUs to the time varying wireless channel (Ali & Yu, 2009). One method to minimize distortion and thus maximize AL QoS is to determine intra refreshing rate together with channel selection for spectrum sensing, spectrum access decision, and intra refreshing rate. The total end-to-end distortion is combination of the quantization distortion and the distortion resulting from channel errors. Keeping in mind the fact that source distortion decreases with decreasing intra refreshing rate whereas channel distortion

Box 1.

$$B(cwnd, RTT, T_0, b, p) = min\left(\frac{cwnd}{RTT}, \frac{1}{RTT.\sqrt{\frac{2bp}{3}} + T_0.min\left(1, 3\sqrt{\frac{3bp}{8}}p(1+32p^2)\right)}\right)$$

decreases with increasing intra refreshing rate, the optimal intra refreshing rate should be found to minimize the total end-to-end distortion.

Along with packet loss and delay jitter lack of bandwidth guarantee is another reason for AL QoS degradation in multimedia applications, because multimedia streaming applications consume significant amount of bandwidth. The problem becomes more severe in case of multi-hop networks because the intermediate network nodes are heterogeneous and may have asymmetric properties in terms of CSI, throughput, power, delay, and packet loss. Faced with bad channel conditions at any hop, a multimedia streaming application might have limited choice for video encoder or streaming algorithm.

Cross-Layer Design for Interference-Limited Spectrum Sharing Systems

To demonstrate the advantage of cross-layer design in systems with underlay access technique, a dynamic spectrum sharing system is considered. The system consists of a primary user, whose licensed spectrum is allowed to be accessed by a SU as long as it does not violate the prescribed interference limit inflicted on the primary user. We aim at maximizing the performance of the SU's link in terms of Average Spectral Efficiency (ASE) and error performance under the specified Packet Error Rate (PER) and average interference limit constraints. To this end, we employ a cross-layer design combining adaptive power and coded discrete M-QAM modulation scheme at the PHY with a truncated Automatic Repeat Request (ARQ) protocol at the DLL, and simultaneously satisfies the aforementioned constraints. The number of retransmissions in ARQ is bounded by a specific maximum $\left(N_r^{\max}\right)$, and the incorrectly received packet after $N_r^{\max}$ transmissions is dropped from the receiver buffer.

System Model

Figure 2 shows the system model used in this subsection. The channel gains from the secondary transmitter to the primary and secondary receivers are respectively denoted by h^{sp} and h^s. Both h^{sp} and h^s are assumed to be stationary and ergodic with pdf, $f_{sp}(h^{sp})$ and $f_s(h^s)$, respectively. Furthermore, both h^{sp} and h^s are assumed to be *i.i.d.* processes, following Nakagami-m fading distribution with unit variance. Moreover, the knowledge of h^{sp} and h^s is assumed to be available at the secondary transmitter.

A variable-power variable-rate transmission scheme at the secondary transmitter, utilizing coded discrete MQAM modulation is considered. It should be noted that the adaptive power and rate of the secondary user are functions of both channel gains, h^{sp} and h^s. This is due to average interference power constraint imposed on secondary transmission, defined by:

$$\int_{h^s}\int_{h^{sp}} P(h^s,h^{sp})h^{sp} f_{sp}(h^{sp}) f_s(h^s) dh^{sp} dh^s \le I_{\max} \tag{1}$$

where $P(h^s,h^{sp})$ is transmit power of the secondary user, and $I_{\max}$ denotes maximum allowed interference at the primary receiver. Using the fact that interference constraint (1) depends on channel gains through the ratio of channel gains (Musavian & Aissa, 2009b; Equation 11), a new random variable $v=h^s/h^{sp}$ with the pdf:

$$f_v(v) = \frac{\rho^{-m^{sp}}}{\beta(m^{sp},m^s)} \frac{v^{m^s-1}}{(v+1/\rho)^{m^{sp}+m^s}} \tag{2}$$

is defined where m^{sp} and m^s denote the Nakagami fading parameters for h^{sp} and h^s, respectively,

Figure 2. System model

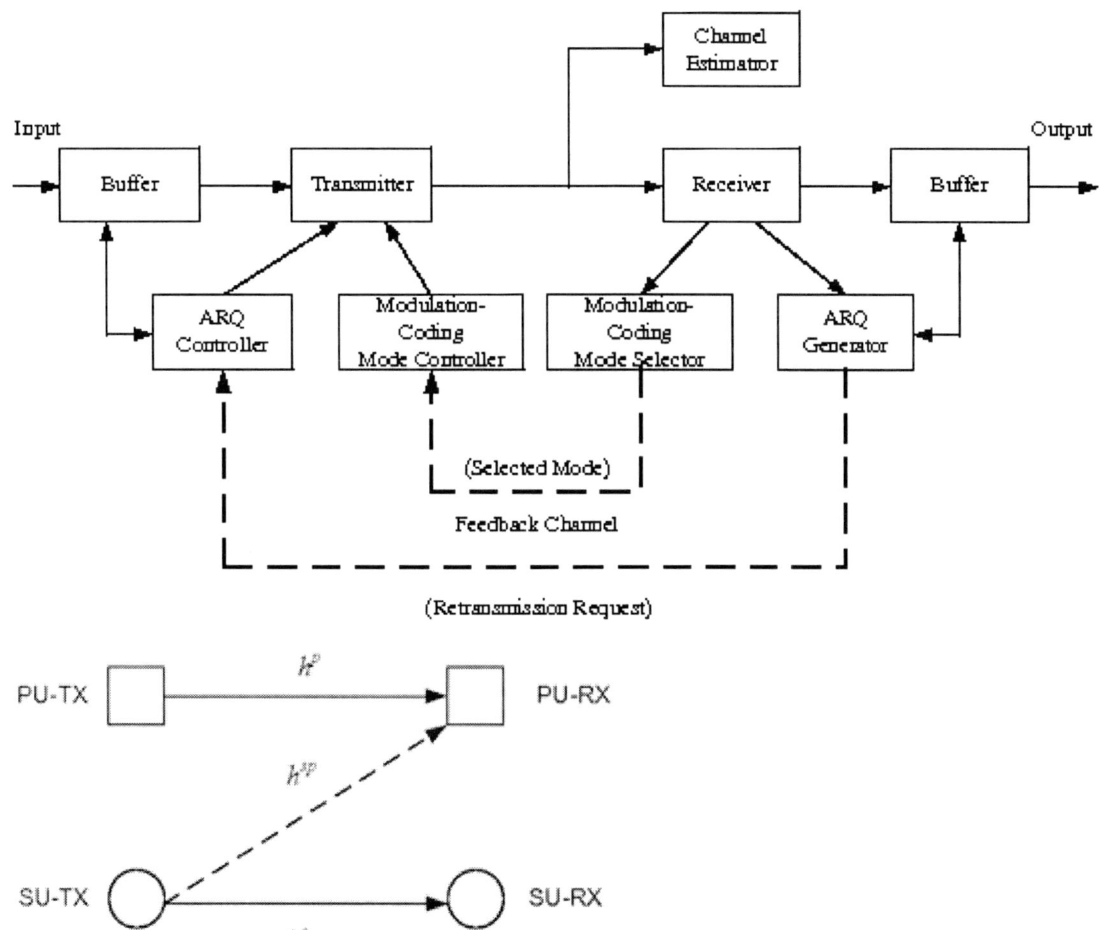

$$\rho = \frac{m^s}{m^{sp}}$$

and

$$\beta(m^{sp}, m^s) = \frac{\Gamma(m^{sp})\Gamma(m^s)}{\Gamma(m^{sp} + m^s)}$$

$\Gamma(\cdot)$ refers to the Gamma function.

Let $\gamma^v = \overline{P}h^{sp}v$ denote the pre-adaptation secondary received SNR with average secondary transmit power $\overline{P}$. In order to perform AMC, γ^v is divided into N+ 1 non-overlapping consecutive intervals

$$[\gamma_n^v, \gamma_{n+1}^v),\ n = 0, 1, \ldots, N$$

where γ_0^v=0, $\gamma_{N+1}^v = \infty$, and N is the number of AMC modes. The AMC mode *n* is chosen when $\gamma^v \in [\gamma_n^v, \gamma_{n+1}^v)$, and transmission takes place with rate $R_n(h^s,h^{sp})$ and power $P_n(h^s,h^{sp})$. Both $P(h^s,h^{sp})$ used in (1) and $P_n(h^s,h^{sp})$ are essentially the same, and we use $P_n(h^s,h^{sp})$ to denote transmit power of the secondary user in subsequent expressions. No transmission occurs when $\gamma^v \in [\gamma_0^v, \gamma_1^v)$, corresponding to the case when h^s is weak compared to h^{sp}. With the purpose of maximizing spectral efficiency, S_{ef} of the secondary user the following expression is used to approximate the secondary channel PER in mode n as a function of post-adaptation received SNR, $\gamma^{eq} = P_n(h^s,h^{sp}) h^{sp}v$:

$$\mathrm{PER}_n(\gamma^v) = \begin{cases} 1, & 0 \leq \gamma^v < \gamma_n^{bnd} \\ a_n \exp(-g_n P_n(h^s, h^{sp}) h^{sp} v) & \gamma_n^{bnd} \leq \gamma^v \end{cases} \tag{3}$$

The parameters a_n, g_n, and γ_n^{bnd} are transmission mode and packet-size dependent, and can be obtained by fitting the PER expression given in (3) to the exact PER obtained through simulation. The PER model corresponding to constant-power allocation at the secondary transmitter has been used and verified in Liu et al. (2004).

Adaptive Coded Rate and Power Allocation with ARQ

In this section, the problem of maximizing spectral efficiency of the secondary user under specified PER and average interference-limit constraints is addressed. The scheme is started by determining the optimal SNR boundaries (v) for AMC mode switching in order to maximize the ASE under the aforementioned constraints.

Adaptive Modulation and Coding at the Physical Layer

The ASE of the secondary link for the discrete AMC case is essentially the sum of data rates of all the modes weighted by the probability of occurrence of the respective mode. Upon selection of the nth mode, each symbol is transmitted with the rate $R_n(h^s,h^{sp})=R_c log_2(M_n)$ associated with QAM constellation size M_n and FEC code rate R_c. Presuming a Nyquist pulse-shaping filter with bandwidth $B=1/T_s$, where T_s corresponds to the symbol rate, ASE achieved at the physical layer without considering possible packet retransmission in the data-link layer is expressed by:

$$\overline{S}_{\mathrm{eff}} = \sum_{n=0}^{N} R_n Pr(n) \tag{4}$$

where Pr(n) is the probability of transmission in n^{th} mode, and is defined by:

$$Pr(n) = \int_{v_n}^{v_{n+1}} f_v(v)dv = \frac{\rho^{m^s}}{m^s \beta(m^{sp}, m^s)} \left[v^{m^s} {}_2F_1 \left(\begin{matrix} [m^s, m^s + m^{sp}]; \\ [1 + m^s]; -v\rho \end{matrix} \right) \right]_{v_n}^{v_{n+1}} \tag{5}$$

where ${}_2F_1$([a,b];[b];c) denotes the Gaussian hypergeometric function (Gel'fand, et al., 2003).

In accordance with the defined random variable v and discrete AMC modes, the average interference constraint given in (1) transforms to:

$$\sum_{n=1}^{N} P_n(h^s, h^{sp}) \int_{v_n}^{v^{n+1}} h^{sp} f_v(v)dv \leq I_{\max}. \tag{6}$$

Using (3), power allocated to the nth mode can be expressed in terms of the PER approximation parameters as:

$$P_n(h^s, h^{sp}) = \frac{1}{g_n v_n h^{sp}} \ln\left(\frac{a_n}{\mathrm{PER}_{\mathrm{ins}}} \right) \tag{7}$$

where:

$$\gamma_n^{bnd} \leq v_n h^{sp} \leq \gamma^v < v_{n+1} h^{sp} \tag{8}$$

where PER_{ins} represents the achievable instantaneous PER, and:

$$\mathrm{PER}_{\mathrm{ins}} \leq P_{\mathrm{tgt}}, \quad \gamma_0^v \leq \gamma^v < \gamma_{N+1}^v \tag{9}$$

where P_{tgt} denotes the target PER. The optimization problem of determining AMC mode switching levels for v is formulated, in order to maximize ASE subject to average interference and instantaneous PER constraints as follows:

$$\begin{aligned} \overline{S}_{\mathrm{eff}} &= \max_{\{v_n\}_{n=1}^{N}, \mathrm{PER}_{\mathrm{ins}}} \sum_{n=0}^{N} R_n \int_{v_n}^{v_{n+1}} f_v(v)dv \\ \text{s.t. C1:} &\quad \sum_{n=1}^{N} S_n \int_{v_n}^{v_{n+1}} \frac{1}{v} f_v(v)dv \leq I_{\max} \\ \text{C2:} &\quad \mathrm{PER}_{\mathrm{ins}} \leq P_{\mathrm{tgt}} \end{aligned} \tag{10}$$

where:

$$S_n = \frac{1}{g_n} \log\left(\frac{a_n}{\mathrm{PER}_{\mathrm{ins}}}\right) \tag{11}$$

Constraint C1 represents the maximum allowed interference constraint imposed on the secondary users, and it is obtained by assuming equality in Equation 8 and replacing Equation 7 in the average interference constraint Equation 6. C2 corresponds to the instantaneous PER constraint.

It can be shown that

$$\sum_{n=0}^{N} R_n \int_{v_n}^{v_{n+1}} f_v(v)dv$$

and the constraints C1 and C2 are convex with respect to *v* (the proof is given in appendix), and Slater's condition holds, so there is no duality gap and the optimal solution is characterized by KKT conditions(Boyd & Vandenberghe, 2004). Proof of fulfillment of Slater's condition is provided in Figure 3. The x-axis corresponds to right hand side of the constraint C1, and Y-axis shows the difference between respective LHS and RHS values of the same. Figure 3 ensures that all the values on Y-axis are negative, therefore there exists *v* for which strict inequality holds in constraint C1.

Figure 3. Proof of Slater's condition qualification

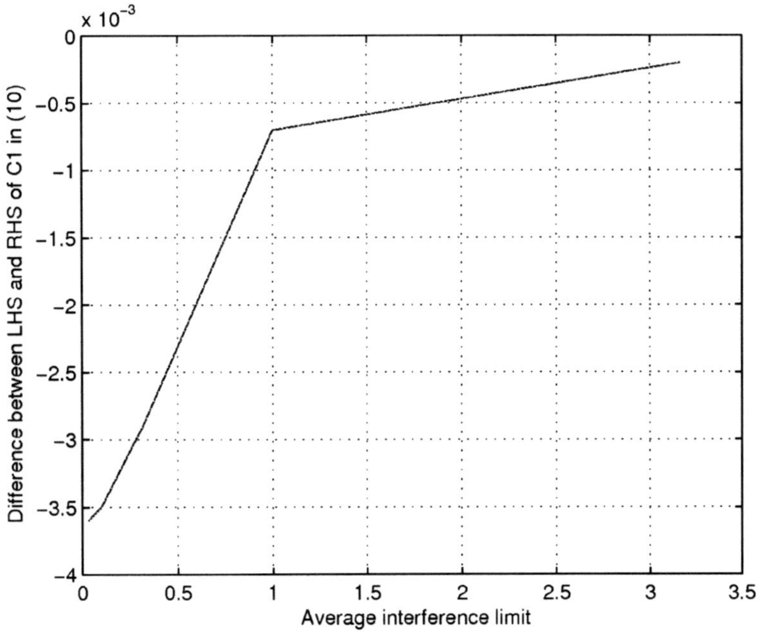

To solve the optimization problem (10) for AMC mode switching levels of v, the Lagrangian for this problem is defined as:

$$L(v_1, v_2, \cdots, v_N, \lambda) = \sum_{n=0}^{N} R_n \int_{v_n}^{v_{n+1}} f_v(v)dv + \lambda \left[\sum_{n=1}^{N} S_n \int_{v_n}^{v_{n+1}} \frac{1}{v} f_v(v)dv - I_{\max} \right] \quad (12)$$

where λ is the Lagrange multiplier. Using KKT conditions, the optimal solution $(v_1, v_2, \ldots, v_N)$ and corresponding Lagrange multiplier λ must satisfy the following conditions:

$$\frac{\partial L}{\partial v_n^*}(v_1^*, v_2^*, \cdots, v_N^*, \lambda^*) = 0 \qquad n = 1, \cdots, N$$

$$\sum_{n=1}^{N} S_n \int_{v_n}^{v_{n+1}} \frac{1}{v} f_v(v)dv = I_{\max}$$

$$\mathrm{PER}_{\mathrm{ins}} \le P_{\mathrm{tgt}}$$

$$\bar{P} h^{sp} v_n^* > \gamma_n^{bnd} \qquad n = 1, \cdots, N \quad (13)$$

Considering (13), the optimal boundary points $\{v_n\}_{n=1}^{N}$ can be obtained as:

$$v_1^* = \frac{\lambda S_1}{R_0 - R_1} = -\frac{\lambda S_1}{R_1}, \text{since } R_0 = 0$$

$$v_n^* = -\frac{\lambda \left(S_{n-1} - S_n\right)}{R_{n-1} - R_n}, \; n = 2, \ldots, N \quad (14)$$

The value of the Lagrangian multiplier λ can be determined by substituting the boundary points in average interference constraint C1 in (10) with the equal sign. By using maximum allowed PER limit $\left(P_{tgt}\right)$ in (11), λ becomes a function of v, its pdf, and $I_{\max}$ (which can also be fixed to a certain value depending upon the interference level allowed by the primary user). As shown in appendix, the optimization problem (10) is a convex optimization problem, therefore the boundary points obtained in Equation 14 are also optimal. Substitution of optimal boundary points v_n, $n=1,\ldots,N$ in (4), and (7) yields the optimum ASE and optimum power allocated in n^{th} transmission mode, respectively. Calculation of adaptive power from (7) requires h^{sp} values along with boundary values of v. Based on initial assumption of availability of h^{sp} at the secondary transmitter, transmit power in the nth mode can be determined.

Power Adaptation and AMC Combined with ARQ

Application of the ARQ protocol at the data-link layer facilitates retransmission of the erroneous packets received at the secondary receiver. However, for practical purposes, it is assumed that the number of retransmissions of a packet with error is bounded by a maximum value $N_r^{\max}$. This is determined by the maximum delay, which can be tolerated in communication between the secondary transmitter and secondary receiver. If a packet is still erroneous after $N_r^{\max}$ retransmissions, it is dropped from the receiver buffer and is considered lost. ASE of the secondary link incorporating ARQ can be expressed as:

$$\bar{S}_{\mathrm{eff}}(N_r^{\max}) = \frac{\bar{S}_{\mathrm{eff}}}{\bar{N}(\overline{\mathrm{PER}}, N_r^{\max})} = \frac{\sum_{n=0}^{N} R_n Pr(n)}{\bar{N}(\overline{\mathrm{PER}}, N_r^{\max})} \quad (15)$$

where $\bar{N}(\overline{\mathrm{PER}}, N_r^{\max})$ is the effective average number of transmissions per packet (defined in

Equation 18). $\overline{PER}$ is the average PER of all modes, and is given by:

$$\overline{\mathrm{PER}} = \frac{\sum_{n=1}^{N} R_n Pr(n) \overline{\mathrm{PER}}_n}{\sum_{n=1}^{N} R_n Pr(n)} \quad (16)$$

$\overline{\mathrm{PER}}_n$ represents the average PER (ratio of the number of incorrectly received packets over those transmitted using mode n) in the nth mode, $\overline{\mathrm{PER}}_n$ is defined by:

$$\overline{\mathrm{PER}}_n = \frac{1}{Pr(n)} \int_{v_n}^{v_{n+1}} \mathrm{PER}_n(v) f_v(v) dv \quad (17)$$

$\bar{N}(\overline{\mathrm{PER}}, N_r^{\max})$ on the secondary link employing joint AMC-ARQ, can be determined from the following equation:

$$\bar{N}(\overline{\mathrm{PER}}, N_r^{\max}) = \frac{1 - \overline{\mathrm{PER}}^{N_r^{\max}+1}}{1 - \overline{\mathrm{PER}}} \quad (18)$$

However, persistence of error in a packet after $N_r^{\max}$ retransmissions results in loss of that packet, and actual packet loss probability, ϕ_{loss} for the considered policy is given as:

$$\phi_{loss} = \overline{\mathrm{PER}}^{N_r^{\max}+1} \leq P_{\mathrm{tgt}}^{N_r^{\max}+1} = \Phi_{loss} \quad (19)$$

where Φ_{loss} is the maximum acceptable packet loss probability.

Truncated ARQ without AMC

This section analyzes performance of the secondary system employing truncated-ARQ without AMC at the physical layer. Therefore, transmit power and transmission mode is not adaptive to CSI. The ASE for the nth transmission mode is calculated with average transmit power $\bar{P}$. Replacing adaptive power in (3) with $\bar{P}$, and using (2), the average PER at physical layer can be obtained from Box 2.

Ei(.) denotes the exponential integral function. The closed form of average PER at the physical layer for $m^s = m^{sp} = 2$ can be expressed as in Box 3.

Average number of transmissions per packet is given by:

$$\bar{N}(\overline{\mathrm{PER}}(n), N_r^{\max+1}) = \frac{1 - \overline{\mathrm{PER}}(n)^{N_r^{\max}}}{1 - \overline{\mathrm{PER}}(n)} \quad (22)$$

Box 2. Average PER at the physical layer

$$\begin{aligned}\overline{\mathrm{PER}}(n) &= \int_0^{\infty} \mathrm{PER}_n(v) f_v(v) dv = \int_0^{\gamma_n^{bnd}} f_v(v) dv + \int_{\gamma_n^{bnd}}^{\infty} a_n \exp(-g_n \bar{P} h^{sp} v) f_v(v) dv \\ &= \frac{\gamma_n^{bnd}}{1+\gamma_n^{bnd}} + \frac{a_n \exp\left(-g_n \bar{P} h^{sp} \gamma_n^{bnd}\right)}{\gamma_n^{bnd}+1} \\ &\quad + a_n g_n \bar{P} h^{sp} \exp(g_n \bar{P} h^{sp}) Ei(-(\gamma_n^{bnd}+1) g_n \bar{P} h^{sp}) \quad [g_n \bar{P} h^{sp} > 0]\end{aligned} \quad (20)$$

for $m^s = m^{sp} = 1$.

Box 3. Closed form of average PER at the physical layer

$$\overline{\mathrm{PER}}(n) = \frac{(\gamma_n^{bnd})^2(\gamma_n^{bnd}+3)}{(\gamma_n^{bnd}+1)^3} - \frac{a_n \exp(g_n \bar{P}h^{sp})}{0.1667} Ei(-g_n \bar{P}h^{sp}(\gamma_n^{bnd}+1))$$
$$\times (g_n \bar{P}h^{sp})^2 \left(\frac{1}{2} + \frac{g_n \bar{P}h^{sp}}{6}\right) + \frac{a_n \exp(-g_n \bar{P}h^{sp}\gamma_n^{bnd})}{0.1667(\gamma_n^{bnd}+1)^2} \qquad (21)$$
$$\left(\frac{1}{2} - \frac{g_n \bar{P}h^{sp}(\gamma_n^{bnd}+1)}{2} - \frac{1}{\gamma_n^{bnd}+1}\sum_{k=0}^{2}\frac{(-1)^k (g_n \bar{P}h^{sp})^k(\gamma_n^{bnd}+1)^k}{n(n-1)\dots(n-k)}\right)\dots[g_n \bar{P}h^{sp} > 0]$$

with $n_r^{\max}$ maximum transmissions.

Packet loss probability $\phi_{loss,n}$ and ASE $\bar{S}_{\mathrm{eff},n}(N_r^{\max})$ can be obtained by following equations:

$$\phi_{loss,n} = \overline{\mathrm{PER}}(n)^{N_r^{\max}+1} \qquad (23)$$

$$\bar{S}_{\mathrm{eff},n}(N_r^{\max}) = \begin{cases} 0 & \bar{P} < \bar{P}_{n,\mathrm{th}} \\ \dfrac{R_n}{\bar{N}(\overline{\mathrm{PER}}(n), N_r^{\max})} & \bar{P} \geq \bar{P}_{n,\mathrm{th}} \end{cases} \qquad (24)$$

where $\bar{P}_{n,\mathrm{th}}$ is the threshold transmit power beyond which $\phi_{loss,n}$ is guaranteed to be not more than the maximum acceptable packet loss probability, that may not be true otherwise for non-adaptive systems. It can be identified numerically that the threshold $\bar{P}_{n,\mathrm{th}}$ exists for mode *n*, which also satisfies interference constraint imposed by primary users. Spectral efficiency of non-adaptive secondary systems employing truncated-ARQ has been plotted in Figures 4 and 5.

The ASE of the secondary link corresponding to different values of $N_r^{\max}$ is plotted against average inflicted interference limit at the primary receiver in Figures 4, 5, 6, and 7 for m=1 and m=2, respectively. It is apparent from Equation 15 that $N_r^{\max}$ =0 corresponds only to the special case of AMC. Figures 4 and 5 show that combined AMC-ARQ at the secondary transmitter provides significant gain in spectral efficiency over AMC only, for both fading scenarios. This is due to the underlying error correcting capability of truncated-ARQ, which depends on the maximum number of retransmissions. The error correcting capability of ARQ increases with $N_r^{\max}$, which benefits physical layer by relaxing the stringent error correction requirements. This lower performance requirement at the physical layer is exploited to increase the transmission rates, which results in overall spectral efficiency improvement (Liu, et al., 2004). Closeness of the spectral efficiency curve corresponding to $N_r^{\max}$ to the respective channel capacity corroborates this. However, increasing $N_r^{\max}$ beyond 2 does not further increase ASE, which is illustrated in Figure 8. Figures 4 and 5 also show the gain in the ASE achieved by physical layer optimization, i.e., AMC only over non-adaptive system with truncated ARQ. Figure 8 depicts the effect of $N_r^{\max}$ on ASE of the secondary user in Rayleigh and Nakagami distributed scenarios. It is clear that spectral efficiency in both fading scenarios becomes almost constant after $N_r^{\max}$ =2. Figure 8 also indicates the spectral efficiency of secondary link following Rayleigh distribution is greater than that following the Nakagami distribution. This is in contrast to when spectral efficiency of fading channel

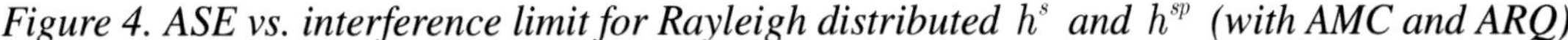

Figure 4. ASE vs. interference limit for Rayleigh distributed h^s and h^{sp} (with AMC and ARQ)

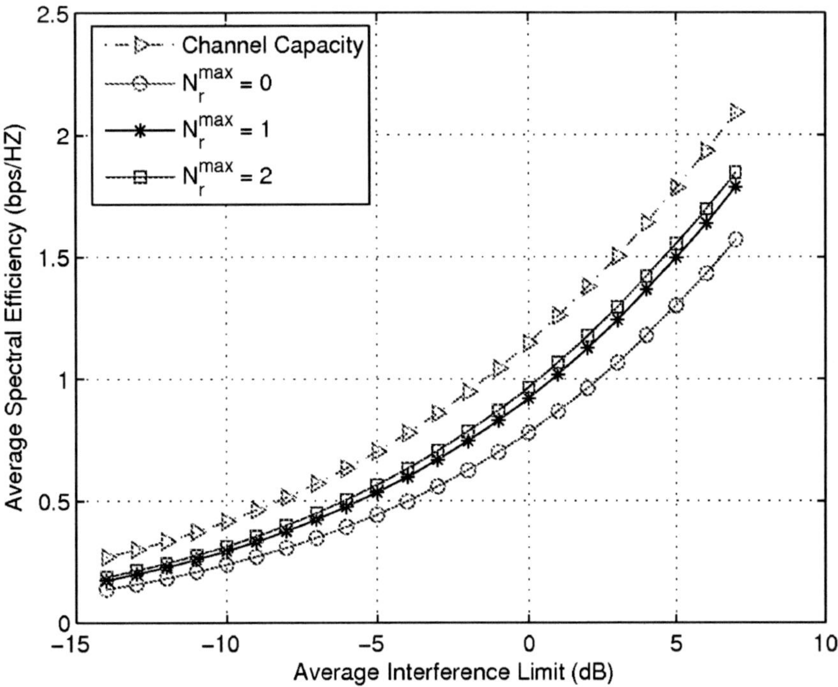

Figure 5. ASE vs. interference limit for Nakagami distributed channel links (with AMC and ARQ)

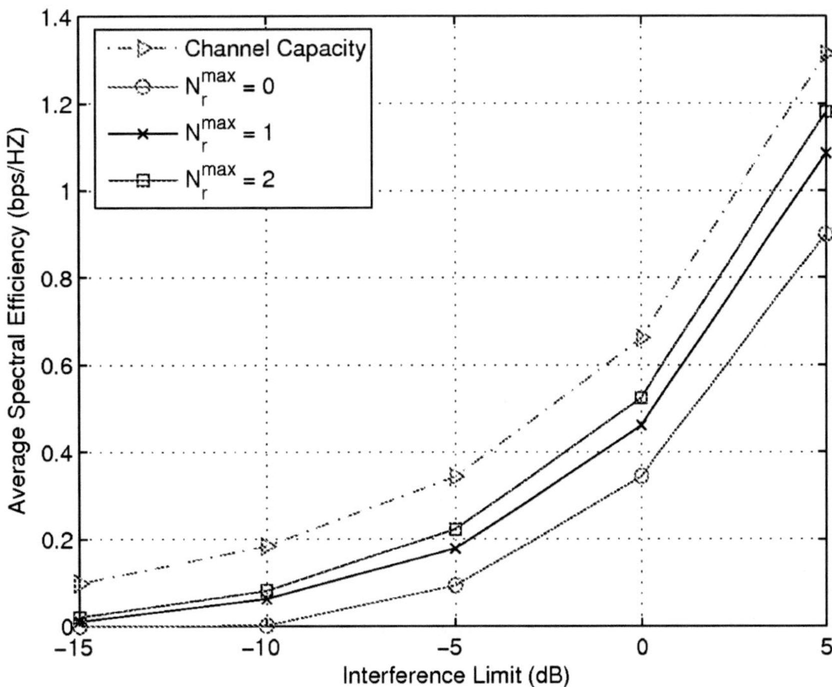

Figure 6. ASE vs. interference limit for Rayleigh distributed h^s and h^{sp} (with ARQ only)

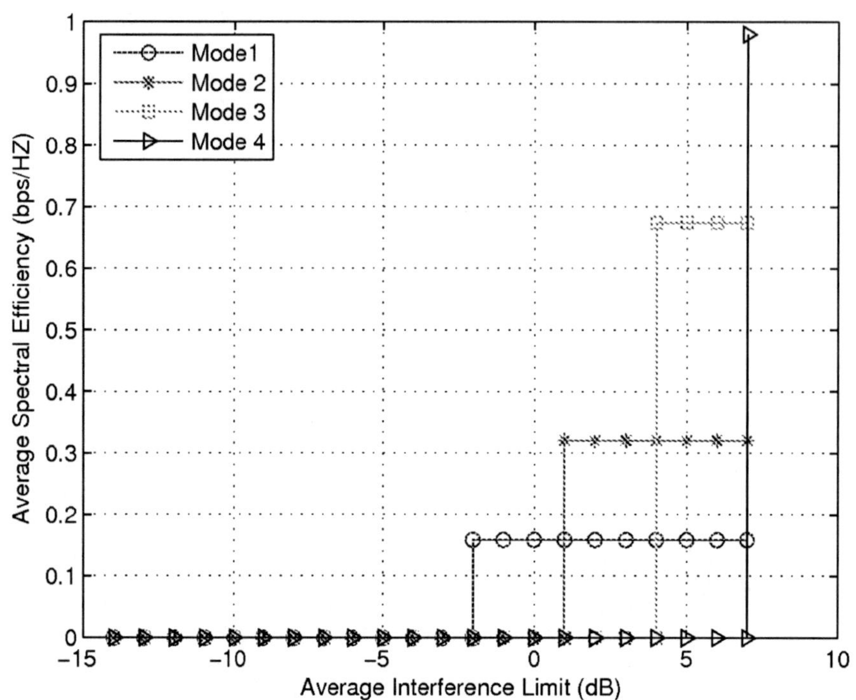

Figure 7. ASE vs. interference limit for Nakagami distributed channel links (with ARQ only)

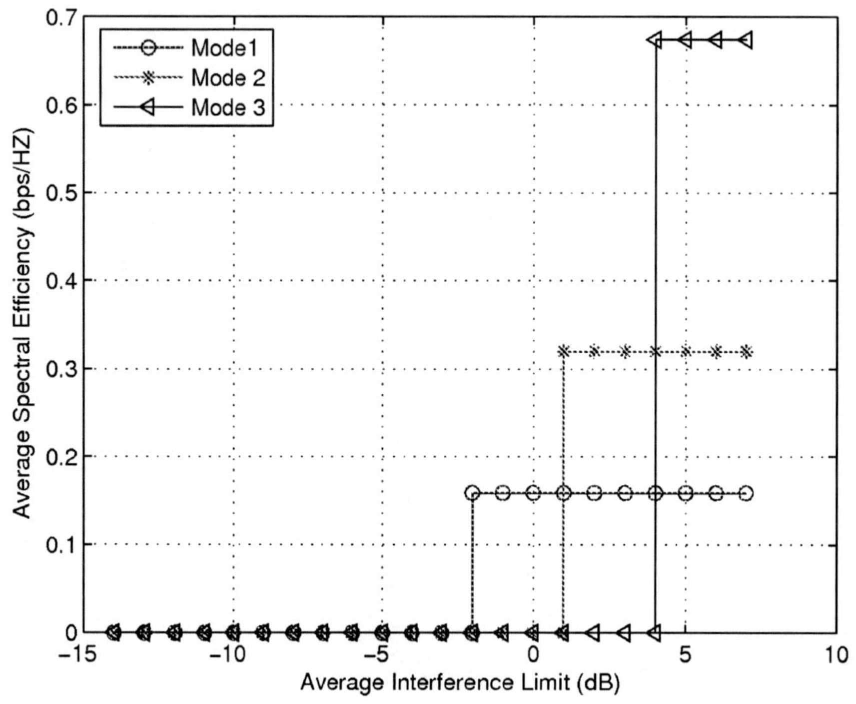

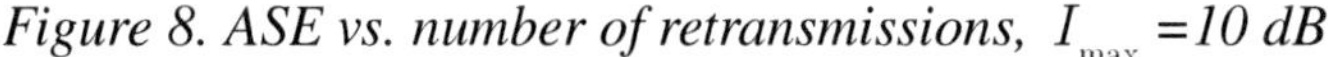

Figure 8. ASE vs. number of retransmissions, $I_{max} = 10$ *dB*

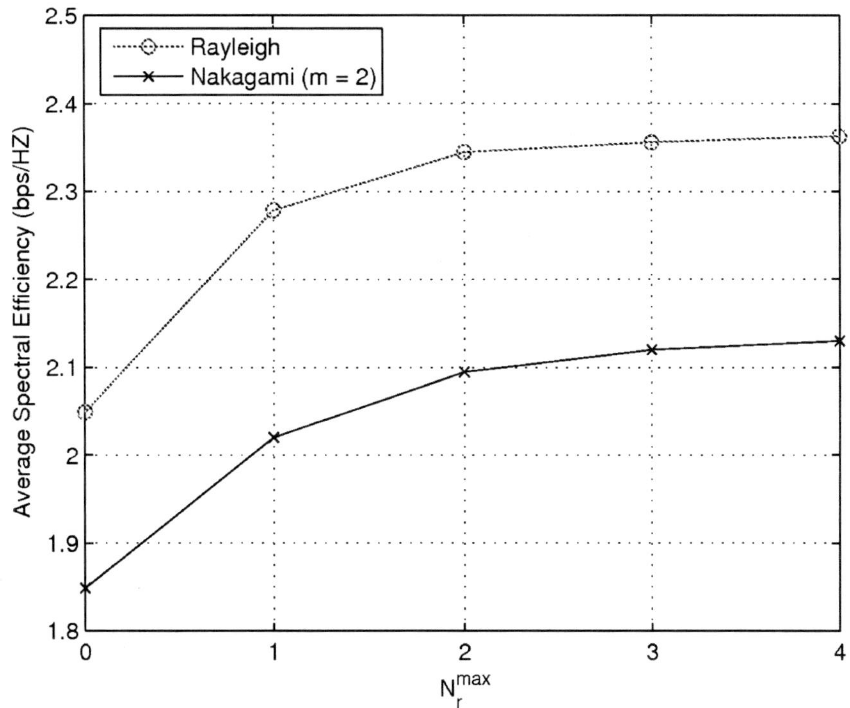

Figure 9. The target PER, $I_{max} = 10$ *dB,* $N_r^{max} = 1$

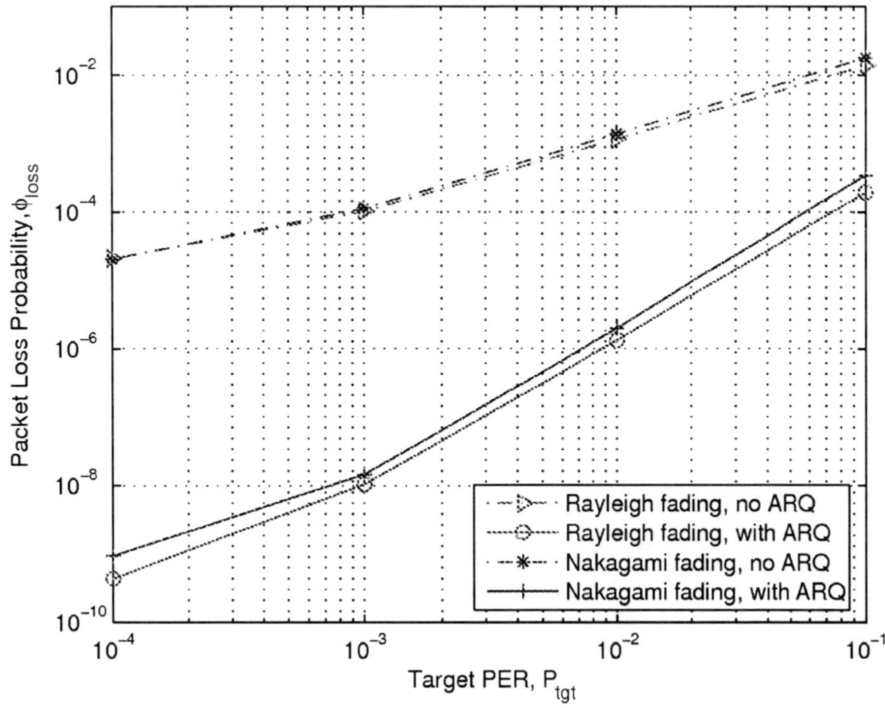

under transmit power constraint increases when the Nakagami parameter increases.

Figure 9 illustrates error performance of the proposed policy by plotting packet-loss probability as a function of target PER, with and without ARQ. It is evident that application of ARQ protocol significantly improves error performance of the secondary link, for both fading scenarios considered. This is generally true for channel links following Nakagami-*m* distribution.

SUMMARY

Cross-layer design in cognitive radio systems has been proved to achieve significant gain in performance over the layered design. There has been an extensive study on joint design of PHY and data link layers, and many studies have also touched upon the higher layers in this direction. However, there are research areas, which still need be explored in more detail such as spectrum mobility, MAC strategies allowing access to SUs in uncertain environments without a priori knowledge about the primary channel availability, and congestion control for both primary and secondary users.

REFERENCES

Akin, S., & Gursoy, M. (2010). Effective capacity analysis of cognitive radio channels for quality of service provisioning. *IEEE Transactions on Wireless Communications*, *9*(11), 3354–3364. doi:10.1109/TWC.2010.092410.090751

Akyildiz, I., Lee, W., Vuran, M., & Mohanty, S. (2006). Next generation/dynamic spectrum access/cognitive radio wireless networks: A survey. *Computer Networks*, *50*(13), 2127–2159. doi:10.1016/j.comnet.2006.05.001

Ali, S., & Yu, F. (2009). Cross-layer qos provisioning for multimedia transmissions in cognitive radio networks. In *Proceedings of the Wireless Communications and Networking Conference, 2009*. IEEE Press.

Bansal, G. (2008). Optimal and suboptimal power allocation schemes for ofdm-based cognitive radio systems. *IEEE Transactions on Wireless Communications*, *7*(11), 4710–4718. doi:10.1109/TWC.2008.07091

Bobarshad, H., van der Schaar, M., & Shikh-Bahaei, M. (2010). A low-complexity analytical modeling for cross-layer adaptive error protection in video over wlan. *IEEE Transactions on Multimedia*, *12*(5), 427–438. doi:10.1109/TMM.2010.2050734

Boyd, S., & Vandenberghe, L. (2004). *Convex optimization*. Cambridge, UK: Cambridge University Press.

Digham, F. (2008). Joint power and channel allocation for cognitive radios. In *Proceedings of the Wireless Communications and Networking Conference, 2008,* (pp. 882–887). IEEE Press.

Etkin, R. (2007). Spectrum sharing for unlicensed bands. *IEEE Journal on Selected Areas in Communications*, *25*(3), 517–528. doi:10.1109/JSAC.2007.070402

Foukalas, F., Gazis, V., & Alonistioti, N. (2008). Cross-layer design proposals for wireless mobile networks: A survey and taxonomy. *IEEE Communications Surveys & Tutorials*, *10*(1), 70–85. doi:10.1109/COMST.2008.4483671

Gastpar, M. (2004). On capacity under received-signal constraints. In *Proceedings of the 42nd Annual Allerton Conference on Communication, Control and Computing*. Allerton.

Gel'fand, I., Gindikin, S., & Graev, M. (2003). *Selected topics in integral geometry*. New York, NY: American Mathematical Society.

Ghasemi, A., & Sousa, E. (2006). Capacity of fading channels under spectrum-sharing constraints. In *Proceedings of the IEEE International Conference on Communications,* (vol. 10, pp. 4373-4378). IEEE Press.

Hu, D., Mao, S., Hou, Y., & Reed, J. (2010). Scalable video multicast in cognitive radio networks. *IEEE Journal on Selected Areas in Communications*, *28*(3), 334–344. doi:10.1109/JSAC.2010.100414

Khan, S., Peng, Y., Steinbach, E., Sgroi, M., & Kellerer, W. (2006). Application-driven cross-layer optimization for video streaming over wireless networks. *IEEE Communications Magazine*, *44*(1), 122–130. doi:10.1109/MCOM.2006.1580942

Le, L., & Hossain, E. (2008). Resource allocation for spectrum underlay in cognitive radio networks. *IEEE Transactions on Wireless Communications*, *7*(12), 5306–5315. doi:10.1109/T-WC.2008.070890

Liang, Y.-C., & Zeng, Y. (2008). Sensing-throughput tradeoff for cognitive radio networks. *IEEE Transactions on Wireless Communications*, *7*(4), 1326–1337. doi:10.1109/TWC.2008.060869

Liu, Q., Zhou, S., & Giannakis, G. (2004). Cross-layer combining of adaptive modulation and coding with truncated arq over wireless links. *IEEE Transactions on Wireless Communications*, *3*(5), 1746–1755. doi:10.1109/TWC.2004.833474

Luo, C., Yu, F., Ji, H., & Leung, V. (2010a). Cross-layer design for tcp performance improvement in cognitive radio networks. *IEEE Transactions on Vehicular Technology*, *59*(5), 2485–2495. doi:10.1109/TVT.2010.2041802

Luo, H., Argyriou, A., Wu, D., & Ci, S. (2009). Joint source coding and network supported distributed error control for video streaming in wireless multihop networks. *IEEE Transactions on Multimedia*, *11*(7), 1362–1372. doi:10.1109/TMM.2009.2030639

Luo, H., Ci, S., Wu, D., & Tang, H. (2010b). Cross-layer design for real-time video transmission in cognitive wireless networks. In *Proceedings of INFOCOM IEEE Conference on Computer Communications Workshops*. IEEE Press.

Musavian, L., & Aissa, S. (2009a). Capacity and power allocation for spectrum-sharing communications in fading channels. *IEEE Transactions on Wireless Communications*, *8*(1), 148–156. doi:10.1109/T-WC.2009.070265

Musavian, L., & Aissa, S. (2009b). Adaptive modulation in spectrum-sharing systems with delay constraints. In *Proceedings of the Communications, 2009*. IEEE Press.

Qing Zhao, S. (2007). A decision-theoretic framework for opportunistic spectrum access. *IEEE Transactions on Wireless Communications*, *14*(4), 14–20. doi:10.1109/MWC.2007.4300978

Rashid, M. (2009). Opportunistic spectrum scheduling for multiuser cognitive radio: A queueing analysis. *IEEE Transactions on Wireless Communications*, *8*(10), 5259–5269. doi:10.1109/TWC.2009.081536

Shi, Y., Hou, Y., Zhou, H., & Midkiff, S. (2010). Distributed cross-layer optimization for cognitive radio networks. *IEEE Transactions on Vehicular Technology*, *59*(8), 4058–4069. doi:10.1109/TVT.2010.2058875

Srinivasa, S. (2008). How much spectrum sharing is optimal in cognitive radio networks? *IEEE Transactions on Wireless Communications*, *7*(10), 4010–4018. doi:10.1109/T-WC.2008.070647

Vu, M., Devroye, N., Sharif, M., & Tarokh, N. (2007). Scaling laws of cognitive networks. In *Proceedings of Cognitive Radio Oriented Wireless Networks and Communications, 2007*. CrownCom. doi:10.1109/CROWNCOM.2007.4549764

Wu, D., & Negi, R. (2003). Effective capacity: A wireless link model for support of quality of service. *IEEE Transactions on Wireless Communications*, *2*(4), 630–643.

Wu, Y., & Tsang, D. (2009). Dynamic rate allocation, routing and spectrum sharing for multi-hop cognitive radio networks. In *Proceedings of Communications Workshops, 2009*. IEEE Press. doi:10.1109/ICCW.2009.5208054

Xin, C., Xie, B., & Shen, C.-C. (2005). A novel layered graph model for topology formation and routing in dynamic spectrum access networks. *Proceedings of New Frontiers in Dynamic Spectrum Access Networks, 2005*, 308–317. IEEE Press.

Yang, K., & Wang, X. (2008). Cross-layer network planning for multi-radio multi-channel cognitive wireless networks. *IEEE Transactions on Communications*, *56*(10), 1705–1714. doi:10.1109/TCOMM.2008.4641901

Yucek, T. (2009). A survey of spectrum sensing algorithms for cognitive radio applications. *IEEE Communications Surveys Tutorials*, *11*(1), 116–130. doi:10.1109/SURV.2009.090109

Zhao, Q., & Sadler, B. (2007). A survey of dynamic spectrum access. *IEEE Signal Processing Magazine*, *24*(3), 79–89. doi:10.1109/MSP.2007.361604

Zhu Han, H. J., & Fan, R. (2009). Replacement of spectrum sensing in cognitive radio. *IEEE Transactions on Wireless Communications*, *8*(6), 2819–2826. doi:10.1109/TWC.2009.080603

APPENDIX

In order to prove the convexity of (10), let $\chi(v_1, v_2, \cdots, v_N)$ denote the objective function in (10), i.e.:

$$\chi(v_1, v_2, \cdots, v_N) = \sum_{n=0}^{N} R_n \int_{v_n}^{v_{n+1}} f_v(v) dv \tag{25}$$

where $f_v(v)$ is given by 2. We introduce the following lemma:

Lemma 1: For a Nakagami-m fading pdf, $v \geq (m^s - 1) / (m^{sp} + 1)\rho$ is a sufficient condition for convexity of both the function χ and constraint function in (10).

Proof: First consider the χ function. Since $\partial^2 \chi / \partial v_i v_j = 0; i, j, \cdots, N, i \neq j$ for χ to be convex, it is sufficient to have $\partial^2 \chi / \partial v_i^2 \geq 0$. We have

$$\frac{\partial^2 \chi}{\partial v_i^2} = (R_{i-1} - R_i) \frac{\partial f_v(v)}{\partial v} |_{v=v_i} \tag{26}$$

Since $R_{i-1} < R_i$, in order to guarantee $\partial^2 \chi / \partial v_i^2 \geq 0$ it is required to have:

$$\begin{aligned} &\frac{\partial f_v(v)}{\partial v} \leq 0, \forall i \\ &\frac{\partial f_v(v)}{\partial v} = \left(\frac{\rho^{-m^{sp}}}{\beta(m^{sp}, m^s)} \right) \left[\frac{\left((m^s - 1) v^{m^s - 2} \right)}{\left((v + 1/\rho)^{m^{sp} + m^s} \right)} - \frac{\left((m^{sp} + m^s)(v + 1/\rho)^{m^{sp} + m^s - 1} v^{m^s - 1} \right)}{\left((v + 1/\rho)^{m^{sp} + m^s} \right)^2} \right] \\ &= \left(\frac{\rho^{-m^{sp}}}{\beta(m^{sp}, m^s)} \right) \left[\frac{\left((m^s - 1) v^{m^s - 2} \right)(v + 1/\rho) - (m^{sp} + m^s) v^{m^s - 1}}{(v + 1/\rho)^{m^{sp} + m^s}} \right] \end{aligned} \tag{27}$$

Since the denominator of the above fraction is always positive, in order to guarantee

$$\frac{\partial f_v(v)}{\partial v} \leq 0,$$

the numerator should be negative. This leads to:

$$v \geq \frac{(m^s - 1)}{(m^{sp} + 1)\rho} \tag{28}$$

which is satisfied for all AMC modes. Now considering the constraint C1 in (10), denoted by:

$$\omega(v_1, v_2, \cdots, v_N) = \sum_{n=1}^{N} S_n \int_{v_n}^{v_{n+1}} \frac{1}{v} f_v(v) dv \leq I_{\max} \tag{29}$$

where

$$S_n = g_n \log\left(\frac{a_n}{P_{\text{tgt}}}\right)$$

It is easy to check that

$$\partial^2 \omega / \partial v_i v_j = 0;\ i, j, \cdots, N, i \neq j,$$

and

$$\frac{\partial^2 \omega}{\partial v_i^2} = \frac{(S_i - S_{i-1})}{v_i^2} f_v(v_i) - \frac{(S_i - S_{i-1})}{v_i} \frac{\partial f_v(v)}{\partial v} |_{v=v_i} . \tag{30}$$

Since $S_i > S_{i-1}$, the first term in the R.H.S. of (30) is positive. From discussion on convexity of χ function we know that when

$$v \geq \frac{(m^s - 1)}{(m^{sp} + 1)\rho},$$

we have

$$\frac{\partial f_v(v)}{\partial v} \leq 0,$$

and therefore the second term in R.H.S. of (30) is also positive and consequently $\partial^2 \omega / \partial v_i^2 \geq 0$, and hence the ω function is convex.

Chapter 6
Interference Suppression Capabilities of Smart Cognitive-Femto Networks (SCFN)

Muhammad Zeeshan Shakir
King Abdullah University of Science and Technology, Saudi Arabia

Rachad Atat
King Abdullah University of Science and Technology, Saudi Arabia

Mohamed-Slim Alouini
King Abdullah University of Science and Technology, Saudi Arabia

ABSTRACT

Cognitive Radios are considered a standard part of future heterogeneous mobile network architectures. In this chapter, a two tier heterogeneous network with multiple Radio Access Technologies (RATs) is considered, namely (1) the secondary network, which comprises of Cognitive-Femto BS (CFBS), and (2) the macrocell network, which is considered a primary network. By exploiting the cooperation among the CFBS, the multiple CFBS can be considered a single base station with multiple geographically dispersed antennas, which can reduce the interference levels by directing the main beam toward the desired femtocell mobile user. The resultant network is referred to as Smart Cognitive-Femto Network (SCFN). In order to determine the effectiveness of the proposed smart network, the interference rejection capabilities of the SCFN is studied. It has been shown that the smart network offers significant performance improvements in interference suppression and Signal to Interference Ratio (SIR) and may be considered a promising solution to the interference management problems in future heterogeneous networks.

DOI: 10.4018/978-1-4666-2812-0.ch006

INTRODUCTION

In this chapter, the interference management capabilities of future generations of wireless communication networks are explored. The contribution of this chapter will create a strong impact on the infrastructure developments for future generations of wireless networks, which facilitate the wireless high quality broadband access to everyone especially to the population residing in sparsely populated areas. Furthermore, the theme of the chapter is fully aligned with the European manifesto which ensures that everyone must have access to good broadband to a minimum of 2 Mb speed service by 2015. The proposal is fully aligned with Race Online 2012 manifesto, along with several other European projects under seventh framework (FP7) such as COGEU, QOSMOS, CROWN, SECARA, ARCOPILIS, WHITESPACECENTER, W-GREEN, and EARTH. The theme of this chapter is also aligned with several projects under COST actions such as IC0804-Energy for distributed systems, IC0905-TERRA, IC0902-HETNETS (Cost Europe, 2012; European Commission, 2012).

In this section, the technology behind the cognitive radio networks is presented, with a deep insight on its applications and trends along with an overview of the several spectrum sensing methods that are currently used. Then, the concept behind heterogeneous networks is introduced with a discussion on Small Cell Networks (SCN), its employment, benefits, problems and challenges.

The rest of the contributions of the chapter is organized as follows. Section 2.1 defines the Smart Cognitive-Femto Networks by recasting the two tier Heterogeneous network into the framework of the Cognitive Enabled Femtocell Network. Section 2.1.1 defines the environment in which the proposed system is being tested. Section 2.1.2 presents the network layout of such system; Section 2.1.4 presents the system model. Later in Section 2.1.5, interference suppression capabilities of the smart network are presented. Next, in Section 2.2, simulation results are presented to show the efficacy of the proposed smart network. Directions for future research are described in Section 3. Finally, conclusions are drawn in section 4.

1. BACKGROUND

1.1. Cognitive Radio Networks

1.1.1. Technology and Forecast

According to the Federal Communications Commission (FCC), the radio spectrum is highly underutilized (Stotas & Nallanathan, 2011), which makes its usage inadequate and its availability limited since one or more users are allocated one fixed band of the spectrum. For instance, in US, the radio spectrum is utilized only 6% most of the time (Cheng, Zhang, & Zhang, 2011) and based on measurements made by Office of Communications (Ofcom) in UK and Spectrum Policy Task Force (SPTF) in USA, many pieces of the licensed spectrum are not utilized for long period of time (Arshad, Imran, & Moessner, 2010). This results in spectrum pieces that are allocated to licensed users but are unused for certain time and at a particular location, and these are referred to as spectrum holes. In addition to being underutilized for most of the time, the radio spectrum has been fully allocated, the fact that initiated an extensive research on how to efficiently reuse the spectrum and resolve the problem of spectrum unavailability. Over the last decade, significant work has been done to explore new radio resources and new technologies which focus more on improving the spectrum and energy efficiency of wireless mobile networks. This highly ambitious goal provides mean for improving end-user data rates, reducing spectrum requirements, and lowering the power consumption/transmission in the network by intelligent utilization of the available spectrum resources. Dynamic Spectrum Access (DSA) is a technique where a fixed band in the spectrum is allocated to one or more licensed users, called Primary Users (PUs), and the access to the allocated band is not exclusively granted to them, unlike the Fixed Spectrum Access (FSA) technique, where licensed users have exclusive access to the licensed spectrum band allocated to them. Although the licensed users or PUs do not have exclusive access to their piece of spectrum, however they have higher priorities than unlicensed users, also called Secondary Users (SUs). These SUs can periodically sense the spectrum to detect any PU activity, and if no PU activity is detected, then SU can utilize the spectrum until the PU accesses it again (Liang, Chen, Li, & Mahonen, 2011). The SU with spectrum sensing capability is called Cognitive Radio user, CR user, where SUs are only allowed to coexist with PUs under the condition of protecting the Quality of Service (QoS). A Cognitive Radio Network or CRN is a network that consists of several CR nodes. CRN can achieve high bandwidth to mobile users by adopting the DSA techniques and by using heterogeneous wireless network architectures. An example of CRN consists of a PU, the TV broadcast network, that is allocated a frequency band around 470-700 MHz. The analog television broadcasting has been converted and replaced by digital television which occupies a significant part of the spectrum, where a number of radio frequency channels previously assigned to analog broadcasting are now available to be used for other purposes such as public safety communications, cellular communications and other broadcast uses. Also, since not the entire allocated spectrum is used all the time by the PU, the SU such as a mobile cellular network that operates in the same band, can utilize the licensed spectrum in order to achieve additional capacity (Sachs, Maric, & Goldsmith, 2010).

In CRN transmission, a single frame consists of a sensing slot of sensing period τ and a data slot of period $T-\tau$ (Li & Cadeau, 2011). These types of CRN are referred to as time-slotted CRN, where a PU can change its status from active to idle or vice versa at the beginning of the frame only and its status is maintained during the whole frame period of T. So, during sensing period τ, SU determines the status of PU, and based on that output, SU will either transmit data or refrain from transmitting it during data transmission period of $T-\tau$ (Cheng, et al., 2011). The time-slotted CRN are referred to as half-duplex systems, and that is because spectrum sensing and data transmission cannot be done simultaneously. With the time-slotted CRN, many interference problems arise. For instance, when PU changes its state from being idle to active after the SU has sensed no PU activity during the sensing period, then in this case SU can cause harmful interference to PU since both SU and PU are accessing the same spectrum band (Cheng, et al., 2011). To solve this problem, it is suggested in Cheng et al. (2011) to use non-time-slotted CRN where an SU can sense the spectrum for the whole frame period instead of just at beginning of the frame, and by doing so, a SU can detect PU state changes at any time and hence, will not cause any interference with PU. Non-time slotted CRN are referred to as full-duplex systems since spectrum sensing and data transmission are done at the same time. CRN can also suffer from jamming attacks that can decrease CRN throughput for up to 70% (Li & Cadeau, 2011). Jamming is the act of transmitting radio signals that aim at interrupting the flow of communications by decreasing the Signal to Noise Ratio (SNR). By transmitting a low signal power, the jammer can cause a CRN to switch channel as the CR user will detect the interfering power signal as a PU activity and will vacate the channel. By switching channel, CRN has to go through series of synchronization, channel setup, handshaking and network setup before the packet can be retransmitted, which means CRN has to spend an additional time for packet retransmission each time there is a jamming attack, the fact that will decrease the CRN throughput dramatically (Li &

Cadeau, 2011). To be less prone to jamming attacks, if a CRN uses larger number of channels, then the probability of the jammer to attack the channel used by CRN becomes low since the jammer will have to randomly select one channel among the many available ones.

1.1.2. Overview of Spectrum Sensing

Spectrum sensing is one of the fundamental components in cognitive radio networks. The SU can access the spectrum in two different ways namely: a) Opportunistic Spectrum Access (OSA), also known as overlay strategy, where an SU can access the spectrum only when no PU activity has been detected, and b) Spectrum Sharing (SS) also known as underlay strategy, where both SU and PU coexist under the condition of protecting the QoS and protecting PU from harmful interference (Stotas & Nallanathan, 2011). By adopting SS, SU will limit its transmit power to very low level power so that their signals will not interfere with those of PU, which means that CR transmitter has to predict the interference power level received at particular location (Liang, et al., 2011). This means that CR interference power level at each PU receiver must be below a certain threshold, which requires knowledge of the channel gain from the CR transmitter to the PU receiver. If a Time-Division Duplex (TDD) mode is used, then using the received PU signal, CR transmitter can estimate the CR to PU channel gain provided that CR knows the PU transmit power. However, in Frequency-Division Duplex (FDD) mode where transmission and reception are done on two different channels, estimation of CR to PU channel gain is not realizable since there is no channel reciprocity between PU and CR (Zhang, 2010). In Zhang (2010), it has been suggested that CR transmits a probing signal to PU receiver, where the latter adapts its transmit rate, and these rate and power adaptations can be observable by the CR. Thus CR would know about PU rate without needing for a dedicated feedback channel from PU. This method only works if CR actively probes the PU receiver, causing the latter to adapt its transmit and power rates. Spectrum sharing can either be done in a centralized or distributed manner. With distributed spectrum sharing, CR nodes opportunistically sense the spectrum and make spectrum access decision individually so that each CR node is responsible for spectrum allocation, while in centralized spectrum sharing, a centralized entity controls the spectrum allocation and access procedures, and is considered impractical with the increase in demand for spectrum which leads to computational complexity at the centralized entity (Lin & Chen, 2010). The spectrum sharing can also result in Adjacent Frequency Interference (AFI) and Co-Channel Interference (CCI) (Ni & Collings, 2009). The AFI is caused by two networks using adjacent frequencies; usually placing a guard band that separates adjacent bands apart helps in reducing AFI. The CCI occurs between cells that are reusing the same band in a single network. Moreover, the CCI can be reduced by assigning orthogonal bands to cells (Ni & Collings, 2009).

A new approach of spectrum sensing is suggested in Stotas and Nallanathan (2011). This approach is referred to as hybrid approach since it combines both OSA and SS approaches. The hybrid approach consists of SU sensing the spectrum and adapting its power level based on the decision made. If no PU activity is detected, then SU communicates with high transmit power, otherwise it will use low transmit power. The frame structure adopted is the non-time-slotted, where data transmission and spectrum sensing can be performed at the same time, which results in better protection of PU since spectrum sensing is done all the time, in addition to a higher throughput and improved detection probability (Stotas & Nallanathan, 2011). Thus, the SU coexists with a PU such that the former will not cause harmful interference to the PU. In addition, by making a decision on whether the frequency band is occupied or not, SU can make more efficient use of the available spectrum instead of assuming that PUs are always active as in the case with OSA strategy, and thus capacity of the hybrid approach is further improved and enhanced (Stotas & Nallanathan, 2011). The SU receiver receives SU transmitter signal, decodes it, and uses the remaining signal to determine the action of cognitive radio system for the next frame.

During the process of sensing spectrum holes, CR transmitter senses the spectrum to detect if there is a PU activity inside the coverage of CR transmitter. Considering *D* as the transmission range of primary transmitter and *R* is the interfering range of a CR transmitter, then with direct spectrum sensing, CR detects nearby primary receivers directly within range *R*. However, with indirect spectrum sensing, CR node detects surrounding primary receivers within a range of *D*+ *R* which is greater than *R* with direct sensing spectrum (Liang, et al., 2011).

Another problem with spectrum sensing arises when a CR node cannot detect the signal from a primary transmitter due to shadowing or fading and this is referred to as hidden node problem (Arshad, et al., 2010; Liang, et al., 2011). If a CR transmitter cannot detect the primary receiver signal due to a high building for example, then the CR transmitter will access the licensed band and may cause harmful interference to the PU. As a solution to this problem, each CR node can independently perform spectrum sensing and transmits its own decision to a fusion center that could be a Wireless Local Area Network Base Station (WLAN BS) or CR BS. The fusion center, in turn, combines the decisions to lead to a final decision about the presence or existence of a PU. This method is referred to as collaborative spectrum sensing or cooperative spectrum sensing (Arshad, et al., 2010; Liang, et al., 2011; Chen, 2010) and can reduce detection error probability and required detection time at each CR node. Each CR transmitter transmits 1 bit decision to the fusion center that can be centralized or distributed. With centralized fusion center, all SUs send their decisions to the fusion center, while with distributed fusion center; all SUs can behave as fusion centers and receive sensing information from neighboring nodes (Arshad, et al., 2010). Eigenvalue based detection schemes to perform spectrum sensing without any prior knowledge are presented in (Shakir, Rao, & Alouini, 2011). The key idea of this method is to infer the absence or presence of a PU from the eigenvalues of the covariance matrix of the received signals. There are three major eigenvalue-based detection techniques studied in the literature: 1) Eigenvalue Ratio Detector (ERD); 2) energy with Smallest Eigenvalue Detection (SED); 3) Largest Eigenvalue Detector (LED). ERD uses the ratio of the largest eigenvalue to the smallest eigenvalue as the test statistic and then gives the decision threshold to determine the presence or absence of the PU. ERD has many advantages over the rest of the sensing methods reported in the literature, because unlike other methods, the decision on presence of the signal can be done irrespective of the knowledge of the signal and the noise properties. Two recently proposed detectors have been presented in (Shakir, et al., 2011), namely (1) the Geometric Mean Detector (GEMD), involving the ratio of the largest eigenvalue to the Geometric mean of the eigenvalues of the received covariance matrix and (2) the Arithmetic Mean Detector (ARMD) involving the ratio of the largest eigenvalue and the Arithmetic mean of the eigenvalues. Thees performance of these detectors outperforms the performance of well-known eigenvalue based detectors (Shakir, et al., 2011).

1.1.3. Application and Trends

Cognitive radios can support a variety of wireless applications and services such as smart grid networks, public safety networks, broadband cellular networks, medical applications, military applications and others.

Using smart grids has many benefits such as energy independence, global warming, and emergency resilience issues. Smart grid consists of a Home Area Network (HAN) that connects smart meters with on-premise appliances, Field Area Network (FAN) that carries information between premises and network gateway, and Wide Area Network (WAN) that acts as a backbone for communication between network gateways. In FAN, wireless meter devices and network gateways can be equipped with CR capabilities to dynamically utilize the unused spectrum to communicate with each other without the need for an infrastructure (Wang, Ghosh, & Challapali, 2011).

In public safety networks, communication plays a critical role in disaster prevention and recovery. Since radio frequencies available for emergency and disaster related issues have become heavily congested and utilized, employing CR capable devices will help in ensuring that emergency responders from police, fire, medical and emergency services have sufficient capacity and means of communications. In this way, emergency responders can access emergency services from voice, messages, emails, and video streaming by exploiting the CR capabilities of utilizing the spectrum that is not widely used. In addition, employing CR for public safety networks will also increase interoperability between devices that operate in different bands or have incompatible wireless interfaces to communicate between each other by using Software Defined Radio (SDR) (Wang, et al., 2011; Maldonado, Le, Hugine, Rondeau, & Bostian, 2005).

In cellular networks, with the introduction of smart phones and availability of social networks and services such as YouTube, Facebook, and Flickr, the cellular networks became overloaded due to limited spectrum resources owned by cellular operators. Using cognitive radios will achieve increased capacity, throughput, and allows cellular operators to access licensed frequency bands (Wang, et al., 2011).

Cognitive radios can also be employed for wireless medical networks. Devices that monitor the patient's blood oxygen, temperature, and other vital signs can be designed using on-body wireless sensors technology. These sensors need to access a frequency band to help wireless patient monitoring, which will result in lower health care costs and better patient comfort since these monitoring devices can be extended to the whole hospital instead of being dedicated to a single patient, in addition they will increase patient mobility and quality of decision making (Wang, et al., 2011). These medical monitoring devices, if equipped with CR capabilities, can use a portion of spectrum to get the benefits of using them instead of sensors connected by wires to a monitor.

Finally, cognitive radios have helped in several military applications such as sensing the interferers, protecting the transmission of information, and recognition of enemy communications. The US Department of Defense (DoD) has invested a lot in the research related to cognitive radios such as next Generation systems (XG), SPEAKeasy radio system, and many others (Maldonado, et al., 2005).

1.2. Heterogeneous Networks

1.2.1. Background

The telecommunications industry is witnessing an explosive increase in data traffic especially with the introduction of wireless modems and smart phones and with the presence of more than one billion wireless subscribers today. The data traffic volume is increasing by a factor of 10 every 5 years (Badic, O'Farrell, Loskot, & He, 2009). Furthermore, the high volume data traffic is leading to an increase of 16% to 20% in energy consumption every 5 years (Badic, et al., 2009), in addition, 20% to 30% of network operational expenses are included in the electricity bills (Ashraf, Boccardi, & Ho, 2011), add to them the 10% increase in carbon emissions of Information and Communication Technology (ICT) despite the extensive researches in creating new energy efficient technologies (Hoydis & Kobayashi, 2011). In addition, current cellular networks have high Operational Expenses

(OPEX) and Capital Expenditures (CAPEX), for instance, in Sweden, the annual cost of a macro BS serving an area coverage of 500 meters to few kilometers is around 20000 Euros with an annual electricity charge of 876 Euros per kilowatt-hour (Tombaz, Vastberg, & Zander, 2011). All these problems that are associated with current cellular networks and that made them reach their capacity limit especially in highly populated areas and that have caused high OPEX and CAPEX have led to the deployment of Small Cell Networks (SCN) or femtocells that eliminate the need for costly cell site acquisition which reduces the CAPEX and OPEX costs. This significantly promising technology will allow the deployment of short-range, low-power, and low-cost user deployed small cells operating in conjunction with the main macro cellular network infrastructure (Lin, Zhang, Chen, & Zhang, 2011; Landstorm, Furuskar, Johansson, Falconetti, & Kronestedt, 2011; Shetty, Parekh, & Walrand, 2009; Femto Forum, 2012; Chandrasekhar, Andrews, & Gatherer, 2008). The resultant wireless network is referred to as two tier Heterogeneous network where the SCN base stations are complementing the macrocell network.

Heterogeneous networks consist of infrastructures with different wireless access technologies, each having different capabilities and functions. A heterogeneous network consists of an inter-operation of macrocells, SCN and relays in order to offer wireless coverage in an environment with several wireless coverage zones, ranging from outdoor areas to hotspot areas such as offices, buildings, and homes. As such, the Heterogeneous networks are envisioned to enable next-generation networks by providing high data rates, allowing offloading traffic from the macrocell network and providing dedicated capacity to homes, enterprises, or urban hot-spots. Furthermore, the substantial improvement in network efficiency will yield large cost savings for mobile operators, i.e. reduction in capital and operating expenditures (CAPEX and OPEX) (Lin, et al., 2011; Shetty, et al., 2009). Although the Heterogeneous network is expected to be the most influential solution to meet the challenges imposed for 4G mobile networks of the future generation of wireless networks however, the management of cross-tier and intra-tier interferences is critical to achieve such gains and of paramount importance for successful operation of the Heterogeneous networks.

A typical heterogeneous network architecture is shown in Figure 1, where a macro BS serves a wide area, while SCN BS or access points serve users in smaller areas such as buildings and airport. Since the largest amount of traffic (50% of the voice calls and more than 70% of data traffic) originates from indoor hotspots such as homes, offices and malls in addition to mobile users, the deployment of SCN will help offloading this huge traffic from the cellular networks, improving the indoor coverage and cell-edge user performance, in addition to enhancing the spectral efficiency per area unit through the spatial reuse (Hoydis & Kobayashi, 2011; Ramanath, Kavitha, & Altman, 2010; Lopez-Perez, Guvenc, Roche, Kountouris, Quek, & Zhang, 2011). Small cells cover femtocells that are used to describe residential premises, picocells that are used for enterprise and business premises, and metrocells for public and outdoor spaces. The main difference between picocells and femtocells is that picocells are operated and managed by the network operator, while the femtocells are installed and powered by end users with less remote management by the network operator. Femtocells look very similar to Wi-Fi access points, but they use a commercial cellular standard and a licensed spectrum. More than 2.3 million femtocells have been deployed worldwide in 2011, with an expectation of reaching 50 million by 2014. Femtocells can interact with the traditional cellular network by performing tasks such as interference management, billing, and authentication (Andrews, Claussen, Dohler, Rangan, & Reed, 2012). Femtocells are characterized namely by, (1) reduced distance between the femtocell and the user, which leads to achieving higher received signal strength, (2) low transmit

Figure 1. Graphical illustration of existence of small cell networks (SCN) in heterogeneous networks

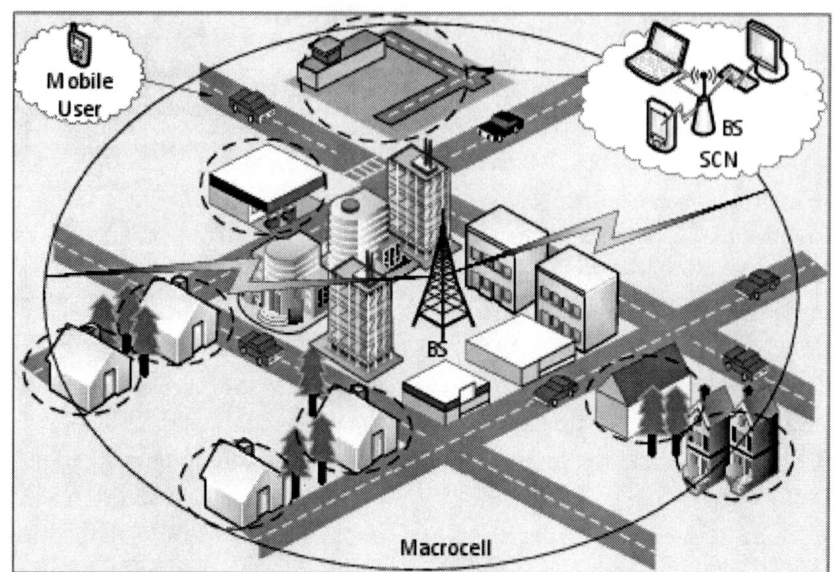

power with less interference from macrocell and femtocell users due to outdoor propagation and penetration losses, (3) serving one to four users, which allows the femtocell to devote a larger portion of its resources from bandwidth and transmit power to each user (Chandrasekhar, et al., 2008).

1.2.2. Characteristics of Small Cell Networks

Although the macrocell BS can provide wide area coverage, however they cannot provide high data rates to very crowded areas and indoor environments. SCN deploy low-cost and low-power access points that are smaller than the macrocell BS and are in close proximity to end users with reduced area coverage of order of ten to several hundreds of meters. There are many differences between the macrocell BS and the SCN BS. The macrocell BS is deployed by the network operator, requires detailed planning and maintenance, and provides open public access and wide area coverage in order of few kilometers, while an SCN BS is a user deployed access point that supports plug-and-play deployment and that can be self-configured by a downloaded software before being operated, can automatically perform node failure recovery, constantly monitor the radio environment to self-optimize its settings in order to achieve higher data rates, greater area coverage and minimum interference (Hoydis & Kobayashi, 2011; Lopez-Perez, et al., 2011). By reducing the distance between transmitter and receiver in SCN, the transmit power is decreased which leads to not only a reduction in energy consumption, but also will lead to overcoming the pathloss resulting from fading and noise and will improve the spatial spectrum reuse and radio link quality. Moreover, a low transmit power is needed to maintain QoS (Lin, et al., 2011; Alouini & Goldsmith, 1999; Harley, 1989).

The SCN BS can operate in three different modes depending on the applications, namely: (1) open access mechanism where operators may install SCN base stations in public areas, e.g. airport, public parks, shopping malls and cafes such that all users in the vicinity may be connected to the SCN access points, (2) closed access mechanism where the SCN base stations are installed in residential places and enterprises for home and office applications such that only a subset of the registered users can access the SCN access points, (3) the hybrid access mode, where only a limited amount of available cell resources are available to non-registered users (Ashraf, et al., 2011). In order to achieve greater energy reduction with the deployment of SCN, the BS can be equipped with SLEEP mode algorithm as described in (Ashraf, et al., 2011), where during idle conditions some of the hardware components of the BS are either completely switched off or operate in low power mode. Equipped with low power sniffer, the BS hardware can detect if an active call is being made by user equipment, if so, all the hardware components of the BS must switched ON. Results showed up to 60% energy savings compared to when there is no SLEEP mode.

1.2.3. Problems and Challenges

With the deployment of SCN, several challenges and limitations have shown to degrade the performance of the network. In this section, a closer look and analysis of SCN reveals many challenges ranging from interference problems to security issues.

The cell size of SCN cause a severe degradation on the network performance and that is because the cell diameter can be around 10 to 100 meters, which means even at a very low speed, users will be switching from one cell to another in just few seconds. As a result, frequent handovers will occur, which will cause too many controlling signals being exchanged between cells, the fact that will impact the services that are offered to end users (Hoydis & Kobayashi, 2011; Ramanath, et al., 2010). Among many of the solutions that have been suggested in the literature to decrease the amount of handovers is user grouping, where static users are served by the small cells, while the active and mobile users are served by the macrocell. Another solution is the concept of virtual cells, where small cells BS cooperate between each other in order to appear to the user as a single BS, thus minimizing handovers since they only occur at boundaries of these virtual cells, and the service of the user traversing the virtual cell will not be affected (Hoydis & Kobayashi, 2011; Ramanath, et al., 2010).

Due to the self-organization and unplanned nature of SCN, unpredictable interference can occur. SCN can be switched on and off and moved from one place to another without any detailed network plan, the fact that makes it hard to the operators to control the number and location of small cells and thus avoiding interference becomes a more difficult and challenging problem. In addition, SCN BSs have a lower antenna height compared to macrocell BS, which makes radio propagation predictions a difficult problem especially in urban areas. One possible solution would be to make every cell have an interference avoidance scheme that can be operated independently and thus intercell interference can be avoided through decentralized interference management (Lopez-Perez, et al., 2011). Another solution is to make the user equipment monitor and detect any potential interfering cells that are in close proximity and report to their servers. As a result, the neighboring cells will know if the transmit power will be set below a specific threshold for downlink transmissions, while for uplink, transmissions can be scheduled to occur in the near future in case the neighboring cells have been informed that the uplink transmissions will cause interference (Lopez-Perez, et al., 2011). Furthermore, intelligent power control schemes can be used, where idle nodes and lightly loaded cells can be temporary shutdown, which will reduce the power density and thus minimize interference affected (Hoydis & Kobayashi, 2011).

The effect of cross-tier and intra-tier interferences significantly depends on the access control mechanism employed by the operator. The challenging problems of open network access mechanism are providing secure access to the network resources; mitigation of cross-tier and intra-tier interferences and excessive signaling for providing handover between adjacent small cells or between small cells and the macrocell in case of high user mobility. On the other hand, the closed access system is vulnerable to cross-tier interference. The cross-tier interference can be managed by employing the overlay or underlay strategy. For overlay strategy, interference can be avoided by employing intelligent resource partitioning schemes (Cheng, Lien, Chu, & Chen, 2011; Guandr, Bayhan, & Alagoandz, 2010). Although such schemes may eliminate cross-tier interference however, they suffer from poor spectrum usage (under or over usage of spectrum) due to static spectrum allocation and poor Signal to Interference Ratio (SIR) of femtocell mobile user due to high macrocell mobile user activity. As a result, there is considerable interest in underlay strategy, where macrocell and femtocells share the common spectrum and interference can be mitigated by employing power control mechanisms (Mo, Quek, & Heath, 2011; Li, Khan, Pesavento, & Ratnarajah, 2011).

The SCN is connected to an IP based backhaul network that is owned and operated by the network operator, which makes the privacy of the user under risk since the traffic must traverse the backhaul network affected (Hoydis & Kobayashi, 2011). SCN must be able to protect user privacy and prevent hackers from getting access to the BS. By using a restricted access policy where only registered users can access network resources, cross-tier interference can occur as the non-registered users who are not able to access the SCN resources have to transmit with high power to compensate for the path losses due to being served by a far located macro BS. Doing so, jamming can occur in the uplink and the downlink (Lopez-Perez, et al., 2011). If the open access mode was used by the small cells, then all users will be able to access the network resources and thus downlink interference can be minimized since users will be connected to the nearest cells all the time. However, with open access strategy, the network resources become available to anyone and thus, user privacy becomes under threat. So a tradeoff can be seen between security and interference.

2. SMART COGNITIVE FEMTO NETWORK (SCFN)

The capability of the femtocell network to become a self-aware entity in the Heterogeneous network can be increased by enabling the cognitive radio technology on femto base stations.

Since the femtocells are characterized by being self-organized due to being randomly deployed based on subscriber requirements, and since there is an absence of direct backhaul coordination between macro and femto networks, inter- and intra-interference management and mitigation become a challenge issue in the deployment of femtocell networks. In addition, some mobile terminals may not be able to connect to the femtocell even if it is close to them, and that is due to restricted association, which result in severe interference from the macrocell transmitter on the femto uplink and from the femto downlink to the macrocell receiver. Moreover, femtocells are connected to the operator's network through subscriber's private Internet Service Provider (ISP), which makes interference coordination more difficult (Rangan, 2010). Using Successive Interference Cancellation (SIC) cannot by itself solve the problem of interference in heterogeneous networks (European Commission, 2012), that is why several papers (Cheng, et al., 2011; Kaimaletu, Krishnan, Kalyani, Akhtar, & Ramamurthi, 2011; Cheng, Ao, & Chen, 2010; Haines, 2010; Costa, Cattoni, Roig, & Mogensen, 2010) discuss the integration of CR technology in femtocells, which is seen to be a promising solution for interference problems in heterogeneous networks.

To mitigate interference of femto-macro cellular networks, femto BS are equipped with CR capabilities and behave as secondary users, while the macro BS behave as primary users. We exploit the term Cognitive-Femto Base Station (CFBS) for the secondary femto base station, which is a low-power network node that utilizes broadband cellular technology with IP backhaul through a local broadband connection, such as DSL, cable, or fiber (Cheng, et al., 2011; Guandr, et al., 2010). This cognitive femtocell idea leads to simpler and easier proliferation of cognitive radio technology into practical systems of future generations of cellular network (Rubaye, Al-Dulaimi, & Cosmas, 2011; Pennanen, Tolli, & Latva-Aho, 2011; Li, Feng, Zhang, Tan, & Tian, 2010). The CFBS will be able to sense the environment and get the required information without the need to communicate with the macro BS, i.e., the CFBS will not interfere with the transmissions of the macro BS and the surroundings femtocells. Thus, a cognitive enabled femto base station can sense spectrum efficiently; manage interference intelligently and allocate resources effectively by learning and adapting accordingly to the communications environment. In Haines (2010), a Cognitive Pilot Channel mechanism is introduced to assist femtocells in avoiding interference and providing them with information about available radio access technologies in the network. Thus, cognitive radio technology constitutes a promising solution to mitigate the interference problems of self-organized femtocells in heterogeneous networks. In Shakir and Alouini (2012), it is shown that the femto base stations can be deployed around the edge of the macrocell in order to increase the area spectral efficiency and reduce the co-channel interferences. In this chapter, by exploiting the cooperation among the CFBS which are arranged in particular configuration, the cross-tier and intra-tier interferences can be managed. This can be considered as a scenario where the CFBS are collaborating to form a triangular antenna array in order to improve desired mobile user SIR, and thereby reduce interferences.

2.1. Smart Cognitive-Femto Network (SCFN)

In this section, the environment setup is first presented; the network layout is then outlined; the random mobile user model is described; and the interference suppression capabilities of the smart cognitive-femto networks is studied.

2.1.1. Environment Setup

In this chapter, a two tier open access Heterogeneous network is considered, where CFBS are underlaid with the macrocell network such that the macrocell network and cognitive radio enabled femtocell network are analogous to primary and secondary network, respectively, as in traditional cognitive radio model. It is known that the cooperation among the base stations can increase the capacity and reduce the Bit Error Rate (BER) under the current network infrastructure (Mo, et al., 2011; Li, et al., 2011). The base station cooperation has been so far mostly based on the assumption that all the base stations in the network are connected to the central processor at the fusion center via links of unlimited capacity (Arshad, et al., 2010; Shakir, et al., 2011). In this case, the set of base stations may effectively act as the multi antenna transmitter (downlink) or receiver (uplink) with the caveat that the antennas are geographically distributed over the region (Li, et al., 2010; Huang, Ding, & Liu, 2007). In this context, by exploiting the cooperation among the CFBS, the multiple CFBS each equipped with single antenna can be considered as a single base station with multiple geographically dispersed antennas. The resultant network is referred to as Smart Cognitive Femto Network (SCFN). A typical SCFN is shown in Figure 2, where it is shown that the CFBS with fixed location are collaborating to form a triangular antenna array. The proposed setting is more applicable to scenarios where operators may install the CFBS in a particular configuration which mandates:

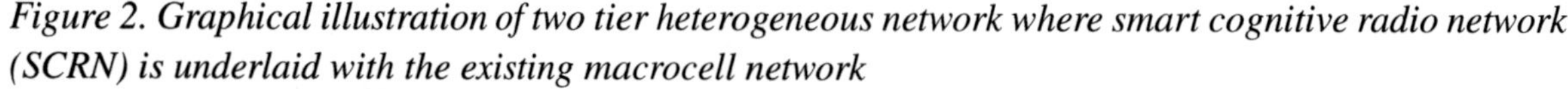

Figure 2. Graphical illustration of two tier heterogeneous network where smart cognitive radio network (SCRN) is underlaid with the existing macrocell network

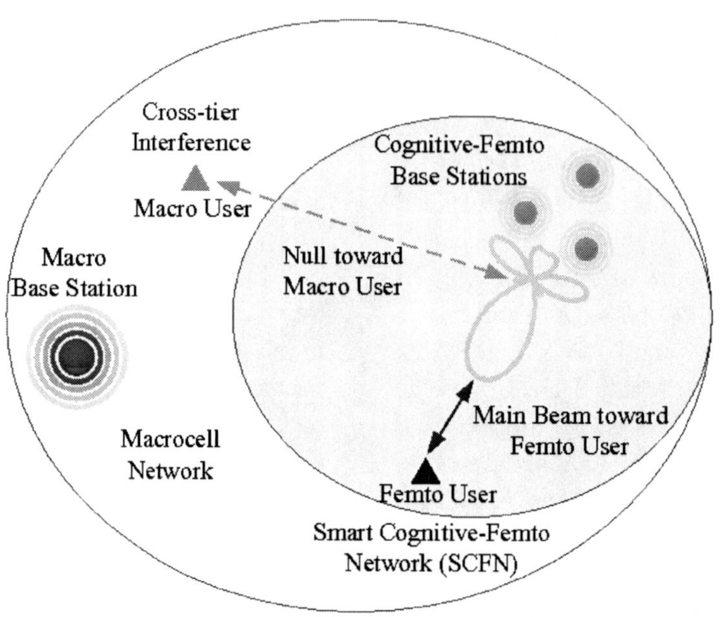

- To reduce the cross-tier and intra-tier interferences by directing the main beam toward the desired femtocell user and creating null toward the cross-tier interference due to macrocell mobile users and thereby facilitate the cross-tier interference management and control.
- To acquire knowledge about its environment without the aid of a macrocell in a decentralized fashion and automatically prevent disturbing the transmission of femtocell and macrocell mobile users.
- To provide connectivity to the femto mobile users which are located in dead zones. The dead zone is defined as the region, which is not covered by femto base station, which might be due to offloading of femto base station.

The proposed setting can be applicable in overlay, underlay or hybrid strategies, which were discussed in Section 1.1.2. The CFBS collaborate to form a beam that is steered towards the desired femtocell mobile user with direction of arrival of ϕ_1 and creating the null toward the interference, which are with direction of arrival ϕ_k.

2.1.2. Network Layout

The first tier of the considered two tier Heterogeneous network comprises of the macrocell network such that the macrocell carrier frequencies are reused throughout the network in the macrocell at a distance D[m]. A circular macrocell of radius R_m [m] with a base station B_m which is equipped with multiple antenna elements is considered. Furthermore, it is considered that there are M macrocell mobile users T_m each equipped with single antenna and are uniformly distributed across each of the macro cells. The second tier of two tier Heterogeneous network which is commonly known as secondary network, comprises of N circular femtocells of radius R_n [m] with low-power low-cost user deployed CFBS B_n which are equipped with an omni-directional antenna and provides back hauling services to the users in its coverage region. Moreover, it is considered that there are F femtocell mobile users T_n, each equipped with single antenna and are uniformly distributed across each of the femtocells.

2.1.3. Mobile User Distribution

In this chapter, it is assumed that mobile users including macrocell users, femtocell mobile users and interfering mobile users are mutually independent and uniformly distributed in their respective cells. Thus, the Probability Density Function (PDF) of the distribution of kth mobile user $T(\cdot)$ which is located at a distance r_k relative to the reference macrocell base station $B(\cdot)$, in polar coordinate (r_k, θ_k) is given by (Shakir & Alouini, 2012):

$$p_{(r_k,\theta_k)}(r_k, \theta_k) = \frac{r_k}{\pi R^2_{(\cdot)}} \quad (1)$$

where $0 \le r_k \le R_{(\cdot)}$ and $0 \le \theta_k \le 2\pi$; $R_{(\cdot)}$ is the radius of the macrocell or femtocell.

2.1.4. System Model

Consider that there are N CFBS which are collaborating to form a triangular array such that the array receives $H=M+NF$ narrowband signals from all mobile users in the Heterogeneous network which are distributed randomly throughout the network as described in previous section. The resultant signal induced at the antenna array can be expressed as:

$$x = \sum_{k=1}^{H}\sum_{n=1}^{N} s_k \, e^{j\frac{2\pi}{\lambda}(r_k - r_{n,\phi_k})} + z \quad (2)$$

where s_k is the signal transmitted by the kth mobile user; λ is the carrier frequency wavelength of the signals; $z=[z_1,z_2,\ldots,z_N]^T$ is $N\times1$ white Gaussian noise vector at the antenna array with zero mean and variance σ^2; $x=[x_1,x_2,\ldots,x_N]^T$ is $N\times1$ vector of measured signal at the antenna array. Moreover, r_k is the reference distance between the origin of the triangular array and k^{th} mobile users which is located at an direction of arrival, ϕ_k and r_{n,ϕ_k} is the distance from nth antenna element to the kth mobile user. In vector notation, Equation 2 can be expressed as:

$$x = \sum_{k=1}^{H} a(\phi_k)s_k + z = A(\phi_k)s + z \quad (3)$$

where

$$A(\phi) = [a(\phi_1), a(\phi_2), \cdots, a(\phi_H)]$$

is an $N\times H$ matrix whose columns are the steering vectors of the mobile users; $s=[s_1,s_2,\ldots,s_N]^T$ is $H\times1$ transmitted signal vector. The $N\times1$ steering vector

$$a(\phi_k) = [a_1(\phi_k), a_2(\phi_k), \cdots, a_N(\phi_k)]^T$$

models the spatial response of the array to an incident plane wave from the direction of arrival ϕ_k such that the array response of the nth element can be expressed as:

$$a_n(\phi_k) = H_n(\phi_k)e^{j\frac{2\pi}{\lambda}(r_k - r_{n,\phi_k})} \quad \text{for } n \in N \quad (4)$$

where $H_n(\phi_k)$ denotes the antenna response of the nth antenna element. For simplicity, the mutual coupling among the cooperative CFBS is neglected and the resultant steering vector becomes:

$$a(\phi_k) = [e^{j\frac{2\pi}{\lambda}(r_k - r_{1,\phi_k})}, e^{j\frac{2\pi}{\lambda}(r_k - r_{2,\phi_k})}, \cdots, e^{j\frac{2\pi}{\lambda}(r_k - r_{N,\phi_k})}]^T \quad (5)$$

2.1.5. Interference Suppression Capability

Let us assume that s_1 is the desired signal of femtocell mobile user 1, T_n^1 with the respective direction of arrival ϕ_1 such that the rest of the signals $\{s_k\}_{k=2}^{H}$ as intra and inter tier interferences arriving at the antenna array from respective direction of arrival. The output of the triangular array is given by:

$$y = w^H x \tag{6}$$

where w is the weight vector which is used to produce a beam pattern with the main beam directed toward the desired femtocell mobile user and $(\cdot)^H$ denotes the Hermitian transpose. It is also assumed that the weights are of equal magnitude while the phases are exploited to steer the direction of the main beam of the array toward the desired femtocell mobile user such that the weight vector may be given as (Durrani & Bialkowski, 2004):

$$w = \frac{1}{\sqrt{N}} a(\phi_1) \tag{7}$$

The desired output at the array can be obtained by substituting Equations 3 and 7 into 6, which may be given as:

$$y = s_1 + \frac{1}{N}\sum_{k=2}^{H} s_k a^H(\phi_1)a(\phi_k) + \frac{1}{N} a^H(\phi_1)z \tag{8}$$

The mean output power of the array is (Durrani & Bialkowski, 2004):

$$\begin{aligned} P &= E\left[|s_1|^2\right] \\ &+ \frac{1}{N^2}\sum_{k=2}^{H} |a^H(\phi_1)a(\phi_k)|^2 \; E\left[|s_k|^2\right] \\ &+ E\left[\frac{1}{N} |a^H(\phi_1)z|^2\right] \\ &= E\left[|s_1|^2\right] + \sum_{k=2}^{H} \alpha(\phi_1,\phi_k) E\left[|s_k|^2\right] + \frac{\sigma^2}{N} \end{aligned} \tag{9}$$

where $E[\cdot]$ is the expectation operator and $\alpha_k(\phi_1,\phi_k)$ is defined as Interference Suppression Measure (ISM) which determines how much undesired power is picked up from the kth interferer and is given by:

$$\alpha_k(\phi_1,\phi_k) = \frac{1}{N^2} |a^H(\phi_1)a(\phi_k)|^2 \tag{10}$$

$a^H(\phi_1)$ is the steering vector of the desired femtocell mobile user with direction of arrival ϕ_1 and $a(\phi_k)$ is the steering vector of the kth interferer with direction of arrival ϕ_k.

By using Equation 9, the mean SIR at the array output of the antenna array $\left(SIR_o\right)$ can be written in terms of input SIR, $\left(SIR_{in}\right)$ as (Durrani & Bialkowski, 2004):

$$SIR_o = \frac{E[|s_1|^2]}{\sum_{k=2}^{H} \alpha(\phi_1,\phi_k) E[|s_k|^2]} = \frac{SIR_{in}}{G_{avg}(\phi_1)} \tag{11}$$

where $G_{avg}(\phi_1)$ is the spatial Interference Suppression Coefficient (ISC). By assuming that the interfering mobile users are uniformly distributed in the range $[0, 2\pi]$ such that the interfering signals are equally probable from any direction in the specified range ϕ_k, the $G_{avg}(\phi_1)$ can be calculated as (Durrani & Bialkowski, 2004):

$$G_{avg}(\phi_1) = \frac{1}{2\pi}\int_0^{2\pi} \alpha_k(\phi_1, \phi_k) d\phi_k \tag{12}$$

where $\alpha_k(\phi_1, \phi_k)$ can be calculated using Equation 10.

The average improvement in SIR at the array output is given by:

$$\begin{aligned}\Delta &= 10\log_{10}\left(\frac{1}{G_{avg}(\phi_1)}\right) \\ &= -10\log_{10}(G_{avg}(\phi_1))\end{aligned} \tag{13}$$

2.2. Performance Analysis and Discussions

2.2.1. Interference Suppression Performance Analysis

In this section, the interference suppression performance analysis of the SCFN is presented by varying the number of collaborating CFBS (*N*). In this context, the beam steering vector is calculated and thereby the interference suppression coefficients for several values of collaborating CFBS in the considered two-tier Heterogeneous network. The geometrical illustration of the triangular array for *N*={3,4,6,9} is shown in Figure 3, where *d*=λ/2. The distances from n^{th} base station to the mobile user can be calculated by exploiting the basics of trigonometry. The distances for *N*=3 and *N*=4 are shown in Figure 3(a) and Figure 3(b) respectively. However, for the sake of simplicity the distances are not shown on the Figure 3(c) and Figure 3(d), which can be similarly calculated. The beam steering vectors and ISM for similar values of *N* are calculated and summarized in Table 1 and Table 2, respectively. Figure 4 shows the variation of the spatial ISC for smart cognitive-femto networks where *N* cognitive-femto base stations are collaborating to form a triangular array and thereby steering the main beam toward the desired femtocell mobile user with direction of arrival of ϕ_1 and creating the null toward the interference which are uniformly distributed in the range $[0, 2\pi]$ with direction of arrival ϕ_k. It can be observed clearly that the interference suppression capability of the proposed smart network has been increased significantly with the increase in number of collaboration cognitive-femto base stations. The improvement in interference suppression due to collaborating femto base stations is considerable as compare with the interference suppression ability of the single non-cooperating femto base station (compare the black curve with red marks with remaining curves where *N* number femto base stations are collaborating). However, it can be seen that the curves have oscillatory behavior for small number of collaborating base stations *N* and these curves start to behave as flat curves overall direction of arrival of desired femtocell mobile user with the increase in *N* and the numerical values for the spatial interference start to decrease.

As an example, the average spatial ISC values for *N*=3 is approximately calculated as 0.39, which reduces to approximately 0.15 for *N*=9 which shows a significant interference suppression improvement when more number of base stations are collaborating in the secondary network. To have a deeper insight into the interference suppression capability of the proposed smart network, the average improvement in SIR is calculated as (13). Figure 5 shows the average improvement in SIR due to collaborating femto base stations versus the direction of arrival ϕ_1 of the desired femtocell mobile user. It can be seen clearly that the SIR of the desired femtocell mobile users improves with the increase in number of collaborating base stations *N*. It can be seen that with the smaller number of Cognitive-femto base stations *N*, the SIR improvement curves show oscillatory behavior. The improvement in SIR is summarized in Table 3 where the maximum SIR improvement is shown as Δ_{max} and the minimum SIR improve-

Figure 3. Geometrical illustration of triangular antenna array where cognitive-femto base station are collaborating to direct the mean beam toward the desired femtocell mobile user and create null toward the interfering mobile users for: (a) N=3; (b) N=4; (c) N=6; and (d) N=9

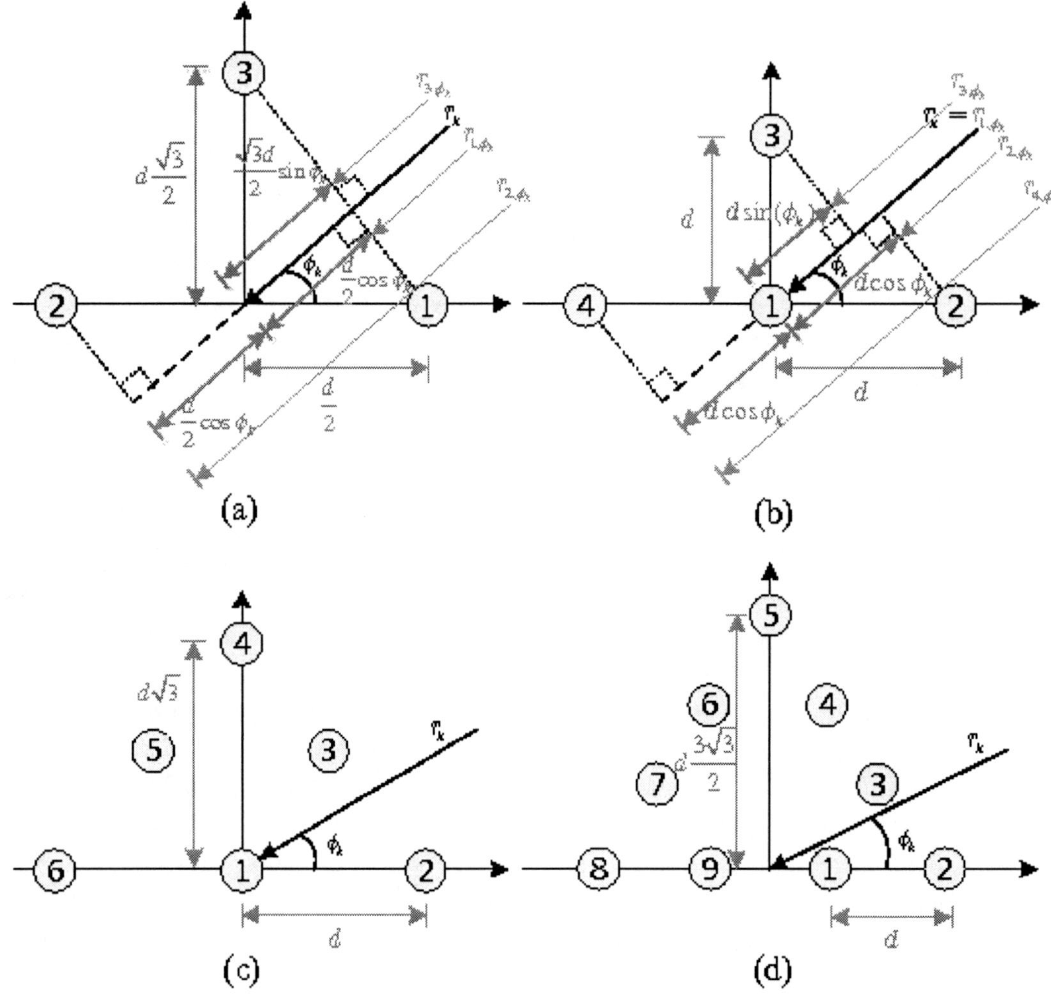

Table 1. Beam steering vectors $a(\phi_k)$ for different number of cognitive-femto base stations (CFBS)

N	Beam Steering Vectors
3	$\boldsymbol{a}(\phi_k) = [e^{jk\frac{d}{2}\cos\phi_k}, e^{-jk\frac{d}{2}\cos\phi_k}, e^{jk\frac{d}{2}\sqrt{3}\sin\phi_k}]^T$
4	$\boldsymbol{a}(\phi_k) = [1, e^{jkd\cos\phi_k}, e^{jkd\sin\phi_k}, e^{-jkd\cos\phi_k}]^T$
6	$\boldsymbol{a}(\phi_k) = [1, e^{jkd\cos\phi_k}, e^{jkd\cos(60^o-\phi_k)}, e^{-jkd\cos\phi_k}, e^{jkd\sqrt{3}\sin\phi_k}, e^{-jkd\cos(60^o+\phi_k)}]^T$
9	$\boldsymbol{a}(\phi_k) = [e^{jk\frac{d}{2}\cos\phi_k}, e^{jk\frac{3d}{2}\cos\phi_k}, e^{jk\frac{\sqrt{7}d}{2}\cos(40.9^o-\phi_k)}, e^{jk\frac{\sqrt{13}d}{2}\cos(73.9^o-\phi_k)},$ $e^{jk3\frac{\sqrt{3}d}{2}\sin\phi_k}, e^{-jk\frac{\sqrt{13}d}{2}\cos(73.9^o+\phi_k)}, e^{-jk\frac{\sqrt{7}d}{2}\cos(40.9^o+\phi_k)}, e^{-jk\frac{d}{2}\cos\phi_k}, e^{-jk\frac{3d}{2}\cos\phi_k}]^T$

Table 2. Interference suppression measure (ISM) for different number of cognitive-femto base stations (CFBS)

N	Interface Suppression Measure (ISM)
3	$\alpha_k(\phi_1,\phi_k) = \frac{1}{N^2}\lvert e^{-j\frac{\pi}{2}(\cos\phi_1-\cos\phi_k)} + e^{-j\frac{\pi}{2}(\cos\phi_k-\cos\phi_1)} + e^{-j\frac{\pi}{2}\cdot\sqrt{3}\cdot(\sin\phi_1-\sin\phi_k)}\rvert^2$
4	$\alpha_k(\phi_1,\phi_k) = \frac{1}{N^2}\lvert 1 + e^{-j\pi(\cos\phi_1-\cos\phi_k)} + e^{-j\pi(\sin(\phi_1)-\sin(\phi_k))} + e^{j\pi(\cos\phi_1-\cos\phi_k)}\rvert^2$
6	$\alpha_k(\phi_1,\phi_k) = \frac{1}{N^2}\lvert 1 + e^{-j\pi(\cos\phi_1-\cos\phi_k)} + e^{-j\pi(\cos(60^o-\phi_1)-\cos(60^o-\phi_k))} + e^{j\pi(\cos\phi_1-\cos\phi_k)}$ $+e^{-j\pi\sqrt{3}(\sin\phi_1-\sin\phi_k)} + e^{-j\pi(\cos(60^o+\phi_k)-\cos(60^o+\phi_1))}\rvert^2$
9	$\alpha_k(\phi_1,\phi_k) =$ $\frac{1}{N^2}\lvert e^{-j\frac{\pi}{2}(\cos\phi_1-\cos\phi_k)} + e^{-j\frac{3\pi}{2}(\cos\phi_1-\cos\phi_k)} + e^{-j\frac{\sqrt{7}\pi}{2}(\cos(40.9^o-\phi_1)-\cos(40.9^o-\phi_k))}$ $+e^{-j\frac{\sqrt{13}\pi}{2}(\cos(73.9^o-\phi_1)-\cos(73.9^o-\phi_k))} + e^{-j\frac{3\sqrt{3}\pi}{2}(\sin\phi_1-\sin\phi_k)} +$ $e^{-j\frac{\sqrt{13}\pi}{2}(\cos(73.9^o+\phi_k)-\cos(73.9^o+\phi_1))}$ $+e^{-j\frac{\sqrt{7}\pi}{2}(\cos(40.9^o+\phi_k)-\cos(40.9^o+\phi_1))} + e^{-j\frac{\pi}{2}(\cos\phi_k-\cos\phi_1)} + e^{-j\frac{3\pi}{2}(\cos\phi_k-\cos\phi_1)}\rvert^2$

Figure 4. Comparison of spatial suppression coefficient (ISC) of smart cognitive-femto base stations with the direction of arrival of the desired femtocell mobile user (ϕ_1) and uniformly distributed interfering mobile users

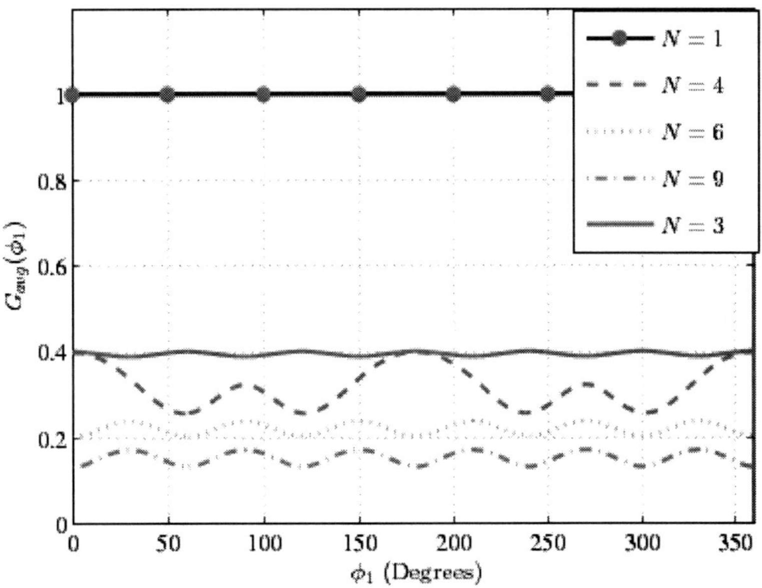

ment as Δ_{min} for each of the considered values of *N*. It is clear from the numerical values also that the SIR improvement reached to the maximum SIR improvement for $\phi_1^{max} \in \phi_1$ and similarly the SIR improvement reached to the minimum improvement for $\phi_1^{min} \in \phi_1$ such that it fluctuates between the range $[\Delta_{max}, \Delta_{min}]$. However, with an increase in *N*, the amplitude of these oscilla-

Figure 5. Comparison of average improvement in signal to interference ratio (SIR) for different number of collaborating cognitive femto base stations with direction of arrival of the desired femtocell mobile user $\left(\phi_1\right)$ and uniformly distributed interfering mobile users

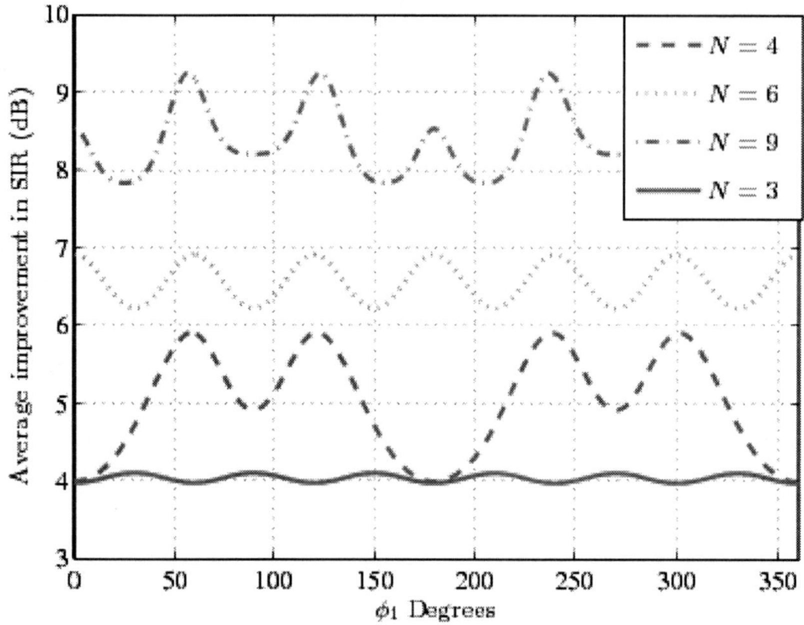

tions start to decrease and leading to less fluctuations in SIR improvement over all direction of arrivals of desired mobile users ϕ_1.

2.2.2. Directivity and Beam Steering Performance Analysis

In this section, the directivity and beam steering performance analysis of the SCFN are presented by varying the number of collaborating CFBS (*N*).

The polar plots for *N* collaborating Cognitive-Femto Base Stations forming a triangular array are shown in Figure 6, where the desired femtocell mobile user is assumed to be located an angle of arrival φ1 = 45°. Looking at these polar plots, and with an increase in *N*, it can be clearly seen that the beam steered towards 45° becomes narrower and thereby suppressing the interferers further, which are uniformly distributed in the range $[0, 2\pi]$ with direction of arrival ϕ_k. In fact, the Half-Power Beam Width (HPBW) in Figure 6(a) is about 260°, while that in Figure 6(d) is approximately reduced to 40°. In communication systems, a narrow beam width is useful to reject unwanted signals from another system, while a wide beam width allows communication with multiple sites distant apart. Overall, it can be clearly seen that the directivity is high when *N* increases.

Table 3. Numeric values of SIR improvement Δ

N	Δ_{max} (dB)	Δ_{min} (dB)
3	4.1	3.9
4	5.9	4.0
6	6.9	6.2
9	9.3	7.8

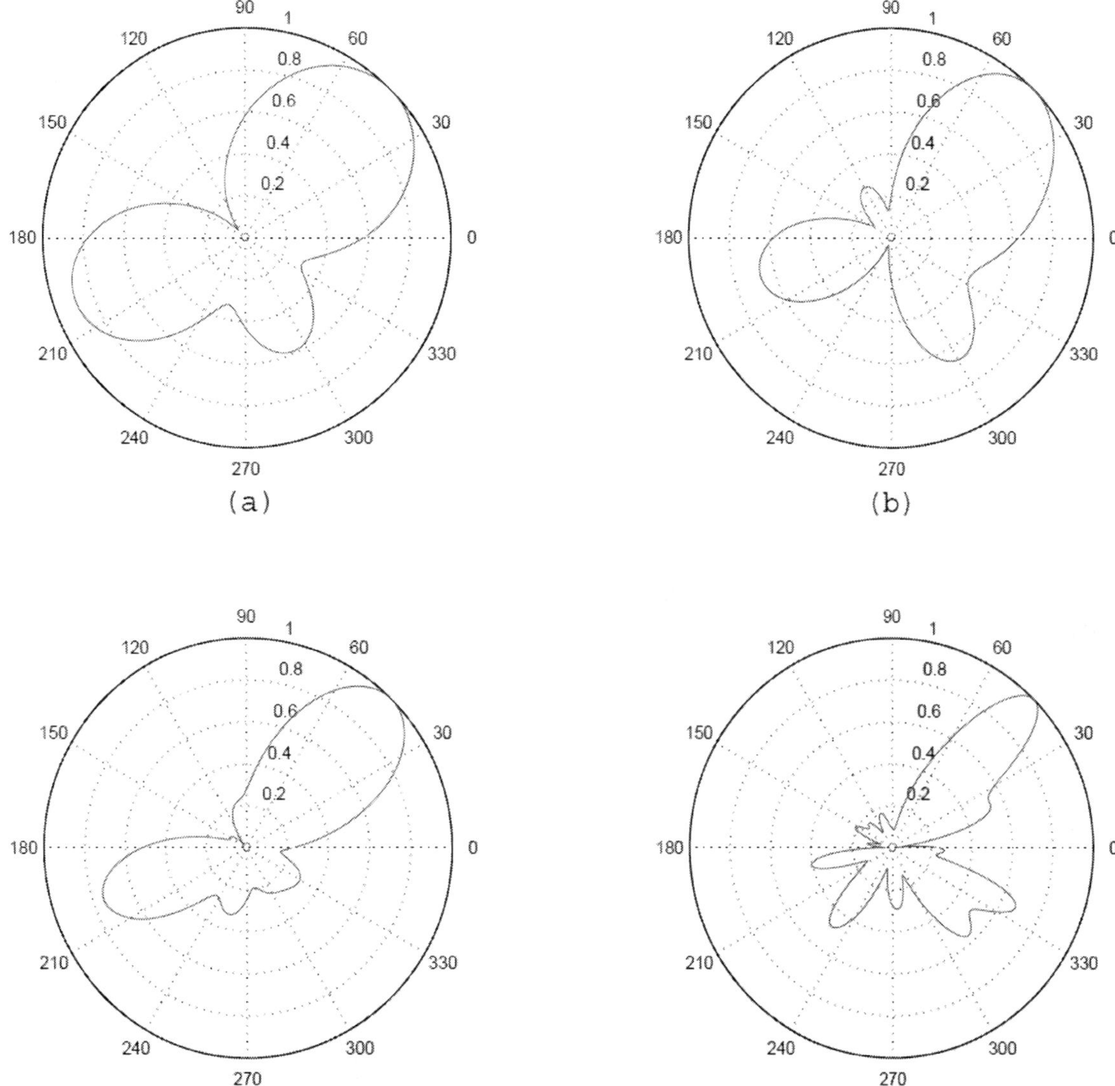

Figure 6. Polar plots of triangular antenna array forming by collaborating cognitive-femto base stations (CFBS) with $\phi_1 = 45^o$ and (a) N=3, (b) N=4, (c) N=6, and (d) N=9

3. FUTURE RESEARCH DIRECTIONS

The smart network idea can be designed for practical cellular systems where CFBS are uniformly random distributed and the subset of the CFBS could collaborate to form a smart antenna array and thereby reduce the cross-tier and the intra-tier interferences and improve the link quality. In this context, the distances among the CFBS are also random however the collaboration among the subset of the CFBS could be possible to facilitate the interference mitigation (Donvito & Kassam, 1979; Lo, 1964).

4. CONCLUSION

In this chapter, a smart cognitive-femto network was proposed where the cognitive enabled femto base station in existing two tier Heterogeneous network collaborate and intelligently adapt accordingly to the communication environment in order to facilitate the interference mitigation.

By exploiting the collaboration among the cognitive enabled femto base station, the main beam can be directed toward the desired femtocell mobile user and create null to the interfering mobile users (may be cross-tier interference and/or inter-femtocell network interference) and thereby improve the SIR of the desired mobile users. It has been shown that the interference suppression capabilities of the smart cognitive-femto network can be improved significantly in comparison to the network without collaboration and coordination among the femto base stations.

REFERENCES

Al-Rubaye, S., Al-Dulaimi, A., & Cosmas, J. (2011). Cognitive femtocell. *IEEE Magazine of Vehicular Technology*, *6*(1), 44–51. doi:10.1109/MVT.2010.939902

Alouini, M. S., & Goldsmith, A. (1999). Area spectral efficiency of cellular mobile radio systems. *IEEE Transactions on Vehicular Technology*, *48*(4), 1047–1066. doi:10.1109/25.775355

Andrews, J. G., Claussen, H., Dohler, M., Rangan, S., & Reed, M. C. (2012). Femtocells: Past, present, and future. *IEEE Journal on Selected Areas in Communications*, *30*(3), 497–508. doi:10.1109/JSAC.2012.120401

Arshad, K., Imran, M. A., & Moessner, K. (2010). Collaborative spectrum sensing optimisation algorithms for cognitive radio networks. *Eurasip International Journal of Digital Multimedia Broadcasting*. Retrieved from http://www.hindawi.com/journals/ijdmb/2010/424036/

Ashraf, I., Boccardi, F., & Ho, L. (2011). SLEEP mode technique for small cell deployments. *IEEE Communications Magazine*, *49*(8), 72–79. doi:10.1109/MCOM.2011.5978418

Badic, B., O'Farrell, T., Loskot, P., & He, J. (2009). Energy efficient radio access architectures for green radio: Large versus small cell size deployment. In *Proceedings of the IEEE 70th Vehicular Technology Conference (VTC 2009 Fall)*. Anchorage, AK: IEEE.

Chandrasekhar, V., Andrews, J., & Gatherer, A. (2008). Femtocell networks: A survey. *IEEE Magazine of Communications*, *46*(9), 59–67. doi:10.1109/MCOM.2008.4623708

Chen, H. (2010). Relay selection for cooperative spectrum sensing in cognitive radio networks. In *Proceedings of the IEEE 2010 International Conference on Communications and Mobile Computing*. Shenzhen, China: IEEE.

Cheng, S. M., Ao, W. C., & Chen, K. C. (2010). Downlink capacity of two-tier cognitive femto networks. In *Proceedings of the IEEE 21st International Symposium on Personal Indoor and Mobile Radio Communications, PIMRC 2010*, (pp. 1303–1308). Istanbul, Turkey: IEEE.

Cheng, S. M., Lien, S. Y., Chu, F. S., & Chen, K. C. (2011). On exploiting cognitive radio to mitigate interference in macro/femto heterogeneous networks. *IEEE Magazine of Wireless Communications*, *18*(3), 40–47. doi:10.1109/MWC.2011.5876499

Cheng, W., Zhang, X., & Zhang, H. (2011). Full-duplex spectrum sensing in non-time-slotted cognitive radio networks. In *Proceedings of the IEEE 2011 Military Communications Conference (MILCOM 2011)*. Baltimore, MD: IEEE Press.

Cost Europe. (2012). *Cost actions*. Retrieved May 5, 2012, from http://www.cost.eu/domains_actions/ict/Actions

Costa, G., Cattoni, A., Roig, V. A., & Mogensen, P. E. (2010). Interference mitigation in cognitive femtocells. In *Proceedings of the 2010 IEEE GLOBECOM Workshops (GC Wkshps)*. Miami, FL: IEEE.

Donvito, M., & Kassam, S. (1979). Characterization of the random array peak sidelobe. *IEEE Transactions on Antennas and Propagation*, *27*(3), 379–385. doi:10.1109/TAP.1979.1142097

Durrani, S., & Bialkowski, M. (2004). Effect of mutual coupling on the interference rejection capabilities of linear and circular arrays in CDMA systems. *IEEE Transactions on Antennas and Propagation*, *52*(4), 1130–1134. doi:10.1109/TAP.2004.825640

European Commission. (2012). *Frame work 7, ICT*. Retrieved May 5, 2012, from http://cordis.europa.eu/fp7/ict/home en.html

Femto Forum. (2012). *An overview of the femtocell concept*. Retrieved May 5, 2012, from http://www.femtoforum.org

Guandr, G., Bayhan, S., & Alagoandz, F. (2010). Cognitive femtocell networks: An overlay architecture for localized dynamic spectrum access. *IEEE Magazine of Wireless Communications*, *17*(4), 62–70. doi:10.1109/MWC.2010.5547923

Haines, R. J. (2010). Cognitive pilot channels for femto-cell deployment. In *Proceedings of the 7th International Symposium on Wireless Communication Systems (ISWCS 2010)*. York, UK: IEEE Press.

Harley, P. (1989). Short distance attenuation measurements at 900 MHz and 1.8 GHz using low antenna heights for microcells. *IEEE Journal on Selected Areas in Communications*, *7*, 5–11. doi:10.1109/49.16838

Hoydis, J., & Kobayashi, M. (2011). Green small-cell networks. *IEEE Vehicular Technology Magazine*, *6*(1), 37–43. doi:10.1109/MVT.2010.939904

Huang, S., Ding, Z., & Liu, X. (2007). Non-intrusive cognitive radio networks based on smart antenna technology. In *Proceedings of the IEEE Conference on Global Communications, GLOBECOM 2007*, (pp. 4862–4867). Washington, DC: IEEE Press.

Kaimaletu, S., Krishnan, R., Kalyani, S., Akhtar, N., & Ramamurthi, B. (2011). Cognitive interference management in heterogeneous femto-macro cell networks. In *Proceedings of the IEEE International Conference on Communications (ICC 2011)*. Kyoto, Japan: IEEE Press.

Landstrom, A., Furuskar, A., Johansson, K., Falconetti, L., & Kronestedt, F. (2011). Heterogeneous networks increasing cellular capacity. *Journal of the Ericson Review*, *89*, 4–9.

Li, L., Khan, F. A., Pesavento, M., & Ratnarajah, T. (2011). Power allocation and beamforming in overlay cognitive radio systems. In *Proceedings of the IEEE 73rd Vehicular Technology Conference, VTC-Spring 2011*. Budapest, Hungary: IEEE Press.

Li, X., & Cadeau, W. (2011). Anti-jamming performance of cognitive radio networks. In *Proceedings of the IEEE 45th Annual Conference on Information Sciences and Systems (CISS)*. Baltimore, MD: IEEE Press.

Li, Y., Feng, Z., Zhang, Q., Tan, L., & Tian, F. (2010). Cognitive optimization scheme of coverage for femtocell using multi-element antenna. In *Proceedings of the IEEE 72nd Vehicular Technology Conference, VTC-Fall 2010*, (pp. 1–5). Ottawa, Canada: IEEE Press.

Liang, Y. C., Chen, K. C., Li, G. Y., & Mahonen, P. (2011). Cognitive radio networking and communications: An overview. *IEEE Transactions on Vehicular Technology*, *60*(7), 3386–3407. doi:10.1109/TVT.2011.2158673

Lin, P., Zhang, J., Chen, Y., & Zhang, Q. (2011). Macro-femto heterogeneous network deployment and management: From business models to technical solutions. *IEEE Magazine of Wireless Communications*, *18*(3), 64–70. doi:10.1109/MWC.2011.5876502

Lin, Y., & Chen, K. C. (2010). Distributed spectrum sharing in cognitive radio networks - Game theoretical view. In *Proceedings of the 7th IEEE Consumer Communications and Networking Conference (CCNC 2010)*. Las Vegas, NV: IEEE Press.

Lo, Y. (1964). A mathematical theory of antenna arrays with randomly spaced elements. *IEEE Transactions on Antennas and Propagation*, *12*(3), 257–268. doi:10.1109/TAP.1964.1138220

Lopez-Perez, D., Guvenc, I., Roche, G., Kountouris, M., Quek, T., & Zhang, J. (2011). Enhanced intercell interference coordination challenges in heterogeneous networks. *IEEE Wireless Communications*, *18*(3), 22–30. doi:10.1109/MWC.2011.5876497

Maldonado, D., Le, B., Hugine, A., Rondeau, T. W., & Bostian, C. W. (2005). Cognitive radio applications to dynamic spectrum allocation: A discussion and an illustrative example. In *Proceedings of the First IEEE International Symposium on New Frontiers in Dynamic Spectrum Access Networks*, (pp. 597–600). Baltimore, MD: IEEE Press.

Mo, R., Quek, T. Q. S., & Heath, R. W. (2011). Robust beamforming and power control for two-tier femtocell networks. In *Proceedings of the IEEE 73rd Vehicular Technology Conference, VTC-Spring 2011*. Budapest, Hungary: IEEE Press.

Ni, W., & Collings, I. (2009). Centralized inter-network spectrum sharing with opportunistic frequency reuse. In *Proceedings of the IEEE Global Telecommunications Conference (GLOBECOM 2009)*. Hawaii, HI: IEEE Press.

Pennanen, H., Tolli, A., & Latva-aho, M. (2011). Decentralized coordinated downlink beamforming for cognitive radio networks. In *Proceedings of the IEEE 22nd International Symposium on Personal Indoor and Mobile Radio Communications, PIMRC 2011*, (pp. 566–571). Toronto, Canada: IEEE.

Ramanath, S., Kavitha, V., & Altman, E. (2010). Impact of mobility on call block, call drops and optimal cell size in small cell networks. In *Proceedings of the IEEE 21st International Symposium on Personal, Indoor and Mobile Radio Communications Workshops (PIMRC Workshops)*. Istanbul, Turkey: IEEE Press.

Rangan, S. (2010). Femto-macro cellular interference control with subband scheduling and interference cancelation. In *Proceedings of the 2010 IEEE GLOBECOM Workshops (GC Wkshps)*. Miami, FL: IEEE Press.

Sachs, J., Maric, I., & Goldsmith, A. (2010). Cognitive cellular systems within the TV spectrum. In *Proceedings of the 2010 IEEE Symposium on New Frontiers in Dynamic Spectrum*. Singapore, Singapore: IEEE Press.

Shakir, M. Z., & Alouini, M. S. (2012). On the area spectral efficiency improvement of heterogeneous network by exploiting the integration of macro-femto cellular networks. In *Proceedings of the IEEE International Conference on Communications, ICC 2012*, (pp. 1–6). Ottawa, Canada: IEEE Press.

Shakir, M. Z., Rao, A., & Alouini, M. S. (2011). Collaborative spectrum sensing based on the ratio between largest eigenvalue and Geometric mean of eigenvalues. In *Proceedings of the International Conference on Global Communications, GLOBECOM 2011*. Houston, TX: IEEE Press.

Shetty, N., Parekh, S., & Walrand, J. (2009). Economics of femtocells. In *Proceedings of the IEEE Conference on Global Communications, GLOBECOM 2009*. Honolulu, HI: IEEE Press.

Stotas, S., & Nallanathan, A. (2011). Enhancing the capacity of spectrum sharing cognitive radio networks. *IEEE Transactions on Vehicular Technology*, *60*(8), 3768–3779. doi:10.1109/TVT.2011.2165306

Tombaz, S., Vastberg, A., & Zander, J. (2011). Energy- and cost-efficient ultra-high-capacity wireless access. *IEEE Wireless Communications*, *18*(5), 18–24. doi:10.1109/MWC.2011.6056688

Wang, J., Ghosh, M., & Challapali, K. (2011). Emerging cognitive radio applications: A survey. *IEEE Communications Magazine*, *49*(3), 74–81. doi:10.1109/MCOM.2011.5723803

Zhang, R. (2010). On active learning and supervised transmission of spectrum sharing based cognitive radios by exploiting hidden primary radio feedback. *IEEE Transactions on Communications*, *58*(10), 2960–2970. doi:10.1109/TCOMM.2010.082710.090412

ADDITIONAL READING

Adhikary, A., Ntranos, V., & Caire, G. (2011). Cognitive femtocells: Breaking the spatial reuse barrier of cellular systems. In *Proceedings of the Workshop on Information Theory and Applications, ITA 2011*, (pp. 1-10). San Diego, CA: IEEE Press.

Akan, O., Karli, O., Ergul, O., & Haardt, M. (2009). Cognitive radio sensor networks. *IEEE Magazine of Networking*, *23*(4), 34–40. doi:10.1109/MNET.2009.5191144

Akoum, S., Kountouris, M., & Heath, R. W. (2011). *Limited feedback over temporally correlated channels for the downlink of a femtocell network*. Retrieved from http://arxiv.org/abs/1101.4477.

Bennis, M., Debbah, M., & Chair, A. (2010). On spectrum sharing with underlaid femtocell networks. In *Proceedings of the IEEE 21st International Symposium on Personal Indoor and Mobile Radio Communications (PIMRC 2010)*, (pp. 185-190). Istanbul, Turkey: IEEE Press.

Bornhorst, N., Pesavento, M., & Gershman, A. B. (2012). Distributed beamforming for multi-group multicasting relay networks. *IEEE Transactions on Signal Processing*, *60*(1), 221–232. doi:10.1109/TSP.2011.2167618

Cao, F., & Fan, Z. (2010). The tradeoff between energy efficiency and system performance of femtocell deployment. In *Proceedings of the International Symposium on Wireless Communications Systems, ISWCS 2010*. York, UK: IEEE Press.

Cheng, S. M., Ao, W. C., & Chen, K. C. (2011). Efficiency of a cognitive radio link with opportunistic interference mitigation. *Transactions on Wireless Communications*, *10*(6), 1715–1720. doi:10.1109/TWC.2011.040411.101503

Dong-Chan, O., Heui-Chang, L., & Yong-Hwan, L. (2010). Cognitive radio based femtocell resource allocation. In *Proceedings of the 2010 International Conference on Information and Communication Technology Convergence (ICTC)*. Jeju, South Korea: IEEE Press.

Du, H., Ratnarajah, T., Pesavento, M., & Papadias, C. B. (2012). Joint transceiver beamforming in MIMO cognitive radio network via second-order cone programming. *IEEE Transactions on Signal Processing*, *60*(2), 781–792. doi:10.1109/TSP.2011.2174790

Gupta, N. K., & Banerjee, A. (2011). Power and subcarrier allocation for OFDMA femto-cell based underlay cognitive radio in a two-tier network. In *Proceedings of the 2011 IEEE 5th International Conference on Internet Multimedia Systems Architecture and Application (IMSAA)*. Bangalore, India: IEEE Press.

Harjula, I., & Hekkala, A. (2011). Spectrum sensing in cognitive femto base stations using welch periodogram. In *Proceedings of the 2011 IEEE 22nd International Symposium on Personal Indoor and Mobile Radio Communications (PIMRC)*. Toronto, Canada: IEEE Press.

Le, Z., Shiyang, C., & Xiaoying, G. (2011). From reconfigurable SDR to cognitive femto-cell: A practical platform. In *Proceedings of the 2011 International Conference on Wireless Communications and Signal Processing (WCSP)*. Nanjing, China: IEEE Press.

Li, B. (2011). An effective inter-cell interference coordination scheme for heterogeneous network. In *Proceedings of the 2011 IEEE 73rd Vehicular Technology Conference (VTC Spring)*. Yokohama, Japan: IEEE Press.

Liqiang, Z., Wan, Z., Wenjie, S., & Hailin, Z. (2011). Cognitive radio CSMA/CA protocol for femto-WLANs. In *Proceedings of the 2011 IEEE 3rd International Conference on Communication Software and Networks (ICCSN)*. Xi'an, China: IEEE Press.

Meerja, K. A., Pin-Han, H., & Bin, W. (2011). A novel approach for co-channel interference mitigation in femtocell networks. In *Proceedings of the 2011 IEEE Global Telecommunications Conference (GLOBECOM 2011)*. Houston, TX: IEEE Press.

Meng, J., Yin, W., Li, H., Hossain, E., & Ha, Z. (2011). Collaborative spectrum sensing from sparse observations in cognitive radio networks. *IEEE Journal on Selected Areas in Communications*, *29*(2), 327–337. doi:10.1109/JSAC.2011.110206

Ngo, D. T., Le, L. B., Ngoc, T. L., Hossain, E., & Kim, D. I. (2012). Distributed interference management in two-tier CDMA femtocell network. *IEEE Transactions on Wireless Communications*, *11*(3). doi:10.1109/TWC.2012.012712.110073

Wajid, I., Pesavento, M., Eldar, Y. C., & Gershman, A. (2010). Downlink beamforming for cognitive radio networks. In *Proceedings of the IEEE Conference on Global Communications, GLOBECOM 2010*, (pp. 1-5). Miami, FL: IEEE Press.

Zhao, Z., Schellmann, M., Boulaaba, H., & Schulz, E. (2011). Interference study for cognitive LTE-femtocell in TV white spaces. In *Proceedings of the 2011 Technical Symposium at ITU Telecom World (ITU WT)*. Geneva, Switzerland: IEEE Press.

Zhe, C., Nan, G., & Qiu, R. C. (2011). Building a cognitive radio network testbed. In *Proceedings of IEEE Southeastcon*, (pp. 91 – 96). Nashville, TN: IEEE Press.

KEY TERMS AND DEFINITIONS

Cognitive-Femto Base Stations (CFBS): Low-power low-cost user deployed base stations or access points in femtocell networks. They are equipped with cognitive radio capabilities so they can sense the environment and get the required information without the need to interfere with the transmissions of the macrocell base stations and the surrounding femtocells.

Cross-Tier Interference: When both macrocell and femtocell users are sharing the same spectrum, there will be a degradation of either the femtocell user or the macrocell user signal quality in downlink and uplink.

Heterogeneous Networks: Consist of infrastructures with different wireless access technologies, each having different capabilities and functions. Such network consists of an interoperation of macrocells, small cell networks and relays that offer several wireless coverage zones, ranging from outdoor areas to hotspot areas such as offices, buildings, and homes.

Interference Suppression Capability: How much signal distortion caused by cross-tier and intra-tier interferences between macrocells and femtocells can be reduced by exploiting the collaboration among multiple CFBS.

Radio Access Technology (RAT): Provides connection to a mobile phone, laptop, *etc*, through a core network.

Signal to Interference Ratio (SIR): Defines how much undesired power is picked up from interferers or other transmitters than the useful signal.

Small Cell Network (SCN): A network that consists of short-range, low-power, and low-cost user deployed small cells, with reduced area coverage of order of ten to several hundreds of meters, operating in conjunction with the main macro cellular network infrastructure.

Chapter 7
A Novel Spectrum Sensing Scheduling Algorithm for Cognitive Radio Networks

Sami H. O. Salih
Sudan University of Science and Technology, Sudan

Maria Erman
Blekinge Institute of Technology, Sweden

Abbas Mohammed
Blekinge Institute of Technology, Sweden

ABSTRACT

Cognitive Radios are recognized as a novel approach to improve the utilization of a precious natural resource of wireless communications: the radio frequency spectrum. Historically, telecom regulators assigned fixed spectrum bands to the licensed wireless network operators. This spectrum management approach guarantees an interference free environment, except for some configuration faults or illegal usage. However, with the increasing demand for more bandwidth in the finite radio spectrum, the spectrum becomes underutilized. Hence, the concept of secondary operators have emerged, but with emphasis not to influence licensed operators. Consequently, the Cognitive Radio Network (CRN) architecture enters the market as an intelligent solution to these issues, with concentration on spectrum sensing procedures to achieve the regulatory constraint. The most successful sensing algorithms are those applying cooperation and scheduling to have better scanning information; however, those algorithms are developed based on the primary network activities, which are good in terms of reducing expected interference, albeit with more computational load on the CRN. In this chapter, a novel sensing scheduler algorithm is proposed. The idea is to utilize the CRN by fairly distributing the sensing task among the sensors and afterwards utilizing the radio spectrum shared with the primary networks.

DOI: 10.4018/978-1-4666-2812-0.ch007

1. INTRODUCTION

Most telecom regulatory authorities adopt the command-and-control approach when managing the Radio Frequency (RF) spectrum. This methodology has been successful in mitigating the effects of harmful interference in multi-operator environments. However, its usage has resulted in inefficient RF spectrum allocation. The Federal Communications Commission (FCC, 2003) has reported that 15% to 85% of the licensed spectrum is idle in various spatial and temporal accesses depending on the customer distribution and their individual usage.

In the future, such rigid assignments will not be able to accommodate the dramatically increasing demands for more spectrum bandwidth in meeting user requests for broadband access. In fact, even the unlicensed spectrum bands, such as the Industrial, Scientific, and Medical (ISM) band, need an overhaul. Congestion, resulting from the coexistence of heterogeneous devices, operating in these bands, is on the rise. Hence, with an increased saturation of wireless devices, the fixed spectrum usage strategy has been shown to strain the available spectrum.

This necessitates a new license regime approach, known as secondary operators, which is more effective in utilizing the RF spectrum. However, there are concerns about a possible decrease in primary network transmission capacity, due to interference from unlicensed operators. In parallel, when increasing demands for more bandwidth and its scaling growth in applications are addressed, the situation becomes even worse. Hence, a scarcity with the inefficient usage of the spectrum further imposes a new spectrum management model to utilize the wireless spectrum resource opportunistically.

Technically, Cognitive Radio (CR) is recognized as an emerging technology to mitigate the unutilized scarce radio frequency spectrum dilemma. Thus, dynamic spectrum access is proposed to share the available spectrum through opportunistic usage of the frequency bands by secondary operators without interfering with the primary networks. When the CR Network (CRNs) approach is used, it will enable the secondary networks to perform the following tasks:

1. **Spectrum Sensing:** Determining instantaneous available spectrum portions and detecting the presence of primary users when they appear.
2. **Spectrum Management:** Coordinating the assignment of radio channels to the CRN clients from the available channel list.
3. **Mobility Management:** Vacating the channel when the licensed user is detected and handoff to another available channel.

Due to the CRN being responsible for detecting the existence of the primary user transmission, no additional protocols are needed in the primary networks. Hence, spectrum sensing accuracy is the key challenge in CRN deployment in order to avoid harmful interference to primary networks. This necessitates intelligent algorithms for sensing and scheduling in the access layer of the CRN architecture (Cabric, Mishara, & Brodersen, 2004). Accurate sensing information facilitates the CRN to efficiently optimize the spectrum access, assist in reducing interference probability, and adapt an instant spectrum slots available from the unutilized spectrum pool. This chapter addresses the issue of spectrum sensing in the CRN and proposes a novel scheduling algorithm to utilize the sensing task schedule in CRN and hence utilizing the radio spectrum.

This chapter is organized as follows: in section 2, an overview of spectrum sensing techniques are presented with a derivation of the performance parameters affecting the operation of the CRN. A novel algorithm for spectrum sensing scheduling is proposed, after which its performance is investigated according to the parameters mentioned in section 3 and 4, respectively, whereof conclusions are presented in section 5.

2. SPECTRUM SENSING OVERVIEW

By definition, CRNs have to continuously sense their surrounding environments, scanning all available spectrum bands, and use the sensing information to allocate vacant wireless channels, most suited to the CR clients' requirements. CR clients are considered as secondary users (unlicensed users) which operate in the RF spectrum, already assigned to primary users (licensed users). As a consequence, the crucial requirement for the CRN is to avoid interference to the primary users in their vicinity. In contrast, primary networks are not related to the opportunistic dynamic spectrum access. Thus, no changes are required in their infrastructure. Therefore, the CRN should independently be able to detect the primary users' presence through continuously scanning the radio spectrum. The techniques used as well as the performance parameters of the spectrum sensing techniques are presented in subsections 2.1-2.3.

2.1. Spectrum Sensing Techniques

Generally, spectrum sensing can be performed at one of the communication ends (transmitter or receiver) as illustrated in Figure 1. Therefore, the spectrum sensing techniques can be classified into two main categories based on the detection end.

2.2. Primary Transmitter Detection

In this approach, secondary users continuously scan the desired frequency bands to detect the primary signal, appearing in the CRN area. The received signal ***X*(t)** can be modeled as:

$$X(t) = AS(t) + n(t) \quad (1)$$

where ***S*(t)** is the transmitted signal, ***n*(t)** is the Additive White Gaussian Noise (AWGN), and ***A*** is the antenna gain. Thus, primary user activity is detected if the power level of the received signal exceeds a predefined threshold. This scheme is referred to as Energy Detection (Akyildiz & Lee, 2006). Other intelligent schemes used to detect primary user presence are known as Matched Filter (Wild & Ramchandran, 2005) and Feature Detection (Yang & Fei, 2008). However, they require additional information about the primary user signal, such as modulation scheme, coding technique, frequency ranges, and/or other features to recognize the primary signal when it appears.

Figure 1. Spectrum sensing techniques

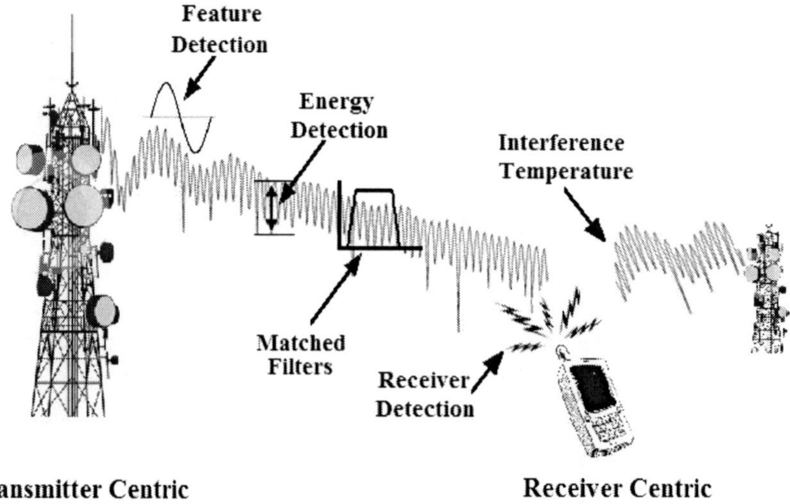

These schemes are more precise than energy detection, even though they consume more computational resources in the CRN. Table 1 provides tabular comparison between transmitter detection sensing techniques.

The main problem with transmission detection schemes is the effect of the hidden node problem (Krenik & Batra, 2005). As shown in Figure 2, when the secondary user is located out of the primary transmitter range, its signal cannot be detected, and hence the CRN assumes that the spectrum is free. However, interference occurs when the secondary user starts to transmit, which affects the primary receiver and decreases the licensed network capacity. Therefore, cooperative transmission detection (Akyildiz, Brandon, & Ravikumar, 2011) schemes are proposed where CR clients cooperatively sense the primary transmitter signal and periodically report to a centralized station. In this scenario, the primary transmitter is detected if one of the CR Customer Premises Equipments (CR-CPEs) notices its signal.

Cooperative detection schemes are more accurate in terms of detecting the primary transmitter activities and consequently these schemes increase the detection probability defined as:

$$\overline{\mathrm{Pd}}^{c} = 1 - \left(1 - \overline{\mathrm{Pd}}\right)^{N} \quad (2)$$

Figure 2. The hidden node problem in CRN

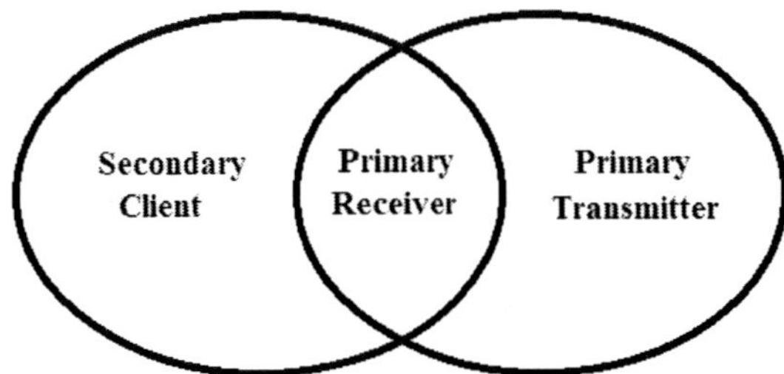

where N is the number of cooperative CR-CPEs.

On the other hand, these schemes will also increase the false alarm probability which is defined as the probability of missed spectrum opportunities:

$$\overline{\mathrm{Pf}}^{c} = 1 - \left(1 - \overline{\mathrm{Pf}}\right)^{N} \quad (3)$$

2.2.1. Primary Receiver Detection

In modern radio receivers, which use multi-stage super heterodyne architecture, the primary receiver emits leakage power from its local oscillators which can be sensed to detect the activity of the primary user (Won-Yeol & Akyildiz, 2008). These schemes shift the sensing spots to the receiver side which in reality is affected by the interference rather than the transmitter. In other words, the

Table 1. Comparison between primary transmitter detection schemes

Sensing Scheme	Energy Detection	Matched Filter	Feature Detection
Implementation Complexity	Easy	Fair	Complex
Primary Signal Properties	Not required	Required	Requires a cyclostationary signal
False Alarms	Highly probable	Less probable	Less probable
Samples Required	$O\,(\mathrm{SNR})^{-2}$	$O\,(\mathrm{SNR})^{-1}$	Primary signal cycle
Noise Effects	Sensitive due to threshold	Less affected	Not affected
Synchronization between the Primary User and the CR-CPEs	Not at all	Required	Not required
Weak Point	Uncertainty in noise power	Dedicated receiver for every primary user	Consumes more computational power
Strong Point	Simplicity	Fast decision	Robustness

sensing task has become receiver-centric instead of transmitter-centric. Adopting a similar approach, the FCC has introduced the Interference Temperature (IT) model (FCC, 2003) to measure interference when its effects take place (i.e. at the receiver). This model, as shown in Figure 3, provides licensed operators with greater certainty regarding the maximum acceptable interference and hence greater protection can be achieved against harmful interference. To the extent when the specified IT limit is not reached, there could be opportunities for secondary transmitters to operate. The main obstacle facing CRNs to adopt this approach is the lack of techniques to measure IT at the primary receiver, taking into account that primary networks do not cooperate with CRNs. For the time being, there is no way to obtain this measurement from the primary receiver, and thus the CRN has to estimate this information independently. For this model to work, limited interaction may be needed between CRNs and primary networks in order to obtain more accurate measurements and save computational resources of the CRNs.

2.3. Scheduling Spectrum Sensing Tasks

The objective of spectrum sensing techniques in the CRN, as described in the cognitive radio cycle proposed by Mitola and Gerald (1999), is to provide accurate spectrum access opportunities without affecting the transmission capacity of the primary network by harmful interference. Thus, the sensing accuracy and precision are the key factors in determining the performance of CRNs.

For the sake of simplicity, portability, and affordability of the CR-CPEs, a single RF front-end circuit is assumed in each node. Consequently, to perform both sensing and data transmission, clients must switch between these two modes in time division manner as in Figure 4, which shows the periodic cycle of the CR-CPE.

Sensing time (t_s) directly influences the sensing accuracy, which is important to guarantee that the interference level (T_I) can be tolerated by the primary network (T_P) (Akyildiz & Lee, 2006). In contrast, transmission time (T_s - t_s) is proportional to the CRN throughput and is defined as:

$$\eta = \frac{T}{T + ts} \tag{5}$$

Figure 3. Interference temperature model

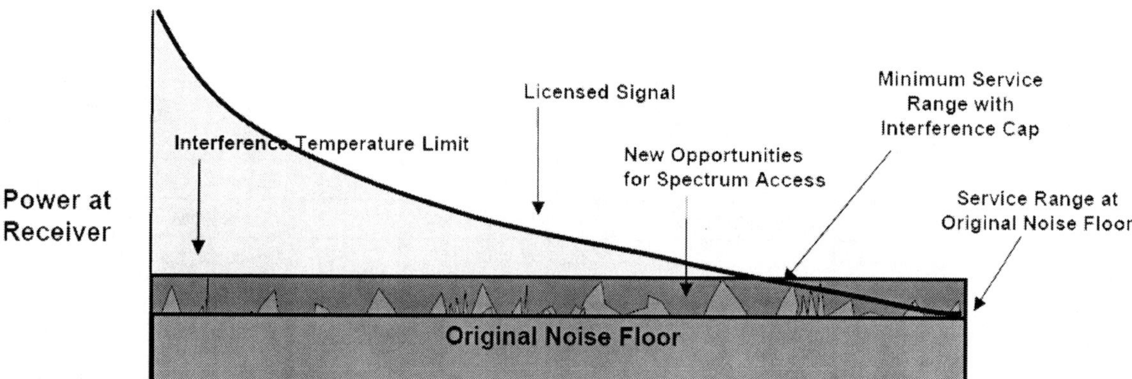

Figure 4. CR-CPE timing structure

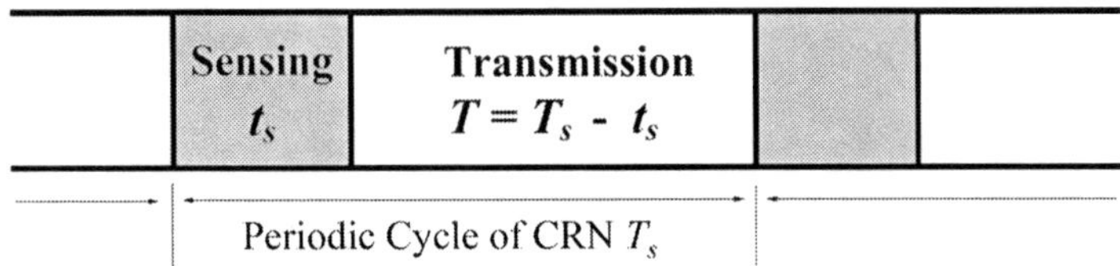

In the cooperative sensing environment, with CR-CPEs sensing the RF spectrum and reporting to a centralized station, all clients have to be synchronized on the same time period for sensing and transmission. Thus, all CR-CPEs should stop their transmission and undertake sensing tasks concurrently to prevent false alarms when another CR-CPE transmits. Preventing false alarms is necessary to increase network performance. However, having all CR-CPEs adjust their transmission period in a fixed manner puts a limitation on the network operators with regards to offering a single service package, taking into account that each application has a specific traffic pattern.

2.4. Performance of Spectrum Sensing

In unwanted cases, CRNs affect the primary network by interference either due to a sensing failure as shown in Figure 4(A) or due to primary transmission starting during the CRN transmission time as in Figure 4(B). In the former case, none of the CR clients detect the presence of the primary user when it is actually transmitting during the sensing period; this type of interference is known as interference during busy period (I_{Busy}). In the latter case, interference has occurred due to the status change of the primary transmitter after the sensing period and during the secondary user transmission; this type is known as interference during idle period (I_{Idle}). Although interference is highly related to the primary network activities in both cases, primary networks are without options to reduce the interference occurrence probability. Hence, it is the responsibility of the CRNs to resolve this issue in accordance with primary network activities. Thus, an estimation of these activities is a very crucial issue in order to examine the spectrum sensing performance (see Figure 5).

As shown in Figure 6, the probabilities of the BUSY and IDLE states of the primary user follow the Birth-Death process with **β** and **α** representing the birth and death rate of the primary user activities, respectively.

$$P_{Busy} = \frac{\beta}{\alpha + \beta} \tag{6}$$

Figure 5. Interference scenarios

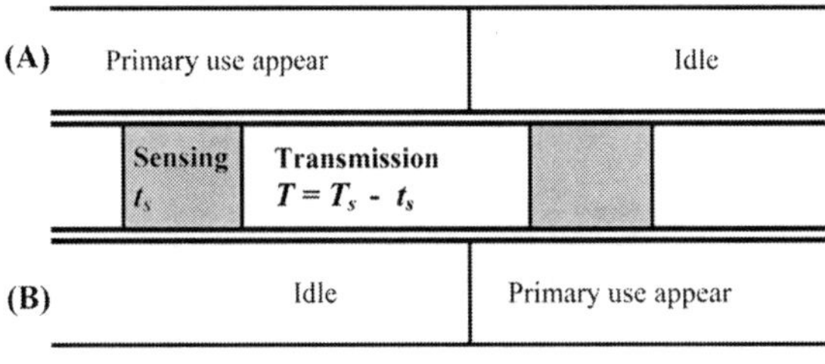

Figure 6. Birth-death processes

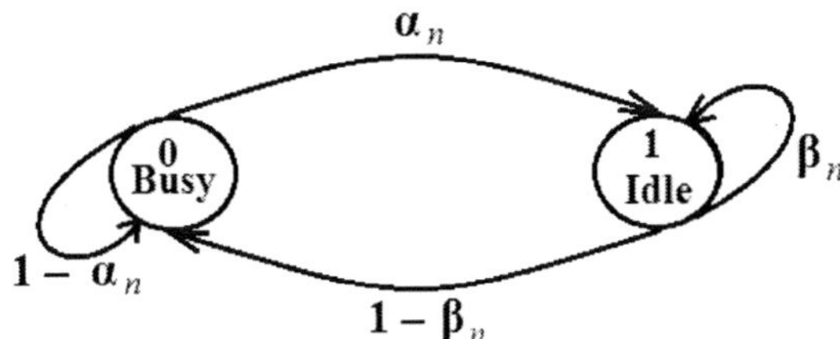

$$P_{Idle} = \frac{\alpha}{\alpha + \beta} \tag{7}$$

The probability of the primary user activity changes during the transmission time T is obtained from the Poisson Arrival Process as $e^{-\alpha T}$ (Chen, Zhao, & Swami, 2008), and consequently the expected interference for the BUSY and IDLE states during the transmission time T are:

$$E\left[I_{Busy}\right] = P_{Idle}\overline{P}_f\left(\frac{\alpha}{\alpha+\beta}e^{-\beta T} + \frac{\beta}{\alpha+\beta}\right)\times T \tag{8}$$

$$E\left[I_{Idle}\right] = P_{Idle}\left(1-\overline{P}_f\right)\left(1-e^{-\beta T}\right)\frac{\beta}{\alpha+\beta}\times T \tag{9}$$

The Interference Avoidance (T_I) is an important parameter to protect the primary network from possible interference to ensure that its capacity will not decrease because of the activities of the CRN. The expected interference ratio is obtained by combining both the expected busy and idle interference during the CRN transmission with the primary user being in the busy state:

$$T_I = \frac{E\left[I_{Busy}\right] + E\left[I_{Idle}\right]}{T \cdot P_{Busy}} \tag{10}$$

Furthermore, another important factor influencing the performance of the CRN is the Lost Spectrum Opportunity (T_L) parameter. Missed spectrum opportunity due to false alarms, will obviously decrease the CRN throughput. This ratio is obtained by combining both the busy and idle interference during CRN transmission with the primary user being in the idle state. The T_L is given as

$$T_L = \frac{E\left[I_{Busy}\right] + E\left[I_{Idle}\right]}{T \cdot P_{Idle}} \tag{11}$$

In the following section 3, a novel scheduling algorithm for spectrum sensing is proposed for the multi-user multi-band network environment.

3. INTERCHANGED SENSING SCHEDULER ALGORITHM (ISS)

Previous sections have explained the performance parameters used to evaluate the coexistence between primary and secondary networks, sharing the radio spectrum resources for single-band/single-user sensing. However, in reality, in order to mitigate the fluctuating nature of the opportunistic spectrum access, the CRN is supposed to exploit multiple spectrum bands. To do so, Wu, Yang, and Huang (2010) describe two different types of sensing strategies can be used: wideband sensing and sequential sensing. The former provides more precise sensing information, but complex arrays of sensors are required, while the later requires less hardware to sense a limited number of spectrum bands on a time division base. In this section, a novel spectrum sensing scheduler is proposed to perform virtual wideband sensing by distributing the sensing task among the CR-CPEs, which reduces the hardware requirement while maintaining precise sensing information.

3.1. Operation Hypothesis

This subsection presents the operating environment of the CRN, as well as the architectures of the CRN, which may either be infrastructure based or ad-hoc based.

3.1.1. CRN Operating Environment

A multiband radio spectrum environment structure is shown in Figure 7. In this environment the CRN is placed in geographical vicinity of various licensed networks, such as TV, GSM, and WiMAX to name a few. Each of these networks utilize different frequency bands with guard bands in between to prevent any possible interference among them as specified in the license issued by the regulatory authority and according to international regimes recommended by the International Telecommunication Union (ITU). Among them, neither spectrum sharing, nor interference is expected; hence each of them serves its customers independently from the others. Moreover, the IT level is always below the specific limit of each sensor (i.e. the receiving antennas). Furthermore, all those licensed networks have exclusive rights to access their assigned spectrum bands.

3.1.2. CRN Architecture

CRNs do not have the authorization to operate at any RF spectrum band, and consequently it can only serve their customers opportunistically without interfering with any licensed operators. The architecture of CRNs can either be infrastructure based or Ad-hoc based.

3.1.2.1. Infrastructure Based CRN (IBCRN)

In this scenario, a centralized architecture for a secondary network is assumed, as shown in Figure 7. A single base transceiver station provides coverage to the area of interest. The other element of the IBCRN is the CR-CPEs, so called CRN clients, each of which is assumed to have several transceivers, and in the case of a single transceiver, the sensing period is defined to have several sensing slots. Finally, a well established error free signaling channel is assumed between the CRN components, and thus synchronized control information is available all of the time.

Figure 7. CRN working environment

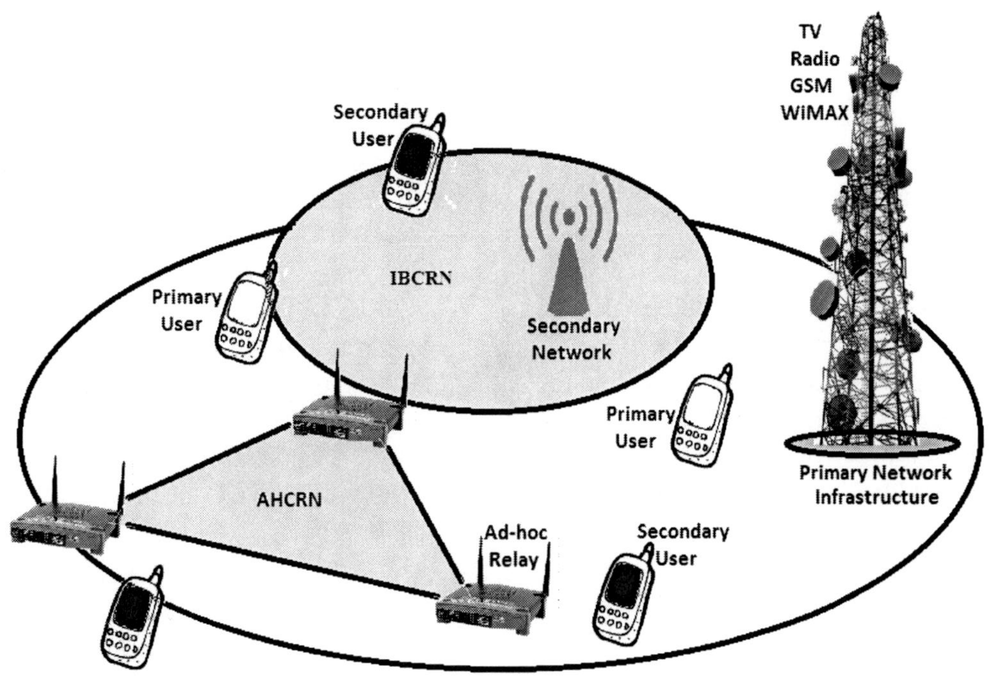

3.1.2.2. Ad-Hoc CRN (AHCRN)

The architecture of an Ad-Hoc CRN (AHCRN), as shown in Figure 7, consists of a mesh Ad-hoc Relay (AhR), which provides services to the secondary users which called in this architecture the Ad-hoc Clients (AhC). This subsection proposes two levels of mesh in accordance with spectrum allocation methodologies.

3.1.2.2.1. Relay Mesh

Mesh architecture is proposed for AhR connectivity. At this level, each access point utilizes the ISM band to setup the wireless link among its neighbors in the access layer (i.e., physical and D-link), and have the whole network topology in the network layer using multi-hop routing protocols such as AODV, OLSR (Sunil & Jyotsna, 2010).

The open and shared ISM band is utilized at this level of the AHCRN to exchange the surrounding environment information as well as handling multi-hop user data transmission. A D-link layer provides next hop information for the AhRs while they are operating independently with their close neighbors. Routing tables are more complex in mesh architectures; however, fortunately, in most scenarios, mesh networks requires a centralized facility (a gateway to access Internet for example). These centralized points help to provide the network topology, which is used to locate the available routes from source to destination in multi-hop communications.

3.1.2.2.2. Local Mesh

The AhRs provide services to their associated AhCs in their local vicinity using star topology. Moreover, each client has direct access to the other clients served by the same access point which perform full mesh architecture. The local mesh uses an AhR as a gateway to access remote clients. In this level of the AHCRN, where the area covered by the AhR is relatively small, the effects of the environment (e.g., multipath propagation and hidden node) can be neglected and accordingly the sensing information may be assumed to be the same in the local clients mesh. Hence, AhCs can perform cooperative sensing and report their corresponding AhRs with the sensing information to allocate vacant spectrum portions to the clients opportunistically. This implies that a control signaling channel should be fully supervised by the AhR to perform spectrum management tasks. Indeed, no data transmission will be handled by the access point, unless the destination is located outside the local mesh. Consequently, the client network is a full mesh network with a maximum of $\binom{n}{2}$ links required. To provide these resources, a local mesh has to utilize multiple spectrum bands and opportunistically allocate vacant spectrum portions by continuously sensing the primary network activities.

3.2. ISS Philosophy

CRN transmission time (***T***) is the key performance parameter to achieve the maximum benefit from the opportunistic approach of the unlicensed CRN. To capitalize on this, a novel scheduling concept is proposed in this section according to the operating environment and the architecture of the CRN presented in 3.1.1 and 3.1.2, respectively.

As illustrated in Section 2.3 and Figure 4, increasing the CRN transmission time (***T***) will efficiently increase the CRN efficiency; however, long CR transmissions may dramatically increase the interference probability to the primary networks, decreasing their capacity. This tradeoff necessitates the designers to calculate the optimal CRN transmission period. Taking into consideration the varying nature of the radio environment, the system will need to recalculate this parameter frequently, which will further increase the computational load on the CRN.

The concept of the proposed scheduling algorithm is to divide the spectrum sensing tasks into groups according to the available spectrum bands. Therefore, each sensor (CR Client) has to sense the presence of the primary users in all bands. Ultimately, the sensing information is collected and processed at the centralized point (base station in the case of IBCRN or AhR in the AHCRN); from the perspective of the CRNs, it is arbitrary which source provides the sensing information. The main objective of this new scheduling algorithm is to enable CRNs to sense the spectrum band from CR client use different band for their actual transmission. Thus, each client senses its own spectrum band while providing sensing information to other CR clients, using other bands interchangeably, hence the suggested name of Interchangeable Sensing Scheduler (ISS) algorithm. Figure 8 demonstrates the operation of the ISS algorithm. With $\boldsymbol{n}$ available spectrum bands, each sensor allocates sensing slots to provide sensing information for the other band while benefiting from the other clients scanning its spectrum. This approach enables CRN clients to transmit and sense simultaneously.

3.3. Performance of ISS

Applying ISS will dramatically decrease the probability of interference. Both $\boldsymbol{I}_{Idle}$ and $\boldsymbol{I}_{Busy}$ will improve due to a virtual decrease in transmission time $\boldsymbol{T}$ from the primary network perspective, by having more sensing slots from the interchanged cooperative sensors during the transmission time of the CRN clients. Consequently, from the point of view of the CRN, $\boldsymbol{T}$ can safely be increased without affecting the performance parameters.

ISS performance is highly dependent on the number of CRN clients involved in the algorithm. Hence, the more users operating in different radio spectrum band, the more interchangeable sensing information, which provides more precise decisions.

Sensing time offset among different bands, as shown in Figure 7, have to be carefully managed by the central controller in order to harmonize the sensing and transmission period for each band. To have equivalent sensing intervals during the transmission period of the CRN clients, the central controller has to synchronize the sensing tasks

Figure 8. Interchangeable sensing scheduler approach

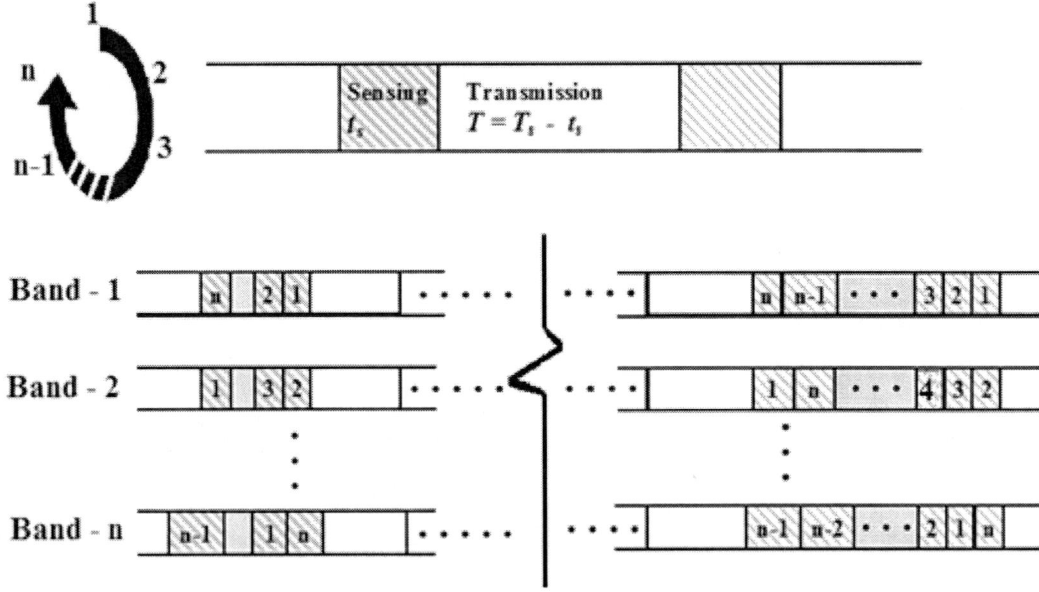

offset time among the clients, and this synchronization can be modeled in matrix as:

$$\begin{pmatrix} S_{11} & \cdots & S_{1m} \\ \vdots & \ddots & \vdots \\ S_{n1} & \cdots & S_{nm} \end{pmatrix}$$

where S_{ij} represents the sensing interval, m is the number of time slots sensed in the same frequency band, and n is the number of spectrum bands utilized by the CRN.

For a specific spectrum band, a group of m CRN clients synchronously senses the presence of primary users. Synchronization is needed to avoid false alarms intended when other clients are transmitting during the sensing period as described in Section 2.3. Therefore, the ordinary cooperative spectrum sensing scheduler is a special case of the proposed ISS when $n=1$. However, the usefulness of ISS is recognized when a multi-band environment is considered as in section 3.1.1 and Figure 7.

Each CRN client contributes to the sensing information matrix by assigning sensing intervals to different spectrum bands than the one used for its own data transmission. These sensing measurements are concatenated to form the sensing information for the whole spectrum (as in wideband sensing). Indeed, this approach provides multi dimension spectrum sensing information instead of just sensing before transmission as in the one dimension sensing algorithms. Thereafter, the central controller constructs the sensing information matrix, which helps the CRN to properly and instantaneously allocate idle spectrum to its clients as needed.

From the central controller point of view, the periodic cycle for each band is identified as in Figure 9. As shown in the figure, sensing intervals conducted by other CR-CPEs will split the transmission time in n sections. This segmentation does not affect CRN transmission efficiency since it is performed by other users.

4. MODELING OF THE ISS

In this section, behavioral simulation is implemented to examine the operation of the proposed algorithm. Essentially, ISS is sensitive to CRN activities as well as the primary network activities. Hence, the sensing algorithm must consider the traffic pattern of both networks. More specifically, the number of sensing slots assigned to CRN clients is dependent on its utilization. For instance, as Figure 10 shows, when the client generates coherent traffic, the base station can assign a single sensing schedule interval in between the transmitting packets. In contrast, for burst traffic, several sensing slots can be fitted between successive bursts.

*Figure 9. Sensing band **i***

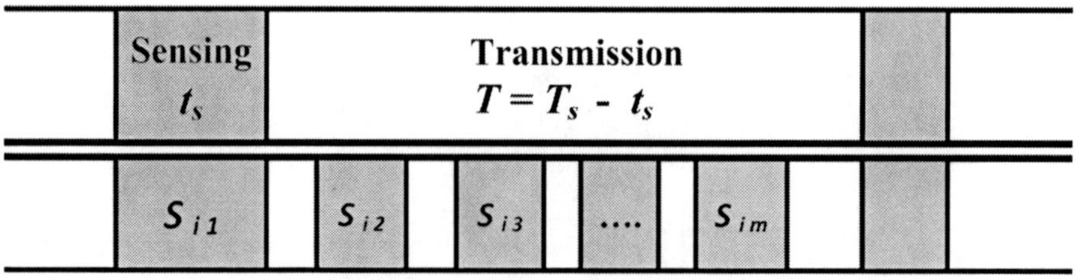

Sensing t_s | Transmission $T = T_s - t_s$

S_{i1} S_{i2} S_{i3} .... S_{im}

Figure 10. Sensing time slot in burst and coherent data traffic pattern

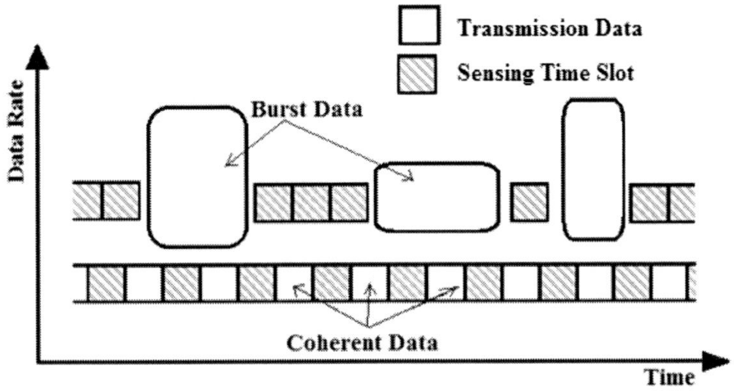

4.1. Theoretical Modeling

Mathematically, the expected interference is given as the linear combination of the interference in the busy (I_{Busy}) and the idle (I_{Idle}) states of the primary user as in Equation 10. Both are highly sensitive to the transmission period of the CR-CPE (***T***) as shown in Equations 8 and 9. The ISS algorithm virtually divides this period into ***n*** splits, where ***n*** is the number of radio spectrum bands as:

$$T_{ISS} = T / n \tag{12}$$

For Equations 8-11, when replacing the transmission time with the new virtual split time after adopting ISS (T_{ISS}) as in Equation 12, the performance parameters, regardless of the other factors, will follow the multiplicative inverse function. Thus, the performance of the algorithm improves based on the number of sensing bands, while the number of cooperative users scanning each band will further improve the accuracy of the sensing performance.

4.2. Behavioral Simulation

To illustrate the operation of the ISS algorithm, a random channel status has been generated according to the Poisson prossess with birth and death rates of 1 and 9, respectively. A behavioral simulation has then been conducted to demonstrate the operation of the spectrum sensing task, while the CRN transmits data in order to increase the throughput of the CRN. In this example, since the transmission time (***T***) is three times as high as the Sensing time (t_s), the CRN effeciency is 15/16= 93.75%. In contrast, this time extension does not affect the primary networks, since the sensing tasks are performed every four bits. Hence, from the point of view of the primary networks, the continuous transmission time is four bits. The complete behavioral simulation output is presented in Figure 11.

Figure 12 represents the first three (out of 1024 slots) time slots for the transmission of the primary users. Each row represents the spectrum band assigned to a specific licensed user, e.g., in the case of this simulation, there are four spectrum bands. From the channel status in Figure 12, it can be seen that the first band has not been utilized in the first and the third time slot, as represented by all the zeros.

Consequently, the sensing matrix is constructed by the ISS algorithm, representing the channel status and determining the operating band of the client which detects primary user activity. As shown in Figure 12, the first and the third time slot of the first band appear vacant and can be utilized by the CRN since none of the clients

Figure 11. Operation of ISS in four spectrum bands

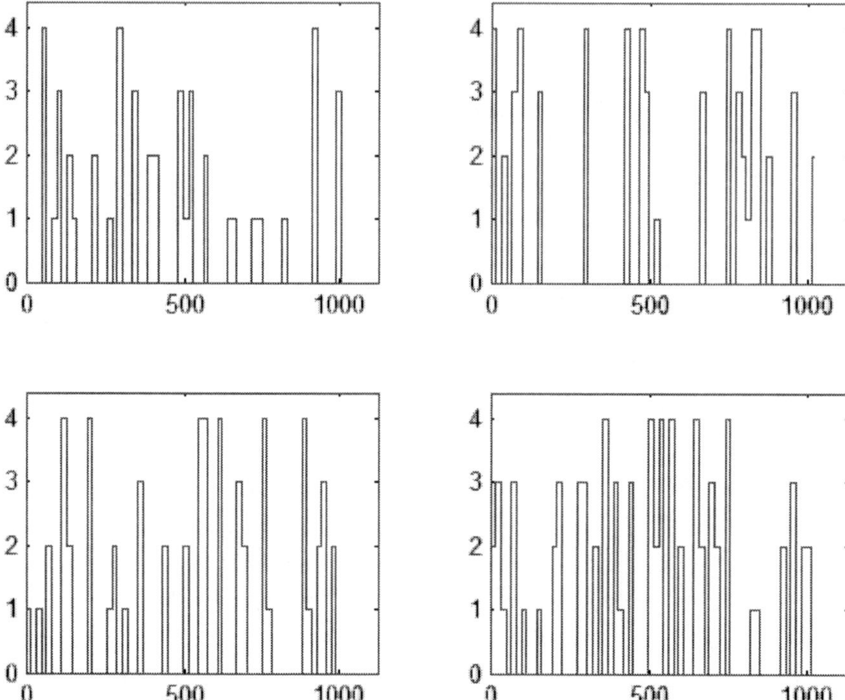

detect the primary signal. As in the second band of the first time slot, primary user activity has been detected by a CR-CPE operating in band four. This is due to the first activity (represented by a logic '1') appearing at the 14th bit, sensed by clients in band four as shown in Figure 13.

As unwanted results from the constructing of the sensing matrix in Figure 12, the CRN will affect the primary network due to some of the activities of the licensed users not being detected by the algorithm. For instance, in the second and third bands of the second time slot, none of the CR-CPEs detect the presence of the primary users because of their appearing on transmission bits, not having been sensed by any of the interchangeable sensing slots. However, considering ISS performance in terms of detecting primary user activity, it is much better to use the same client for sensing and transmission, as illustrated in the next subsection.

4.3. Performance Evaluation

The graph in Figure 14 demonstrates the contribution of ISS with regards to the interference probability. This result is obtained by comparing the interference probability before and after applying ISS. As can be seen in the figure, with two interchangeable bands, ISS decreases the expected interference by 50% when the primary network activities are very low. The improvement is increased to about 75% of decreased interference, when using four bands with interchangeable scheduling. As the primary network increases its spectrum utilization, ISS can still perform about a 30% of decrease over the legacy cooperative approaches. A numerical analysis has been conducted 1000 times for 1024 bits per spectrum band for the CNR transmission.

Figure 12. Construction of the sensing matrix

```
********************************************************
Channel Status
********************************************************
1st Time Slot
  0  0  0  0  0  0  0  0  0  0  0  0  0  0  0  0
  0  0  0  0  0  0  0  0  0  0  0  0  0  1  0  0
  0  0  1  0  0  0  0  0  0  0  0  0  0  0  1  0
  0  0  0  0  0  0  0  1  0  0  0  0  1  0  0  0

2nd Time Slot
  0  0  0  1  0  0  0  0  0  0  0  0  0  0  0  0
  0  0  1  0  0  0  0  0  0  0  0  0  0  0  0  0
  0  0  0  1  0  0  0  0  0  0  0  0  0  0  0  0
  0  0  0  0  0  0  0  0  0  0  1  1  0  0  1  0

3rd Time Slot
  0  0  0  0  0  0  0  0  0  0  0  0  0  0  0  0
  0  0  0  0  0  1  1  0  0  1  0  0  0  1  0  0
  0  0  1  0  0  0  0  0  0  0  0  0  0  0  0  0
  0  1  0  1  0  0  0  0  0  0  0  0  0  0  1  0

************************************************************

Sensing Matrix
************************************************************
1st Time Slot
  0  0  0  0  0  0  0  0  0  0  0  0  0  0  0  0
  4  4  4  4  4  4  4  4  4  4  4  4  4  4  4  4
  1  1  1  1  1  1  1  1  1  1  1  1  1  1  1  1
  2  2  2  2  2  2  2  2  2  2  2  2  2  2  2  2

2nd Time Slot
  0  0  0  0  0  0  0  0  0  0  0  0  0  0  0  0
  0  0  0  0  0  0  0  0  0  0  0  0  0  0  0  0
  0  0  0  0  0  0  0  0  0  0  0  0  0  0  0  0
  3  3  3  3  3  3  3  3  3  3  3  3  3  3  3  3

3rd Time Slot
  0  0  0  0  0  0  0  0  0  0  0  0  0  0  0  0
  2  2  2  2  2  2  2  2  2  2  2  2  2  2  2  2
  1  1  1  1  1  1  1  1  1  1  1  1  1  1  1  1
  1  1  1  1  1  1  1  1  1  1  1  1  1  1  1  1
```

5. IMPLEMENTATION ISSUES OF ISS

The positive impacts of the proposed scheduling algorithm in reducing the expected interference and increasing the opportunities to find vacant spectrum slots have been shown in the previous section. This section demonstrates other advantages of the ISS algorithm and how they pertain to implementation. Indeed, the performance of the proposed scheduling algorithm depends on several factors which form research challenges listed below.

5.1. Supports CPE Diversity

The proposed scheduling algorithm pertains both to the CRN activities as well as the primary network activities. Hence, the CRN first distributes the sensing tasks among its CRN clients according to their computational power and the type of

Figure 13. Sensing matrix legend

Packet	1	2	3	4	5	6	7	8	9	10	11	12	13	14	15	16
1st Band	1				2				3				4			
2nd Band		1				2				3				4		
3rd Band			1				2				3				4	
4th Band				1				2				3				4

Figure 14. Performance of ISS in multiband environment

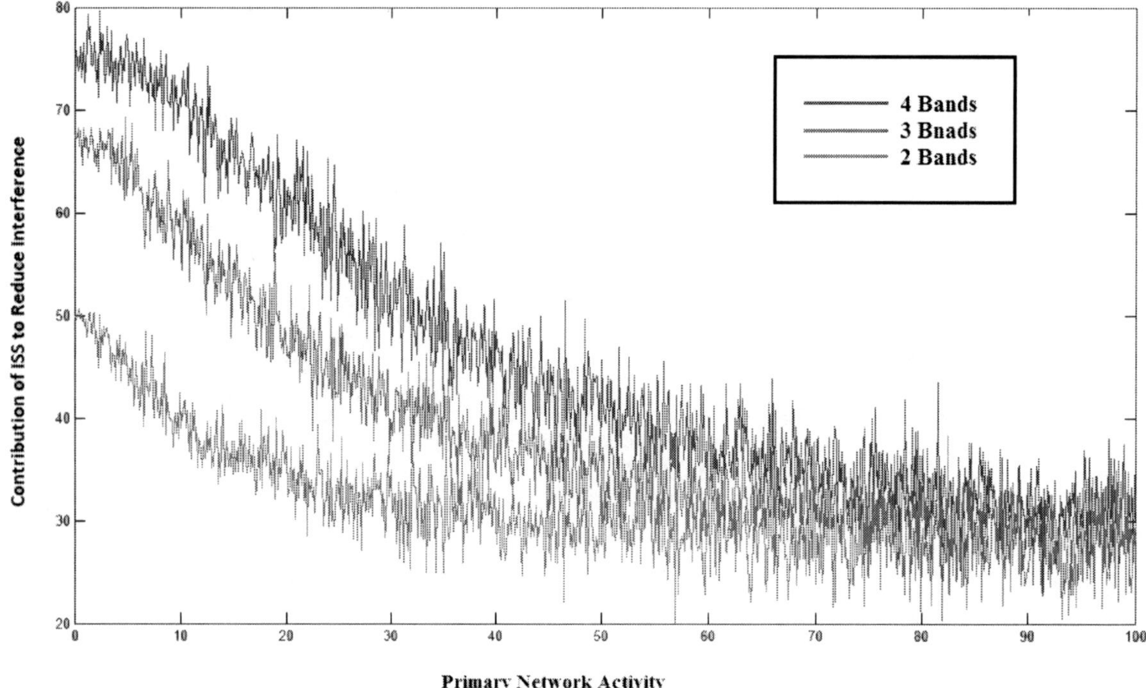

application used. Further, its aim is to utilize the shared radio spectrum with the primary networks. The amount of sensing tasks assigned to the CR clients is dependent on the instantaneous ability and the utilization of their transponders. Therefore, different customer devices can be involved in the CRN with no emphasis on their capabilities. Consequently, the central controllers have to manage this diversity which offers, from a marketing point of view, a wide variety of customer devices according to their affordability and needs.

5.2. Inter-Band Softer Handover

When idle spectrum slots are registered in a central database at the centralized controller, the controller then can decide on the communication parameters used for the secondary network depending on the sensing information sent by the CR-CPEs. Therefore, the controller can assign a primary channel on a specific band and other auxiliary channels depending on availability. Subsequently, in case of a primary user being detected, the controller can perform Inter-Band Softer Handover (IBSHO) to one of the auxiliary channels. By doing so, CRN can guarantee a certain Quality of Service (QoS) level to its customers, particularly for operators providing real-time applications which are very sensitive to the stability of the communication channel. Therefore, the adaptation process not only depends on the access layer conditions, but also the requirements of the upper layers.

5.3. Hybrid Sensing Schemes

As mentioned in Table 1, there are several schemes to detect primary user activities, each of which has its own specific pros and cons. Thus, according to the situation of the CRN, any one of the schemes can be applied for each sensing slot; hence, the central controllers may assign different sensing schemes in different sensing slots, forming Hybrid Sensing Schemes (HSS) to maximize the CRN utilization.

5.4. Optimal Number of CR Clients

Even though Figure 14 shows a decrease in the expected interference by increasing the number of CR-CPEs, a continuous increase of CR-CPEs serviced under a single controller, will eventually lead to an over-utilization of the RF spectrum in the vicinity. Therefore, the size of the network should be carefully adjusted to serve the optimal number of CR clients. Furthermore, considering the varying nature of primary networks and CRNs, the optimal number of clients changes according to the instantaneous traffic in both networks. Hence, when implementing ISS in a geographic area covered by a cluster of central controllers, the cell size of each controller can be adaptively modified to balance the traffic amount, which can be performed by Adaptive Cell Breathing (Bejerano & Seung-Jae, 2009).

5.5. CRN Cross-Layer Considerations

The proposed algorithm combines CRN activities with primary user activities. Thus, ISS utilizes the CRN first and then shared spectrum with the primary networks. Therefore, for the CR system to fit in accordance with the variations of the primary networks, it is better for the CRN to deliver various applications with diverse traffic patterns, which can be accomplished by offering diverse packages to the customers. This diversity will further support the harmonization of the CRN operation, and therefore the upper layer data flows will affect, and be affected, by the performance of the CRN. Consequently, simulation platforms are needed to examine the cross layer influence on the ISS performance.

6. CONCLUSION

The development of the CR architecture has opened the doors for communication networks to be implemented with a minimum of available radio spectrum resources. Consequently, the command-and-control spectrum management policies will no longer apply and the time has come to develop intelligent networks which can efficiently utilize the resources and fairly allocate radio spectrum on a usage basis.

This chapter has addressed the spectrum sensing issue as a crucial routine for the unlicensed CRNs to co-allocate with licensed networks. Indeed, the CRNs share the spectrum bands in an opportunistic manner with the primary networks. So far, the primary networks do not cooperate with the CRNs and regard their signals as noise. This hypothesis puts the coexistent load on the CRNs, which consume considerable computational resources. From another perspective, this hypothesis facilitates a fast emergence of CRNs, without any modifications to the existing networks. Thus, this approach is helpful in the early stages of deploying the primary/secondary spectrum sharing paradigm. However, in the near future when the concept matures, it would beneficial for the primary networks to share information with the CRNs in order for them to gain a better understanding of the surrounding environment. In addition, CRNs would then have to enroll in a light regulatory framework such as a *General Authorization* and compensate for some of the cost of the individual licensed operators, encouraging the cooperation of the primary network operators with the CRNs.

This chapter has also proposed a novel spectrum sensing scheduling algorithm, named Interchangeable Sensing Schedule (ISS). The impact of this algorithm in reducing the interference with the primary networks while increasing the opportunity of the AHCRN to detect a vacant spectrum slot has been shown. Furthermore, other implementation benefits of ISS have been considered and compared with their operational necessities. Since this algorithm presents new scheduling concepts, various open issues remain for researchers in order to implement ISS to gain a further understanding of its behavior in a real world environment. Moreover, ISS introduces new CRN procedures such as IBSHO and HSS, which need a further discussion.

REFERENCES

Akyildiz, I. F., Brandon, F., & Balakrishnan, R. (2011). Cooperative spectrum sensing in cognitive radio networks: A survey. *Journal of Physical Communication*. Retrieved January 16, 2012, from http://www.elsevier.com/locate/phycom

Akyildiz, I. F., & Lee, W. Y. (2006) NeXt generation/dynamic spectrum access/cognitive radio wireless networks: A survey. *Computer Networks Journal*. Retrived March 21, 2012, from http://www.sciencedirect.com

Bejerano, Y., & Han, S.-J. (2009). Cell breathing techniques for load balancing in wireless lans. *IEEE Transactions on Mobile Computing*, *8*(6). doi:10.1109/TMC.2009.50

Cabric, Mishra, & Brodersen. (2004). Implementation issues in spectrum sensing for cognitive radios. In *Proceedings of the Thirty-Eighth Asilomar Conference on Signal, System, and Computer,* (Vol. 1, pp. 272-276). Berkeley, CA: Berkeley Wireless Research Center.

Chen, Y., Zhao, Q., & Swami, A. (2008). Joint design and separation principle for opportunistic spectrum access. *IEEE Transactions on Information Theory*, *54*(5), 2053–2071. doi:10.1109/TIT.2008.920248

Federal Communications Commission. (2003a). *Establishment of an interference temperature metric. FCC 03-289*. Washington, DC: FCC.

Federal Communications Commission. (2003b). *ET docket no 03-222 notice of proposed rule making and order*. Washington, DC: FCC.

Krenik, W., & Batra, A. (2005). Cognitive radio techniques for wide area networks. In *Proceedings of Design Automation Conference 42nd*, (pp. 409-412). IEEE.

Kumar, S., & Sengupta, J. (2010). AODV and OLSR routing protocols for wireless ad-hoc and mesh networks. In *Proceedings of the International Conference on Computer & Communication Technology, ICCCT,* (pp. 402-407). ICCCT.

Lee, W.-Y. Kausbik, Cbowdbury, & Mebmet. (2008). Spectrum sensing algorithms for cognitive radio networks. In Y. Xiao & F. Hu (Eds.), *Cognitive Radio Networks*. New York, NY: Auerbach Publications.

Lee, W.-Y., & Akyildiz, I. F. (2008). Optimal spectrum sensing framework for cognitive radio networks. *IEEE Transactions on Wireless Communications*, *7*(10), 2845–3857.

Mitola, J. III, & Maguire, G. Q. Jr. (1999). Cognitive radio: Making software radios more personal. *IEEE Personal Communications Journal*, *5*(4), 13–18. doi:10.1109/98.788210

Wild, B., & Ramchandran, K. (2005). Detecting primary receivers for cognitive radio applications. In *Proceedings of DySPAN First IEEE International Symposium*, (pp. 124-130). IEEE Press.

Wu, A., Yang, C., & Huang, D. (2010). Cooperative sensing of wideband cognitive radio: A multiple-hypothesis-testing approach. *IEEE Transactions on Vehicular Technology*, *59*(4), 1835–1846. doi:10.1109/TVT.2010.2043967

Chapter 8
MAC Layer Protocols for Cognitive Radio Networks

Lokesh Chouhan
ABV-Indian Institute of Information Technology and Management (ABV-IIITM), India

Aditya Trivedi
ABV-Indian Institute of Information Technology and Management (ABV-IIITM), India

ABSTRACT

In the last few decades, the Cognitive Radio (CR) paradigm has received huge interest from industry and academia. CR is a promising approach to solve the spectrum scarcity problem. Moreover, various technical issues still need to be addressed for successful deployment of CRNs, especially in the MAC layer. In this chapter, a comprehensive survey of the Medium Access Control (MAC) approaches for CRN is presented. These MAC technologies under analysis include spectrum sharing, multiple antenna techniques, cooperation, relays, distributed systems, network convergence, mobility, and network self-optimization. Moreover, various classifications of MAC protocols are explained in this chapter on the basis of some parameters, like signaling technique, type of architecture, sharing mode, access mode, and common control channel. Additionally, some case studies of 802.11, 802.22, and Mobile Virtual Node Operator (MVNO) are also considered for the case study. The main objective of this chapter is to assist CR designers and the CR application engineers to consider the MAC layer issues and factors in the early development stage of CRNs.

1. INTRODUCTION

Frequency spectrum is a limited resource for wireless communications and may become congested owing to a need to accommodate the diverse types of air interface used in cognitive radio networks. However, since conventional wireless communications systems also utilize the frequency bands allocated by the Telecom Regulatory Authority of India (TRAI) and Federal Communications Commission (FCC) in a static manner, they lack adaptability in the existing framework (Telecom Regulatory Authority of India, 2012). Also, several studies show that while some frequency bands in the spectrum are heavily used, other bands are largely unoccupied most of time. These latent

DOI: 10.4018/978-1-4666-2812-0.ch008

vacant spectrums result in the under-utilization of available frequency bands. To defeat the over-crowding, different governing and non governing agencies and organization such as TRAI and FCC have been investigating new ways to manage Radio Frequency (RF) resources (FCC, 2003).

Today's the fixed spectrum assignment policy is used to characterize wireless networks. However, a huge portion of the allocated spectrums are used rarely and the environmental variation in the utilization of assigned spectrum ranges from 15% to 85%. The recent years have been seen major and remarkable development in the field of Cognitive Radio Network (CRN) technologies. Cognitive radio is a revolutionary technology; they guarantee to enhance the utilization of radio frequencies and make room for new and additional commercial data, emergency and military communication services etc. (Haykin, 2005). The influence of CRN's functions on the performance of the upper layer protocols such as routing and transport are complex and open research issues in these areas are also challenging in CRN (Cesana, 2011). Therefore, there is always requirement of CR-MAC (Cognitive Radio Medium Access Control) protocol to provide efficient, fair, and seamless services to user. One of the important factors which should be considered during design process of CRN is MAC layer issue (Chouhan, 2011).

1.1. Layered Architecture of Cognitive Radio Network

The introduction summarized above necessitates new communication protocols to be designed for awareness of the spectrum in CRN. This direct association requires the new MAC layer design in the entire CR networking protocol stack. The effects of the preferred spectrum bands and the variations due to spectrum mobility need to be carefully monitored in the design of these MAC protocols. The rapidly changing radio environment, more radio channels to utilize, number of factors to select during decisions taken by MAC and routing protocols, etc., makes design of CRN very challenging (Chouhan, 2011).

Typically, CRN consists of two types of users, i.e., Primary User (PU) and Secondary User (SU). PU or licensed user is the legitimate user who has more priority. SU is opportunistic user, who has less preference than the primary users. SU does not interrupt the transmission of the PU. So there is always requirement of dynamic algorithm to create separation between the transmission of the PU and SU. In the Figure 1, five layer architecture of CRN is shown. This layered architecture, is based on OSI reference model which represents bridge between the existing network architecture and the new CRN. Unlike from the existing models, spectrum sensing, spectrum sharing, scheduling, reconfiguration, Quality of Service (QoS), and mobility are very critical and challenging in the CRN (Akyildiz, Lee, Vuran, & Mohanty, 2006; Haykin, 2005). Therefore, CRN architecture requires spectrum management and spectrum mobility functions additionally (represented by vertical blocks in the Figure 1). The Second layer of this architecture is the MAC layer, which comprises two important functions: spectrum sensing and spectrum sharing of the CRN. Moreover, this architecture is primarily focused on the four basic functions (Akyildiz, Lee, Vuran, & Mohanty, 2006; Haykin, 2005):

1. **Spectrum Sensing:** Sensing is to identify vacant spectrum bands and distributing this spectrum with the tolerable interference to the PUs.
2. **Spectrum Management:** Selecting the best offered spectrum band to achieve high Quality of Service (QoS) requirements.
3. **Spectrum Mobility:** Preserving seamless and faultless network requirements during switch to better spectrum.
4. **Spectrum Sharing:** Offering the flexible and fair spectrum scheduling scheme simultaneously among SUs.

Figure 1. Layered architecture for cognitive radio networks

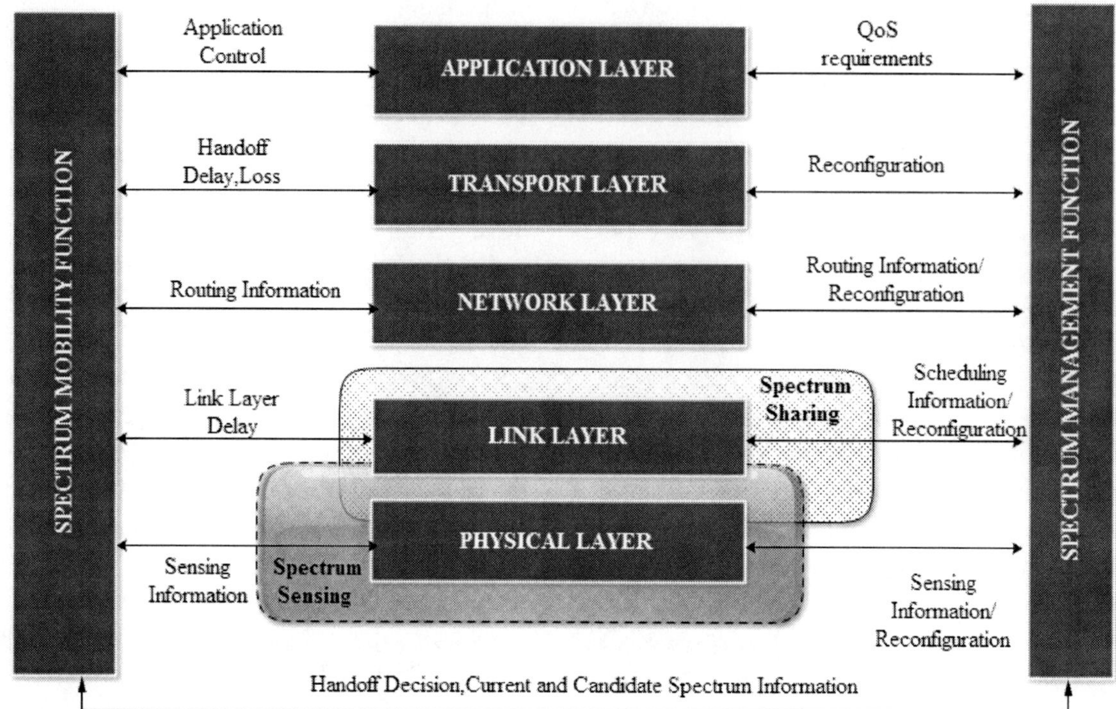

Moreover, sensing information, link layer delay, routing information, handover delay, and application control are the key issues of mobility management for physical, link, network, transport, and application layer respectively. Whereas, scheduling information, routing information, reconfiguration, and QoS requirements are important issues of spectrum management (Dubey, 2012)(Rajni Dubey, 2012). These issues are also discussed in the second and third sections of this chapter.

1.2. Organization of Chapter

Organization of this chapter as follows. In the second section, MAC layer protocol architecture is elaborated. Various issues and challenges of MAC layer are discussed in the third section. Classifications of various MAC schemes are described in the section four. In the fifth section, priority based MAC modelling are explained. Some of the case studies are described for CR-MAC scheme in the sixth section. Chapter is concluded in the last section.

2. SPECTRUM SENSING IN COGNITIVE RADIO NETWORK

Key factors of the CRN are the cognitive capability, sensing, availability of spectrum and transmission power, bandwidth, modulation techniques, learn, and awareness of radio environment and characteristics of the channel, user requirements and applications, available networks (infrastructures) and nodes, local policies and other operating restrictions. In the spectrum sensing, the radio scene and thereby estimating the power spectra of incoming RF stimuli, we can broadly classifying the spectrum into three types (Haykin, 2005; Hu & Yang, 2012):

1. **Black Space:** Spectrum bands which are engaged by PUs (high-power local interferers) some of the time.
2. **Grey Space:** Spectrum bands which are partially engaged by PUs (low-power interferers).
3. **White Space or Spectrum Hole:** Spectrum bands which are not occupied by any PUs, and free of RF interferers excluding the ambient noise.

Cognitive cycle is the four-stage process, i.e., spectrum sensing, spectrum analysis, spectrum decision, and spectrum mobility (see Figure 2). CR observes the offered spectrum bands, captures their information, and then identifies the white spaces (Akyildiz, Lee, Vuran, & Mohanty, 2006).

Spectrum sensing is a useful concept for discovering spectrum opportunities for secondary spectrum usage in real-time. Main objective of the sensing is the identification of unused spectrum (detection of spectrum holes) or to detect the primary user's transmission. When any two users want to communicate, then transmitter and receiver are responsible for sensing operation; following operations are performed by them:

- Source and destination choose a set of channels to sense.
- Source and destination estimate channel availability.
- Channel filtering is performed, and a communication path is set up.

Both reactive (on-demand) and proactive sensing may be exploited in a cognitive network. Sensing performance is limited by hardware and physical constraints. For instance, secondary nodes with a single transceiver cannot transmit and sense simultaneously. Moreover, users usually only observe a partial state of the network to limit sensing overhead. There is a fundamental trade-off between the undesired overhead and spectrum holes detection effectiveness: the more bands are

Figure 2. Cognitive radio cycle

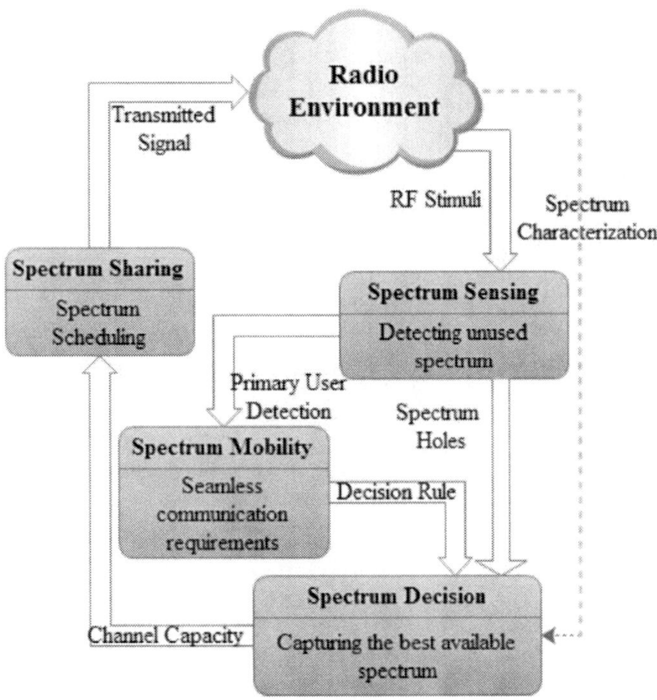

sensed, the higher the number and quality of the available resource. In the next section, classification of spectrum sensing is described.

2.1. Classification of Spectrum Sensing

In Arslan (2009), most justified classification is described by the authors. In this classification, authors have not only classified the sensing but also mentioned various approaches, enabling algorithms, issues, and challenges related to the spectrum sensing, see Figure 3. Broadly, spectrum sensing can be classified into three parts: transmitter detection, cooperative detection, and interference based detection, see Figure 4 (Akyildiz, Lee, Vuran, & Mohanty, 2006). Let us discuss each classification.

2.1.1. Transmitter Detection

In this type of the of detection technique, SU locally observes weak signal from a PU's transmitter. This detection model can also be defined by two hypotheses (Akyildiz, Lee, Vuran, & Mohanty, 2006):

$$r(t) = \begin{cases} n(t), & \mathrm{H}_0 \\ hs(t) + n(t), & \mathrm{H}_1 \end{cases}$$

where *r(t)* is received signal by SU, s(t) is the transmitted signal by PU, *n(t)* is the noise and *h* is the amplitude gain of the channel. H_0 is the null hypothesis, which states that there is no PU signal in the particular spectrum. Alternative hypothesis defined by H_1, which specifies existence of some

Figure 3. Various aspects of spectrum sensing

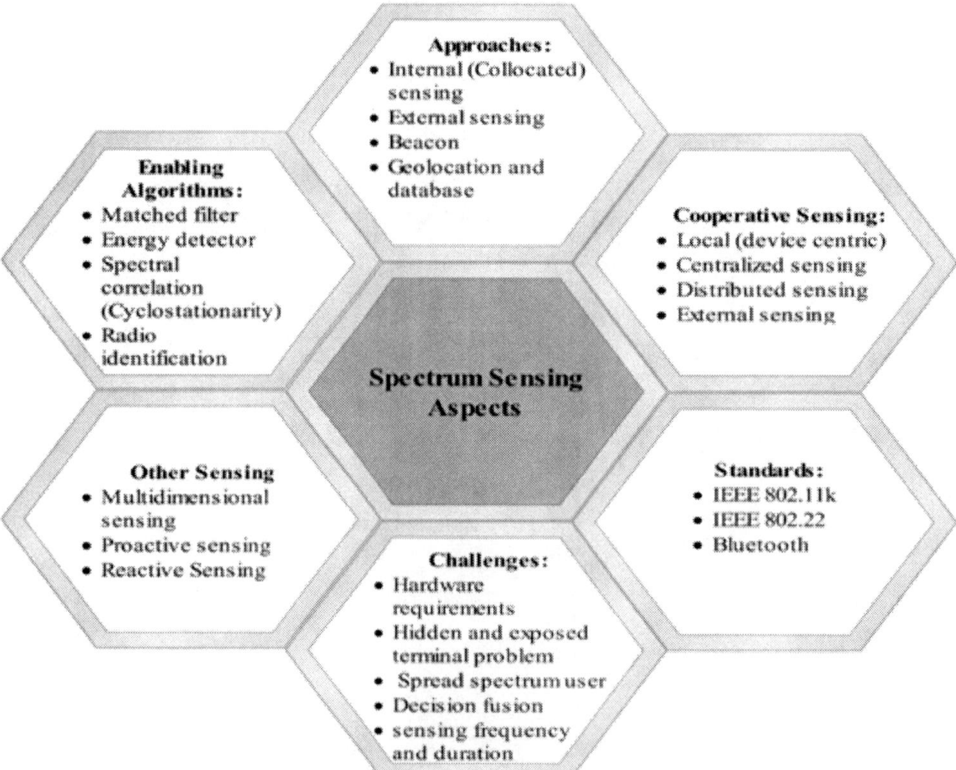

Figure 4. Spectrum sensing classification

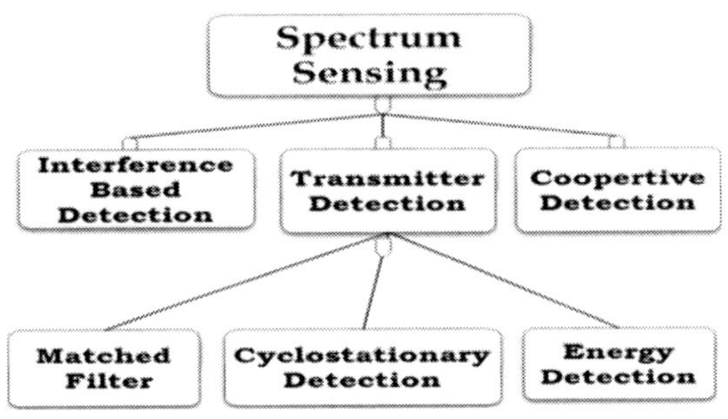

PU in the certain spectrum band. This transmitter detection can be implemented by three ways: matched filter detection, energy detection and cyclostationary feature detection, see Figure 4.

Among various spectrums sensing schemes such as matched filter detection (Sahai, 2004) and the cyclostationary feature detection (Sutton, 2008), energy detection is the most popular method discussed in the literature (Urkowitz, 1967), which is also called the conventional energy detector. The conventional energy detector measure square of the magnitude of received sample. The energy detector is a non coherent device and performance of conventional energy detector (Urkowitz, 1967), can be improved by replacing the squaring operation of the received signal amplitude with an arbitrary power operation p (Chen, 2010). In Singh (2011), performance of cooperative spectrum sensing with an improved energy detector was discussed. In this work, optimal number selection method of cooperative CRs needed for optimum detection of the PU for the optimization of the sensing threshold and the energy detector in order to minimize the total error rate are discussed.

2.1.2. Cooperative Detection

In the previously explained detection scheme, the SU detects the PU's transmission separately by their local observations. Whereas in case of cooperative detection scheme, information is collected from multiple SUs and this shared information incorporate to detect the PU's transmitter signal. This scheme can be implemented either by the distributed or by the centralized manner.

Centralized detection gathers all sensing information by the central unit (SU's base station), and each of the associated SU sends the information to this base station about the PU's presence. Sensing information is collected by the base station of SU from cognitive devices. Consequently, SU determines the available spectrum, and broadcasts this information to other CR users or directly manage the CRN traffic.

On the other hand, in the distributed detection scheme, an appropriate handshake operation is required among various SUs. This procedure includes observation information from the each of the participant SU. Although cognitive nodes share information with each other, they generate their individual choice as to which portion of the spectrum band they can utilize. Distributed sensing is superior to centralized sensing, because distributed detection is infrastructure-less and is also relatively cheaper. Distributed sensing solutions are also described in Gandetto (2005a, 2005b, 2007).

Cooperative spectrum sensing is more successful when associated SUs separately monitors shadowing or fading. The performance reduction due to correlated shadowing is examined in Pawełczak (2006) and Sousa (2007) with respect to missing the opportunities. This paper showed that it is more beneficial to have the same amount of users collaborating over a large area than over a small area. In order to reducing shadowing, directional antennas and beamforming can also be applied (Cabric, 2006). Cooperation with every user in the CRN does not necessarily obtain the optimal throughput and SUs with the highest PU's Signal to Noise Ratio (SNR) are preferred for the cooperation (Liang, 2007). Moreover, constant detection rate and constant false alarm rate are also used for optimally choosing the users for cooperative detection scheme (Liang, 2007). Following are the advantages of cooperative sensing scheme (Akyildiz, Lee, Vuran, & Mohanty, 2006; Jha, Rashid, Bhargava, & Despins, 2011; Arslan, 2009):

- This can solve various problems that occur in other the spectrum detection methods due to noise uncertainty, fading, and shadowing. Consequently, it can significantly reduce the probabilities of false alarm and misdetection.
- Theoretically, cooperative detection scheme among SUs is more precise since the uncertainty and ambiguity in a single user's detection can be reduced.
- Cooperation not only solves the hidden primary user problem, but can also reduce the sensing time.

However cooperative detection offers very precise sensing performance, it also causes bad effect on limited resources networks, because of the additional overhead and traffic. Moreover, lack of the primary receiver location knowledge influences the PU's receiver uncertainty problem which is still unresolved in the cooperative detection scheme (Cormio & Chowdhury, 2010; Lee & Wong, 2011).

2.1.3. Interference-Based Detection

One of the most important factors that affect the performance of sensing is the interference tolerance by the PU. Consider the example, when a SU is utilizing opportunities in the TV band, then sensing should be done as frequently as possible in order to prevent any interference. Therefore, the FCC proposes the model for measuring interference, also known as interference (FCC, 2003). This model represents the signal of a radio station considered to work in the range at which the received power nearly to the level of the noise floor. The interference temperature model handles interference at the destination through the interference temperature limit, which is defined by the summations of the new interference that the destination could accept. Because, interference is typically originated in a source-centric and this interference can be managed through location of individual transmitters, the out-of-band emissions, and the radiated power. However, the interference actually takes place at the receiver's side.

The interference temperature model also comprises several limitations in measuring the interference. In Brown (2005), the interference is recognized as the expected number of PUs interrupted by the SU's actions and operations. This scheme has investigated various aspects, i.e., transmission power, antennas, capability to detect licensed channels, the type of SU's modulation, and activation level of SUs and PUs. In addition, this model also explained the interference interrupted by a single SU and did not conceal the multiple SUs effect. Moreover, if SUs are not able to detect the position of nearby PUs, the accurate interference cannot be determined by applying this scheme.

2.2. Challenges of Spectrum Sensing

Some of the important challenges of spectrum sensing are hardware requirements, hidden and exposed problem of PU, spread spectrum users, decision fusion, sensing frequency, and security, also depicted in Figure 3 (Arslan, 2009). Let us discuss each challenge one by one.

- **Hardware Requirements:** Spectrum sensing for CR applications necessitates huge sampling rate, better resolution Analog to Digital Converters (ADCs) with large dynamic range, and high speed signal processors. Noise discrepancy estimation techniques have been commonly applied for optimal receiver designs like the channel estimation, soft information generation *etc.*, as well as for better handover, power control, and channel assignment techniques. Spectrum sensing can be applied on three architectures: single-radio, dual-radio, and multi-radio. In the single radio scheme, only a specific time slot is assigned for the spectrum sensing. Due to this limited sensing duration, only small precision can be assured for spectrum sensing determination. However, the spectrum efficiency is minimized as various part of the offered time slot is used for sensing instead of the transmission of data. The observable benefit of single radio architecture is its simplicity and cheaper cost. In the dual radio sensing, dedicated single radio is specifically used for the data transmission, while the other one is used for controlling and managing the spectrum. The disadvantage of such a scheme is the high power consumption and cost of hardware.

- **Hidden and Exposed Problem of Primary User:** Multichannel hidden and exposed terminal problems are already explained in Section 4.2.1. Cooperative detection scheme is proposed in this chapter for handling hidden PU problem. We have already discussed cooperative sensing in previous subsection.
- **Decision Fusion:** In CRN, a SU can select the bandwidth, mode of the transmission, and data rate of the transmission. Then, the suitable spectrum band is selected as per the spectrum characteristics and quality of service. This is very critical challenge for the cooperative type of sensing. Because, in the cooperative sensing, soft or hard decision about the presence of primary transmission take place from the shared information which is collected from various cognitive users.
- **Sensing Frequency:** In the CRN, PUs can arrive to their spectrum bands anytime, no matter SU is effective on their bands or not. CR should capable enough to detect the existence of PUs as quickly as possible and should leave the band instantly to prevent the transmission interference to and from PUs. Therefore, sensing scheme must able to identify the existence of PUs within definite time interval. This condition produces a boundary on the performance and efficiency of the sensing algorithm and it also makes a challenge for CR design.
- **Spectrum Decision over Heterogeneous Spectrum Bands:** Presently, some spectrum bands are already allocated to various purposes while some bands are still unlicensed and unoccupied. So, the spectrums used by CRN are generally being an amalgamation of exclusively accessed spectrum and unlicensed spectrum. In the case of licensed bands, the SU requires to think about of the PU's operations in the spectrum analysis and decision phase, in order not to consume the bandwidth of the PU's transmission. However, superior and well defined spectrum sharing schemes are required in the unlicensed bands, since all the SU have same spectrum access privileges. In order to choose the best spectrum band over this heterogeneous environment, the CRN should assist spectrum decision activities on both the licensed and the unlicensed bands by allowing these various characteristics.
- **Security:** Due to the unique characteristics of CRN, such network is highly vulnerable to security attacks compared to wireless network or infrastructure based wireless Network. Cognitive Radios facilitates in the secondary usage of the licensed spectrum (when not in use). If primary user is using, then the secondary user cannot use it. Therefore, accurate spectrum occupancy information needs to be maintained by a secondary user. This minimizes the interference. A malicious user can try to falsify the spectrum occupancy information,

which may cause interference. The misbehaving nodes can be categorized as:

- ◦ **Selfish Node:** Seeks to maximize its own gains at the expense of others.
- ◦ **Malicious Node:** Acts to degrade the system or individual node performance with no explicit intention to maximize its own gains, and acts as a primary and transmit false information to the secondary user.

Some other important challenges interference temperature measurement, spectrum sensing in multi-user networks, and detection capability are also investigated by authors (Akyildiz, Lee, Vuran, & Mohanty, 2006).

3. ISSUES IN DESIGNING A MAC PROTOCOL FOR CRN

Similar to conventional wireless networks, CRNs also have various MAC issues and challenges. Moreover channel definition, channel availability; channel heterogeneity, channel quality, common control channel, and multichannel hidden and exposed terminal are the major challenges for designing MAC protocol for CR. Some of the issues and challenges are explicit to the wireless nature of CRNs, and few issues are specific to the ad hoc nature of the cognitive radio networks, also known as cognitive radio ad hoc networks (Zhang, 2010) (Yan Zhang, 2010). Issues related to both are discussed below.

3.1. Issues Related to Wireless Nature

In the past few decades, wireless network research has become one of the hot spot in emerging technologies. With the appearance of increasingly large-scale size of the networks, wireless networks become more and more difficult to describe complex phenomenon due to lack of dynamic and effective methods. Moreover, the capacity of signaling information exchanged in a CR network is significantly larger as compare to the conventional wireless network. In this section, issues related to the wireless nature are discussed.

3.1.1. Interference Temperature Measurement

Most important aspect in the designing of the MAC scheme of the CR system is the broadcast nature of the radio channel, i.e., transmission by any user always gives the interference to all the types of users within the radio range. The problem of this receiver detection model lies in effectively measuring the interference temperature. It also causes collision of packets, packet drop, and failure of transmission of the CR system, as result degradation in the performance of the CR system. Since multiple users contended for the channels concurrently, chances of the collision will be much high. Therefore, CR designers always try to minimize the interference, packet drops, collision rate, and error rate in the early of the development process (Dubey, 2012)(Rajni Dubey, 2012).

Various solutions are suggested to reduce these factors for the CR systems. In Qaraqe, Ekin, Agarwal, and Serpedin (2011), authors have examined the capacity of CRMAC over the dynamic fading environment. Nash equilibrium game based algorithm for SIR (Signal-to-Interference Ratio) with power control was described and a novel adaptive scheme was also proposed based on a space-time block coding MC-CDMA (STBC MC-CDMA) system in Cheng and Yang (2008). In Ma, Li, and Juang (2009), signal processing aspects of cognitive radio are described by the authors.

3.1.2. Need of Common Control Channel (CCC)

The capacity of signalling information exchanged in a CR network is significantly larger as compare to the conventional wireless network. Therefore, many CR solutions either centralized or distributed uses explicit or out of band MAC schemes to exchange the control information. These channels are physically separated from the data channels and responsible for the transmission of actual data, also known as common control channels. CR users share dedicated common control channels to exchange the signaling, synchronisation, sensing, decision, and sharing information about the spectrum bands.

Various design and solutions are proposed for the common control channel. In Cormio and Chowdhury (2010), common control channel design CRAHNs is proposed, also called Adaptive Multiple Rendezvous Control Channel (AMRCC), and is based on frequency hopping. The author has also compared the performance from the classical CCC shames. Design of an out of band CCC based on an Ultra Wideband pulse shaped signal is proposed by author in Baldini and Pons (2011). Some authors have also proposed a distributed spectrum sharing scheme which is based on individual spectrum decisions, priority and messaging mechanism between CR users (Uyanik & Oktug, 2011).

3.1.3. Mobility of the Users

Another important factor which affects the performance of the CR system is the mobility. Various solutions are proposed for the mobility effect of the CR systems. Spectrum mobility management architecture and requirements and solutions for the heterogeneous CRNs are discussed by (Damljanovi, 2010). The Channel Assigning Agent (CAA) was proposed by Al-Dulaimi, Al-Rubaye, and Cosmas (2011), at the mobile Internet Protocol (IP) layer to allow CR mobile users to use their channel as long as possible, to avoid additional interruption loss due to unnecessary channel adaptations. Some solutions are also provided for the Long Term Evolution (LTE) (Chen, Cho, You, & Chao, 2011). In this article, authors have proposed cross-layer protocol of spectrum mobility (layer-2) and handover (layer-3) in cognitive LTE networks.

3.1.4. Sensing in Multi-User Networks

Generally CRNs consist of two types of users, i.e., primary and secondary users. Moreover, number of primary or secondary users always gives the big impact on the performance of the CR system. Multi-user environment creates it more difficult to sense the primary users and to estimate the actual interference. Lots of solutions are proposed to solve this problem. Multi-carrier DS CDMA modulation over a frequency-selective fading channel was proposed by authors to solve the multiuser problem in Cognitive Radio Ad Hoc Networks (CRAHNs) in Qu, Milstein, and Vaman (2008). In Ban, Choi, Jung, and Sung (2009), authors have examined the effects of multi-user diversity in a spectrum sharing system. An efficient multiuser based algorithm is proposed to maximize the sum capacity of the cognitive OFDM systems while maintain the proportional rate by Yuan and Du (2012). However, cooperative sensing is the most popular for this problem. Cooperative solution is discussed in the forth section of this chapter.

3.1.5. Insecure Operational Environment

The operating environments where CRNs are used may be vulnerable. One of the important applications of such networks is in battlefields. In such applications, users may go in and out of hostile and protected adversary region, where they would be extremely susceptible to security attacks (Dubey, 2012).

3.1.6. Limited Resource Availability

Resources such as bandwidth, battery power, and computational power are scarce in CRNs. Moreover, composite cryptography-based security mechanism is hard to design and implement in such networks.

3.1.7. Synchronization

The MAC protocol must consider the synchronization between various types of user nodes in the CRNs. CRN requires precise information to exchange the control signals. These control signals should be synchronized to achieve better throughput and performance of the network. Lack of synchronization causes more collision of packets and imperfect information about the networks (Wyglinski, 2010).

3.1.8. Multichannel Hidden and Exposed Terminal Problems

The multichannel hidden terminal problem can significantly reduce the performance of the CRN systems. This problem needs to be deal at the MAC layer by proper synchronization and the signaling scheme. The hidden terminal problem is defined as the collision of packets at the destination caused by the concurrent transmission of those users, which are within the direct transmission range of the receiver.

In the Figure 5, node *A* wants to send a packet to node *B* and starts the RTS/CTS handshake on the control channel (channel 1). After negotiation, channel 2 is selected and node *A* starts communication. Node *C* does not hear, however, the RTS/CTS messages because it is listening to channel 3, and decides to initiate a transmission on channel 2, causing a collision.

It can be addressed better in a multi-transceiver MAC as shown in a few CRN MAC schemes (i.e., Kondareddy & Agrawal, 2008; Su & Zhang, 2008). However, the presence of multiple transceivers at each node makes the SU node more complex and expensive. The single transceiver cognitive MAC seems to be a more practical solution to keep the node complexity lower. However, current single-transceiver MAC solutions either do not address the issue at all or handle the issue very inefficiently due to high control overhead.

Figure 5. Multichannel hidden node problem

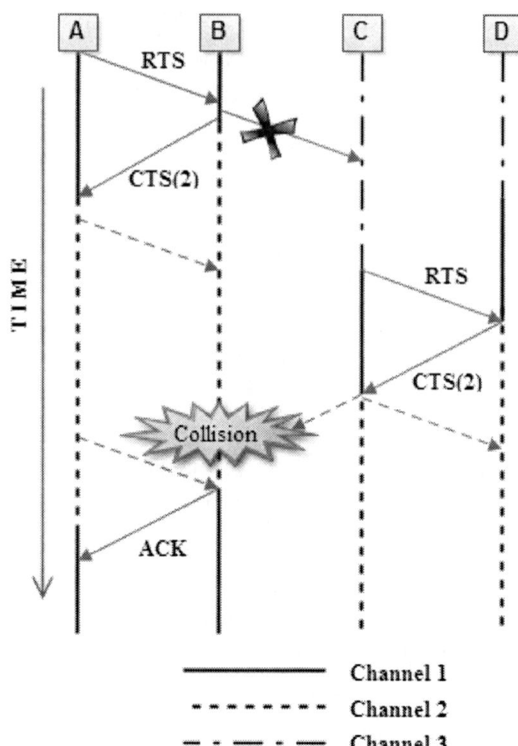

3.1.9. Quality of Service (QoS)

Quality of Service (QoS) is the performance indicator of a service accessible at the CRN to their users. The goal of QoS provisioning is to accomplish more deterministic network performance, so that information transferred by the CRN can be better transported and network resources can be better utilized. A network or a service provider can propose various services to its users. A QoS support is very critical, due to the intrinsic nature of CRNs, where two types of mobile users most of the time. QoS is very important factor when we deal with the cognitive radio networks, because availability of the secondary users always give the higher impact on the performance of the cognitive radio networks. Moreover, QoS is very challenging at the presence of the primary users.

Various QoS solutions are proposed for cognitive radio. QoS for the cross layer based MAC is described in Su and Zhang (2008). In this article author has proposed cross-layer based opportunistic multi-channel Medium Access Control (MAC) protocols, to enhance the delay-QoS provisioning for the CRNs. Multicast QoS routing is explained by the authors in Xie and Zhou (2012). Cognitive MAC scheme is proposed in Wang and Chen (2007), specifically for the QoS of the wireless ad hoc networks.

3.2. Issues Related to Cognitive Radio Ad Hoc Networks (CRAHNs)

3.2.1. Lack of Central Authority and Association

In wired networks and infrastructure-based wireless networks, the base station acts as the central authority or coordinator and it allocates channels to the various users in the system. So, extra overhead is always present when CR system size is large. Since Cognitive Radio Ad Hoc Networks (CRAHNs) do not have any such central points, these mechanisms cannot be applied in CRN ad hoc networks.

3.3. Design Objectives of MAC Protocols for CRNs:

The following are the important objectives while designing a MAC protocol for the CRNs (Dubey, 2012; Chouhan, 2012):

- The protocol should be suitable for both the types of users, i.e., primary and secondary users.
- The protocol should ensure fair allocation (same or weighted) of spectrum bands/ channels.
- The protocol should effectively overcome the problem of hidden and exposed terminal problem.
- **Adaptive Algorithms:** In the sensing and detection operation, the CR can sense its surroundings, flexible to its policy and configuration constraints, and agreements with peers to best utilize the radio spectrum and meet the QoS.
- **Distributed Collaboration:** In CRN, CRs will exchange existing and accessible information on their local environment, user demand, and radio performance among themselves on the regular basis. CR will use their local information and peer information to decide their operating settings. Hence, operations of the MAC protocols should be distributed.
- **Average Delay:** The access delay or average delay of the MAC scheme should be low. Control overhead should also be kept low.
- **Bit Error:** Bit Error Rate (BER), collision rate, packet drop rate, and packet loss rate should be less for MAC protocols.
- **QoS:** The MAC scheme should support QoS and real time traffic for better performance and throughput of the CR systems.
- **Security:** CR will connect and leave wireless networks. CRNs require mechanisms to authenticate, authorize, and protect information flows of participants.
- **Power Control Mechanism:** Effective power control mechanism is required to utilize the energy of the system, because all users are mobile in CR system and they have very limited energy resources.

- **Synchronization:** Since synchronization among various users is very important for the bandwidth reservation, so protocol should provide synchronization between various users and their resources.

4. MAC PROTOCOLS FOR COGNITIVE RADIO NETWORKS

4.1. Classification of MAC

MAC layer solutions can be mainly classified on the basis of following features: driven based, duplexing technique, type of architecture, spectrum access technique, spectrum allocation behaviour, access mode, spectrum sharing mode, scheduling based, and priority based as shown in Figure 6. Let discuss each category one by one (Zhang, 2010).

In the first classification, two types of protocols are possible: (1) Source driven or initiated protocol, which includes three stages: in the first stage, a sender user gathers the position information of all the targeted users; second stage, the sender creates a multicast tree with the help of position information; and third, the sender transmit data to down on this multicast tree. (2) Destination initiated or driven protocol which permits receivers to build their individual data delivery path to the sender, and ultimately, multicast tree is automatically build by assembling the data delivery paths, and then, the sender node transmit data to them down on this multicast tree (Lee, Lee, Park, Park, & Kim, 2010).

Duplexing techniques are characterized under the both the bottom layers. Time Division Duplex (TDD) and Frequency Division Duplex (FDD) are two important duplexing techniques. In an FDD system, the uplink and downlink channels are situated on individual frequency bands. A deterministic time frame is used for both uplink and downlink transmissions. In TDD, the uplink and downlink transmissions share the identical frequency but they occur at different times. A TDD frame has a fixed duration and encloses one downlink and one uplink sub frame. Presently, combination of both TDD and FDD schemes are more popular.

There are three types of architectures in the cognitive radio MAC systems: (1) Centralized MAC where each user is controlled by the central entity, also known as base station or the access point. IEEE 802.22 and dynamic spectrum access protocols are the examples; (2) Distributed MAC protocols in which no central controlled entity and each primary and secondary user have equal privilege. Therefore, coordination and synchronization are the main problem in these types of protocols; and (3) Hybrid MAC protocols in which both the types of architectures are used.

Contentions based and contentions free are two main approaches in access based MAC techniques. In contentions based MAC, users contend a channel for transmission opportunities, like in Carrier Sense Multiple Access with Collision Avoidance (CSMA/CA). However, users access the spectrum as per the time slot available in the frame structure in case of contention free MAC protocols. Channel partitioning, coordinated access and spread spectrum techniques are example of contention free MAC schemes.

According to the spectrum sharing cooperative and non-cooperative MAC are two main schemes. In cooperative MAC, secondary user coordinates with the other users to maximize the throughput of overall system. On the other hand, non-cooperative MAC is selfish scheme where each secondary user works independently to maximize its individual profit.

Regarding to modes of spectrum sharing, overlay (out of band) and underlay (in band) are two popular major MAC schemes. In overlay approach, a node accesses the network using a portion of the spectrum that has not been used by licensed

Figure 6. Classification of various MAC protocols for cognitive radio networks

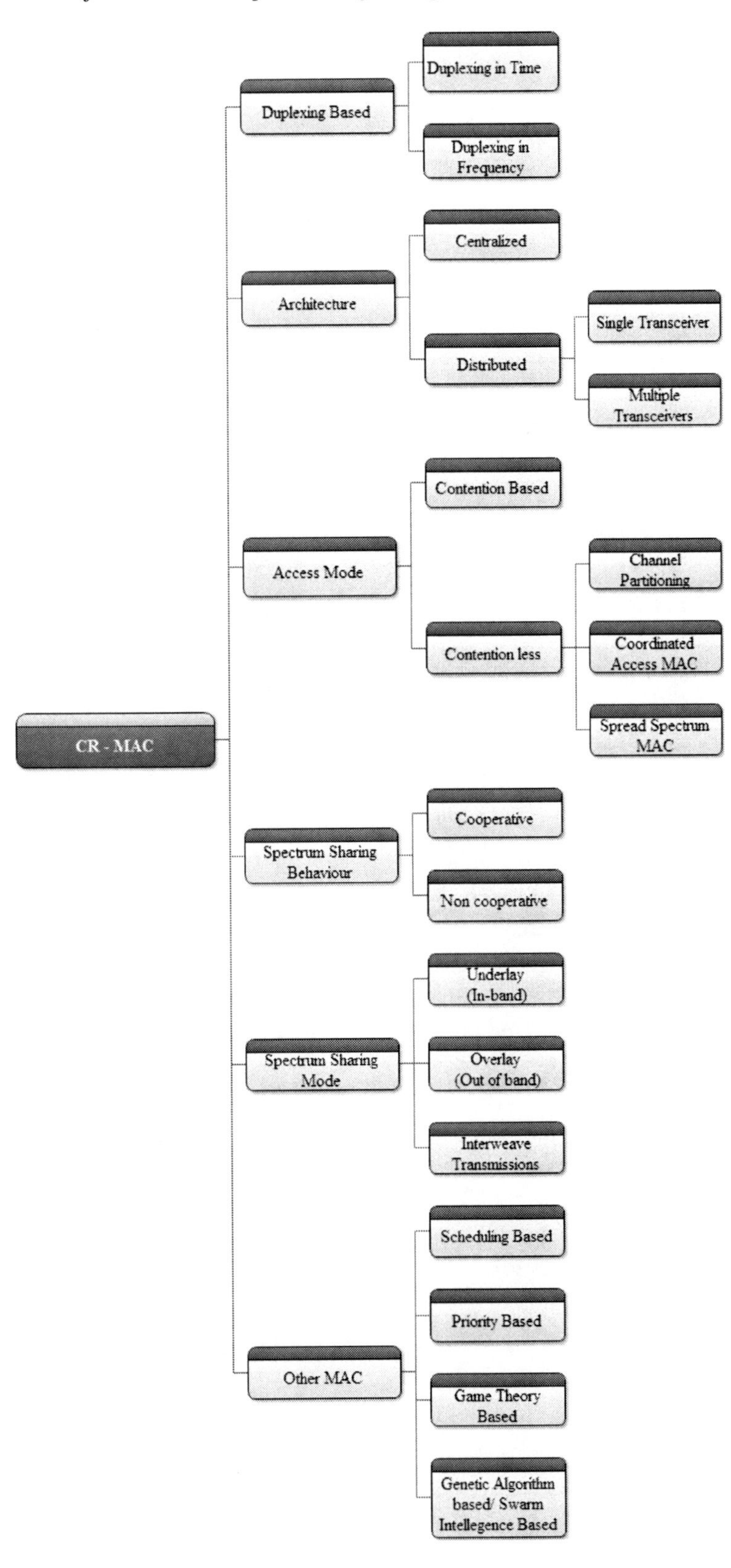

users. While in the case of underlay, secondary user shares the same spectrum, which is assigned to primary user up to tolerable interference.

Random access and reservation based are also two important classifications. Random access (sometimes called direct access) is the ability to access a channel at an arbitrary position in a sequence in equal time, independent of sequence size.

Moreover, various priority based MAC models are proposed. Partial Observed Markov Decision Process (POMDP) model and queuing theory based MAC models are the examples.

Various time division based approaches are proposed for the cognitive radio (Wang, 2011; Iyer & Lim, 2011; Hu & Yang, 2012; Ansari, Zhang, Achtzehn, Petrova, & Mahonen, 2011).

4.2. Duplexing-Based MAC

The most significant work of the MAC protocol is to control the usage of the medium, and this is done through the channel access scheme. The channel access scheme is the method to distribute the resources between the users and the radio channel, by flexible use of the spectrum. Protocols guide each user when it can forward and when it is predictable to receive transmission. The channel access scheme is also the center of the MAC protocol. In this section, we examine TDMA, CSMA, and polling schemes which are the three key classes of channel access methods for radio (MAC Layer, 2004(The MAC level (link layer), 2004).

4.2.1. Duplexing in Time

TDMA (Time Division Multiple Access) is very straightforward. In this scheme, a particular user or the base station has responsibility to manage and synchronize the users of the network. The time on the channel is distributed into time slots, which are generally of fixed and equal size. Every user of the network is assigned a fixed number of slots where it can transmit. Slots are generally prearranged in the frame, which is repetitive on a regular basis.

In this type of system, the central entity specifies in the beacon (a management frame) to manage and organize of the frame. Every user needs to obey blindly the instruction of this central entity. Very often, the frame is ordered as downlink (base station to user) and uplink (user to base station) slots, and all the transmission goes through the central entity. A service slot permits a user to ask for the allocation of a connection, by transmitting a connection request message in it. In various standards, uplink and downlink frames are one special frequency, and the service slots might be a separate channel as well (MAC Level, 2004), see Figure 7.

Figure 7. TDMA frame

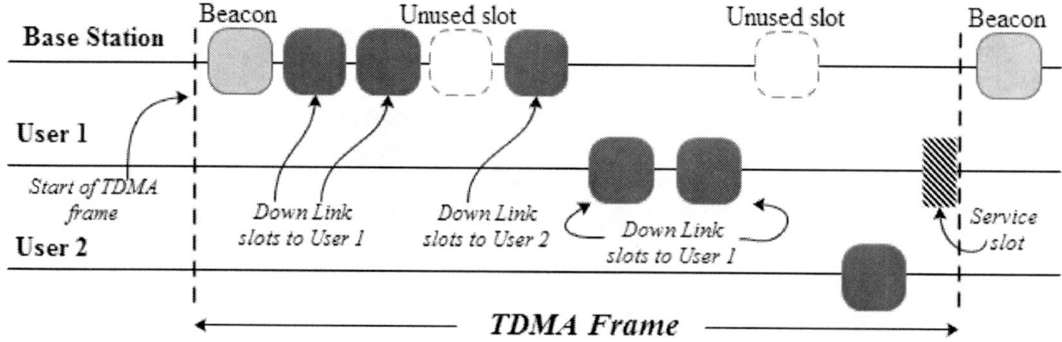

4.2.2. Duplexing in Frequency

Frequency Division Duplexing (FDD) defined that the sender user and destination user operates at various carrier frequencies. The term is frequently used in ham radio operation, where a worker is trying to contact a repeater station. This station must be capable enough to transmit and receive a communication simultaneously, and does so by slightly varying the frequency at which it transmits and receives. This mode of operation is referred to as duplex mode or offset mode.

Uplink and downlink sub bands are supposed to be separated by the frequency offset. Frequency division duplexing can work well and effective in the case of symmetric traffic. In this case, time division duplexing lean to waste bandwidth for the period of the switch over from sending to receiving, has better intrinsic latency, and may need more composite and complex circuitry.

One more benefit of frequency division duplexing is that it constructs radio preparation and planning easier, faster and more well-organized, since central entity do not "listen" every one (as they send and receive in various spectrum bands) and consequently it will usually not interfere with each other. Unlikely with time division duplexing systems, it concerns to preserve guard bands between adjacent base stations (which minimizes spectral efficiency) or to coordinate base stations, so that they will send and receive simultaneously (which enhances network complexity and thus cost, and decrease bandwidth distribution flexibility as all base stations and segments will be forced to utilize the same uplink/downlink ratio). Most popular examples of frequency division duplexing systems are:

- Asymmetric Digital Subscriber Line (ADSL) and very high bit rate digital subscriber line (VDSL or VHDSL).
- Most cellular systems, including the Universal Mobile Telecommunications System (UMTS), Wideband Code Division Multiple Access (WCDMA) and the Code Division Multiple Access (cdma2000) system.
- IEEE 802.16 WiMAX (worldwide interoperability for microwave access).

Qu, Milstein, and Vaman (2008), Tu, Chen, and Prasad (2009), Zhang and Su (2011), and Kandeepan, Sierra, Campos, and Chlamtac (2010) are popular examples of these kinds of MAC schemes in the CR systems.

4.3. Architecture-Based MAC

A neat and specific explanation of the CRN is very complex and critical for the development of communication protocols. A complete architecture of the CRN is explained in Section 1.2. As explained previously, there are three kinds of MAC architecture possible, i.e., centralized, distributed, and hybrid (Chouhan, 2011; Cormio & Chowdhury, 2010), see Figure 8.

4.3.1. Distributed MAC

Distributed MAC is more robust since it does not depend on the central entity (as well specify as cluster head or cluster leader), while in centralized MAC schemes, a single node coordinates control information exchange and radio access. In these protocols, secondary users share observation data and independently take decisions regarding interference, and resource availability and access is based on the global or local policies (Akyildiz, Lee, Vuran, & Mohanty, 2006). Several terminologies for this sharing paradigm might be identified in the literature, such as Dynamic Spectrum Access (Hossain, 2008) or Opportunistic Spectrum Access (Jia, Zhang, & Shen, 2008). A classification of distributed MAC is depicted in the Figure 8. Common Control Channel (CCC) is used to create categories in the classification.

Various distributed MAC strategies are proposed for CRNs. Some of the proposed MAC

Figure 8. Classification of existing cognitive MAC (CCC: common control channel)

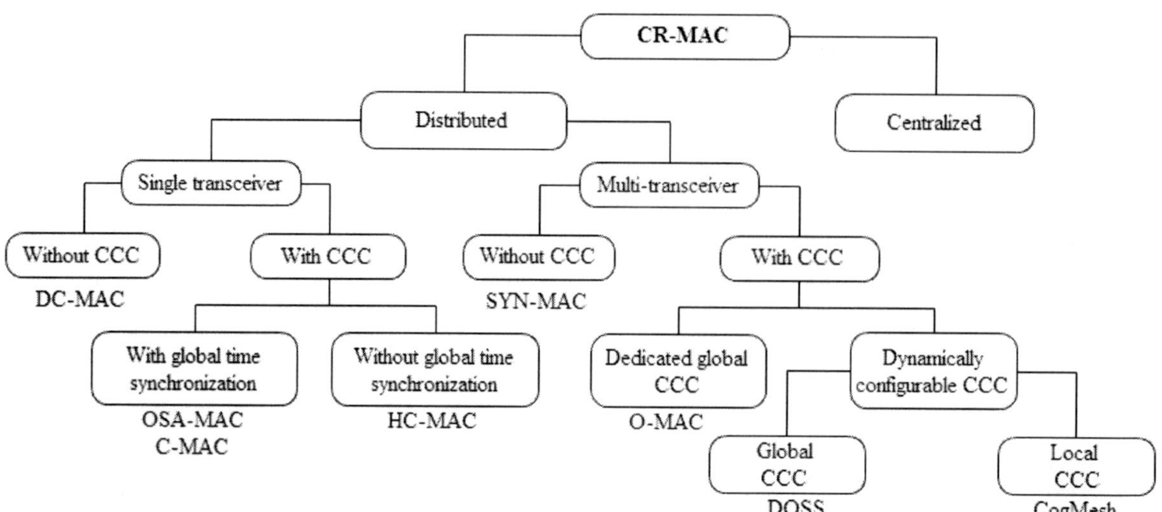

protocols (e.g., DC-MAC (Zhao, Tong, Swami, & Chen, 2007), OSA-MAC (Hossain, 2008), HCMAC (Jia, Zhang, & Shen, 2008), and C-MAC (Cordeiro & Challapali, 2007)) work well with a single transceiver, while others (e.g., SYN-MAC (Kondareddy & Agrawal, 2008), O-MAC (Su & Zhang, 2008), DOSS (Ma, Han, & Shen, 2005), and CogMesh (Chen, Zhang, Maggio, & Chlamtac, 2007)) need multiple transceivers at SU nodes. In Jha, Rashid, Bhargava, and Despins (2011) performance of various distributed MAC strategies are compared.

In Table 1, various distributed MAC protocols are compared on the basis of the network topology. Three important parameters are used in this table, i.e., global CCC, local CCC, and without a dedicated CCC. Different MAC properties are compared with the above parameters, like network coordination, spectrum sensing policy, spectrum allocation, spectrum access, power control and Interference Mitigation (IM), time synchronization, mobility, reconfiguration overheads, control and communication overheads, multi-hop network, and QoS (Quality of Service).

4.3.2. Centralized MAC

In this type of architecture a central entity controls, allocates, and share the spectrum bands to various users. A distributed algorithm is implemented at the central entity (base station or access point) to crate spectrum allocation map, which imply additional signaling overhead. A centralized spectrum manager is used in this architecture that must have knowledge about the spectrum usage and requirements of shared spectrum band at that location (Chouhan, 2011).

Distributed and centralized architectures are compared by Salami, Durowoju, Attar, Holland, Tafazolli, and Aghvami (2011). In this article they have surveyed two paradigms, Centralized Dynamic Spectrum Allocation (DSA) and Distributed Dynamic Spectrum Selection (DSS) that aim to address this problem, whereby they use DSS (distributed) as universal term for a range of terminologies for decentralized access, such as Opportunistic Spectrum Access (OSA) and Dynamic Spectrum Access.

Various other architectures have been proposed in the literature for such centralized spectrum management MAC. In Table 2, a comparison of

Table 1. Various distributed MAC protocols

Properties	Network Topology			
	Global CCC	**Local CCC**	**Configurable CCC**	**Without a Dedicated CCC**
MAC Protocol	ILP based MAC (Krishna, 2009), DOSS (Ma, 2005), and HC-MAC (Jia, Zhang, & Shen, 2008)	HD-MAC (Zhao, 2005) and CogMesh Scheme (Chen, Zhang, Maggio, & Chlamtac, 2007)	DUB-MAC (Adamis, 2007) and C-MAC (Cordeiro, Challapali, Birru, & Sai Shankar, 2005)	DC-MAC (Zhao, 2005) and SRAC (Ma, 2007)
Coordination and Synchronization	extreme	medium	medium	low
Sensing Policy	available only in HC-MAC	not addressed	not addressed	not addressed
Spectrum Assignment	assigned for all users	assigned for users in the cluster only	repetitively assigned in the cluster	not guaranteed
Spectrum Access	guaranteed access in ILP based MAC, random access in DOSS, combined access in HC-MAC	joint access	combined access	random access
Power Control and Interference Mitigation (IM)	IM addressed only in ILP based MAC	not available	not available	available in SRAC, and Haykin (2005) (Haykin, 2005)
Synchronization	strictly necessary	necessary	necessary in C-MAC	not necessary
Mobility	available only in ILP based MAC	not available	addressed only in C-MAC	not available
Reconfiguration Overheads	low	medium	slightly high	extreme
Control Message Overheads	medium in DOSS. High in ILP based MAC and HC-MAC	medium	medium in DUB-MAC. High in C-MAC	low in (Zhao, 2007). High in (Ma, 2007). Not addressed in Haykin (2005)
Multi-hop Network	available in DOSS and ILP based MAC	possible	not possible	not guaranteed
QoS	available only in HC-MAC	not available	not available	not available

centralized MAC protocols is presented which based on the network topology. Four important centralized protocols are compared in this table, i.e., WRAN (802.22), DIMSUMNet, DSAP, and On-demand CPC. Different MAC properties are used to compare above protocols, like network topology, operating spectrum, spectrum allocating entity, spectrum sensing, control channel, spectrum access, power control, pricing plans. A dynamic spectrum allocation procedure in a multi-operator environment is depicted in the Figure 9 (De Domenico, 2010).

4.4. Access Mode

MAC protocols can also organize into two categories: contention based and contention free. Contention based or random access MAC protocols are not necessarily required synchronization between the users accessing the band. However, contention free MAC requires high coordination between their users for this synchronization. Let discuss each protocol one by one.

Table 2. Various centralized MAC protocols

Properties	WRAN	DIMSUMNet	DSAP	On-Demand CPC
Network Topology	Centralized spectrum access network	Centralized spectrum sharing network	Centralized spectrum sharing network	Centralized spectrum sharing network
Effective Spectrum	TV spectrum (UHF and VHF)	CAB (Cellular, PCS and TV)	ISM band (2.4 GHz)	not specified
Spectrum Assignment Entity	Base stations (BS)	RANMAN and BS	DSAP server	Cognitive Pilot Channel (CPC) transmitter
Sensing Type	distributed sensing	distributed sensing	not available	not available
Control Channel	Each channel has its MAC super-frames	reserved in CAB	available in ISM bad	CPC
Spectrum Access	complete support	partial support	not available	not available
Power Control	defined	defined	defined	not available
Pricing Plans	defined	defined	not defined	not defined

Figure 9. Dynamic spectrum allocation procedures in a multi-operator environment

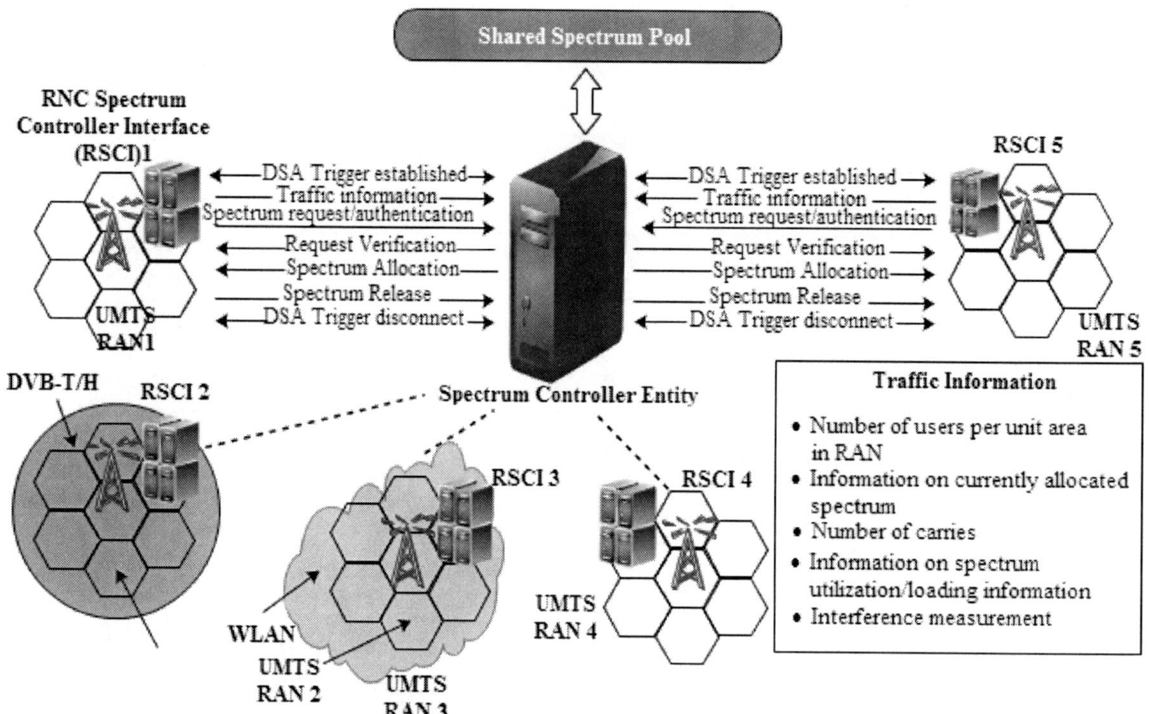

4.4.1. Contention-Based MAC

A Contention-Based (CBP) or random-access protocol is the MAC protocol for operating wireless telecommunication devices that allows several users to utilize the same radio band excepting synchronization and prior coordination.

The "sense before transmit" working system in the IEEE 802.11 is the most popular contention based protocol. FCC has also defined these protocols under rule Section 90.7 of Part 90, as FCC (2005):

A procedure that permits several nodes to distribute the similar spectrum band through defining the activities that must happen when two or more senders try to concurrently access the similar band creating policy by which a sender offers realistic opportunity for other senders to utilize. Such a protocol may contain of methods for starting new transmissions, schemes for deciding the status of the channel (accessible or not-accessible), and techniques for super-visioning and managing retransmissions in case when the channel is busy.

In the CRN, initially the sender and receiver exchange their sensing outcomes, and then this pair evaluates available resource and negotiates the channel for communication. This complete process is also known as Channel Filtering Sender Receiver (CFSR) handshake.

4.4.2. Contention-Free MAC

Contention approaches when two nearby users try to access the same transmission band. Contention make up signal and data collisions, which are particularly to arise when the communication overhead is extreme and interrelated, and it can minimize the lifetime of this CR system. A MAC protocol is defined as contention free, if the protocol disallows any collisions. All proposed solution for contention free protocols supposes that the users are synchronized in various manners. This is generally extremely critical and difficult when the large number of users in CRN. Contention free protocols can also be divided into channel partitioning, coordinated access MAC, and spread spectrum MAC techniques.

4.4.2.1. Channel Partitioning

Various solutions are proposed on channel partitioning MAC scheme. Meko and Chaporkar (2009) have proposed channel partitioning and relay positioning technique so as to better the number of nodes entered in the CR region. In this technique, Signal to Interference and Noise Ratio (SINR) acknowledged by the central entity and Fixed Relay Nodes (FRN) are examined with concern on nodes at worst case SINR. Some pre-partitioning based approaches for CR are also proposed by Jiang, Grace, and Mitchell (2009). In this scheme, authors have proposed reinforcement learning based spectrum sharing technique to minimize the potential action space of users by randomly partitioning the spectrum for different users.

4.4.2.2. Coordinated Access MAC

A variety of coordinated access MAC protocols available in the CRN by Lee and Wong (2011) and Wong, Chen, and Hsu (2009). Lee and Wong (2011) have proposed synchronized non-sensing Medium Access Control (MAC) scheme in the Distributed Coordinated Dynamic Spectrum Reservation (DCDSR) protocol, which is not based on the CR systems. In this proposed scheme, all PUs and SUs have two transceivers; first one operating in a common control channel and another one for licensed or primary data channels to access control and data packets in the particular control and data frames. PU informs SUs which channels will be engaged in next data frame and then SU captures one of the remaining channels by competing with each other in Contention Interval (CI) period of control frame. The existing channels which can be captured by PUs will depend on the spectrum utilization of PU and additional variations in the efficiency of SUs. When, SU successfully captures band in the CI, it will begin access the spectrum band in subsequent data frame (Wong, Chen, & Hsu, 2009).

4.4.2.3. Spread Spectrum MAC

Spread spectrum schemes are techniques by which a message (e.g. an electrical, electromagnetic, or acoustic signal) produced in the specific spectrum are intentionally spread in the frequency domain, follow on in the message with a larger bandwidth. Spread spectrum MAC protocols are applied for various reasons, incorporating the formulization of protected transmissions, enhancing resistance to expected interference, noise and jamming, to avoid detection, and to bound power flux density, e.g. in the case of satellite downlinks (Spread Spectrum, 2012).

This technique is message configuration schemes that employ direct sequence, frequency hopping, or a hybrid of these, which can be applied for multiple access and/or multiple purposes. Spread spectrum is more suitable in the case of CRN where bandwidth and resources are very limited. Chirp Spread Spectrum (CSS) (Shen, Zhang, & Kwak, 2008; Salous, 2010), Direct-Sequence Spread Spectrum (DSSS) (Deng, Shen, Bao, Su, Lin, & Wang, 2011; Zamat & Natarajan, 2009), Frequency-Hopping Spread Spectrum (FHSS) (Yao, 2009; Vigneron, 2006) and Time-Hopping Spread Spectrum (THSS) (Shao & Beaulieu, 2011; Ying-Jie, 2011) are the popular spread spectrum techniques for CRNs.

4.5. Spectrum Sharing Behavior

Most important classification of the CRN-MAC is based on its spectrum sharing mode. Two important types of sharing modes are cooperative and noncooperative MAC.

4.5.1. Cooperative MAC

Primary user detection efficiency is compromised by noise uncertainty, less information about the primary receiver location, fading, and shadowing effects. So, precisely detection of primary user is especially critical and harder in CRNs. In the cooperative sharing MAC, the interference information is shared among all the SUs to detect the existence of the PUs in the particular spectrum band. This approach of MAC is very popular for the designing of the CRNs.

Various solutions are proposed in this category. Cooperative transmission investigates to be a promising scheme for enhance the performance of SU by accumulating the spatial diversity and spectrum diversity (Zhang, 2009)(Qian Zhang, 2009). This approach can take advantage of inherent multi-user spatial diversity to increase detection, and minimize the missing and false alarm probabilities (Sousa, 2005). Increased performance arises at the cost of amplified latency and communication overhead. Cooperation can also resolve the hidden terminal problem as well as minimize the sensing observation time and bandwidth. Moreover, it also allows decreasing the effects of malicious sensing nodes (De Domenico, 2010). However all the centralized schemes can be considered as cooperative methods, there also exist various distributed cooperative schemes (Akyildiz, Lee, Vuran, & Mohanty, 2006).

4.5.2. Non-Cooperative or Selfish Sharing MAC

On the other hand, non-cooperative sharing MAC approach is self centric based scheme. In this scheme, SU always try to maximize its own profits which results degradation the performance of the CR system. This scheme is also known as selfish or non-collaborative spectrum sharing MAC. Various solutions are proposed for these types of MAC approaches (Sankaranarayanan, 2005; Zhen, 2005; Zhao, 2005).

4.6. Spectrum Sharing Mode

PU is licensee holder of the spectrum resource and, opportunistic users should not interfere with their transmissions. Three different cognitive transmission access paradigms are presented: underlay, overlay, and interweave (De Domenico, 2010).

4.6.1. In Band (Underlay) MAC

In underlay MAC scheme, SUs are allowed to operate while generated interference stays below a given threshold. Underlay MAC spectrum allocation scheme utilizes the spread spectrum techniques developed for conventional wireless networks (Huang, 2005). Since the associated interference constraints, the underlay technique is appropriate for the short-range communications. This approach necessitates complex spread spectrum schemes and can be operated on enhanced bandwidth as compared to overlay techniques.

4.6.2. Out of Band (Overlay) MAC

In overlay sharing mode, SUs utilizes the spectrum band or part of the spectrum band which has not been used by the PUs, so that the interference to the PU's system is reduced. In this scheme, SUs exploit the knowledge of PUs messages to either stop or diminish interference at both PU and SU's side.

4.6.3. Interweave Transmissions MAC

In interweave transmissions, opportunistic radios transmit only in spectrum holes; if during in-band sensing a secondary user detects a licensed one, it vacates its channel to avoid harmful interference. Due to the less information on PU's receivers, currently most of CR protocols are developed according to the interweave transmission paradigm. SUs avoid contention with incumbent primary nodes by performing periodically sensing on the occupied channels. If an incumbent is detected, the channel is vacated, transmission is interrupted, and a communication link is set up on a different channel (De Domenico, 2010).

4.7. MAC Protocols with Scheduling Mechanisms

Scheduling is the technique by which threads; processes or data flows are certain access to the system resources (e.g. processing time, transmission bandwidth). This is generally achieved to weight balance the CRN efficiently or accomplish the intentioned QoS. The requirement for this scheduling scheme occur from the requirement for present wireless systems to execute multitasking (perform more than one task at the same time) and multiplexing (transmit several flows concurrently). The scheduler is mainly dealt with the following terms (Scheduling, 2012):

- **Throughput:** Number of successful transmission per time unit.
- **Latency:** Specially:
    - **Turnaround:** Overall time from start the transmission to its completion.
    - **Response Time:** Total time that obtains from when a signal was requested until the first reply is transmitted.
- **Fairness/Waiting Time:** Same transmission time slot to each user (or normally suitable period according to each transmission priority).

A resource schedule, also generally called a "shared resource schedule," is the time plan of activities and resources. Distributed resources are accessible at given times and actions are considered throughout these period. The timetable preserves partition among users of these resources. The procedure of generating the plan and timetable is called scheduling (Schedule, 2012)

Various scheduling based solutions are proposed for the CRN-MAC. Li, Dai, and Zhang (2008) have proposed cross layer scheme for combined link scheduling and power control in CRNs, in order to build more efficient and resourceful. Zhang, Bai, Liu, and Tang (2010) have combined OVErwater Radio-Time Scheduling (OVERTS) with the cognitive radio to solve the communication problem in the overwater. Adaptively spectrum sensing and data transmission scheduling are also proposed to minimize the harmful impacts on the performance of the CRNs (Hoang, Liang, & Zeng, 2010).

4.8. Other MAC Protocols

4.8.1. Game Theory-Based MAC Protocols

Interaction between cognitive radios can be represented as a game. Game theory efficiently models the dynamics of a cognitive network: adaptation and recursive interactive decision procedures are naturally modeled by the repetitive game. Moreover, with game theory each player may adopt a different utility function to pursue specific goals. Interactive behaviors among CR are represented as a game $\Gamma = [N, \{S_i\} \{U_i\}]$. N is the set of game players, each sender receiver pair is an element of this set; S_i represents the strategy space (modulation and coding schemes, transmission power, antenna parameters, etc) of player *i*; U_i is the local utility function that models the scope of player *i* (Wang, 2011). Various game theory based solutions are also discussed in De Domenico (2010), Jha, Rashid, Bhargava, and Despins (2011), Zhang (2010), and Zhao (2005).

4.8.2. Graph Theory-Based MAC Schemes

CRN can be modeled as a graph $G=(V,E)$ where V and E indicate the vertex vs. the edge sets. Two kinds of representations are available: Node Contention Graph (NCG) and Link Contention Graph (LCG). In NCG, cognitive users are represented by nodes while edges indicate that two nodes are in the interfering range of each other. In LCG, the vertex set represents active flows, while edges represent a contention between different flows.

4.8.3. Genetic Algorithm-Based MAC Schemes

These are adaptive search algorithms based on the evolutionary ideas of natural selection. An iterative process starts with a randomly generated set of solutions called population. Best individuals are selected through the utility function (called here fitness function). Then, starting from this subset, a second population is produced through genetic operators: crossover and/or mutation. The new population shares many of the characteristics of its parents, and it hopefully represents a better solution. The algorithm typically terminates when it converges to the optimal solution or after a fixed number of iterations. Genetic algorithms are chosen to solve resource allocation problems due to their fast convergence and the possibility of obtaining multiple solutions.

4.8.4. Swarm Intelligence-Based MAC Algorithms

Inspired by the collective behaviour of social biological individuals, Swarm Intelligence (SI) algorithms model network users as a population of simple agents interacting with the surrounding environment. Each individual has comparatively slight intelligence; however, the collaborative behaviour of the population leads to a global intelligence, which permits to solve complex tasks. For instance, in social insect colonies, various actions are often performed by those individuals that are better equipped for the operation. This phenomenon is called division of labour. SI algorithms are scalable, fault tolerant and moreover, they adapt to changes in real time (De Domenico, 2010; Krishna, 2009).

5. PRIORITY-BASED MAC PROTOCOLS

There are various related studies on performance analysis of CRNs. Performance of open spectrum access in licensed bands has analyzed with proposed Continuous-Time Markov Chain (CTMC) model, is discussed in Tung (2008). In Zhang (2008), forced termination probability can be greatly minimized through the proposed channel reservation scheme for the PU's spectrum sharing system. A finite queue has introduced to store the newly arriving SU if there are all busy channels, which are able to considerably decrease the SU's blocking probability, and non-completion probability is described in Zhu (2007).

Foh (2009) have proposed the loss/failure model for CR spectrum access within their user population, the impact of licensed band loads and spectrum band sharing on the delay performance of SUs are examined. In Zhao (2007), Partial Observed Markov Decision Process (POMDP) framework based MAC protocol is illustrated for the distributed CRN. Jah (2011) have defined higher priority for QoS supplies of delay sensitive applications. In Wang (2011), the queuing theory based CRN framework is deliberated with the preemptive resumption priority M/G/1 system.

5.1. POMDP

Several of the logical frameworks for opportunistic spectrum sharing are proposed based on the theory of POMDP. This decision theoretic scheme incorporates the design of spectrum access technique at the MAC layer with spectrum sensing at the physical layer and communication overhead statistics determined by the application layer of the primary network. Moreover, POMDP permits simple integration of spectrum sensing error and also limitation on the probability of colliding with the PUs. In addition, authors have proposed cognitive MAC protocols that maximize the efficiency/performance of SUs though restrictive the obstruction served by PUs. A suboptimal strategy with minimized complexity, however comparable performance is also explained. The suggested decentralized protocols guaranteed synchronous hopping in the spectrum among the sender and the destination in the existence of collisions and spectrum sensing errors devoid of extra control signals exchange among the SU's transmitter and receiver (Zhao, 2007).

Various architectures and models are also proposed which are based on the POMDP framework. Ma, Hsu, and Feng (2009) has developed MAC protocol based on POMDP to control handoffs in the CRNs. Hossain (2008) has proposed Opportunity Spectrum Access (OSA) strategy for multiuser based on POMDP. In this scheme, OSA strategy has proposed with randomized POMDP framework. It targets at maximizing the throughput of the entire cognitive radio network. The OSA also consider the service orders of the CR users and the CR user with high service order will have more chance to access the idle channels.

5.2. Queuing Theory-Based MAC

Interruption from PUs is a key factor impacting the performance of SUs in CRNs. Based on a simplified model of primary user interruptions (Markov chain), queuing analysis is carried out for two-server-single-queue and single-server-two-queue cases (Li, 2009).

6. CASE STUDY OF VARIOUS CRN STANDARDS

Various architectures and standards are proposed for the CRN-MAC. In this section we examine 802.11, Wireless Regional Area Network (WRAN) or 802.22, and mobile virtual node operator based MAC protocols for CRNs.

6.1. 802.11-Based MAC Protocols

IEEE 802.11 is the group of standards for designing the Wireless Local Area Network (WLAN) for computer communication in the 2.4, 3.6, and 5 GHz frequency bands. These standards are formed and preserved by the IEEE LAN/MAN Standards Committee (IEEE 802). The basic information of the standard IEEE 802.11-2007 has had successive revisions. IEEE 802.11 offers the foundation for wireless network products by utilizing the Wi-Fi brand (IEEE 802.11, 2012).

In the CRN, various MAC solutions are proposed on the 802.11 standards. Cognitive MAC protocols permits SUs to detect and utilize the offered frequency band without interference to the PUs. Zhao, Hu, and Shen (2010) have proposed a Carrier Sense Multiple Access with Collision Avoidance (CSMA/CA) MAC protocol to permits the SUs dynamically so that they can utilize the offered band, and also analyses its performance.

Wang, Huang, Lau, and Lin (2011) proposed a MAC protocol for CRs that aims to "borrow" the link bandwidth from IEEE 802.11 networks. This protocol allows CR users to opportunistically exchange data on an idle 802.11 channel. However, the protocol proposed in selects an 802.11 channel for data exchange in a random manner. This random channel selection process is time-consuming for CR users to start a data transmission.

CR-MAC protocol is proposed for packet scheduling in wireless networks (Benslimane, Ali, Kobbane, & Taleb, 2009). In this MAC protocol, every SU is equipped with two transceivers. First transceiver is used for control signals while the other one sporadically senses and dynamically exploit the vacant data channel. The SUs inform the state of the channels on control channel and negotiate on the preferred data channel itself for forward data transmission. Each channel is utilized by various groups of SUs. Unlike to existing protocols, data transmission takes place in the time slot in which spectrum opportunity is discovered.

Modified IEEE 802.11 based CRN where each PU occupies on a slot-by-slot basis and SUs attempt to utilize the time unused by PUs. To improve the throughput for SUs, proposed IEEE 802.11-based Opportunistic Spectrum Access (OSA) allows SUs to transmit packets with variable length depending on the remaining time of the current slot (Bae, Alfa, & Choi, 2010).

TDMA based energy efficient CR multichannel Medium Access Control (MAC) protocol called ECR-MAC has proposed for ad hoc networks. ECRMAC necessitates simply a single half duplex radio transceiver on each user that incorporates the spectrum sensing at Physical (PHY) layer and the packet scheduling at MAC layer (Kamruzzaman, 2010).

6.2. Wireless Regional Area Networks (802.22)

IEEE 802.22 standards will offer broadband sharing to large region around the globe and achieve consistent and safe and protected high speed communications to underserved and unnerved population. This novel paradigm for Wireless Regional Area Networks (WRANs) acquires benefit of the constructive communication properties of the VHF and UHF-TV bands to offers broadband wireless sharing over a huge area up to 100 km from the sender. Each WRAN will deliver up to 22 Mbps per channel devoid of interfering with reception of existing TV broadcast stations, using the so-called spectrum holes among the occupied TV channels. This contemporary approach is particularly supportive for helping fewer densely populated areas, such as rural areas, and developing countries where large number of vacant TV channels can be discovered (IEEE 802.22, 2011).

IEEE 802.22 integrates advanced CR capabilities with dynamic spectrum sharing, present database access, precise geo-location system, spectrum sensing, authoritarian area dependent strategy, spectrum propriety, and coexistence for best possible utilization of offered spectrum band.

The IEEE 802.22 Working Group (WG) is employed by the improvement of a CR based Wireless Regional Area Network (WRAN), Physical (PHY), and Medium Access Control (MAC) layers for use by license excepted devices in the spectrum band that is currently payable to the Television (TV) service. Since 802.22 is necessary to reuse the empty TV spectrum band devoid of causing any destructive interference to incumbents (i.e., the TV receivers), CR approaches are of key significance with the purpose of sense and measure the spectrum and detect the presence/absence of incumbent signals (Cordeiro, Challapali, Birru, & Sai Shankar, 2005).

6.3. Mobile Virtual Node Operator (MVNO)

The primary network preserves to offer a lease network through permitting opportunistic based access to its licensed spectrum band with the deal to a moderator/third-party devoid of compromising the QoS for the PU. Such as, the primary network can lease its spectrum access right to a Mobile Virtual Network Operator (MVNO). In addition, the primary network can offer its spectrum access privileges to a local community to make use of broadband access (Akyildiz, Lee, Vuran, & Mohanty, 2006).

A Mobile Virtual Network Operator (MVNO) is the cell phone service provider that offers services directly to their own consumers but does not own key network resources such as a licensed frequency assignment of radio spectrum and the base station infrastructure. On the other hand, these assets are leased from a mobile service provider in the area where the MVNO works.

MVNOs are approximately comparable to the "switchless resellers" of the fixed landline telephone market. An MVNO operates separately of the host MNO and locates its personal trading price arrangements to its clients. Conversely, the MVNO should also compensate a comprehensive payment to the host in quantity of the usage by their clients.

Furthermore, the MVNO does not generally own core network infrastructure, they so often do include a Home Location Register (HLR). This permits more flexibility and rights of the subscriber's mobile phone number (MSISDN). Some MVNOs implement their own billing and customer care solutions known as Business Support Systems (BSS) (Mobile Virtual Network Operator, 2012).

7. CONCLUSION

The issues and challenges to enhance spectrum efficiency have been studied in this chapter, the main focus being on Dynamic Spectrum Allocation (DSA). This Chapter has reviewed the state of art in the area of spectrum sensing, sharing and management under the paradigms of CRN and DSA. Moreover, the major challenges, solutions, and open research issues have been examined in each case.

This chapter investigated the MAC layer behavior, activities, properties, and the performance of the CRNs. These activities can go beyond what is done, and when the existing network communicates. Various classifications of the MAC layer protocols are also explored with functionality, advantage, disadvantage of each type of protocol. However, some of the researchers also prefer hybrid or heterogeneous protocols to achieve more adaptability and computability for the design of the CRNs. Some of the MAC solutions are also proposed with game theory, graph theory, swarm intelligence algorithm, priority based algorithms, and genetic algorithms to achieve high throughput and optimized performance.

Table 3 summarizes and tries to cover all aspects of this chapter. In this table, various MAC protocols are compared with architecture, spectrum sharing mode, common control chan-

Table 3. Summary of features

MAC	Quality Provision	Synchronization	MC Problem	Access Mode	No. of Tranceivers	Cooperation	Common Control Channel	Spectrum Sharing Mode	Architecture
IEEE 802.22	Yes	Yes	No	Contention free	2	Possible	No	Overlay	Centralized
DSAP				Contention based	1 (Server req. 2)	Possible	Yes	Overlay	Centralized
HC-MAC	Yes	Yes	No	Contention based	1	Possible	Yes	Overlay	Distributed
DC-MAC	Not addressed	No	No	Contention based	≥1	Yes	No	Overlay	Distributed
SCA-MAC				Contention based	≥1	Possible	Yes	Overlay	Distributed
OS-MAC	Not addressed	Yes	No	Contention based	1	Possible	Yes	Overlay	Distributed
C-MAC	Yes	Yes	No	Contention free	1	Possible	No	Overlay	Distributed
SYN-MAC	No	Yes	No	Contention based	2	Possible	No	Overlay	Distributed
O-MAC	Yes	Yes	No	Contention based	2	No	Yes	Overlay	Distributed
CREAM-MAC				Contention based	1 (additional sensors)	Possible	Yes	Overlay	Distributed
CogMesh	Yes	Yes	No	Contention free	1	Yes	Yes (Local)	Overlay	Distributed
CSMA-CMAC	Yes	Yes	No	Contention based	1	Possible	No	Underlay	Distributed
DOSS	No	No	No	Contention free	3	Yes	Yes	Underlay	Distributed
MMAC-CR	Yes	Yes	No	Contention free	2	Yes	Yes (Global)	Underlay	Distributed
COMAC	No	No	No	Contention free	≥2	Yes	Yes	Underlay	Distributed
COMNET	Not addressed	Yes	No	Contention free	1	Yes	No	Overlay	Centralized
HDMAC	Not addressed	Yes	Yes	Contention free	1	Yes	Yes (Local)	Overlay	Distributed
Ghaboosi	No	No	No	Contention based	2	No	Yes	Overlay	Distributed
Su	Yes	Yes	No	Contention based	2	No	Yes	Overlay	Distributed
Zuo	Not addressed	Yes	No	Contention free	1	Yes	No	Overlay	Centralized

nel, cooperation, number of transceivers, access mode, MC problem, synchronization, and quality provision parameters.

Most of CR literature deals with opportunistic networks. However, the impacts of cognitive paradigms on cellular networks still need to be explored. Hence, further analysis and research on MAC based protocols is expected to improve the spectral occupancy and better performance for both PU and SU networks. Therefore, CRN would form objective for green next generation wireless networks.

SUMMARY

In this chapter, MAC layer requirements for cognitive radio network are highlighted. In the first section brief introduction about the cognitive radio, its behavior and requirement of MAC layer are discussed. Moreover, five layered architecture (see, Figure 1) is also illustrated on the basis of OSI layered model, in this section. MAC layer covers spectrum sensing, sharing and the scheduling properties. Spectrum sensing, spectrum sharing, spectrum mobility, spectrum management are four important functions.

Spectrum sensing is the ability to detect the spectrum hole. Two important terms in the sensing are: cognation capability and reconfiguration. Spectrum sensing mainly classified into three parts: transmitter detection, cooperative detection, interference temperature measurement, see Figure 4. In addition, various sensing issues are challenges of spectrum sensing are discussed in second section, see Figure 3.

Issues, challenges, and some properties are described with wireless nature and *ad hoc* nature of the CRN, in the third section. Problem statement of MAC layer is also highlighted in this section. In the Figure 5, multichannel hidden terminal problem is investigated.

In the fourth section, classifications of the MAC protocols are described on the basis of duplexing technique (time and frequency), architecture (centralized, distributed, and hybrid), access mode (contention based and contention less), spectrum sharing behavior (cooperative and non-cooperative), spectrum access mode (underlay, overlay, and interweave), and some other types of protocols which are based on the scheduling, game theory, priority, and genetic and swarm intelligence based (see Figure 6).

Priority based MAC protocols and modeling, i.e., POMDP and queuing theory are explained in the fifth section.

802.11-based MAC, 802.22 (WRAN), and MVNO are described as a case study in the sixth section.

In the last section, conclusion is summarized, which is evidently demanding of specific MAC requirement and new framework architecture for successful deployment of cognitive radio networks in the future. In the Table 3, this chapter is concluded and summarized with various MAC solutions by different features and parameters.

REFERENCES

Adamis, K. M. (2007). A new MAC protocol with control channel auto-discovery for self-deployed cognitive radio networks. In *Proceedings of the Program for European Wireless 2007, EW 2007*. EW.

Akyildiz, I. F., Lee, W.-Y., Vuran, M. C., & Mohanty, S. (2006). NeXt generation/dynamic spectrum access/cognitive radio wireless networks: A survey. *Computer Networks*, *50*(13), 2127–2159. doi:10.1016/j.comnet.2006.05.001

Al-Dulaimi, A., Al-Rubaye, S., & Cosmas, J. (2011). Adaptive congestion control for mobility in cognitive radio networks. *Proceedings of the Wireless Advanced (WiAD), 2011*, 273–277. doi:10.1109/WiAd.2011.5983268

Ansari, J., Zhang, X., Achtzehn, A., Petrova, M., & Mahonen, P. (2011). A flexible MAC development framework for cognitive radio systems. *Proceedings of the, WCNC-2011*, 156–161. IEEE Press.

Bae, Y. H., Alfa, A., & Choi, B. D. (2010). Performance analysis of modified IEEE 802.11-based cognitive radio networks. *IEEE Communications Letters*, *14*(10), 975–977. doi:10.1109/LCOMM.2010.082310.100322

Baldini, G., & Pons, E. C. (2011). Design of a robust cognitive control channel for cognitive radio networks based on ultra wideband pulse shaped signal. In *Proceedings of the 4th International Conference on Multiple Access Communications.* (pp. 13-23). Berlin, Germany: Springer.

Ban, T. W., Choi, W., Jung, B. C., & Sung, D. K. (2009). Multi-user diversity in a spectrum sharing system. *IEEE Transactions on Wireless Communications*, *8*(1), 102–106. doi:10.1109/TWC.2009.080326

Benslimane, A., Ali, A., Kobbane, A., & Taleb, T. (2009). A new opportunistic MAC layer protocol for cognitive IEEE 802.11-based wireless networks. In *Proceedings of 2009 IEEE 20th International Symposium on Personal, Indoor and Mobile Radio Communications,* (pp. 2181 - 2185). Tokyo, Japan: IEEE Press.

Brown, T. (2005). An analysis of unlicensed device operation in licensed broadcast service bands. In *Proceedings of DySPAN,* (pp. 11–29). IEEE Press.

Cabric, A. T. (2006). Spectrum sensing measurements of pilot, energy, and collaborative detection. In *Proceedings of Military Communication* (pp. 1–7). Washington, DC: IEEE Press. doi:10.1109/MILCOM.2006.301994

Cesana, C. E. (2011). Routing in cognitive radio networks: Challenges and solutions. *Ad Hoc Networks*, *9*(3), 228–248. doi:10.1016/j.adhoc.2010.06.009

Chen, T., Zhang, H., Maggio, G., & Chlamtac, I. (2007). CogMesh: A cluster-based cognitive radio network. In *Proceedings of 2nd IEEE International Symposium on New Frontiers in Dynamic Spectrum Access Networks, 2007,* (pp. 168-178). IEEE Press.

Chen, Y. (2010). Improved energy detector for random signals in Gaussian noise. *IEEE Transactions on Wireless Communications*, *9*(2), 558–563. doi:10.1109/TWC.2010.5403535

Chen, Y.-S., Cho, C.-H., You, I., & Chao, H.-C. (2011). A cross-layer protocol of spectrum mobility and handover in cognitive LTE networks. *Simulation Modelling Practice and Theory*, *19*(8), 1723–1744. doi:10.1016/j.simpat.2010.09.007

Cheng, S., & Yang, Z. (2008). Adaptive power control algorithm based on SIR in cognitive radios. *Journal of Electronics & Information Technology*, *1*, 15.

Chouhan, A. T. (2011a). Cognitive radio netwoks: Application and implemetation issues. In *Proceedings of the 26th M.P. Young Scientist Congress,* (p. 65). Jabalpur, India: MPCOST.

Chouhan, A. T. (2011b). Cognitive radio networks: Implementation and application issues in India. In *Proceedings of the Seminar on Next Generation Networks – Implementation and Implications.* New Delhi, India: Telecom Regulatory Authority of India (TRAI). Retrieved from http://www.trai.gov.in/WriteReadData/trai/upload/misc/174/Lokesh_Chauhan.pdf

Chouhan, L. (2012). *Design and analysis of mac protocols for cognitive radio networks.* (Unpublished Dissertation). ABV-Indian Institute of Information Technology and Management. Gwalior, India.

Cordeiro, C., & Challapali, K. (2007). C-MAC: A cognitive MAC protocol for multi-channel wireless networks. In *Proceedings of 2nd IEEE International Symposium on New Frontiers in Dynamic Spectrum Access Networks, 2007,* (pp. 147-157). IEEE Press.

Cordeiro, C., Challapali, K., Birru, D., & Sai Shankar, N. (2005). IEEE 802.22: The first worldwide wireless standard based on cognitive radios. In *Proceedings of the First IEEE International Symposium on New Frontiers in Dynamic Spectrum Access Networks, 2005,* (pp. 328 - 337). Baltimore, MD: IEEE Press.

Cormio, C., & Chowdhury, K. R. (2010). Common control channel design for cognitive radio wireless ad hoc networks using adaptive frequency hopping. *Ad Hoc Networks, 8,* 430–438. doi:10.1016/j.adhoc.2009.10.004

Damljanovi, Z. (2010). Mobility management strategies in heterogeneous cognitive radio networks. *Journal of Network and Systems Management, 18,* 4–22. doi:10.1007/s10922-009-9146-0

De Domenico, E. C.-G. (2010). A survey on MAC strategies for cognitive radio networks. *IEEE Communications Surveys & Tutorials, 14*(1), 1–24.

Deng, Z., Shen, L., Bao, N., Su, B., Lin, J., & Wang, D. (2011). Autocorrelation based detection of DSSS signal for cognitive radio system. In *Proceedings of the International Conference on Wireless Communications and Signal Processing (WCSP), 2011.* WCSP.

Dubey, S. S. (2012). Security for cognitive radio networks. In Lin, M.-L. K.-C. (Ed.), *Cognitive Radio and Interference Management: Technology and Strategy.* Hershey, PA: IGI Global. doi:10.4018/978-1-4666-2005-6.ch013

FCC. (2003). *Notice of proposed rulemaking and order, ET Docket No 03-222.* Washington, DC: The Federal Communications Commission (FCC).

FCC. (2005). *Rules for wireless broadband services in the 3650-3700 MHz band (FCC 05-56).* Washington, DC: Federal Communications Commission (FCC).

Foh, E. W. (2009). Analysis of cognitive radio spectrum access with finite user population. *IEEE Communications Letters, 13*(5), 294–296. doi:10.1109/LCOMM.2009.082113

Gandetto, A. F. (2005). A distributed approach to mode identification and spectrum monitoring for cognitive radios. In *Proceedings of the SDR Forum Technical Conference.* Orange County, CA: SDR.

Gandetto, A. F. (2005). Distributed cooperative mode identification for cognitive radio application. In *Proceedings of the International Radio Science Union (URSI).* New Delhi, India: URSI.

Gandetto, A. F. (2007). Spectrum sensing: A distributed approach for cognitive terminals. *IEEE Journal on Selected Areas in Communications, 25*(3), 546–557. doi:10.1109/JSAC.2007.070405

Haykin, S. (2005). Cognitive radio: Brain-empowered wireless communications. *IEEE Journal on Selected Areas in Communications, 23*(2), 201–220. doi:10.1109/JSAC.2004.839380

Hoang, A., Liang, Y.-C., & Zeng, Y. (2010). Adaptive joint scheduling of spectrum sensing and data transmission in cognitive radio networks. *IEEE Transactions on Communications, 58*(1), 235–246. doi:10.1109/TCOMM.2010.01.070270

Hossain, L. L. (2008). OSA-MAC: A MAC protocol for opportunistic spectrum access in cognitive radio networks. *Proceedings of IEEE WCNC, 2008,* 1426–1430. IEEE Press.

How, K.-C., & Ma, M. (2011). Routing and qos provisioning in cognitive radio networks. *Computer Networks, 55*(1), 330–342. doi:10.1016/j.comnet.2010.09.008

HP. (2004). *MAC level (link layer)*. Retrieved February 28, 2012, from http://www.hpl.hp.com/personal/Jean_Tourrilhes/Linux/Linux.Wireless.mac.html

Hu, Y. Y.-D., & Yang, Z. (2012). Cognitive medium access control protocols for secondary users sharing a common channel with time division multiple access primary users. *Wireless Communications and Mobile Computing*, *12*(4), 20–36.

Huang, R. B. (2005). Spectrum sharing with distributed interference compensation. *Proceedings of IEEE DySPAN*, *2005*, 88–93. IEEE Press.

IEEE. (2011). *IEEE 802.22 working group on wireless regional area networks*. Retrieved March 10, 2012, from http://www.ieee802.org/22/

IEEE 802.11. (2012). *Wikipedia*. Retrieved March 1, 2012, from http://en.wikipedia.org/wiki/IEEE_802.11

Iyer, G., & Lim, Y. C. (2011). Efficient multi-channel MAC protocol and channel allocation schemes for TDMA based cognitive radio networks. In *Proceedings of the International Conference on Communications and Signal Processing (ICCSP), 2011,* (pp. 394-398). Kerla, India: IEEE Press.

Jha, S., Rashid, M., Bhargava, V., & Despins, C. (2011). Medium access control in distributed cognitive radio networks. *IEEE Wireless Communications*, *18*(4), 41–51. doi:10.1109/MWC.2011.5999763

Jha, U. P. (2011). Design of omc-mac: An opportunistic multi-channel mac with qos provisioning for distributed cognitive radio networks. *IEEE Transactions on Wireless Communications*, *10*(10), 3414–3425. doi:10.1109/TWC.2011.072511.102196

Jia, J., Zhang, Q., & Shen, X. (2008). HC-MAC: A hardware-constrained cognitive MAC for efficient spectrum management. *IEEE Journal on Selected Areas in Communications*, *26*(1), 106–117. doi:10.1109/JSAC.2008.080110

Jiang, T., Grace, D., & Mitchell, P. D. (2009). Improvement of pre-partitioning on reinforcement learning based spectrum sharing. In *Proceedings of IET International Communication Conference on Wireless Mobile and Computing (CCWMC 2009),* (pp. 299-302). IET.

Kamruzzaman, S. (2010). An energy efficient multichannel MAC protocol for cognitive radio ad hoc networks. *International Journal of Communication Networks and Information Security*, *2*(2), 112–119.

Kandeepan, S., Sierra, A., Campos, J., & Chlamtac, I. (2010). Periodic sensing in cognitive radios for detecting UMTS/HSDPA based on experimental spectral occupancy statistics. In *Proceedings of the Wireless Communications and Networking Conference (WCNC),* (pp. 1-6). Sydney, Australia: IEEE Press.

Kondareddy, Y., & Agrawal, P. (2008). Synchronized MAC protocol for multi-hop cognitive radio networks. In *Proceedings of the International Conference on Communications, 2008,* (pp. 3198-3202). Beijing, China: IEEE Press.

Le, L., & Hossain, E. (2008). A MAC protocol for opportunistic spectrum access in cognitive radio networks. In *Proceedings of the Wireless Communications and Networking Conference, 2008,* (pp. 1426-1430). IEEE Press.

Lee, B.-H., & Wong, C.-M. (2011). Coordinated non-sensing MAC protocol in dynamic spectrum access networks. *Wireless Personal Communications*, *58*, 867–887. doi:10.1007/s11277-010-9999-2

Lee, J., Lee, E., Park, S., Park, H., & Kim, S.-H. (2010). Destination-initiated geographic multicasting protocol in wireless ad hoc sensor networks. In *Proceedings of the IEEE 71st Vehicular Technology Conference (VTC 2010-Spring)*, (pp. 1 - 5). IEEE Press.

Li, D., Dai, X., & Zhang, H. (2008). Cross-layer scheduling and power control in cognitive radio networks. In *Proceedings of the 4th International Conference on Wireless Communications, Networking and Mobile Computing, 2008*, (pp. 1 - 3). IEEE Press.

Li, Z. H. (2009). Queuing analysis of dynamic spectrum access subject to interruptions from primary users. In *Proceedings of the 5th International ICST Conference on Cognitive Radio Oriented Wireless Networks and Communications (CROWNCOM)*, (pp. 1-5). IEEE Press.

Liang, E. P.-C. (2007). Optimization for cooperative sensing in cognitive radio networks. In *Proceedings of the Wireless Communications and Networking Conference*, (pp. 27-32). Hong Kong, China: IEEE Press.

Ma, C.-C. S. (2007). Single-radio adaptive channel algorithm for spectrum agile wireless ad hoc networks. *Proceedings of the DySPAN, 2007*, 547–558. IEEE Press.

Ma, J., Li, G., & Juang, B. H. (2009). Signal processing in cognitive radio. *Proceedings of the IEEE, 97*(5), 805–823. doi:10.1109/JPROC.2009.2015707

Ma, L., Han, X., & Shen, C.-C. (2005). Dynamic open spectrum sharing MAC protocol for wireless ad hoc networks. In *Proceedings of the First IEEE International Symposium on New Frontiers in Dynamic Spectrum Access Networks, 2005*, (pp. 203-213). IEEE Press.

Ma, R.-T., Hsu, Y.-P., & Feng, K.-T. (2009). A POMDP-based spectrum handoff protocol for partially observable cognitive radio networks. In *Proceedings of the Wireless Communications and Networking Conference, 2009*, (pp. 1-6). IEEE Press.

Ma, X. H.-C. (2005). Dynamic open spectrum sharing MAC protocol for wireless ad hoc networks. *Proceedings of the IEEE DySPAN, 2005*, 203–213. IEEE Press.

Meko, S. F., & Chaporkar, P. (2009). Channel partitioning and relay placement in multi-hop cellular networks. In *Proceedings of the 6th International Symposium on Wireless Communication Systems, ISWCS 2009*, (pp. 66-70). Piscataway, NJ: IEEE Press.

Mobile Virtual Network Operator. (2012). *Wikipedia.* Retrieved March 12, 2012, from http://en.wikipedia.org/wiki/MVNO

Murthy, C. B. (2004). *Ad hoc wireless networks: Architectures and protocols.* Upper Saddle River, NJ: Prentice Hall PTR.

Pawełczak, G. J. (2006). Performance measures of dynamic spectrum access networks. In *Proceedings of the Global Telecommunications Conference (Globecom).* San Francisco, CA: IEEE Press.

Qaraqe, K., Ekin, S., Agarwal, T., & Serpedin, E. (2011). Performance analysis of cognitive radio multiple-access channels over dynamic fading environments. *Wireless Personal Communications.* Retrieved from http://www.qscience.com/doi/abs/10.5339/qfarf.2012.CSPS2

Qu, Q., Milstein, L., & Vaman, D. (2008). Cognitive radio based multi-user resource allocation in mobile ad hoc networks using multi-carrier CDMA modulation. *IEEE Journal on Selected Areas in Communications, 26*(1), 70–82. doi:10.1109/JSAC.2008.0801007

Sahai, N. H. (2004). Some fundamental limits on cognitive radio. In *Proceedings of the Allerton Conference on Communication, Control and Computing*, (pp. 1662–1671). IEEE Press.

Salami, G., Durowoju, O., Attar, A., Holland, O., Tafazolli, R., & Aghvami, H. (2011). A comparison between the centralized and distributed approaches for spectrum management. *IEEE Communications Surveys Tutorials*, *13*(2), 274–290. doi:10.1109/SURV.2011.041110.00018

Salous, S. (2010). Chirp sounder measurements for broadband wireless networks and cognitive radio. In *Proceedings of the 7th International Symposium on Communication Systems Networks and Digital Signal Processing (CSNDSP), 2010,* (pp. 846-851). IEEE Press.

Sankaranarayanan, P. P. (2005). A bandwidth sharing approach to improve licensed spectrum utilization. In *Proceedings of IEEE DySPAN,* (pp. 279–288). IEEE Press.

Schedule. (2012). *Wikipedia.* Retrieved from http://en.wikipedia.org/wiki/Schedule_%28resource%29

Scheduling. (2012). *Wikipedia.* Retrieved from http://en.wikipedia.org/wiki/Scheduling_algorithm

Shao, H., & Beaulieu, N. (2011). Direct sequence and time-hopping sequence designs for narrowband interference mitigation in impulse radio UWB systems. *IEEE Transactions on Communications*, *59*(7), 1957–1965. doi:10.1109/TCOMM.2011.060911.100581

Shen, H., Zhang, W., & Kwak, K. S. (2008). Cognitive implementation of chirp waveform in UWB system. *IEICE Transactions on Communications, E91.B*(1), 147-150.

Singh, M. B. (2011). Cooperative spectrum sensing with an improved energy detector in cognitive radio network. In *Proceedings of the National Conference on Communications (NCC),* (pp. 1-5). IEEE Press.

Sousa, A. G. (2005). Collaborative spectrum sensing for opportunistic access in fading environments. In *Proceedings of the Symposium on Dynamic Spectrum Access Networks (DySPAN 2005),* (pp. 131–136). Baltimore, MD: IEEE Press.

Sousa, A. G. (2007). Asymptotic performance of collaborative spectrum sensing under correlated log-normal shadowing. *IEEE Communications Letters*, *11*(1), 34–36.

Spread Spectrum. (2012). *Wikipedia.* Retrieved from http://en.wikipedia.org/wiki/Spread_spectrum

Su, H., & Zhang, X. (2008). Cross-layer based opportunistic MAC protocols for QoS provisionings over cognitive radio wireless networks. *IEEE Journal on Selected Areas in Communications*, *26*(1), 118–129. doi:10.1109/JSAC.2008.080111

Sutton, K. N. (2008). Cyclostationary signatures in practical cognitive radio applications. *IEEE Journal on Selected Areas in Communications*, *26*(1), 13–24. doi:10.1109/JSAC.2008.080103

Telecom Regulatory Authority of India. (2012). *Telecom regulatory authority of India (TRAI) online.* Retrieved from http://www.trai.gov.in

Tu, S.-Y., Chen, K.-C., & Prasad, R. (2009). Spectrum sensing of OFDMA systems for cognitive radio networks. *IEEE Transactions on Vehicular Technology*, *58*(7), 3410–3425. doi:10.1109/TVT.2009.2014775

Tung, A. H. (2008). Dynamic spectrum access with prioritization in open spectrum wireless networks. In *Proceedings of the 11th IEEE Singapore International Communication Systems, 2008,* (pp. 1026 –1030). IEEE Press.

Urkowitz, H. (1967). Energy detection of unknown deterministic signals. *Proceedings of the IEEE, 55*(4), 523–531. doi:10.1109/PROC.1967.5573

Uyanik, G., & Oktug, S. (2011). A priority based cooperative spectrum utilization considering noise in common control channel in cognitive radio networks. In *Proceedings of the 7th International Wireless Communications and Mobile Computing Conference (IWCMC), 2011,* (pp. 477-482). IEEE Press.

Vamsi Krishna, A. D. (2009). A survey on MAC protocols in OSA networks. *Computer Networks, 53*(9), 1377–1394. doi:10.1016/j.comnet.2009.01.003

Vigneron, B. C. (2006). Multiband frequency hopping for high data-rate communications with adaptive use of spectrum. In *Proceedings of the IEEE 63rd Vehicular Technology Conference, 2006,* (pp. 251-255). IEEE Press.

Wang, C. W.-T. (2011). A queueing-theoretical framework for qos-enhanced spectrum management in cognitive radio networks. *IEEE Wireless Communications, 18*(6), 18–26. doi:10.1109/MWC.2011.6108330

Wang, K. R. (2011). *Cognitive radio networking and security: A game theoreitic view*. Cambridge, UK: Cambridge University Press.

Wang, L.-C., & Chen, A. (2007). A cognitive MAC protocol for QoS provisioning in overlaying ad hoc networks. In *Proceedings of the 4th IEEE Consumer Communications and Networking Conference, CCNC 2007,* (pp. 1139-1143). IEEE Press.

Wang, L. N. (2011). Timeslot allocation scheme for cognitive satellite networks. *Advanced Materials Research, 230-232*, 40–43. doi:10.4028/www.scientific.net/AMR.230-232.40

Wang, S.-Y., Huang, Y.-M., Lau, L.-C., & Lin, C.-C. (2011). Enhanced MAC protocol for cognitive radios over IEEE 802.11 networks. In *Proceedings of the Wireless Communications and Networking Conference (WCNC),* (pp. 37 - 42). Cancun, Mexico: IEEE Press.

Wong, C.-M., Chen, J.-D., & Hsu, W.-P. (2009). Coordinated non-sensing MAC protocol for dynamic spectrum access networks. In *Proceedings of the IEEE 20th International Symposium on Personal, Indoor and Mobile Radio Communications, 2009,* (pp. 471-475). IEEE Press.

Wyglinski, M. N. (Ed.). (2010). *Cognitive radio communications and networks: Principels and practice*. New York, NY: Elsevier Academic Press.

Xie, J. X., & Zhou, K. (2012). QoS multicast routing in cognitive radio ad hoc networks. *International Journal of Communication Systems, 25*, 30–46. doi:10.1002/dac.1285

Yan Zhang, J. Z.-H. (Ed.). (2010). *Cognitive radio networks*. Boca Raton, FL: CRC Press. doi:10.1201/EBK1420077759

Yao, N. Y. (2009). Joint design of frequency and power adaptation in FHSS systems based on cognitive radio. In *Proceedings of the 5th International Conference on Wireless Communications, Networking and Mobile Computing, WiCom 2009.* IEEE Press.

Ying-Jie, M. A. (2011). Cognitive UWB adaptive pulse design for interference suppression. *Journal of Beijing University of Posts and Telecommunications, 4*, 61–69.

Yuan, H. F., & Du, S. (2012). Resource allocation for multiuser cognitive OFDM networks with proportional rate constraints. *International Journal of Communication Systems, 25*, 254–269. doi:10.1002/dac.1272

Zamat, H., & Natarajan, B. (2009). Practical architecture of a broadband sensing receiver for use in cognitive radio. *Physical Communication*, *2*(12), 87–102. doi:10.1016/j.phycom.2009.02.005

Zhang, J. X. (2010). Medium access control in cognitive radio networks. In Zhang, J. X., & Yan Zhang, J. Z.-H. (Eds.), *Cognitive Radio Netwoks* (pp. 89–120). Boca Raton, FL: CRC Press. doi:10.1201/EBK1420077759

Zhang, Q. (2009). Cooperative relay to improve diversity in cognitive radio networks. *IEEE Communications Magazine*, *47*(2), 111–117. doi:10.1109/MCOM.2009.4785388

Zhang, W., Bai, S., Liu, Y., & Tang, J. (2010). Cognitive radio scheduling for overwater communications. In *Proceedings of the 53rd IEEE Global Communications Conference, GLOBECOM 2010,* (pp. 1-5). Miami, FL: IEEE Press.

Zhang, X., & Su, H. (2011). Opportunistic spectrum sharing schemes for CDMA-based uplink MAC in cognitive radio networks. *IEEE Journal on Selected Areas in Communications*, *29*(4), 716–730. doi:10.1109/JSAC.2011.110405

Zhang, Y. (2008). Dynamic spectrum access in cognitive radio wireless networks. In *Proceedings of the IEEE International Conference on Communications,* (pp. 4927–4932). IEEE Press.

Zhao, C., Hu, J., & Shen, L. (2010). A MAC protocol of cognitive networks based on IEEE 802.11. In *Proceedings of the 2010 12th IEEE International Conference on Communication Technology (ICCT),* (pp. 1133 - 1136). IEEE Press.

Zhao, H. Z.-H. (2005). Distributed coordination in dynamic spectrum allocation networks. *Proceedings of the IEEE DySPAN*, *2005*, 259–268. IEEE Press.

Zhao, L. T. (2005). Decentralized cognitive MAC for dynamic spectrum access. *Proceedings of IEEE DySPAN*, *2005*, 224–232. IEEE Press.

Zhao, L. T. (2007). Decentralized cognitive mac for opportunistic spectrum access in ad hoc networks: A pomdp framework. *IEEE Journal on Selected Areas in Communications*, *25*(3), 589–600. doi:10.1109/JSAC.2007.070409

Zhao, Q., Tong, L., Swami, A., & Chen, Y. (2007). Decentralized cognitive MAC for opportunistic spectrum access in ad hoc networks: A POMDP framework. *IEEE Journal on Selected Areas in Communications*, *25*(3), 589–600. doi:10.1109/JSAC.2007.070409

Zheng, C. P. (2005). Collaboration and fairness in opportunistic spectrum access. *Proceedings of the IEEE, ICC*, 3132–3136. IEEE Press.

Zhu, L. S. (2007). Analysis of cognitive radio spectrum access with optimal channel reservation. *IEEE Communications Letters*, *11*(4), 304–306. doi:10.1109/LCOM.2007.348282

KEY TERMS AND DEFINITIONS

3G Wireless: The third generation of technology of the mobile wireless industry. Third generation (3G) systems use wideband digital radio technology as compared to 2nd generation narrowband digital radio. For third generation cordless telephones, 3G wireless describes products that use multiple digital radio channels and new registration processes allowed some 3rd generation cordless phones to roam into other public places.

Base Station: The equipment on the network side of a wireless communications link. The base station contains the tower, antennas and radio equipment needed to allow wireless communications devices to connect with the network.

Carrier-to-Interference Ratio: The ratio of power in an RF carrier to the interference power in the channel.

Channel: A general term used to describe a communications path between two systems. They may be either physical or logical depending on the application.

Cognitive Radio Network (CRN): A kind of two-way radio that automatically changes its transmission or reception parameters, in a way where the entire wireless communication network of which it is a node communicates efficiently, while avoiding interference with licensed or licensed exempt users.

Federal Communications Commission (FCC): The regulatory body governing communications technologies in the US and established by the Communications Act of 1934, as amended, and regulates interstate communications (wire, radio, telephone, telegraph and telecommunications) originating in the United States.

Global System for Mobile Communications (GSMC): A wide area wireless communications system that uses digital radio transmission to provide voice, data, and multimedia communication services. A GSM system coordinates the communication between a mobile telephones (mobile stations), base stations (cell sites), and switching systems.

Medium Access Control (MAC): In the seven-layer OSI model of computer networking, Media Access Control (MAC) data communication protocol is a sub layer of the data link layer, which itself is layer 2. The MAC sub layer provides addressing and channel access control mechanisms that make it possible for several terminals or network nodes to communicate within a multiple access network that incorporates a shared medium, e.g. Ethernet. The hardware that implements the MAC is referred to as a medium access controller.

Multiple In Multiple Out (MIMO): The combining or use of two or more radio or telecom transport channels for a communication channel. The ability to use and combine alternate transport links provides for higher data transmission rates (inverse multiplexing) and increased reliability (interference control).

Open Systems Interconnection (OSI): A standard description or "reference model" for how messages should be transmitted between any two points in a telecommunication network. Its purpose is to guide product implementers so that their products will consistently work with other products.

Operator: A company that operates a telephone or cellular network, for example AT&T, Vodafone, or Airtel.

Orthogonal Frequency Division Multiplex (OFDM): A modulation technique that transmits blocks of symbols in parallel by employing a large number of orthogonal subcarriers. The data is divided into blocks and sent in parallel on separate sub-carriers. By doing this, the symbol period can be increased and the effects of delay spread are reduced.

Primary User (Licensed): Has a license to operate in a certain spectrum band. This access can only be controlled by the primary base station and should not be affected by the operations of any other unlicensed users.

Secondary User: Has no spectrum license, hence additional functionalities are required to share the licensed spectrum band.

Session: A virtual connection between two hosts by which network traffic is passed.

Wideband Code Division Multiple Access (WCDMA): A 3rd generation mobile communication system that uses Code Division Multiple Access (CDMA) technology over a wide frequency band to provide high-speed multimedia and efficient voice services.

Wireless Local Area Network (WLAN): Allows computers and workstations to communicate with each other using radio propagation as the transmission medium. The wireless LAN can be connected to an existing wired LAN as an extension, or can form the basis of a new network.

WiMAX: The 802.16A wide area broadband wireless industry standard.

Wireless Fidelity (Wi-Fi): Another name for the 802.11 WLAN system.

Chapter 9
MAC Protocols for Cognitive Radio Ad Hoc Networks:
Sensing Error-Aware and Spectrum Access Strategies

Abdullah Masrub
Brunel University, UK & University of Al-Mergib, Libya

ABSTRACT

In contrast to infrastructure-based networks, in wireless ad hoc networks nodes can discover and communicate with each other directly without involving central access points. In this mode of multi-hop networks, all nodes have equal right to access the medium. Hence, the performance of wireless ad hoc networks is mostly limited by traffic congestion. To alleviate such a problem, Cognitive Radio (CR) technology can be used. In this chapter, a CR-based Medium Access Control (MAC) layer for wireless ad hoc networks is investigated. The authors focus on Cognitive MAC protocols for an unlicensed user, which can be enabled to access the large amount of unused spectrum allocated for a licensed user in an intelligent way without causing any harmful interference. They propose a cognitive MAC protocol based on the theory of the Partially Observed Markov Decision Process (POMDP), which sense the radio spectrum, detect the occupancy state of different primary channels, and then opportunistically communicate over unused channels. The objective is to benefit as much as possible from the available spectrum opportunities by making efficient decisions on which channels to access, which ensures maximization of the throughput of the secondary user.

DOI: 10.4018/978-1-4666-2812-0.ch009

1. INTRODUCTION

1.1. Wireless Ad Hoc Networks

Wireless ad hoc networks are multi-hop systems in which nodes assist each other in transmitting and receiving packets across the network. In contrast to infrastructure-based networks, in an ad hoc network nodes can discover and exchange information with each other directly without involving central access points. A node may join or leave the network at any time. In ad hoc networks, all nodes have equal right to access the medium. To be able to establish communication with each other, each node needs to be able to see the others. If a node wishes to communicate outside its range, another node within the same range operates as a gateway and forwards the contact in a multi-hop fashion. The recent work related to ad hoc networks focus on many issues such as network architecture and network capacity.

1.1.1. Capacity of Wireless Ad Hoc Networks

The problem of capacity is widely investigated and most researchers showed that the capacity can be increased as the size of the network is increased. In fact, capacity of wireless ad hoc networks is based on the traffic behaviour at the Medium Access Control (MAC) layer. As each node in wireless ad hoc networks has to transmit relayed data as well as its own, so we need to discuss the issue of fairness. In addition, in wireless ad hoc networks as the traffic might be directed to the gateways (e.g. Mesh and Sensor Wireless Networks) which are connected to external networks, these gateways would pose a bottleneck problem. In this case, we need to reduce the bottleneck wireless links along the path to the gateway. Other factors that influence the capacity of the network can also be considered, such as interference between simultaneous transmissions, fading, and environmental noise.

1.1.2. Channel Assignment

To address such above-mentioned problems, the existing MAC and routing protocols can be developed and enhanced to be convenient for wireless ad hoc networks. Multiple channels technique, for example, can be assigned between nodes to multiple radios at the same time, so such problems can be minimized and more data can be sent between nodes increasing the overall throughput of the network. In addition, different MAC protocols based on modification of the existed standards (e.g. IEEE 802.11, CDMA, and OFDM) were proposed for utilizing multiple channels. The notion of “soft” channel reservation, for example, was proposed to give preference to the channel that was used for the last successful transmission. Schemes that negotiate channels dynamically were also proposed to enable clients to communicate in the same region simultaneously. Other techniques such as single or multiple transceivers for each node have been widely discussed.

1.2. Cognitive Radio

With the rapid growth in wireless applications, the radio spectrum becomes of fundamental importance. Recent reports made by spectrum regulators such as Federal Communications Commission (FCC) in the United States (US) have shown that almost all the available spectrum has been allocated (Federal Communications Commission, 2003). However, extensive measurements made by Office of Communications (Ofcom) in the UK and Spectrum Policy Task Force (SPTF) in the US indicate that a large amount of licensed spectrum remains unused at a specific time or slot level (FCC, 2002; Office of Communications, 2005). As a result, in recent years, the FCC has been considering more flexible and comprehensive uses of the available spectrum (FCC, 2003). This phenomenon accelerated the emergence of Opportunistic Spectrum Access (OSA) concepts leading to a Cognitive Radio (CR) technology (Mitola, 2000; Wang & Liu, 2011). In CR Networks, licensed (primary) users have high priority to use their spectrum. Meanwhile, to mitigate the spectrum scarcity and improve their service quality, unlicensed (secondary) users are allowed to opportunistically access the spectrum and use temporarily the spectrum spaces unused by the licensed users, which are also known as spectrum holes or white spaces as illustrated in Figure 1. If the licensed users further use these bands, CR users should have the ability to vacate the bands and move dynamically to other spectrum holes or they might stay in the same band but change the transmission power level to avoid interference. To ensure robust system, spectrum management has to be done efficiently in CR Networks. Spectrum management can be composed of four major steps: sensing, decision-making, sharing, and mobility (Haykin, 2005). Cognitive radios have the ability to perform spectrum sensing continuously to recognize the status of the radio spectrum environment. Once white spaces (unused spectrum) are identified, the cognitive radio can change its transmission parameters, such as carrier frequency, bandwidth, power efficiency, and modulation schemes, according to the interactions with the environment in which it operates.

1.2.1. Spectrum Sensing

Spectrum sensing can be considered as the main issue that has to be considered to enable the CR users to explore vacant spectrum opportunities and then to avoid interference with the primary users. In general, three approaches for spectrum sensing techniques are proposed in the literature, primary transmitter detection, primary receiver detection, and interference temperature management (Akyildiz, Lee, Vuran, & Mohanty, 2008). While the primary transmitter detection technique is based on the detection of the weak signal from a primary transmitter through the local observations of CR users, the primary receiver detection aims at finding the primary users that are receiving data within the communication range of a CR user. On the other hand, signal detection approaches can be classified into three categories: (1) energy detection (Poor, 1994), (2) matched filter detection (Kay, 1998), and (3) cyclostationary detection (Enserink & Cochran, 1994). Moreover, different developed algorithms based on the mentioned approaches are proposed in the literature, for example, eigenvalue-based detection (Bianchi, Najim, Maida, & Debbah, 2009; Zeng, Koh, & Liang, 2008; Zeng & Liang, 2009b); and covariance-based detection (Zeng, & Liang, 2007,

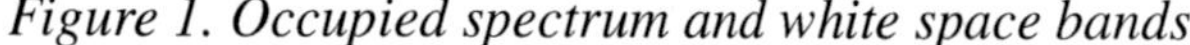

Figure 1. Occupied spectrum and white space bands

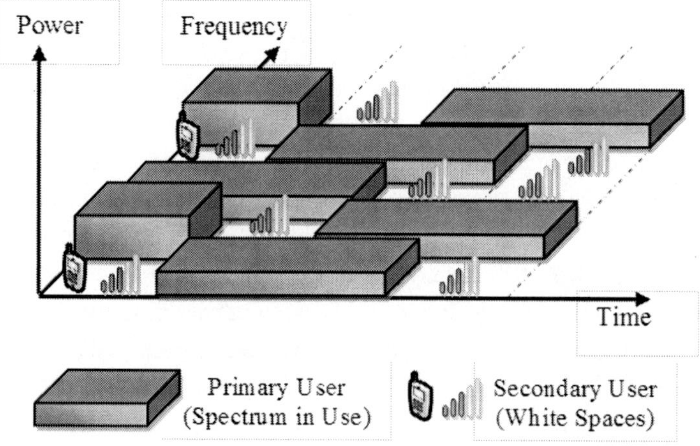

2009a). A survey of various spectrum sensing techniques can be found in (Haykin, Thomson, & Reed, 2009) and the references therein. However, as shown in Figure 2, some factors such as shadowing, hidden terminal problem may significantly affect the detection process of primary signal.

1.2.2. Spectrum Access

Spectrum access techniques have responsibility of enabling CR users to share the spectrum resource by determining which user will access the channel, and when a user can access the channel (Akyildiz, Lee, Vuran, & Mohanty, 2006; Zhao & Sadler, 2007). Different CogMAC protocols based on different architectures and radio technologies have been investigated. In general, MAC protocols for CR-based networks for both infrastructure and mobile ad hoc networks can be classified into three categories: (1) random access protocols, (2) time slotted protocols, and (3) hybrid protocols (Akyildiz, Lee, & Chowdury, 2009). In addition, the number of radio transceivers can be applied in this classification. In the random access approach, the access to the medium is based on the Carrier Sensing Multiple Access and Collision Avoidance (CSMA/CA) scheme, where the time slotted approach aims to divide the time into multiple slots for both control and data transmission. Meanwhile, Hybrid protocols are assumed to be a combination of random access and time slotted approaches. This type of protocols uses a partially slotted channel scheme and tries to access the channel randomly. More details concerning the previous work related to spectrum access techniques can be found in the next section.

2. BACKGROUND AND RELATED WORK

In this section, we focus on the most famous techniques that have been identified as one of the key techniques for designing cognitive MAC in the field of research. Unlike infrastructure-based CR MAC protocols, CR MAC protocols for ad hoc networks do not have a central entity to control and operate the network. To enhance the performance of spectrum usage, CR users, which should be aware of the primary user's activities within the same transmission area, need to determine and control the spectrum sensing, sharing, and access robustly. All these tasks necessitate efficient cooperation with the neighboring nodes located in the same transmission range. Therefore, maintaining time synchronization across the network and obtaining valuable information from surrounding nodes are some of the major factors that need to be considered in the sensing and access protocols design.

Figure 2. Illustration of hidden primary user in CR networks

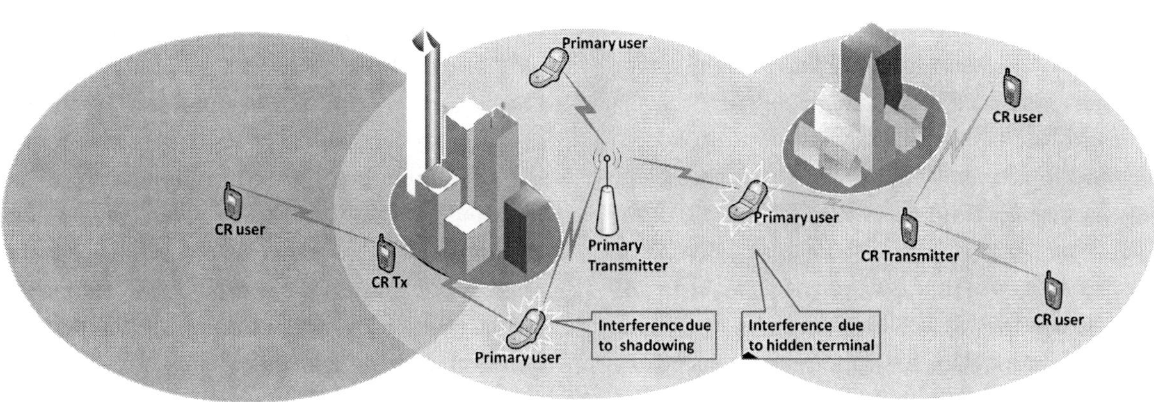

2.1. System Models Overview

Different spectrum sensing, sharing, and access approaches are proposed in the literature. A graph coloring theory based, game theory based, and Markov decision process theory based are the most known approaches proposed for sensing, sharing, and accessing the medium. A bidirectional graph G=(V,L,E), for example, can be used as a framework, where V is a set of vertices denoting the users, L is the color list at each vertex denoting the available channels, and E is a set of undirected edges between vertices representing interference between any two vertices (Wang, Huang, & Jiang, 2009). However, the coloring scheme is constrained by the problem of color-sensitive graph coloring when a specific colored edge cannot be assigned to any two vertexes simultaneously if that color already exists between the two vertexes.

On the other hand, in game theory approach, various strategies are also developed to analyze competitions between the players, which represent the CR users and the primary users, to achieve optimal solutions. A game theoretic approach, for example, can be formulated as follows: The players in the game are the secondary users and the primary user. The primary user is willing to share some portion of the spectrum with the secondary users. The primary user charges a secondary user for the spectrum at a rate of the total size of spectrum available for sharing. The payoff for each player is the profit of the secondary user. the Nash equilibrium is also considered as the solution of this game (Niyato & Hossain, 2008). However, this assumption may not be realistic in some cognitive radio systems.

In addition, various spectrum sensing and access techniques are proposed and developed based on Markov decision process theory. A Partially Observed Markov Decision Process (POMDP), for example, was used for dynamic spectrum access in wireless ad hoc networks. An analytical POMDP framework for opportunistic spectrum access was also developed to allow CR users to access the spectrum by using a decentralized cognitive MAC strategy (Zhao, Tong, Swami, & Chen, 2007). The decision-theoretic approach aims at identifying optimal policies to optimize the performance of the CR network.

2.2. Related Work

In general, previous work on scheduling of CogMAC protocols has considered two main scenarios: primary network scenario and secondary network scenario. For the primary network scenario, the spectrum can be slotted into small bands or un-slotted (e.g. Kim & Shin, 2008a, 2008b; Lee & Akyildiz, 2008). For the secondary network scenario, many models based on POMDP, Game Theory or other approaches have been proposed. All of these approaches offer different solutions and aspects that aim to increase spectrum efficiency by allowing cognitive users to exploit the existence of the huge unused spectrum. In this chapter, we focus only on time slotted approaches.

In Srinivasa, Jafar, and Jindal (2006) the authors presented a scenario for accessing the medium in which the secondary transmitter can sense all the available primary channels, whereas the receiver does not participate in any sensing action. After sensing the medium, the transmitter decides which channel to access. However, a cross-layer approach was presented in Zhao, Tong, Swami, and Chen (2007). Considering the hardware limitation and energy constrains, the authors suppose that the secondary transmitter can only sense a subset of the possible primary channels at the beginning of each time slot. Mehanna, Sultan, and El Gamal (2009) preferred to optimize the on-line learning capabilities of the secondary transmitter and ensure perfect synchronization between the secondary transmitter and receiver. The authors also propose a scenario where only a limited number of channels are sensed by the secondary user at every time slot before making a decision on which cannel to access.

On the other hand, a myopic approach based on the obtained occupancy state estimation for independent and dependent channels has been addressed in Chen, Zhao, and Swami (2008), Zhao, Luo, Yue, and He (2009), and Zhao and Krishnamachari (2007). While Zhao and Krishnamachari (2007) pointed out obvious throughput improvement of myopic sensing as time increases under independent channels, Zhao, Luo, Yue, and He (2009) demonstrated that it is not the case under dependent channels. Chen, Zhao, and Swami (2008) introduced two heuristic approaches with three integrated components to exploit channel correlation: (1) a spectrum sensor at the Physical (PHY) layer as "PHY layer approach," (2) a spectrum sensing strategy at the MAC layer as "MAC layer approach," and (3) a spectrum access strategy as "MAC layer approach." It has been shown that exploiting channel correlation at the PHY layer is more effective than at the MAC layer. In addition, the performance of the PHY layer spectrum sensor can improve over time by incorporating the MAC layer sensing and access decisions. Furthermore, Zhao, Luo, Yue, and He (2009) have demonstrated how sensing errors at the PHY layer affect MAC design and how incorporating MAC layer information into PHY layer leads to a cognitive spectrum sensor whose performance improves over time by learning from accumulating observations.

In addition, alternative approaches that improve sensing reliability are proposed. Zhang and Su (2011) proposed a cognitive radio-enabled transceiver and multiple channel sensors to exploit the vacant spaces in the spectrum. The authors also introduced the four-way handshakes scheme. They have demonstrated that both the traditional and multi-channel hidden terminal problems can be solved. Furthermore, Quan, Cui, Poor, and Sayed (2008), Quan, Cui, and Sayed (2008), and Quan, Cui, Sayed, and Poor (2009) have focused on the development of cooperative sensing schemes among multiple CR users. Quan, Zhang, Shellhammer, and Sayed (2011) proposed a spectral feature detector for spectrum sensing. Using the asymptotic properties of Toeplitz and circular matrices, the authors have demonstrated that this spectral feature detector is asymptotically optimal at very low SNR.

3. SYSTEM MODEL

3.1. Markov Decision Process Strategy

POMDP-based approach is a dynamic system that can be used to achieve sensing and access strategies leading to identification of optimal policies to optimize the CR network. Our POMDP-based design is first proposed in Masrub, Al-Raweshidy, and Abbod (2011). It specified by a finite set of states *S*, set of control actions *A*, a transition probability *P*, and a reward function *R*. The main goal behind using this strategy is to maximize the throughput of unlicensed users. To achieve this, we need to maximize the expected sum of rewards:

$$R^T = E\left\{\Sigma_{t=1}^{T} R\left(t\right)\right\} \tag{1}$$

3.2. Primary Network Model

We consider a spectrum consisting of *N* independent channels, each channel with bandwidth $B_n (n=1,\ldots,N)$. These *N* channels are licensed to PUs, which have an authority to communicate over it according to a synchronous slot structure. The presence and absence of the PUs in each channel of the network represents the traffic strategy of the primary network and can be modeled as alternative time intervals of busy and free states. At a particular time and due to the absence of the primary users, some of these N channels might be free and available for opportunistic transmission by secondary users, which seek free spectrum spaces. We consider each channel to be divided into *T* slots and the network state in a slot t $(t=1,\ldots,T)$ is

Figure 3. The primary network model: N independent channels with T time slots

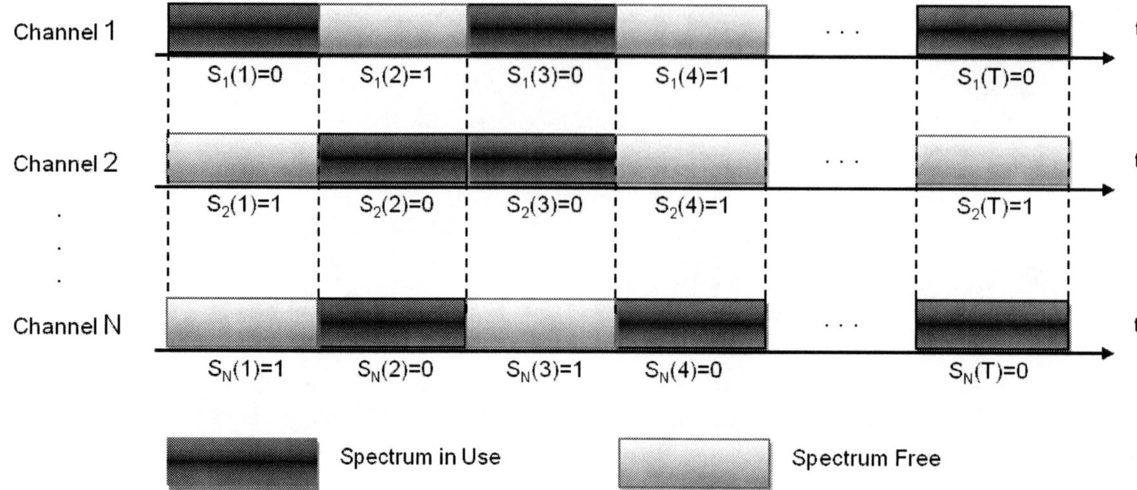

given by $\{S_1(t),...,S_N(t)\}$, where $S_n(t)=1$ when the channel is free and 0 when the channel is busy. The slotted channels diagram is illustrated in Figure 3.

3.3. Secondary Network Model

We consider the following components forming the core of our proposed model strategy:

- **States:** We define a set of states as follows:

$$S=\begin{bmatrix} S_1(1) & S_1(2) & \dots & S_1(T) \\ S_2(1) & S_2(2) & \dots & S_2(T) \\ \vdots & \vdots & \ddots & \vdots \\ S_N(1) & S_N(2) & \dots & S_N(T) \end{bmatrix} \qquad (2)$$

  where a row vector $[S_n(1), S_n(2),..., S_n(T)]$ is a vector state of channel *n*, and the network state in channel *n* and slot t is represented by $S_n(t)$ So with *N* channels and *T* slots for each channel we can get $S=2^{N\times T}$ different states.
- **Sub-States:** To reduce the computational complexity problem, we define a sub-set of states as a column vector state of slot *t*:

$$S_t = [S_1(t), S_2(t),..., S_N(t)]^T$$

where the network with *N* channels is represented by $S_t = 2^N$ different sub-states for each time slot. At the start of each time slot the SUs will be invited to sense the medium for the *N* channels of the system. Example: N=2, so we can get $2^2=4$ possible sub-states (see Table 1).

For any SU, checking the availability of channels at the slot time *t* can be indicated by the column vector by performing the following hypothesis test:

$$\begin{cases} H_0 : S_n(t) = 1 & \text{if the channel is free} \\ H_1 : S_n(t) = 0 & \text{otherwise} \end{cases} \qquad (3)$$

Table 1. The possible sub-states for n=2

	Channel 1	Channel 2
Sub-State 1	0	0
Sub-State 2	0	1
Sub-State 3	1	0
Sub-State 4	1	1

- **Actions:** We now define a set of actions which consist of 2^N different possible actions.
- **Example:** N=2, so we can get $2^2=4$ possible actions: (1) transmit nothing, (2) transmit through channel one, (3) transmit through channel two, and (4) choose one of the available channels to transmit through and send a packet to other SUs informing them that there are other available channels to be used.
  We assume that at some point of time, the SU sensed the current state as:

$$S = \begin{bmatrix} 0 & 0 \\ 0 & 0 \end{bmatrix}$$

Taking action "sensing $S_{n,1}$" will lead to four possible next states:

1. There are no available slots:

$$S' = \begin{bmatrix} 0 & 0 \\ 0 & 0 \end{bmatrix}$$

2. There is only one slot available at channel one $S_1(1)$:

$$S' = \begin{bmatrix} 1 & 0 \\ 0 & 0 \end{bmatrix}$$

3. There is only one slot available at channel two $S_2(1)$:

$$S' = \begin{bmatrix} 0 & 0 \\ 1 & 0 \end{bmatrix}$$

4. There are two available slots $S_1(1)$ and $S_2(1)$:

$$S' = \begin{bmatrix} 1 & 0 \\ 1 & 0 \end{bmatrix}$$

The SU, which acts as a decision maker, has to select which channel to be accessed based on the early observations.

- **Information State:** We refer to the belief state $\pi(t)$ as the summary statistic of all past decisions and observations:

$$\pi(t) = [\lambda_1(t), \lambda_2(t), \ldots, \lambda_N(t)]$$

where $\lambda_n(t)$ denotes the conditional probability that channel n is available in slot t.

- **Reward:** We finally define the transition reward. We assume that each used channel will give an amount of reward:

$$R(t) = \Sigma_{n=1}^{N} S_n(t) B_n(t) \quad (4)$$

where $S_n(t) \in \{0, 1\}$ is the state of channel n in slot t, and at the beginning of each time slot using the belief state $\pi(t)$, the secondary user decides to access the channel:

$$n*(t) = \arg\max_{n=1,N} [\lambda_n(t) B_n(t)] \quad (5)$$

4. ERROR FREE SPECTRUM SENSING

We consider that the information state is updated after each action and observation with the application of Bayes' rule as follows:

$$\pi(t+1) = [\lambda_1(t+1), \lambda_2(t+1), \ldots, \lambda_N(t+1)]$$

We also consider that the conditional probability $\lambda_n(t+1)$ depends on the channel transition probabilities P^{01} and P^{11}. We assume that the channel transition probabilities P^{01} and P^{11} are random variables that update their values after sensing all the primary channels at the start of each time slot. To update the channel transition probabilities P^{01} and P^{11}, the secondary user should keep track of the history of all transition states as described in Mehanna, Sultan, and El Gamal (2009):

- Number of state transitions from busy to busy:

$$S_n^{00}(t) = \Sigma_{i=1}^{t-1}\left(1 - S_n(i)\right)\left(1 - S_n(i+1)\right) \tag{6}$$

- Number of state transitions from busy to free:

$$S_n^{01}(t) = \Sigma_{i=1}^{t-1}\left(1 - S_n(i)\right)S_n(i+1) \tag{7}$$

- Number of state transitions from free to busy:

$$S_n^{10}(t) = \Sigma_{i=1}^{t-1}S_n(i)\left(1 - S_n(i+1)\right) \tag{8}$$

- Number of state transitions from free to free:

$$S_n^{11}(t) = \Sigma_{i=1}^{t-1}S_n(i)S_n(i+1) \tag{9}$$

Depending on the previous state transitions, P^{01} is updated as follows:

$$P^{01}(t) = \frac{S^{00}(t) + S^{01}(t)}{t} \tag{10}$$

and P^{11} is updated as follows:

$$P^{11}(t) = \frac{S^{10}(t) + S^{11}(t)}{t} \tag{11}$$

We consider that the initial belief state is given as the stationary distribution of the channel occupancy state:

$$\pi(0) = \frac{P^{01}}{1 + P^{01} - P^{11}} \tag{12}$$

Since the belief state $\pi(t) = [\lambda_1(t), \lambda_2(t), \ldots, \lambda_N(t)]$, we will consider two cases of formulation for the conditional probability of the availability of a channel:

1. In the first strategy, we consider that the conditional probability $\lambda_n(t+1)$ depends on the channel transition probabilities P^{01} and P^{11} as follows:

$$\lambda_n(t+1) = \lambda_n(t)P^{11} + \left(1 - \lambda_n(t)\right)P^{01} \tag{13}$$

2. In the second strategy, we consider that the conditional probability $\lambda_n(t+1)$ depends only on the last belief state $\pi(t)$ as follows:

$$\lambda_n(t+1) = \lambda_n(t)\pi(t) \tag{14}$$

3. In the last strategy, we consider that the initial belief state is generated randomly:

$$0 < \pi(0) < 1$$

5. IMPACT OF SENSING ERRORS ON MAC PERFORMANCE

5.1. Problem Formulation

Obviously, spectrum sensing is a critical function of CR networks; it allows secondary users to detect spectral holes and to opportunistically use under-utilized frequency bands without causing harmful interference to legacy systems. Because of low computational complexity, the energy detector approach has been considered as one of the most common ways for spectrum sensing. The spectrum sensing problem can be formulated as follows:

$$y(n) = \begin{cases} w(n) & \mathcal{H}_0(white\ space) \\ s(n) + w(n) & \mathcal{H}_1(occupied) \end{cases} \tag{15}$$

In general, the performance of the detection algorithm can be summarized with two probabilities: probability of false alarm P_F and probability of detection P_D. P_F is the probability that the test incorrectly decides that the considered frequency is occupied when it actually is not, and it can be written as:

$$P_F = P_r\left(T(y) > \lambda | \mathcal{H}_0\right) \tag{16}$$

P_D is the probability of detecting a signal on the considered frequency when it truly is present. Thus, a large detection probability is desired. It can be formulated as:

$$P_D = P_r\left(T(y) > \lambda | \mathcal{H}_1\right) \tag{17}$$

and the missed detection as:

$$P_{MD} = 1 - P_D \tag{18}$$

Let us assume that under the hypotheses H_0 when the primary user is absent, the received signal has the following simple form:

$$y(n) = w(n) \tag{19}$$

Under the hypotheses H_1 when the primary user is active, the received signal has the following form:

$$y(n) = s(n) + w(n) \tag{20}$$

where $s(n)$ is the primary user's transmitted signal, and $w(n)$ is the Additive White Gaussian Noise (AWGN) sample. For M independent measurements is taken at the beginning of each slot, the decision metric for the energy detector is given by:

$$T(y) = \frac{1}{M} \Sigma_{n=0}^{M} \left|y(n)\right|^2 \tag{21}$$

Let $\sigma_{n,0}^2 = \sigma_w^2$ and $\sigma_{n,1}^2 = \sigma_s^2$ denote the noise and the primary signal power, respectively, in channel n. The white noise can be modeled as a zero-mean Gaussian random variable with variance $\sigma_{n,0}^2$ as:

$$w(n) = \mathcal{N}\left(0, \sigma_{n,0}^2\right)$$

We also consider that the signal can be modelled as a zero-mean Gaussian random variable with variance $\sigma_{n,1}^2$ as:

$$s(n) = \mathcal{N}\left(0, \sigma_{n,1}^2\right)$$

The decision metric $T(y)$ is normally distributed under both the mentioned hypotheses as follows:

$$T(y) \sim \begin{cases} \mathcal{N}\left(0, \sigma_{n,0}^2\right) & \mathcal{H}_0\left(S_n = 0\right) \\ \mathcal{N}\left(0, \sigma_{n,1}^2 + \sigma_{n,0}^2\right) & \mathcal{H}_1\left(S_n = 1\right) \end{cases} \tag{22}$$

5.2. Joint Sensing and Scheduling

We recognize that it is hard to distinguish between a white spectrum and a weak primary signal attenuated by noise and deep fading. However, we develop our proposed model to investigate the effect of sensing errors on the MAC layer performance. In this section, we present the joint sensing and scheduling scheme. We consider $\delta_{n,t}$ as a threshold which can be determined from the summary statistic of the tth time slot in the nth channel. For each time slot, the decision on the presence or absence of primary signals can be made by CR users according to the threshold δ_t which can be written as a row vector state of the slot t: $\delta_t=[\delta_{1,t}, \delta_{2,t},..., \delta_{N,t}]^T$. Globally, the statistics across the network can be represented in matrix form as follows:

$$\delta = \begin{bmatrix} \delta_{1,1} & \delta_{1,2} & \cdots & \delta_{1,T} \\ \delta_{2,1} & \delta_{2,2} & \cdots & \delta_{2,T} \\ \vdots & \vdots & \ddots & \vdots \\ \delta_{N,1} & \delta_{N,2} & \cdots & \delta_{N,T} \end{bmatrix} \tag{23}$$

For the energy detector, the probability of false alarm is given by:

$$P_F = 1 - \gamma\left(\frac{M}{2}, \frac{\delta}{\sigma_{n,0}^2}\right) \tag{24}$$

The probability of missed detection is given by:

$$P_{MD} = \gamma\left(\frac{M}{2}, \frac{\delta}{\sigma_{n,1}^2 + \sigma_{n,0}^2}\right) \tag{25}$$

where:

$$\gamma\left(m, a\right) = \frac{1}{r\left(m\right)} \int_0^a t^{m-1} e^{-t} dt \tag{26}$$

is the incomplete gamma function.

To ensure certain values for P_F and P_{MD}, the required number of measurements M is given by:

$$M = 2\left[Q^{-1}\left(P_F\right) - Q^{-1}\left(1 - P_{MD}\right)\sqrt{1 + 2SNR}\right]^2 SNR^{-2} \tag{27}$$

where SNR denotes the ratio of the primary signal power to the noise power, i.e.

$$SNR = \frac{\sigma_{n,1}^2}{\sigma_{n,0}^2} \tag{28}$$

6. PERFORMANCE EVALUATION

In this section, based on the POMDP framework proposed in the previouse section, we present simulation results for different scenarios of POMDP-based cognitive MAC protocols based on greedy sensing approach. In the first scenario we assume that the number of primary channels are three independent channels, each with bandwidth B=1 and number of slots T=30. In this scenario we assume that sensing errors are ignored and we only focus on one secondary user. We first investigate the throughput achieved by the SU using three different strategies.

Figure 4 shows the throughput comparison between the different spectrum occupancy statistics after applying (13), with prior knowledge about the channel transition probabilities $\{P^{01}, P^{11}\}$. In Case 1 and Case 2 with assumed large values for the probability of the channel state remain unchanged from the idle state $\{P^{11}=0.9$

Figure 4. Throughput comparison using Equation 4 with different parameters (bandwidth B=1, number of channels N=3, and different transition probabilities {P01, P11})

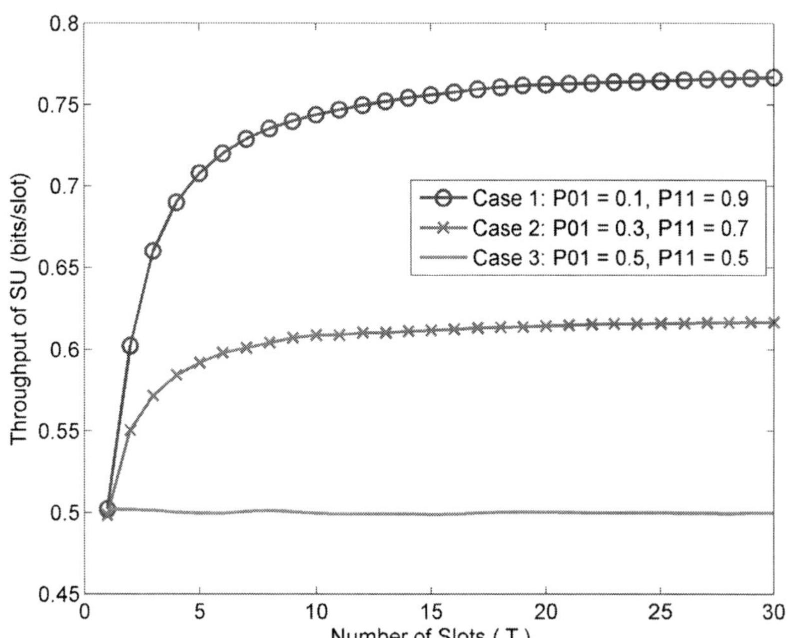

and 0.7, respectively}, we observe an increase in the aggregate throughput of SU over time for both cases. On the other hand, the comparison of Case 1 with Case 2 shows that whenever the probability of channel unchanged from idle state is bigger, the throughput achieved by the SU is higher. This corresponds to large message length and large inter-arrival time. However, in Case 3 where the values of transition probabilities are equals, 0.5 for each, (which means the channel is as likely to change the state or remain at the current state), we notice no change in the throughput of SU over time.

Applying (14) on the same scenario by ignoring the channel transition probabilities, Figure 5 shows the throughput comparison between the different spectrum occupancy statistics where the updated information state depends only on the last belief state $\pi(t)$. In other words, we keep using the channels transition probabilities $\{P^{01}, P^{11}\}$ to generate the initial belief state in each case and ignoring it in updating the belief state. It is obvious from Figure 5 that the three cases achieved high throughput over time. Comparing Figure 5 to Figure 4, Figure 5 shows sharp improvement in the throughput of SU especially for case 3 and case 2. These results reflect the usage of (14) where the SU always expects the channel to stay available. We further proceed by applying the same scenario with some changes. We consider that the initial belief is generated randomly and we do not need to use the channel transition probabilities any more. By using (14) to update the belief state, we plot the aggregate throughput by repeating the simulation three times. As shown in Figure 6, we observe that the three plots are almost congruent achieving the highest aggregate throughput.

In the second scenario, we consider one of the above mentioned three cases. We set the channel transition probabilities at a specific values $\{P^{01}=0.1, P^{11}=0.9\}$, as case 1 performs the best throughput in Figure 4 and Figure 5. We apply this scenario to a different number of primary

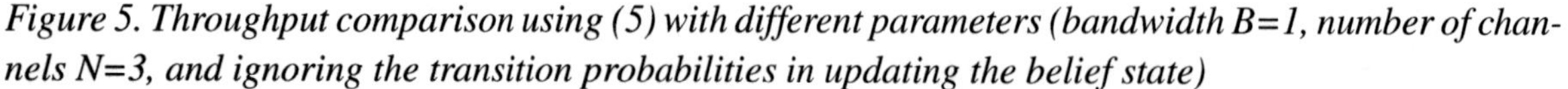

Figure 5. Throughput comparison using (5) with different parameters (bandwidth B=1, number of channels N=3, and ignoring the transition probabilities in updating the belief state)

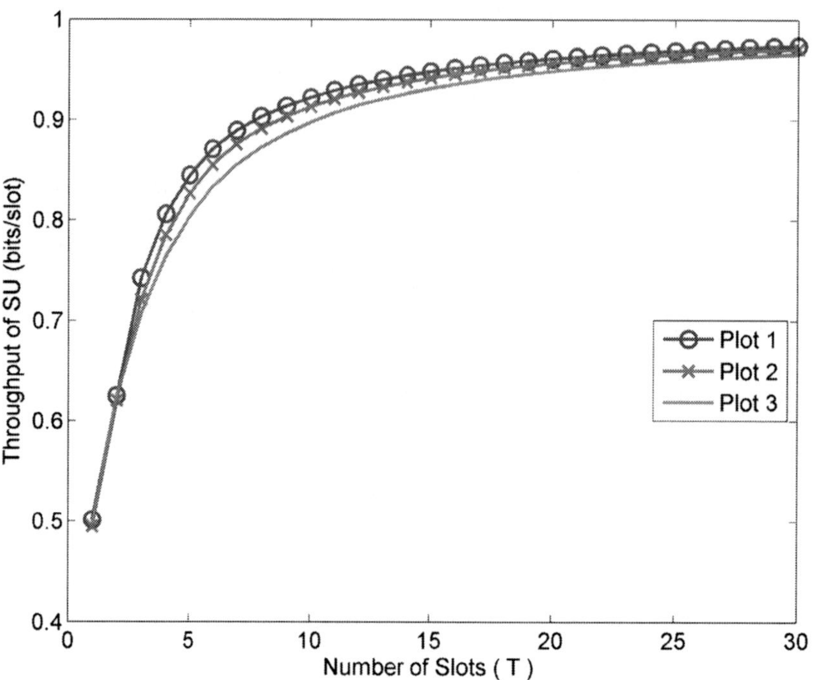

Figure 6. Throughput comparison using Equation 5 with different parameters (bandwidth B=1, number of channels N=3, and generating the initial belief state randomly)

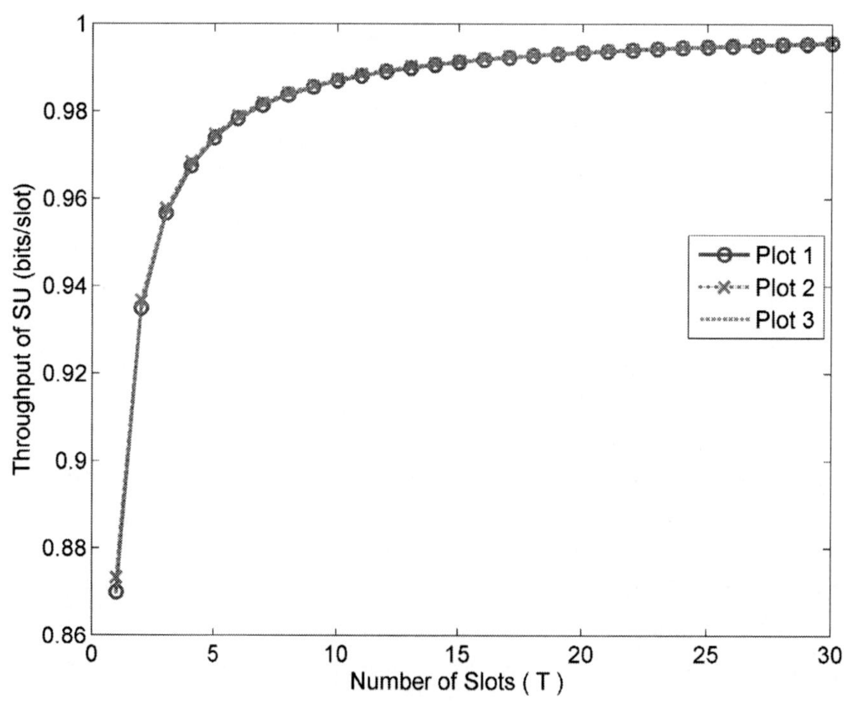

channels {N=2, 3, 4, and 5}, with bandwidth B=1 and number of slots T=30. After applying this scenario, firstly considering the channel transition probabilities by using (13), secondly ignoring the channel transition probabilities in updating the information state by using (14), and finally by generating the initial belief state randomly, Figures 7, 8, and 9 show the throughput comparison of the greedy approach. In general, we observe that the aggregate throughput of SU increases as the number of channels increases in the three strategies. While Figure 7 and Figure 8 show similar increase in the aggregate throughput for all cases {N=2, 3, 4, and 5}, Figure 9 shows that the achievable throughput converges asymptotically, achieving the highest throughput. These results again reflect the usage of (14) where the SU always expects the channel to stay available. In addition, applying this strategy is shown to achieve a throughput very close to the upper bound; as a result, a convergence of throughput is observed.

Now we present the third scenario. We consider the sensing errors and we investigate the impact of false alarm and missed-detection probabilities on the MAC layer performance. We consider a scenario that maximizes the aggregate throughput under the constraint that the false-alarm probability does not exceed a specified level. In our simulation, we adapt the threshold level depending on the value of P_{MD}. Figure 10 shows that the throughput of the secondary user increases with the increase of SNR. Notice that when the P_{MD} is small, the detection threshold increases with the increase of SNR, and hence the P_F decreases with the increase of SNR. Therefore, when SNR is small, the throughput of the secondary user is limited by the large P_F. On the other hand, when SNR is large, the P_F is reduced at the expense of less transmission time in each slot, which also leads to high throughput. In other words, the Figure 10 shows that small P_{MD} gives a better performance at high SNR, whereas large P_{MD} gives high throughput at low SNR. The fact behind that can be illustrated in Figure 11. Figure 11 illustrates that high spectrum efficiency can be obtained at high SNR whenever the P_{MD} is small.

Figure 7. Throughput comparison using Equation 4 with different parameters (bandwidth B=1, transition probabilities {P01=0.1, P11=0.9}, and different number of channels N)

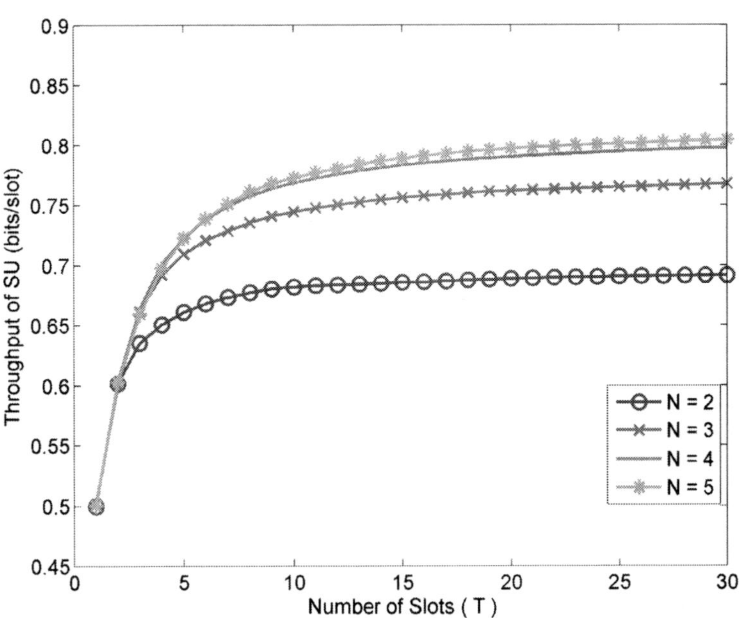

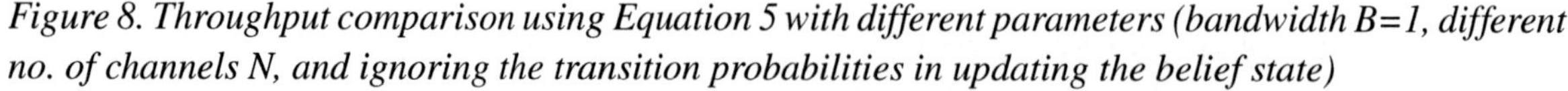

Figure 8. Throughput comparison using Equation 5 with different parameters (bandwidth B=1, different no. of channels N, and ignoring the transition probabilities in updating the belief state)

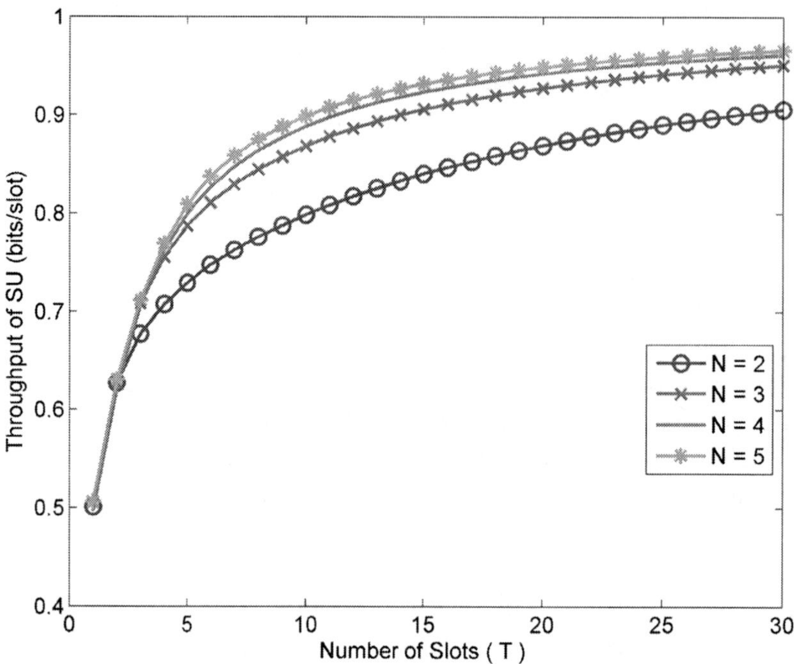

Figure 9. Throughput comparison using Equation 5 with different parameters (bandwidth B=1, different number of channels N, and generating the initial belief state randomly)

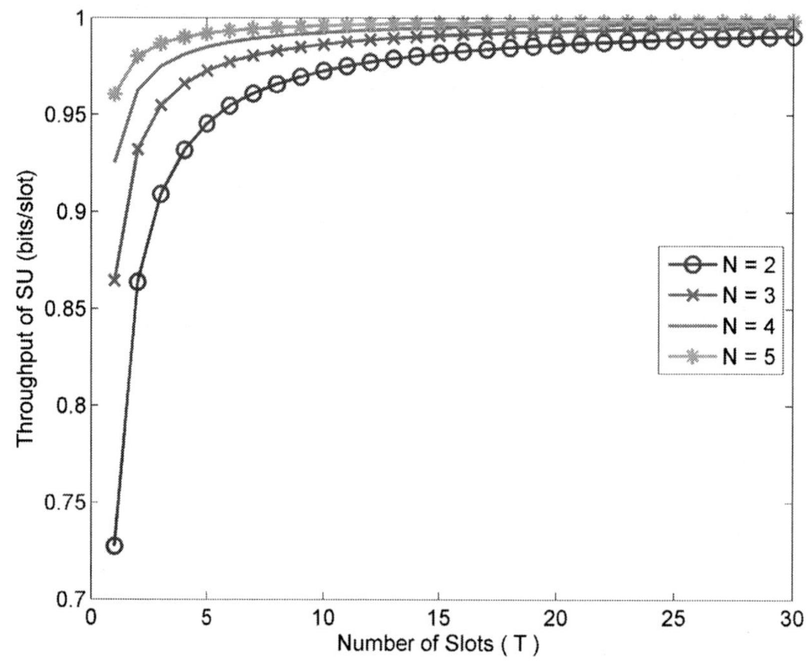

Figure 10. Throughput comparison with different parameters (bandwidth B=1, number of channels N=3, and different values of missed detection probability)

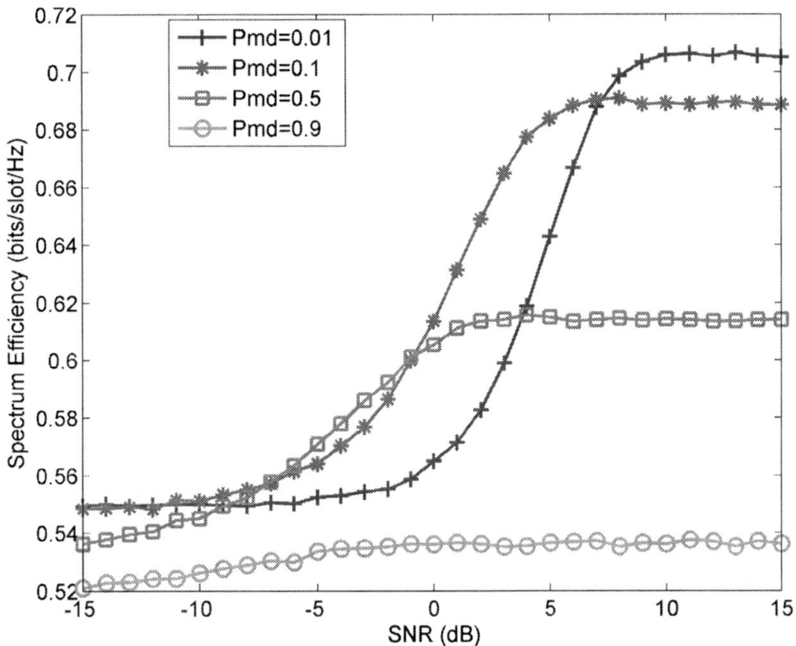

Figure 11. Spectrum efficiency comparison with different parameters (bandwidth B=1, number of channels N=3, and different values of missed detection probability)

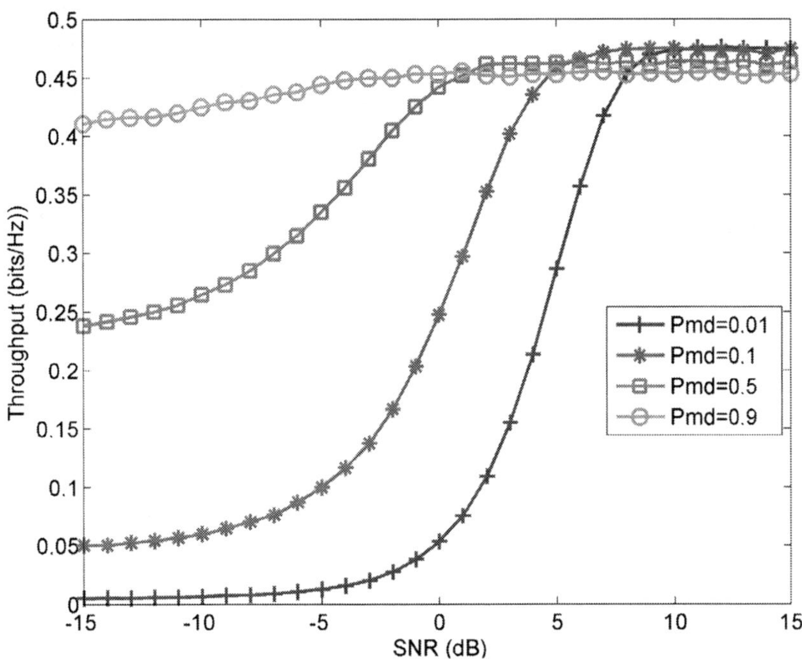

7. CONCLUSION

In this chapter, we have studied a POMDP-based cognitive MAC protocols for multi-channel wireless ad hoc networks. Using multiple channels and assuming a time slotted strategy for the primary network, we propose that the unlicensed user makes efficient decisions for sensing and accessing the medium based on the channel transition probabilities and the belief state. By updating the belief state that summarizes the knowledge of the network state based on all past decisions and the channel transition probabilities or previous belief states, the secondary user perform more spectrum efficiency. Using three different strategies, our results demonstrate an improvement in the aggregate throughput using greedy strategy. Furthermore, by generating the belief state randomly, we present an alternative way to improve the aggregate throughput of the SU. In addition, we demonstrate that the aggregate throughput can be improved whenever the number of channels increases. In context of considering the sensing errors, we investigated the impact of false-alarm and missed-detection probabilities on the MAC layer performance. We have shown that the throughput of the secondary user can be improved with the increase of SNR by decreasing the P_F which can be adapted by the threshold level. Our results demonstrates that whenever the P_{MD} is small, the detection threshold increases with the increase of SNR. As aresult, the P_F decreases and the high throughput is achieved.

REFERENCES

Akyildiz, I. F., Lee, W. Y., & Chowdury, K. R. (2009). CRAHNs: Cognitive radio ad hoc networks. *Journal of Ad Hoc Networks*, *7*(5), 810–836. doi:10.1016/j.adhoc.2009.01.001

Akyildiz, I. F., Lee, W. Y., Vuran, M. C., & Mohanty, S. (2006). NeXt generation/ dynamic spectrum access/cognitive radio wireless networks: A survey. *Elsevier Computer Networks Journal*, *50*(13), 2127–2159. doi:10.1016/j.comnet.2006.05.001

Akyildiz, I. F., Lee, W. Y., Vuran, M. C., & Mohanty, S. (2008). A survey on spectrum management in cognitive radio networks. *IEEE Communications Magazine*, *46*(4), 40–48. doi:10.1109/MCOM.2008.4481339

Bianchi, P., Najim, J., Maida, M., & Debbah, M. (2009). Performance analysis of some eigen-based hypothesis tests for collaborative sensing. In *Proceedings of the IEEE Workshop Datatistical Signal Processing*, (pp. 5–8). Cardiff, UK: IEEE Press.

Chen, Y., Zhao, Q., & Swami, A. (2008). Joint design and separation principle for opportunistic spectrum access in the presence of sensing errors. *IEEE Transactions on Information Theory*, *54*(5), 2053–2071. doi:10.1109/TIT.2008.920248

Enserink, S., & Cochran, D. (1994). A cyclostationary feature detector. In *Proceedings of the 28th Asilomar Conerence on Signals, Systems, and Computers*, (vol. 2., pp. 806–810). Pacific Grove, CA: Asilomar.

Federal Communications Commission. (2002). *Spectrum policy task force report*. Washington, DC: Federal Communications Commission (FCC).

Federal Communications Commission. (2003). *Notice of proposed rule making and order*. Washington, DC: Federal Communications Commission (FCC).

Haykin, S. (2005). Cognitive radio: Brain-empowered wireless communications. *IEEE Journal on Selected Areas in Communications*, *23*(2), 201–220. doi:10.1109/JSAC.2004.839380

Haykin, S., Thomson, D. J., & Reed, J. H. (2009). Spectrum sensing for cognitive radio. *Proceedings of the IEEE*, *97*(5), 849–877. doi:10.1109/JPROC.2009.2015711

Kay, S. M. (1998). *Fundamentals of statistical signal processing: Detection theory*. Englewood Cliffs, NJ: Prentice-Hall.

Kim, H., & Shin, K. (2008). Efficient discovery of spectrum opportunities with MAC-layer sensing in cognitive radio networks. *IEEE Transactions on Mobile Computing*, *7*(5), 533–545. doi:10.1109/TMC.2007.70751

Kim, H., & Shin, K. G. (2008). Fast discovery of spectrum opportunities in cognitive radio networks. In *Proceedings of the 3rd IEEE Symposia on New Frontiers in Dynamic Spectrum Access Networks*, (pp. 1-12). Chicago, IL: IEEE Press.

Lee, W., & Akyildiz, I. F. (2008). Optimal spectrum sensing framework for cognitive radio networks. *IEEE Transactions on Wireless Communications*, *7*(10), 3845–3857. doi:10.1109/TWC.2008.070391

Masrub, A., Al-Raweshidy, H., & Abbod, M. (2011). Cognitive radio based MAC protocols for wireless ad hoc networks. In *Proceedings of the 4th International Conference on Developments in eSystems Engineering*, (pp. 465-469). Dubai, UAE: IEEE.

Mehanna, O., Sultan, A., & El Gamal, H. (2009). Blind cognitive MAC protocols. In *Proceedings of the IEEE International Conference on Communications,* (pp. 1-5). Dresden, Germany: IEEE Press.

Mitola, J. (2000). *Cognitive radio: An integrated agent architecture for software defined radio*. (Doctoral Dissertation). KTH Royal Institute of Technology. Stockholm, Sweden.

Nasipuri, A., Zhuang, J., & Das, S. R. (1999). A multichannel CSMA MAC protocol for multihop wireless networks. In *Proceedings of the IEEE Wireless Communications and Networking Conference,* (vol. 3, pp. 1402-1406). New Orleans, LA: IEEE Press.

Niyato, D., & Hossain, E. (2008). Competitive spectrum sharing in cognitive radio networks: A dynamic game approach. *IEEE Transactions on Wireless Communications*, *7*(7), 2651–2660. doi:10.1109/TWC.2008.070073

Office of Communications. (2005). *Technology Research Programme*. Washington, DC: Office of Communications.

Poor, H. V. (1994). *An introduction to signal detection and estimation*. New York, NY: Springer-Verlag.

Quan, Z., Cui, S., Poor, H. V., & Sayed, A. H. (2008). Collaborative wideband sensing for cognitive radios. *IEEE Signal Processing Magazine*, *25*(6), 60–73. doi:10.1109/MSP.2008.929296

Quan, Z., Cui, S., & Sayed, A. H. (2008). Optimal linear cooperation for spectrum sensing in cognitive radio networks. *IEEE Journal on Selected Topics in Signal Processing*, *2*(1), 28–40. doi:10.1109/JSTSP.2007.914882

Quan, Z., Cui, S., Sayed, A. H., & Poor, H. V. (2009). Optimal multiband joint detection for spectrum sensing in cognitive radio networks. *IEEE Journal Transaction on Signal Processing*, *57*(3), 1128–1140. doi:10.1109/TSP.2008.2008540

Quan, Z., Zhang, W., Shellhammer, S. J., & Sayed, A. H. (2011). Optimal spectral feature detection for spectrum sensing at very low SNR. *IEEE Journal Transaction on Communications*, *59*(1), 201–202. doi:10.1109/TCOMM.2010.112310.090306

Srinivasa, S., Jafar, S., & Jindal, N. (2006). On the capacity of the cognitive tracking channel. In *Proceedings of the IEEE International Symposium on Information Theory*, (pp. 2077-2080). Seattle, WA: IEEE Press.

Wang, B., & Liu, K. J. (2011). Advances in cognitive radio networks: A survey. *IEEE Journal on Selected Topics in Signal Processing*, *5*(1), 5–23. doi:10.1109/JSTSP.2010.2093210

Wang, J., Huang, Y., & Jiang, H. (2009). Improved algorithm of spectrum allocation based on graph coloring model in cognitive radio. In *Proceedings of the International Conference on Communication and Mobile Computing*, (vol. 3, pp. 353-357). Yunnan, China: IEEE.

Zeng, Y., Koh, C. L., & Liang, Y. C. (2008). Maximum eigenvalue detection: Theory and application. In *Proceedings of the IEEE International Conference on Commununications*, (pp. 4160–4164). Beijing, China: IEEE Press.

Zeng, Y., & Liang, Y. C. (2009b). Eigenvalue-based spectrum sensing algorithms for cognitive radio. *IEEE Transactions on Communications*, *57*(6), 1784–1793. doi:10.1109/TCOMM.2009.06.070402

Zeng, Y. H., & Liang, Y. C. (2007). Covariance based signal detections for cognitive radio. In *Proceedings of the IEEE Dynamic Spectrum Access Networks*, (pp. 202–207). Dublin, Ireland: IEEE Press.

Zeng, Y. H., & Liang, Y. C. (2009a). Spectrum sensing algorithms for cognitive radio based on statistical covariance. *IEEE Transactions on Vehicular Technology*, *58*(4), 1804–1815. doi:10.1109/TVT.2008.2005267

Zhang, X., & Su, H. (2011). CREAM-MAC: Cognitive radio-enabled multi-channel MAC protocol over dynamic spectrum access networks. *IEEE Journal of Selected Topics in Signal Processing*, *5*(1), 110–123. doi:10.1109/JSTSP.2010.2091941

Zhao, H., Luo, T., Yue, G., & He, X. (2009). Myopic sensing for opportunistic spectrum access using channel correlation. In *Proceedings of the International Conference on Wireless Communications and Mobile Computing*, (pp. 512-516). Leipzig, Germany: IEEE.

Zhao, Q., & Krishnamachari, B. (2007). Structure and optimality of myopic sensing for opportunistic spectrum access. In *Proceedings of the IEEE International Conference on Communications*, (pp. 6476-6481). Glasgow, UK: IEEE Press.

Zhao, Q., & Sadler, B. M. (2007). A survey of dynamic spectrum access. *IEEE Signal Processing Magazine*, *24*(3), 79–89. doi:10.1109/MSP.2007.361604

Zhao, Q., Tong, L., Swami, A., & Chen, Y. (2007). Decentralized cognitive MAC for opportunistic spectrum access in ad hoc networks: A POMDP framework. *IEEE Journal on Selected Areas in Communications*, *25*(3), 589–600. doi:10.1109/JSAC.2007.070409

KEY TERMS AND DEFINITIONS

Channel: A transmission medium used to convey information from user to other.

Cognitive Radio: A transmission system allows the unlicensed users to adjust its transmission parameters to avoid interference with the licensed users.

Medium Access Control (MAC): A sub-layer acting as an interface between the upper sub-layer LLC and the physical layer, so users can communicate with each other.

Primary User: A licensed user who has the priority to use a certain bands of the spectrum.

Secondary User: An unlicensed user who has a permission to use the spectrum without causing any harmful interference with the primary user who licensed that spectrum.

Throughput: The average rate of the amount of information sent successfully over a communication channel.

Wireless Network: A type of computer networks that uses radio waves to connect between a group of terminals or nodes.

Chapter 10
MAC Layer Spectrum Sensing in Cognitive Radio Networks

Mohamed Hamid
University of Gävle, The Royal Institute of Technology (KTH), Sweden

Abbas Mohammed
Blekinge Institute of Technology, Sweden

ABSTRACT

Efficient use of the available licensed radio spectrum is becoming increasingly difficult as the demand and usage of the radio spectrum increases. This usage of the spectrum is not uniform within the licensed band but concentrated in certain frequencies of the spectrum while other parts of the spectrum are inefficiently utilized. In cognitive radio environments, the primary users are allocated licensed frequency bands while secondary cognitive users can dynamically allocate the empty frequencies within the licensed frequency band, according to their requested quality of service specifications. In this chapter, the authors investigate and assess the performance of MAC layer sensing schemes in cognitive radio networks. Two performance metrics are used to assess the performance of the sensing schemes: the available spectrum utilization and the idle channel search delay for reactive and proactive sensing schemes. In proactive sensing, the adapted and non-adapted sensing period schemes are also assessed. Simulation results show that proactive sensing with adapted periods provides superior performance at the expense of higher computational cost performed by network nodes.

1. INTRODUCTION

In cognitive radio environments the primary users are allocated licensed frequency bands while secondary cognitive users can be dynamically allocated the empty frequencies within the licensed frequency band, according to their requested Quality of Service (QoS) specifications. Spectrum sensing is commonly recognized as the most fundamental task in cognitive radio based on dynamic spectrum access due to its important role in discovering the spectrum holes (Yucek & Arslan, 2009; Haykin, 2005). To achieve this goal the unlicensed user should monitor the licensed channels to identify the spectrum holes and to properly utilize them. In order to adopt the spectrum-agile features required

DOI: 10.4018/978-1-4666-2812-0.ch010

by cognitive radios, enhancements in physical and MAC (Medium Access Control) layers are needed (Ying, Kwang, Li, & Mahonen, 2011).

In cognitive radio networks, MAC protocols can be categorized as Direct Access Based (DAB) or Dynamic Spectrum Allocation (DSA) (De Domenico, Strinati, & Di Benedetto, 2012). In DAB protocols, each transmitter-receiver pair aims to optimize its local optimization goal and no global network optimization is considered since the computational cost and latency are restricted by the simple protocol architecture. DSA protocols, on the other hand, aim at maximizing the global optimization goal for the whole network in an adaptive manner using complicated algorithms.

MAC layer protocols play important roles in cognitive radio networks. The main functionalities of MAC layer in cognitive radios includes: spectrum mobility, spectrum sensing, resource allocation, and spectrum sharing (Akyildiz, Lee, Vuran, & Mohanty, 2006). Spectrum mobility lets the secondary user to jump from one channel when it is re-occupied by its primary user to another vacant channel. Spectrum sensing is the process of detecting primary user's existence and maintaining a picture of the available channels from secondary usage. Resource allocation is responsible for resources assignment for secondary users according to their QoS requirements. Spectrum sharing handles the coordination among primary and secondary users to avoid harmful interference.

The task of sensing in the physical layer is to decide on the existence or absence of licensed user's signals on different channels. There are several proposed physical layer detection methods for performing this task such as energy detection (Kim, Xin, & Rangarajan, 2010), matched filtering detection (Kapoor, Rao, & Singh, 2011), feature detection (Liu & Zhai, 2008), and Eigen-values based spectrum detection (Yonghong & Ying-Chang, 2007). The channel-sensing outcome could be one of the following possibilities:

- The channel is idle.
- The channel is occupied by a licensed user but can be utilized by the unlicensed user with some power constrains so that the Quality of Service (QoS) of the licensed user transmission is not degraded to unacceptable level by the interference from the unlicensed user.
- The channel is not available to the unlicensed user.

For the aforementioned possible outcomes of physical layer sensing to be available, some important fundamental questions arise: when the available radio channels for the unlicensed usage should be sensed and in which order and how frequent? These tasks are the responsibility of MAC layer. Thus, in this chapter we address the potential need of MAC layer sensing schemes in cognitive radio networks aiming to find a clear policy describing how the available spectrum should be utilized to achieve as low idle channel search delay as possible. Both reactive and proactive sensing methods are considered. In proactive sensing the adapted and non-adapted sensing periods, schemes are also assessed. The assessment of these sensing schemes will be validated via two performance metrics: available spectrum utilization and idle channel search delay. According to the best of our knowledge, proactive and reactive sensing performance has not been extensively studied before.

This chapter is structured as follows. In Section 2 we briefly introduce the channel usage model and utilization factor definition. The MAC layer sensing modes (reactive and proactive) are presented in Section 3. In Section 4, the system model, simulation scenario and parameters used in the simulations are presented. In addition, the simulation results showing the performance of each scheme are also analyzed in this section. Finally, the conclusions and directions for future work are drawn in Section 5.

2. CHANNEL USAGE PATTERN

2.1. Markov Process Channel Usage Model

Communication channels can be modeled as ON/OFF Markov process model or 0/1 state; 0 for free channel and 1 for occupied channel by either licensed or other unlicensed user under the assumption that there are no priority considerations among the unlicensed users. This 0/1 alternating model is referred to as channel usage pattern where unlicensed users can utilize only portions of the OFF periods to communicate with other nodes. The channel usage pattern is shown in Figure 1.

Let us assume a radio system with N channels where each channel is addressed as $i(i=1,2,3,\ldots,N)$; the lengths of ON(Y^i) and OFF(X^i) periods are described by their corresponding random probability density functions (pdf), $f_Y^i(y)$ and $f_x^i(x)$, respectively. If we assume these ON and OFF periods to be exponentially distributed with means of $E_{Yi}(y)$ and $E_{Xi}(x)$, then we will end up with distributions of ON and OFF periods as stated in Equations 1 and 2:

$$f_{Y^i}(y) = \lambda_{y^i} e^{-\lambda_{y^i} y} \tag{1}$$

$$f_{X^i}(x) = \lambda_{x^i} e^{-\lambda_{x}^{i} x} \tag{2}$$

where:

Figure 1. Alternating ON/OFF channel usage pattern

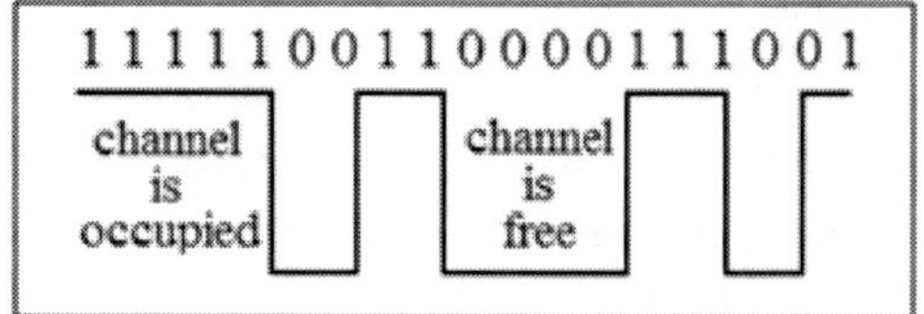

$f_{Y^i}(y)$: pdf of the ON periods of channel *i*,

$f_{X^i}(x)$: pdf of the OFF periods of channel *i*,

$$\lambda_{y^i} = 1 / E_{y^i}(y)$$

$$\lambda_{x^i} = 1 / E_{x^i}(x)$$

2.2. Channel Utilization Factor

Channel utilization factor of channel $i(u^i)$ is defined as the fraction of time (t) in which the channel has been utilized by its licensed users throughout long enough time period (i.e., $t \to \infty$). From this definition we can derive a relationship between the channel utilization and its random distribution parameters as follows

$$u^i = \frac{\int_0^{\infty} \lambda_{y^i}{}^{y} dy}{\int_0^{\infty} \lambda_{y^i} e^{-\lambda_{y^i y}} dy + \int_0^{\infty} \lambda_{x^i} e^{-\lambda_{x^i x}} dx} \tag{3}$$

where

$$u^i = \frac{E_{yi}(y)}{E_{yi}(y) + E_{xi}(x)} \tag{4}$$

which can be expressed in terms of λ_{yi} and λ_{xi} as

$$u^i = \frac{\lambda_{xi}}{\lambda_{yi}(y) + \lambda_{xi}} \tag{5}$$

3. MAC LAYER SENSING

From the MAC layer point of view, the availability of a particular channel to the unlicensed user can be sensed either reactively or proactively (Runsheng, Zhongyu, & Tao, 2008; Wang & Wang, 2008).

3.1. Reactive Sensing

Reactive sensing is an on demand-sensing scheme where the available channels are sensed when the unlicensed user has a packet to be sent or received; otherwise, the unlicensed user sleeps. During sensing, if any idle channel is found then it will be utilized and the wireless link between the unlicensed user and the other entity will be established. If no idle channel is found after completing the sensing of all channels then the unlicensed user will sleep for a short period of t seconds, and then resume sensing until finding an idle channel to utilize. According to the channels utilization factors the channel are sensed in a random order since there is no prior knowledge about channels utilization factors. During utilizing any channel, the unlicensed user use Listen-Before-Talk mechanism to check licensed users presence, and if any licensed user appearance is detected then the channel should be released and restart the sensing procedure from the beginning. The procedure of reactive sensing is illustrated in Figure 2.

3.2. Proactive Sensing

In this type of sensing unlicensed user periodically sense the channels besides the on demand sensing when communication is needed. The purpose of the periodic sensing is to estimate the channel usage pattern in order to determine the most desirable sensing order for on demand sensing. This most desirable sensing order is governed by the estimated channel utilization factors order aiming to reduce the idle channel search delay. Hence, the

Figure 2. System flow diagram of reactive sensing procedure

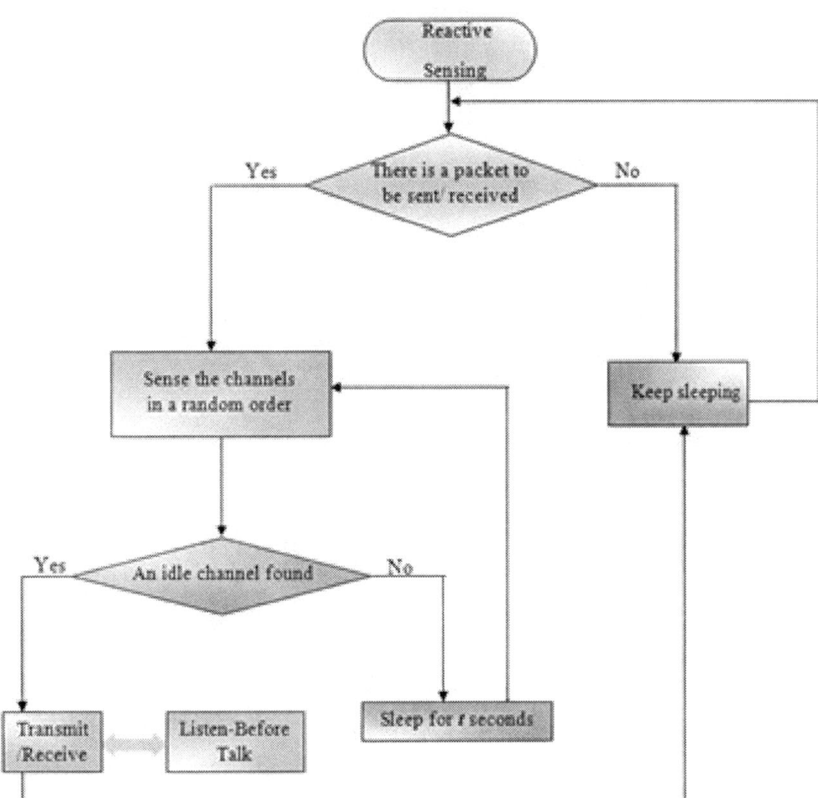

on demand sensing part is the same as the one on reactive sensing except that the channels are sensed in a specific order rather than in a random order. Figure 3 shows the proactive sensing procedure.

Wireless communication channels are not generally stationary and so when they are proactively sensed their sensing periods may need to be adapted according to the channel variations. The change we mean here is the change in channel utilization factor. In fact, the adaptation of sensing periods is an additional computational overhead, which may be traded off with the benefits from this adaptation represented by the achieved spectrum utilization and idle channel search delay. The choice of which proactive sensing method to be used (adapted sensing periods or with non-adapted sensing periods) is thus dependent on the degree of channel stationarity. Proactive sensing with adapted sensing periods is shown in Figure 4.

3.3. Unexplored Opportunities and Sensing Overhead

In proactive sensing, the channel is sampled discretely in time and thus it is not possible to identify when an opportunity (spectrum hole) begins and ends exactly which may result in missing some opportunities. These missed opportunities increase with increasing of the sensing periods; however, reducing the sensing periods blindly is not desirable either since it will increase the sensing

Figure 3. System flow diagram of proactive sensing procedure

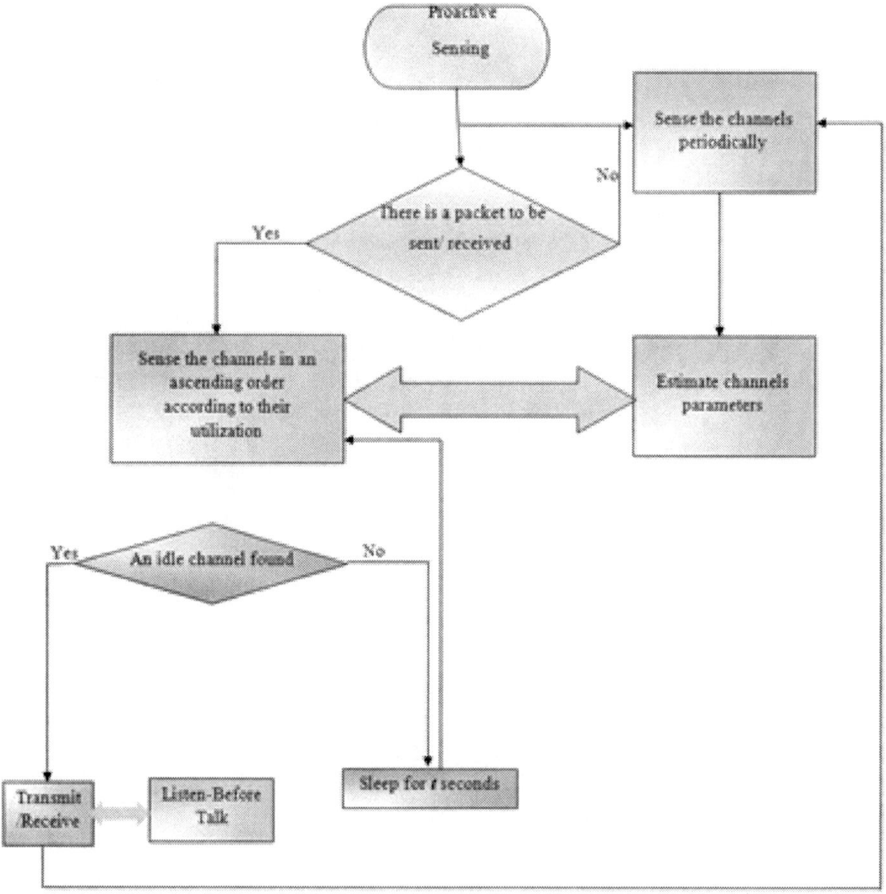

Figure 4. System flow diagram of proactive sensing with sensing periods adaptation procedure

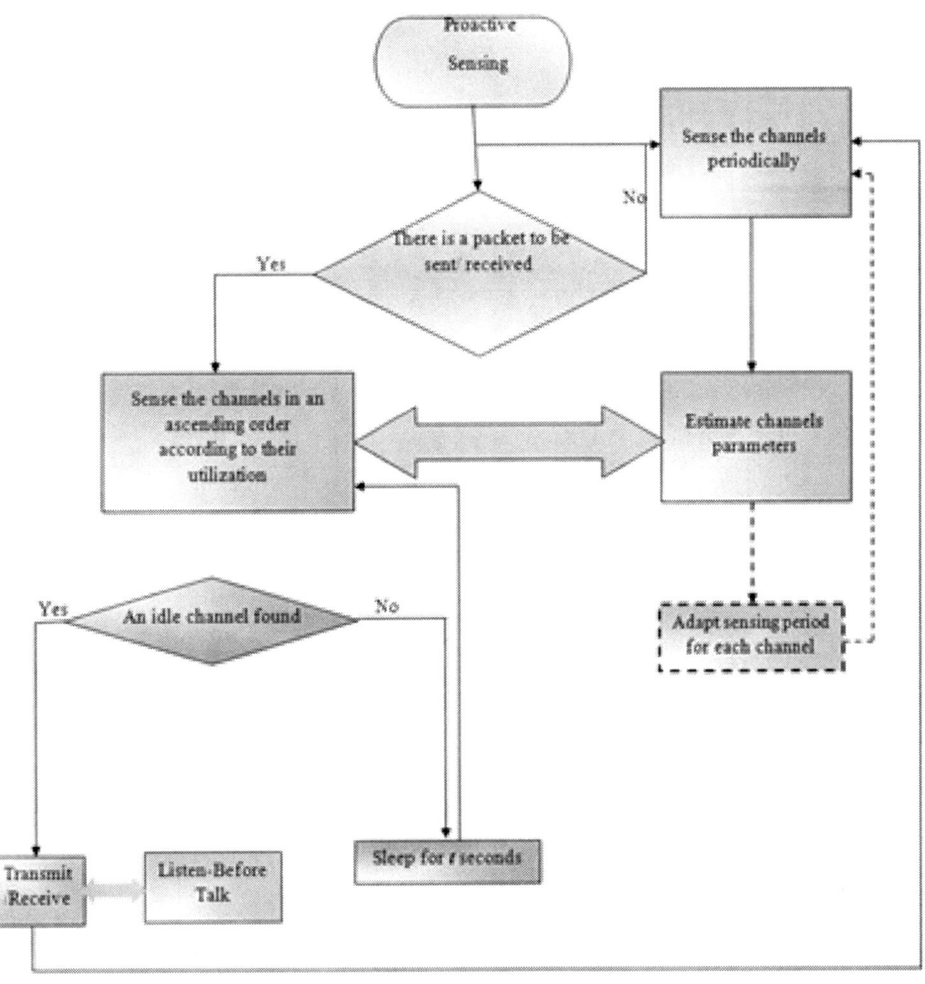

overhead. Therefore, we need to tradeoff between these two parameters since they impact the value of sensing period. Here we introduce two terms: unexplored opportunities and sensing overhead.

- **Unexplored Opportunities** ***(UOPP*** ^{i} ***):*** The fraction of time during which channel *i*'s opportunities are not discovered.
- **Sensing Overhead** ***(SSOH*** ^{i} ***):*** The average fraction of time during which channel *i*'s discovered opportunities cannot be utilized due to sensing of other channels, assuming that the secondary user must stop utilizing a discovered channel while it is sensing one of the other channels. Figure 5 describes the concept of SSOH graphically for two channels.

3.4. Optimization of Sensing Periods in Adapted Sensing Periods Proactive Sensing

Let us assume that the time needed to sense channel *i*, which is referred to as the listening interval, to be (T_I^i) and the sensing period of channel *i* is (T_P^i). T_I^i is determined by physical layer sensing since it depends on the used modulation scheme, sample duration, sample energy and other physical

Figure 5. Example of unexplored opportunities (UOPP) and sensing overhead (SSOH) in two channels system

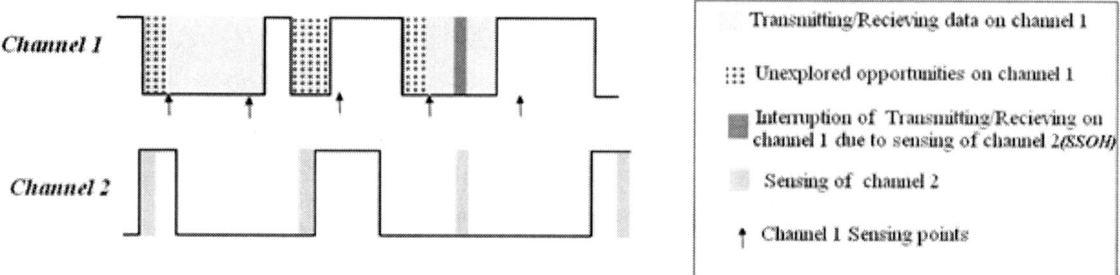

layer characteristics. Thus, our task in MAC layer is to optimize T_P^i in order to utilize the available spectrum as much as possible. In Figure 6, T_I^i and T_P^i are shown for two channels.

For a radio system with *N* channels our objective function is

$\mathbf{T_P}: = (T_P^1, T_P^2, \ldots, T_P^N)$

Then our task is to find T_P such that

$$T_p^* = \arg\max_{Tp}\left\{\sum_{i=1}^{N}\left\{\left(1-u^i\right) - SSOH^i - UOPP^i\right\}\right\} \tag{6}$$

where T_p^* is the optimal sensing periods vector.

Since $(1\text{-}u^i)$ is not related to T_p^*, so Equation 6 can be rewritten as

$$T_p^* = \arg\min_{Tp}\left\{\sum_{i=1}^{N}\left\{SSOH^i + UOPP^i\right\}\right\} \tag{7}$$

To derive a mathematical expression for $SSOH^i$ and $UOPP^i$ a couple of assumptions should be dealt with to simplify the problem:

- In case there are simultaneous opportunities on multiple channels, unlicensed users can assign themselves simultaneously to one or more data links using noncontiguous OFDM technique (Gao, Wang, & Li, 2010).
- Each unlicensed user performs consistent transmission. That is, there is always an incoming/outgoing packet from/to any unlicensed node. So, in this case, every discovered idle channel is assigned to one of the

Figure 6. Example of sensing period $\left(T_P^i\right)$ and listening interval $\left(T_I^j\right)$ in two channels radio system

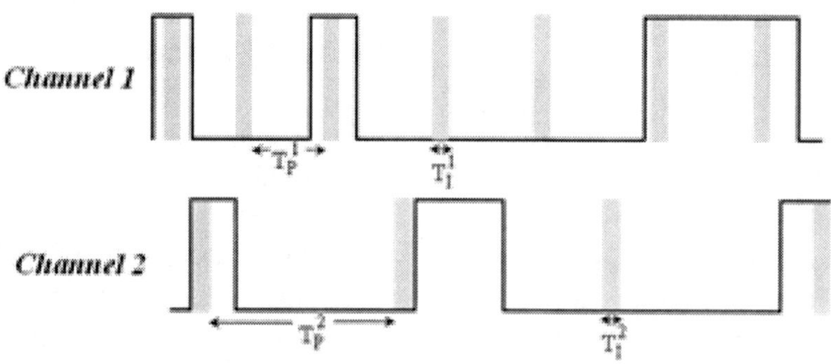

data links and is utilized until its current idle period ends.

3.4.1. Analysis of UOPP^i

Let $T_d^i(t)$ to be the average opportunities on channel i through a period between t and $t+t_s$, where t_s is the sensing point and d is either 0 or 1 given that a sample d is captured at time t_s. t_s can be an end or start of an idle period, so in this case we use $\tilde{T}_d^i(t)$ instead of $T_d^i(t)$; then we have four possible cases: $\tilde{T}_0^i(t)$, $\tilde{T}_1^i(t)$, $T_0^i(t)$, and $T_1^i(t)$ as illustrated in Figure 7.

Let $\tilde{X}^i$ to be the remaining time of an OFF period at the sensing time t_s. The distribution of $\tilde{X}^i$ is given by (Cox, 1967):

$$f_{\tilde{X}^i}(x) = \frac{\mathcal{F}_X(\tilde{X}^i)}{E(x^i)} \tag{8a}$$

where

$$\mathcal{F}_X(\tilde{X}^i) = 1 - F_x(\tilde{X}^i)$$

Similarly, for an ON period we get

$$f_{\tilde{Y}^i}(x) = \frac{\mathcal{F}_Y(\tilde{Y}^i)}{E(y^i)} \tag{8b}$$

where

$$\mathcal{F}_Y(\tilde{Y}^i) = 1 - F_y(\tilde{Y}^i)$$

Since we are interested in calculating $UOPP^i$ then we need to find both $\tilde{T}_0^i(t)$ and $\tilde{T}_1^i(t)$ respectively. To achieve this task we can apply the renewal theory concepts (Cox, 1967) to get:

$$T_0^i(t) = t\int_t^{\infty} \frac{\mathcal{F}x^i(x)}{E(x^i)}dx + \int_0^t \frac{\mathcal{F}x^i(x)}{E(x^i)}\left(x + \tilde{T}_1^i(t-x)\right)dx \tag{9}$$

$$T_1^i(t) = \int_0^t \frac{\mathcal{F}y^i(y)}{E(y^i)}\tilde{T}_0^i(t-y)\,dy \tag{10}$$

$$T_0^i(t) = t\int_t^{\infty} f_{x^i}(x)dx + \int_t^{\infty} f_{x^i}(x) + \left(x + \tilde{T}_1^i(t-x)\right)dx \tag{11}$$

$$T_1^i(t) = \int_0^t f_{y^i}(y)\tilde{T}_0^i(t-y)\,dy \tag{12}$$

By applying Laplace transforms, we get:

Figure 7. $\tilde{T}_d^i(t)$

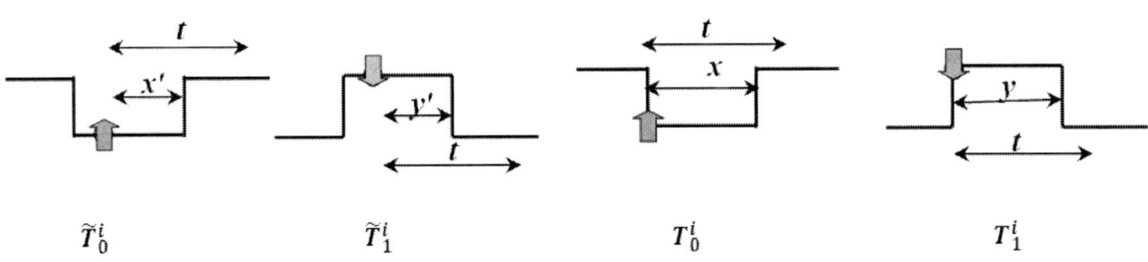

$$E\left(x^{i}\right)T_{0}^{i*}\left(s\right)=\frac{\mathcal{F}x^{i*}\left(0\right)-\mathcal{F}x^{i*}\left(s\right)}{s^{2}}+\mathcal{F}x^{i*}\left(s\right)\tilde{T}_{1}^{i}\left(s\right) \quad (13)$$

$$E\left(y^{i}\right)T_{1}^{i*}\left(s\right)=\mathcal{F}y^{i*}\left(s\right)\tilde{T}_{0}^{i}\left(s\right) \quad (14)$$

$$T_{1}^{i*}\left(s\right)=f_{y^{i*}}\left(s\right)\tilde{T}_{0}^{i}\left(s\right) \quad (15)$$

$$\tilde{T}_{0}^{i*}\left(s\right)=\frac{f_{x^{i*}}\left(0\right)-f_{x^{i*}}x^{i*}\left(s\right)}{s^{2}}+f_{x^{i*}}\left(s\right)\tilde{T}_{1}^{i}\left(s\right) \quad (16)$$

This leads to:

$$T_{0}^{i*}\left(s\right)=\frac{1}{E\left(x^{i}\right)s^{2}}\left[\mathcal{F}_{x^{i*}}\left(0\right)-\mathcal{F}_{x^{i*}}\left(s\right)\frac{1-f_{x}^{*}i\left(0\right)f_{y}^{*}i\left(s\right)}{1-f_{x}^{*}i\left(s\right)f_{y}^{*}i\left(s\right)}\right] \quad (17)$$

$$T_{1}^{i*}\left(s\right)=\frac{\mathcal{F}_{y^{i*}}\left(s\right)}{E\left(y^{i}\right)s^{2}}\left[\frac{f_{x}^{*}i\left(0\right)f_{x}^{*}i\left(s\right)}{1-f_{x}^{*}i\left(s\right)f_{y}^{*}i\left(s\right)}\right] \quad (18)$$

As introduced, $UOPP^i$ is defined as the average fraction of time during which utilizable opportunities on channel *i* are not discovered taking into account that the maximum value of $UOPP^i$ is $(1\text{-}u^i)$. Furthermore, an opportunity means *0* in our channel usage pattern, which makes it mathematically possible to express $UOPP^i$ as the 'length' of undiscovered zeroes in the channel usage pattern between the last captured *1* and next captured symbol as a ratio of $(1\text{-}u^i)$. Hence, $UOPP^i$ is found to be:

$$UOPP^{i}=\left(1-u^{i}\right)\left[\frac{1}{T_{p}^{i}}\int_{0}^{T_{p}^{i}}\frac{\mathcal{F}_{x^{i}}\left(x\right)}{E\left(x^{i}\right)}\tilde{T}_{1}^{i}\left(T_{p}^{i}-x\right)dx\right] \quad (19)$$

If we consider a radio system with exponentially distributed values of the ON and OFF periods as in Equations 1 and 2, then $UOPP^i$ can be expressed as:

$$UOPP^{i}=\left(1-u^{i}\right)\left[1+\frac{1}{\lambda_{x^{i}}T_{p}^{i}}\left(e^{-\lambda_{x^{i}}T_{p}^{i}}-1\right)\right] \quad (20)$$

Thus, $UOPP^i$ and T_p^i are related as shown in Figure 8.

3.4.2. Analysis of SSOH^i

SSOH^i is entire radio system dependant as it depends on sensing of the other channels rather than channel *i* itself as depicted in Figure 5. Since the unlicensed user has no way to detect whether the channel is free or not continuously, then any unlicensed user constructs its own channel usage pattern extracted from the discrete sensing procedure introduced earlier. This new constructed channel usage pattern is referred to as "observed channel usage pattern," as shown in Figure 9.

Using the observed channel usage pattern, there will be a new value of the channel utilization factor $\tilde{u}^i$ as

$$\tilde{u}^{i}=u^{i}+UOPP^{i} \quad (21)$$

From this new value of the channel utilization factor $\tilde{u}^i$ the $SSOH^i$ can be calculated as

$$SSOH^{i}=\left(1-\tilde{u}^{i}\right)\Sigma_{\substack{j=1\\ j\neq i}}^{N}\left(\tilde{u}^{i}\frac{T_{I}^{j}}{T_{P}^{j}}\right) \quad (22)$$

3.5. Performance Metrics

Two performance metrics are considered to assess the performance of the sensing schemes: the spectrum utilization factor and idle channel search delay.

Figure 8. Relationship between $UOPP^i$ and T_p^i with respect to u^i

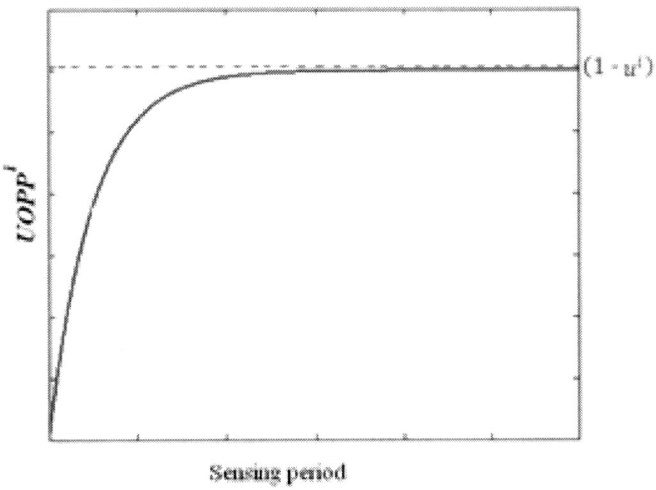

Figure 9. Actual and observed channel usage pattern

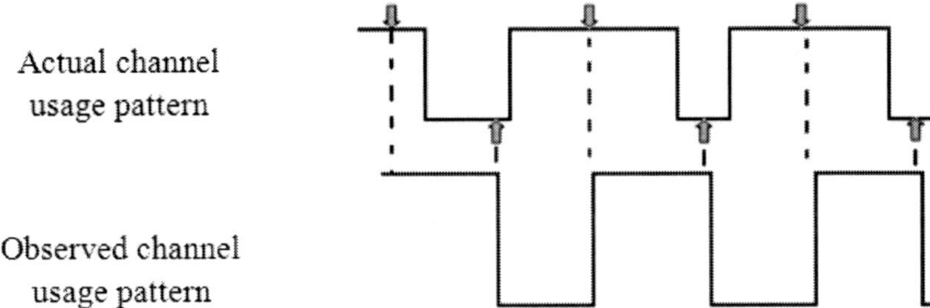

3.5.1. Spectrum Utilization Factor

This is the portion of the available spectrum that has not been utilized by the licensed users and then can be utilized by the unlicensed user. The Spectrum Utilization Factor (*SUF*) is considered for proactive sensing as reactive sensing is an on demand sensing. For an exponentially distributed ON/OFF periods channel with parameters shown in Equations 1 and 2 the spectrum utilization factor can be evaluated as

$$SUF = \left(1 - u^i\right) - \left(SSOH^i + UOPP^i\right) \qquad (23)$$

where both $SSOH^i$ and $UOPP^i$ are defined in Equations 20 and 22, respectively. Substituting these equations into Equation 23 yields

$$\begin{aligned} SUF = \left(1 - u^i\right)\left[\frac{1}{\lambda_x T_p^i}\left(e^{-\lambda_x i T_p^i} - 1\right)\right] \\ - \left(1 - \tilde{u}^i\right)\sum_{\substack{j=1 \\ j \neq i}}^{N}\left(\tilde{u}^i \frac{T_I^j}{T_p^j}\right) \end{aligned} \qquad (24)$$

where $\tilde{u}^i$ is defined in Equation 21.

3.5.2. Idle Channel Search Delay

Idle channel search delay for proactive sensing $\left(T^{P}_{idle}\right)$ and reactive sensing $\left(T^{R}_{idle}\right)$ is defined as the time required for the unlicensed user to locate the first free channel (Kim & Shin, 2008).

3.5.2.1. Idle Channel Search Delay in Proactive Sensing

Proactive sensing sorts the channels in an ascending order according to their channel utilization factors. Then we can state the following relation:

$(1-u^1) \geq (1-u^2) \geq \ldots \geq (1-u^N)$

Therefore, channel 1 is sensed for a time of T_I^1 and if it is free, with a probability of $(1-u^1)$, then it will be assigned to a data link; if not, channel 2 will be sensed for a time of T_I^2 and if it is free then it will be assigned to a data link, with a probability of $(u^1)(1-u^2)$ (i.e., channel 1 is occupied and channel 2 is free). This process will go on through all channels; if all channels are occupied then the packet will be buffered and sent later, with a probability of $u^1u^2\ldots u^N$. Consequently, $\left(T^{P}_{idle}\right)$ can be expressed as:

$$\begin{aligned} T^{P}_{idle} &= T_I^1\left(1-u^i\ldots u^N\right)+T_I^2\cdot u^i\left(1-u^2\ldots u^N\right) \\ &\quad +\ldots T_I^N\cdot u^i\ldots u^{N-1}\left(1-u^N\right) \\ &\quad +\sum_{i=1}^{N} T_I^i\cdot u^i\ldots u^{N-1} \\ &= T_I^1+u^1T_I^2+u^1u^2T_I^3+\ldots+u^1\ldots u^{N-1}T_I^N \\ &= T_I^1+\sum_{i=2}^{N}\left\{\left(\prod_{j=1}^{i-1}u^j\right)T_I^i\right\} \end{aligned} \tag{25}$$

3.5.2.2. Idle Channel Search Delay in Reactive Sensing

In the case of reactive sensing, according to their channel utilization factors, channels are sensed randomly. Hence, $N!$ possible orders of channels should be considered with equal probabilities. Let s_m be the mth set of ordered channels from the total of $N!$ possible sets, $s_m(i)$ to be the channel number i in s_m. As all the $N!$ sets can be chosen equally likely with probability of $1/N!$ then $\left(T^{P}_{idle}\right)$ can be expressed as:

$$T^{R}_{idle} = \sum_{m=1}^{N!}\frac{T^{s_m}_{idle}}{N!} \tag{26}$$

where

$$T^{s_m}_{idle} = T_I^{s_m(1)}+\sum_{k=2}^{N}\left\{\left(\prod_{i=1}^{k-1}u^{s_m(i)}\right)T_I^{s_m(k)}\right\} \tag{27}$$

Then,

$$T^{s_m}_{idle} = \frac{1}{N!}\sum_{m=1}^{N!}\left\{T_I^{s_m(1)}+\sum_{k=2}^{N}\left\{\left(\prod_{i=1}^{k-1}u^{s_m(i)}\right)T_I^{s_m(k)}\right\}\right\} \tag{28}$$

3.6. Channel Parameters Estimation

In order to calculate the performance metrics for any sensing scheme and to apply the formulas for calculating $UOPP^i$, $SSOH^i$ and the optimal sensing period vector we need to estimate the channels parameters: channel utilization factor, u^i and channel exponential ON/OFF periods' distribution parameters λ_x^i and λ_y^i for each channel in the system.

3.6.1. Channel Utilization Factor Estimation

Suppose we have collected r^i symbols $(b_1, b_2, \ldots, b_r i)$ from the channel usage pattern of channel *i*. To estimate $u^i \Rightarrow \tilde{u}^i$ we can use the sample mean estimator method which gives us:

$$\tilde{u}^l = \frac{1}{r^i} \Sigma_j^{r^i} b_j \tag{29}$$

3.6.2. Channel Exponential ON/OFF Period's Distribution Parameters λ_x^i and λ_y^i Estimation

We can estimate either λ_x^i or λ_y^i, and then use it to estimate the other by knowing $\tilde{u}^l$. Suppose we estimate λ_x^i, then we can use Equation 30 derived from Equation 5 as:

$$\hat{\lambda}_y^l = \hat{\lambda}_x^l \left(\frac{1 - \hat{u}^l}{\hat{u}^l} \right) \tag{30}$$

The Maximum Likelihood Estimator (MLE) is a suitable method to be used for estimating λ_x^i (Kim & Shin, 2008). Suppose we have k^i OFF periods among the collected r^i symbols with lengths of L_j (j=1,2,…,k^i), then the estimated $\lambda_x^i \Rightarrow \hat{\lambda}_x^l$ is given by:

$$\hat{\lambda}_y^l = \frac{k^i}{\Sigma_{j=1}^{k^i} L_j} \tag{31}$$

After estimating the channel parameters, we can adapt the sensing periods if we are employing proactive sensing with adapted sensing periods method. The whole process is illustrated in Figure 10.

Figure 10. Complete process for proactive sensing with adapted sensing periods

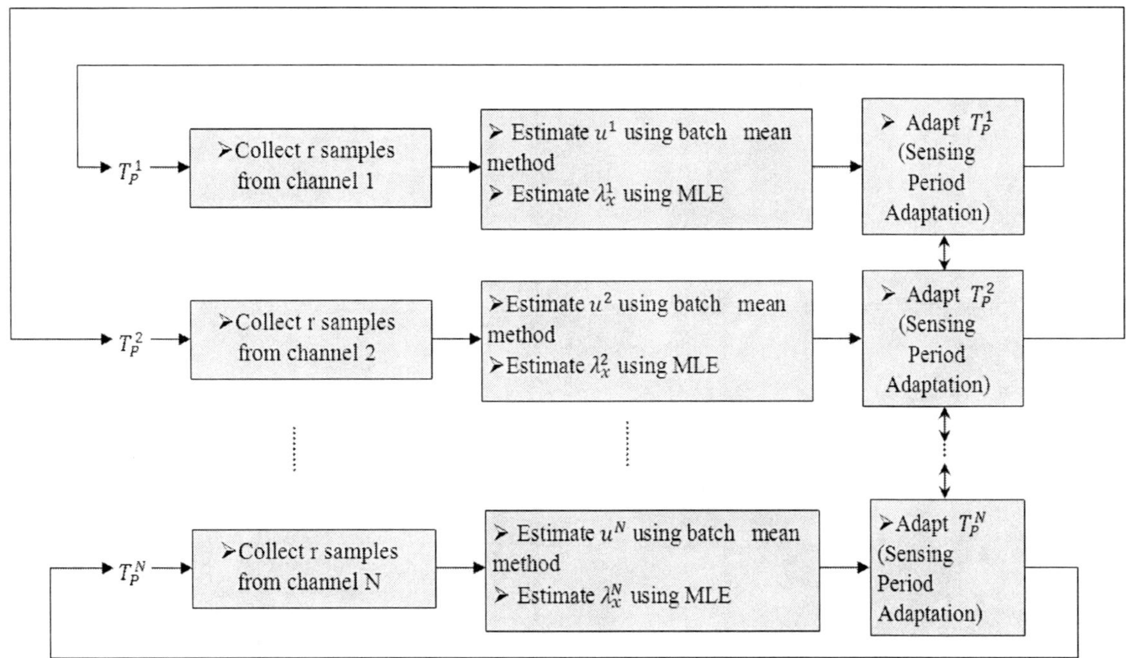

4. SIMULATIONS

This section shows the structure of the cognitive radio network used in the simulations. In addition, the simulation parameters to apply the MAC layer sensing aspects are specified and simulation results and performance comparisons are presented.

4.1. Network Model

A wireless ad-hoc network supporting data transfer among its nodes is considered to represent the unlicensed users' network. The network consists of a group of nodes and the licensed radio network to be shared spectrum with has *N* channels.

Figure 11 shows the topology of the unlicensed network where an unlicensed node N_0 is surrounded by *M* neighbors $N_1, N_2, \ldots, N_M$. A total number of *N* licensed channels can be utilized by N_0 when they are unoccupied by their licensed users. A data link L_j (j=1,2,…, M) is assigned for communication between N_0 and N_j. When N_0 wants to communicate with any other node N_j, both N_0 and N_j should exchange their sensing results via control channels and then assign the appropriate channels to data links with the aid of *Unlicensed Users Coordination mechanisms.*

Figure 11. Simulated cognitive radio ad-hoc network topology

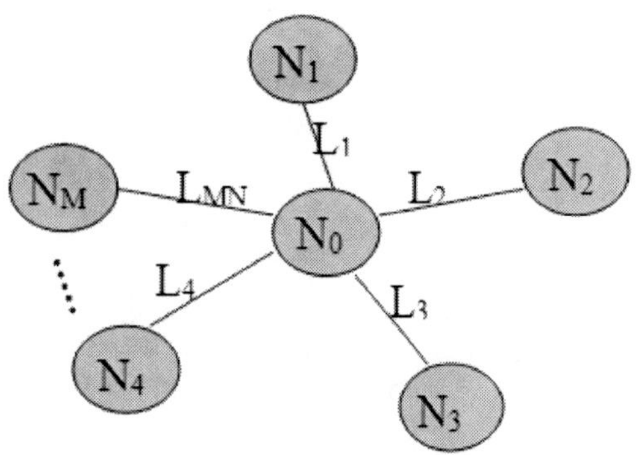

4.2. Simulation Parameters

Table 1 shows the fundamental simulation parameters while Table 2 shows the channels parameters used in the simulations.

4.3. Simulation Results

The obtained results will be classified into four categories:

1. *UOPP+SSOH* for the channels, which reflects the wasted available spectrum due to sensing.
2. Channels parameters estimation.
3. Adapted sensing periods and the impact on the achieved spectrum utilization factor for each channel.
4. Comparison of sensing modes:
    a. Comparison between reactive and proactive sensing modes in term of idle channel search delay.

Table 1. Fundamental simulation parameters

Parameter	Notation	Value
N	Number of channels	5 channels
T_I^i	Listing Interval (for all channels)	20 ms
r^i	Collected symbols in each estimation cycle for each channel	10^4 symbol

Table 2. Channel parameters used in simulation

Channel ⇓ Parameter ⇒	$E_{x^i}(x)$	$E_{y^i}(y)$
Channel 1	5.00s	1.00s
Channel 2	2.50s	1.25s
Channel 3	1.67s	0.50s
Channel 4	1.25s	0.75s

b. Comparison between adapted and non-adapted sensing periods proactive sensing in term of the achieved spectrum utilization factor.

4.3.1. The Wasted Available Spectrum Due to Sensing

Figure 12 illustrates the obtained values of *UOPP+SSOH* for channel 1 and 5 as an example.

The optimal sensing period for channels 1 and 5 can be extracted from Figure 12, and accordingly the value of SUF can be calculated as in Table 3 for all channels.

4.3.2. Channels Parameters Estimation

Here are the estimated values of channels utilization factors $\boldsymbol{u}$ and the exponential distribution parameter, $\lambda_{x,}$ for OFF periods for each channel.

4.3.2.1. Channels Utilization Factor Estimation

Figure 13 shows the estimated values for the channels utilization, the dashed lines illustrate the values calculated from the 'injected' means $E_{x^i}\left(x\right)$ and $E_{y^i}\left(y\right)$. Figure 13 insures that the utilization estimator we have used (i.e., sample mean estimator) is unbiased where the estimated utilizations $\hat{u}$ follow the actual values represented by the dashed lines closely for all channel.

4.3.2.2. Channels OFF Periods Distribution Parameter, $\hat{\lambda}_x$ Estimation

Figure 14 shows the estimated values for the channels distribution parameter $\lambda_{x,}$, the dashed lines illustrate the value calculated from the 'injected' means $E_{x^i}\left(x\right)$ and $E_{y^i}\left(y\right)$. Figure 14 reflects the estimated values of $\hat{\lambda}_x$ which follow the actual values represented by the dashed lines, even though one can observe the biasness of the used MLE; that is the estimation accuracy differ from channel to channel since it is best in channel 1 and worst case is faced with channel 5.

4.3.3. Adapted Sensing Periods and Achieved Spectrum Utilization Factor for Each Channel

During operation, the adapted sensing periods and the corresponding achieved spectrum utilization factors for channels 2 and 4 as examples are shown in Figure 15 and Figure 16, respectively.

From Figures 15 and 16 it is observed that the achieved spectrum utilization is almost consistent in the two channels and in both cases it is more than 80%. This consistent achieved spectrum utilization is an advantage of the adapted sensing periods' mode over the non-adapted mode, which will be discussed and explained in more details later in this section.

4.4. Sensing Modes Comparison and Tradeoffs

4.4.1. Reactive vs. Proactive Sensing Regarding Idle Channel Search Delay

We define three cases concerning the radio environment congestion:

1. Uncongested radio environment where **0.5>u^i>0.1**
2. Congested radio environment where **0.7>u^i>0.4**
3. Highly congested radio environment where **0.9>u^i>0.5**

The above scenarios are plotted in Figures 17, 18, and 19, respectively. The following comments can be noted:

1. In all cases, the idle channel search delay in proactive sensing is affected much less with the number of channels than in reactive sensing.

Figure 12. (a) UOPP+SSOH for channel 1; (b) UOPP+SSOH for channel 5

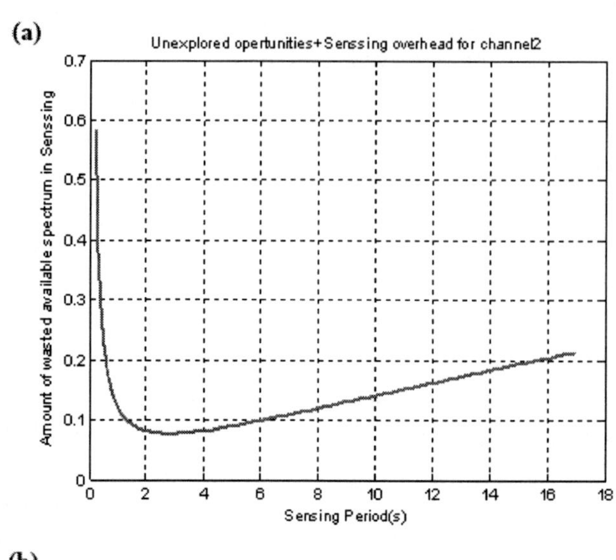

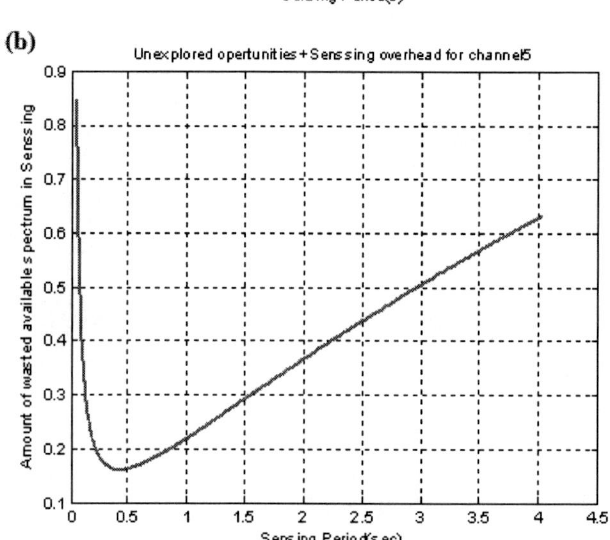

Table 3. Optimum sensing periods and spectrum utilization factors

Channel	Optimum Sensing Period	$\frac{UOPP + SSOH}{(1-u)}$	SUF
1	1.51 sec	5.8%	94.2%
2	2.56 sec	7.8%	92.2%
3	0.43 sec	9.7%	90.3%
4	4.00 sec	8.7%	91.3%
5	0.48 sec	8.1%	91.9%

Figure 13. Estimated utilization factors $\hat{u}$ for the 5 channels

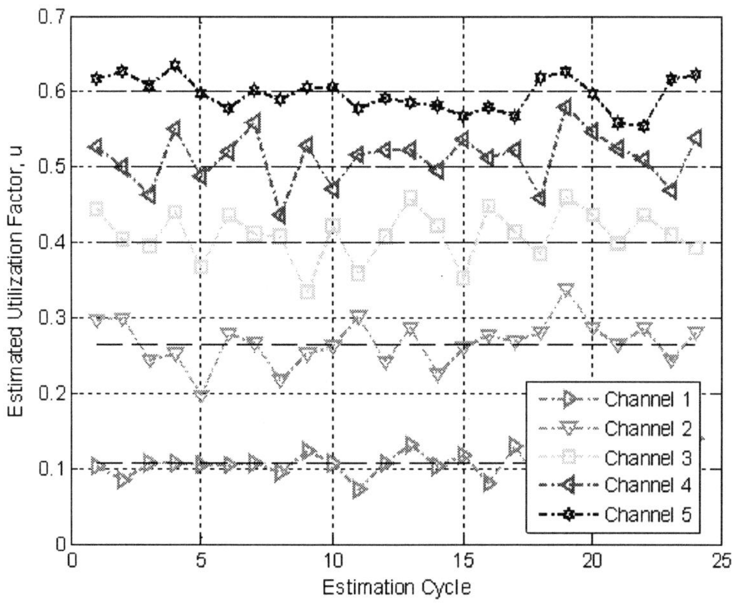

Figure 14. Estimated $\hat{\lambda}_x$ for the 5 channels

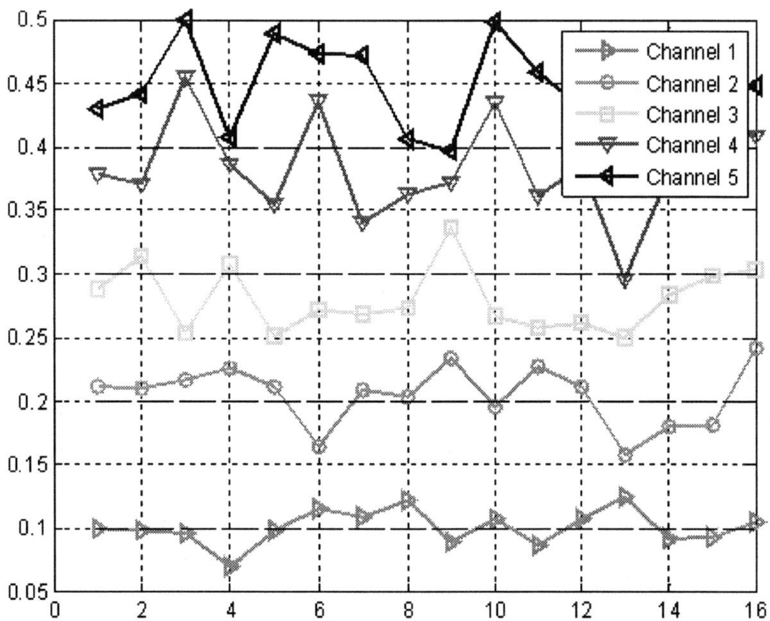

Figure 15. (a) Adapted sensing periods; (b) corresponding achieved spectrum utilization factor for channel 3 during operation

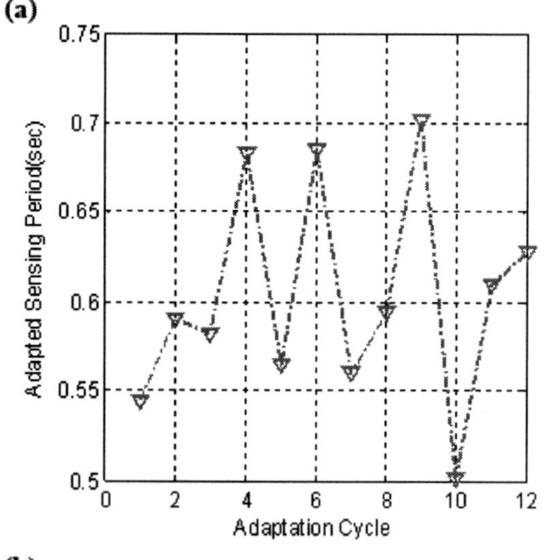

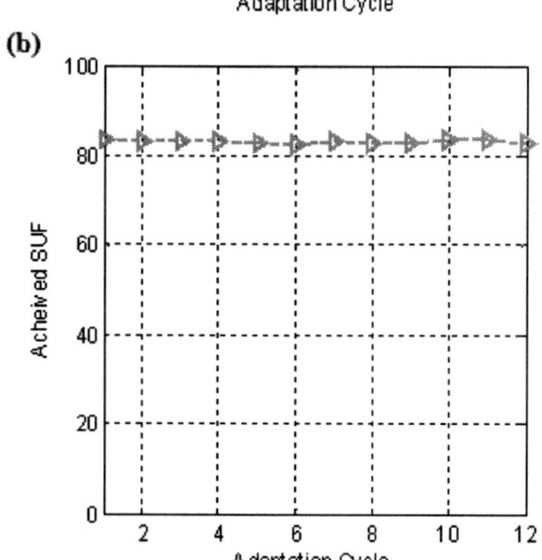

Figure 16. (a) Adapted sensing periods; (b) corresponding achieved spectrum utilization factor for channel 4 during operation

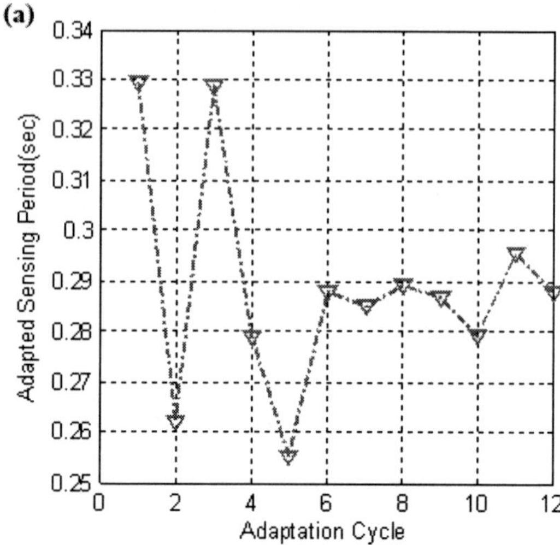

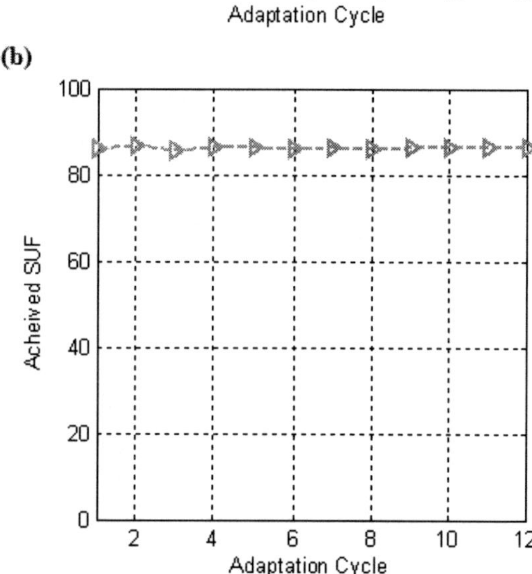

2. Reactive sensing in uncongested radio environments is better to be used since the idle channel search delay is not much higher than proactive sensing. Consequently, trading off between the idle channel search delay and simplicity would end up with selecting reactive sensing for such environment.
3. With the increase of congestion in our radio environment, using of proactive sensing becomes more desirable to decrease the idle channel search delay.

4.4.2. Impact of Sensing Periods' Adaptation on Achieved Spectrum Utilization Factor in Proactive Sensing

Figure 20 demonstrates the achieved Spectrum Utilization Factor (*SUF*) for the whole system, which reflects the amount of the available spectrum on all channels the unlicensed users can utilize in both adapted and non-adapted sensing periods of proactive sensing method. We can conclude from Figure 20 that the adapted sensing periods

Figure 17. Idle channel search delay in proactive and reactive sensing in an uncongested environment

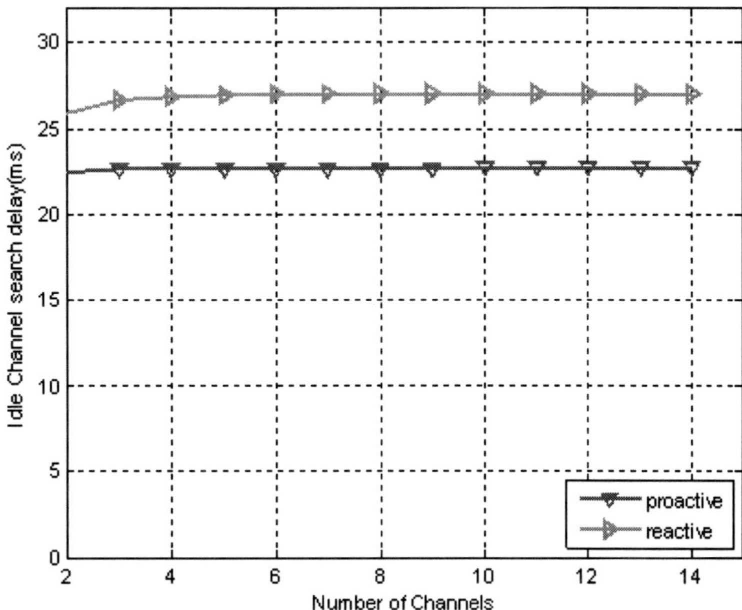

Figure 18. Idle channel search delay in proactive and reactive sensing in a congested environment

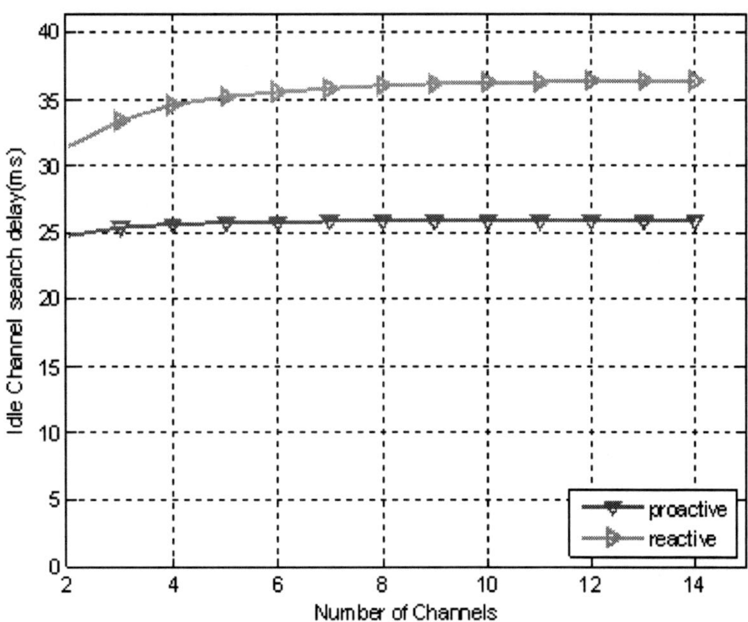

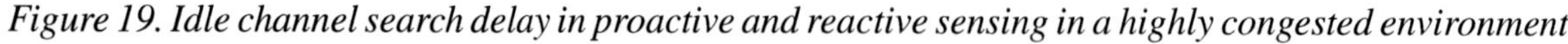

Figure 19. Idle channel search delay in proactive and reactive sensing in a highly congested environment

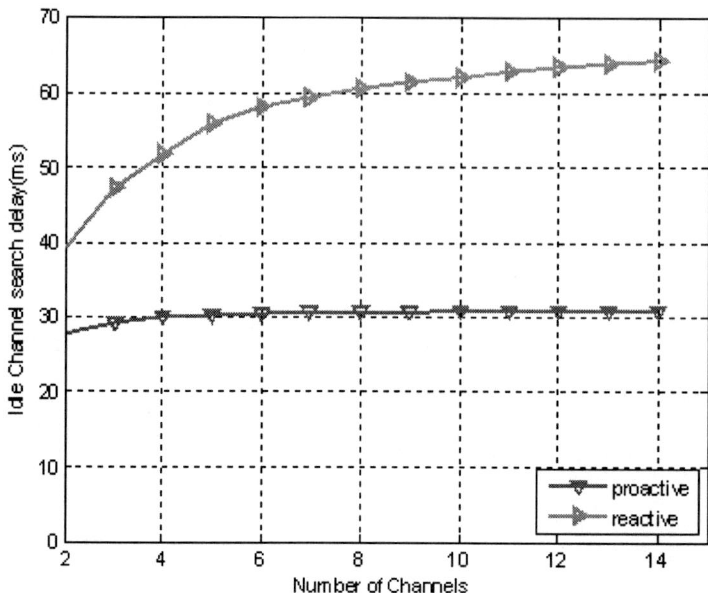

Figure 20. SUF for the whole system in adapted and non-adapted sensing periods of proactive sensing

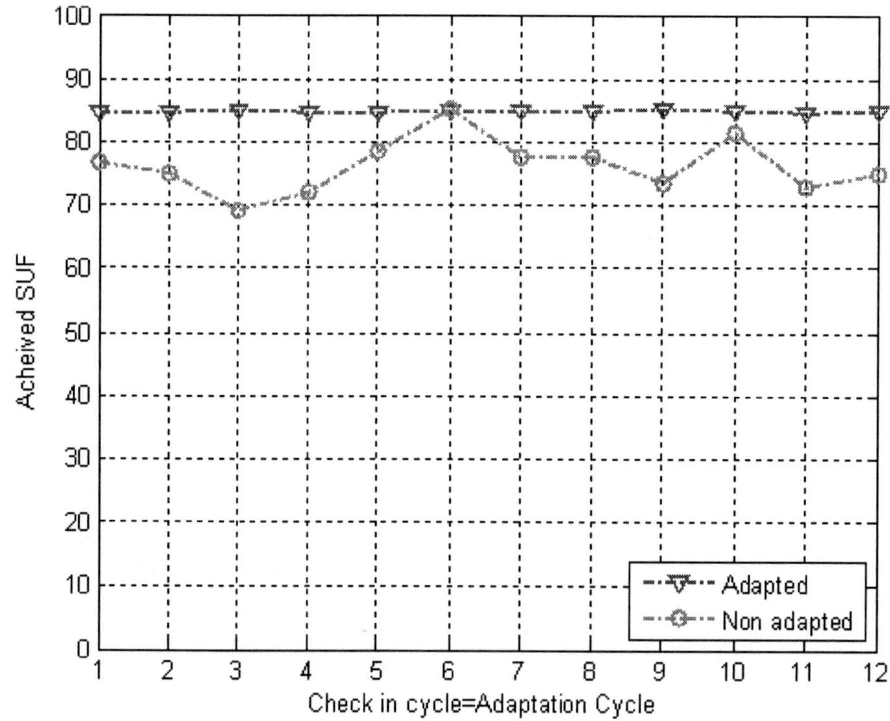

mode is more robust in this context and has more consistent SUF.

5. CONCLUSION AND FUTURE RESEARCH DIRECTIONS

In this chapter we investigated the MAC layer sensing schemes in cognitive radio to find a clear policy describing how the available spectrum should be utilized to achieve as low idle channel search delay as possible. The results show that to grantee higher spectrum utilization and lower idle channel search delay as possible, proactive sensing with adapted sensing periods is the best candidate to be used. However, proactive sensing with adapted sensing periods is costly in terms of nodes complexity needed to perform the required computational tasks. Hence, it is a matter of tradeoffs which is governed by the nature of the licensed and unlicensed networks as described in the analysis of the results.

For future work, it is interesting to investigate various licensed and unlicensed networks and to study the impact of the network nature and characteristics on the sensing modes performance. For instance, studying the inter-arrival packet rate and packet departure rate effects of on the performance of different sensing modes is an interesting subject. In addition, more complicated networks can be considered such as prioritized nodes networks and networks with centralized units. Furthermore, the design and implementation of a transceiver supports one or more of the MAC layer sensing modes would be an interesting and important study. Finally, it is important for future studies to merge the area covered in this chapter with other related issues in cognitive radio such as to consider unlicensed users coordination problems.

REFERENCES

Akyildiz, I. F., Lee, W. Y., Vuran, M. C., & Mohanty, S. (2006). NeXt generation dynamic spectrum access in cognitive radio wireless networks: A survey. *Computer Networks Journal*, *50*(13), 2127–2159. doi:10.1016/j.comnet.2006.05.001

Berlemann, L., Dimitrakopoulos, G., & Moessner, K. (2005). Cognitive radio and management of spectrum and radio resources in reconfigurable network. *Wireless World*. Retrieved from http://www.wireless-world-research.org/fileadmin/sites/default/files/about_the_forum/WG/WG6/White%20Paper/WG6_WP7.pdf

Cox, D. (1967). *Renewal theory*. London, UK: Butler & Tanner Ltd.

De Domenico, A., Strinati, E. C., & Di Benedetto, M. (2012). A survey on MAC strategies for cognitive radio networks. *IEEE Communications Surveys & Tutorials*, *14*(1), 21–44. doi:10.1109/SURV.2011.111510.00108

Gao, P., Wang, J., & Shaoqian, L. (2010). Non-contiguous CI/OFDM: A new data transmission scheme for cognitive radio context. In *Proceedings of the IEEE International Conference on Wireless Information Technology and Systems (ICWITS)*, (pp. 1-4). IEEE Press.

Haykin, S. (2005). Cognitive radio: Brain-empowered wireless communications. *IEEE Journal on Selected Areas in Communications*, *23*(2), 201–220. doi:10.1109/JSAC.2004.839380

Kamal, K., & Mohammed, A. (2012). Game theory for cognitive radio. In *Cognitive Radio and Interference Management: Technology and Strategy*. Hershey, PA: IGI Global. doi:10.4018/978-1-4666-2005-6.ch007

Kapoor, S., & Singh, G. (2011). Opportunistic spectrum sensing by employing matched filter in cognitive radio network. In *Proceedings of the International Conference on Communication Systems and Network Technologies (CSNT)*. CSNT.

Kim, H., & Shin, K. (2008). Efficient discovery of spectrum opportunities with MAC-layer sensing in cognitive radio networks. *IEEE Transactions on Mobile Computing*, *7*(5), 33–45.

Kim, K., Xin, Y., & Rangarajan, S. (2010). Energy detection based spectrum sensing for cognitive radio: An experimental study. In *Proceedings of the Global Telecommunications Conference (GLOBECOM)*, 1-5. IEEE.

Liang, Y., Chen, K., Li, G., & Mahonen, P. (2011). Cognitive radio networking and communications: An overview. *IEEE Transactions on Vehicular Technology*, *60*(7), 3386–3407. doi:10.1109/TVT.2011.2158673

Liu, K. (2011). Cognitive radio and game theory. *IEEE Spectrum*. Retrieved from http://spectrum.ieee.org/telecom/wireless/cognitive-radio-and-game-theory/0

Liu, X., & Zhai, X. (2008). Feature detection based on multiple cyclic frequencies in cognitive radios. In *Proceedings of the China-Japan Joint Microwave Conference (CJMW 2008)*, (pp. 290-293). Piscataway, NJ: CJMW.

Mitola, J. III. (1999). Software radio architecture: A mathematical perspective. *IEEE Journal on Selected Areas in Communications*, *17*(4), 514–538. doi:10.1109/49.761033

Mitola, J. (2000). *Cognitive radio: An integrated agent architecture for software defined radio.* (PhD Dissertation). Royal Institute of Technology (KTH). Stockholm, Sweden.

Mitola, J., & Maguire, G. (1999). Cognitive radio: Making software radios more personal. *IEEE Personal Communications*, *6*(4), 13–18. doi:10.1109/98.788210

Park, J., Park, S., Kim, D., Cho, P., & Cho, K. (2003). Experiments on radio interference between wireless LAN and other radio devices on a 2.4 GHz ISM band. In *Proceedings of the 57th IEEE Semiannual Vehicular Technology Conference,* (vol. 3, pp. 1798-1801). IEEE Press.

Runsheng, G., Zhongyu, H., & Tao, S. (2008). Adaptive CRN spectrum sensing scheme with excellence in topology and scan scheduling. In *Proceedings of the 3rd International Conference on Sensing Technology,* (pp. 384-391). IEEE.

Sherman, M., Mody, A. N., Martinez, R., Rodriguez, C., & Reddy, R. (2008). IEEE standards supporting cognitive radio and networks, dynamic spectrum access and coexistence. *IEEE Communications Magazine*, *46*(7), 72–79. doi:10.1109/MCOM.2008.4557045

Wang, L., & Wang, C. (2008). Spectrum handoff for cognitive radio networks: Reactive-sensing or proactive-sensins? In *Proceedings of the IEEE International Performance, Computing and Communications Conference 2008 (IPCCC 2008)*, (pp. 343-348). IEEE Press.

Yonghong, Z., & Ying-Chang, L. (2007). Maximum-minimum Eigenvalue detection for cognitive radio. In *Proceedings of the 2007 IEEE 18th International Symposium on Personal, Indoor and Mobile Radio Communications*, (pp. 1165-1169). Piscataway, NJ: IEEE Press.

Yucek, T., & Arslan, H. (2009). A survey of spectrum sensing algorithms for cognitive radio applications. *IEEE Communications Surveys Tutorials*, *11*(1), 116–130. doi:10.1109/SURV.2009.090109

Chapter 11
A Survey of High Performance Cryptography Algorithms for WiMAX Applications Using SDR

Rafidah Ahmad
Universiti Sains Malaysia, Malaysia

Widad Ismail
Universiti Sains Malaysia, Malaysia

ABSTRACT

As wireless broadband technology has become very popular, the introduction of Worldwide Interoperability for Microwave Access (WiMAX) based on IEEE 802.16 standard has increased the demand for wireless broadband access in the fixed and the mobile devices. This development makes wireless security a very serious concern. Even though the Advanced Encryption Standard (AES) has been popularly used for protection in WiMAX applications, still WiMAX is exposed to various classes of wireless attack, such as interception, fabrication, modification, and reply attacks. The complexity of AES also produces high power consumption, long processing time, and large memory. Hence, an alternative cryptography algorithm that has a lower power consumption, faster and smaller memory, is studied to replace the existing AES. A Software Defined Radio (SDR) is proposed as a different way of proving the performance of the cryptography algorithm in real environments because it can be reprogrammed, which leads to design cost and time reductions.

INTRODUCTION

Cryptography is known as the science of using mathematics to encrypt and decrypt data. With cryptography, it enables users to store sensitive information or transmit it across insecure networks such as Internet. The information can only be read by the intended recipient. There are many types of cryptography algorithm which had been evaluated in the recent research works. These algorithms will be studied and analyzed in the literature review, in terms of battery power consumption, memory, and speed for mobile Worldwide Interoperability for Microwave Access (WiMAX).

DOI: 10.4018/978-1-4666-2812-0.ch011

As for WiMAX, it is a Wireless Metropolitan Area Network (WMAN) communications technology that is largely based on the wireless interface defined in IEEE 802.16 standard (Scarfone, et al., 2010). Nowadays, WiMAX is gaining popularity in many regions due to wide coverage and high data rate of multimedia transmission rather than WiFi which has a coverage and speed limitations. A new validation approach for WiMAX communication using Software Defined Radio (SDR) is proposed in this chapter.

SDR is an Information Transfer System (ITS) that combines technologies from historically separated fields of computers and radios (Xu, et al., 2006). Emerging from military applications, SDR has gained much attention among researchers and practitioners working in the wireless communication area. SDR technique has been considered as an important technique to enhance the flexibility and usability of many popular communication standards such as GSM, WiFi and 3G. Since WiMAX has never been verified through SDR platform, this can be one of new method to proof the SDR capability. An overview about WiMAX and SDR are presented in the next section, followed by a survey on cryptography algorithms and SDR implementations in the literature review.

BACKGROUND

WiMAX is used for variety of purposes including, but not limited to, fixed last-mile broadband access, long-range wireless backhaul, and access layer technology for mobile wireless subscribers operating on telecommunications networks (Scarfone, et al., 2010). The WiMAX Forum has estimated that new WiMAX equipment will be capable of sending 40 Mbps data over 10 km in a Line-Of-Sight (LOS) fixed environment (Khan & Zaman, 2009). Therefore, WiMAX technology continues to adapt to market demands and provide enhanced user mobility. IEEE 802.16e-2005 was an amendment that enabled mobile WiMAX. This standard was built on Orthogonal Frequency Division Multiple Access (OFDMA). Most countries have allocated the bands for the wireless access between 3.4 and 3.6 GHz but the United States, Mexico, Brazil, and some Southeast Asian nations, have chosen instead the bands between 2.5 and 2.7 GHz (Ahson & Ilyas, 2008).

The mobile WiMAX system also has more enhanced security features than the existing IEEE 802.16-2004-based WiMAX network system. However, the mobile WiMAX system, which uses Advanced Encryption Standard (AES) scheme (Airspan, 2007; Yuksel, 2007), is still not able to guarantee the reliability of the whole mobile WiMAX systems and network architecture (Joseph, 2011). In this short period of their existence, various weaknesses have emerged. Some of the possible threats are similar to the ones that WiFi faced: this observation stresses on the importance of the WiFi threat analysis and the prevention measures that can be taken for WiMAX (Trimintzios & Georgiou, 2010). WiMAX has security vulnerabilities in both Physical (PHY) and Medium Access Control (MAC) layer, exposing to various classes of wireless attack including interception, fabrication, modification, and reply attacks (Jha & Dalal, 2010).

As in MAC layer, the threats are examined with respect to confidentiality and authentication. As shown in Figure 1(a) based on Mishra and Glore (2008), a MAC layer Protocol Data Unit (PDU) consists of a MAC header, a payload and an optional CRC. The payload may consist of user traffic or management messages. MAC headers are not encrypted and all MAC management messages shall be sent in without protection. Therefore, MAC layer is exposed to eavesdropping, man in the middle and Denial of Service (DoS) attacks. Eavesdropping of management messages may reveal network topology to the eavesdropper, posing a critical threat to Sub-Stations (SSs) as well as the WiMAX system (Ahson & Ilyas, 2008). Weaknesses in management messages authentication also open the door to aggressions such as the man-in-the-middle attack. The impact of this type of attack can be high because it might affect the operation of the communications. The risk is at least major for all cases and it might be safe to allow a second line of defense against this attack (Barbeau, 2005). The DoS attack is created by the authentication operations of devices, users and messages that trigger the execution of long procedures. A DoS attack can be executed by flooding a victim with a high number of messages to authenticate. The impact is medium on a system, but could be high on a user because of lower power resources available for handling a large influx of invalid messages (Barbeau, 2005).

In view of the limited processing power and battery life considerations of mobile devices, the authentication and encryption protocols have been designed so that most of the computational processes take place in base stations rather than mobile stations (Kumar, 2008). However, due to the fact that was mentioned in Ahson and Ilyas (2008), encryption that is taken place in base-station only applied to the MAC PHY Data Unit (PDU) payload but not on MAC management messages, with the purpose to increase the efficiency of network operations (Scarfone, et al., 2010). Therefore, an attacker, as a passive listener of the WiMAX channel, can retrieve valuable information from unencrypted MAC management messages. This is why WiMAX requires device-level authentication and encryption, as an alternative or extra defense to tackle this problem rather than base-station level. Since batteries provide a limited amount of energy, it is important for mobile devices to have an efficient power-saving mechanism. Hence, this chapter will survey various types of cryptography algorithms as an alternative and enhancement to the existing AES algorithm in WiMAX applications. This chapter will also help the researchers to decide the most power-saving cryptography algorithm for any wireless communication standard.

Most of the previous research works that will be discussed in the next section, have shown that the cryptography algorithms were only modeled and simulated on software. A further work on validation in the real communication standard needs to be done. Therefore, SDR is one of potential platform to verify these algorithms. The term 'software defined radio' was coined in 1991

Figure 1. Description of WiMAX MAC layer and SDR concept: (a) MAC PDU; (b) the ideal SDR transmitter and receiver

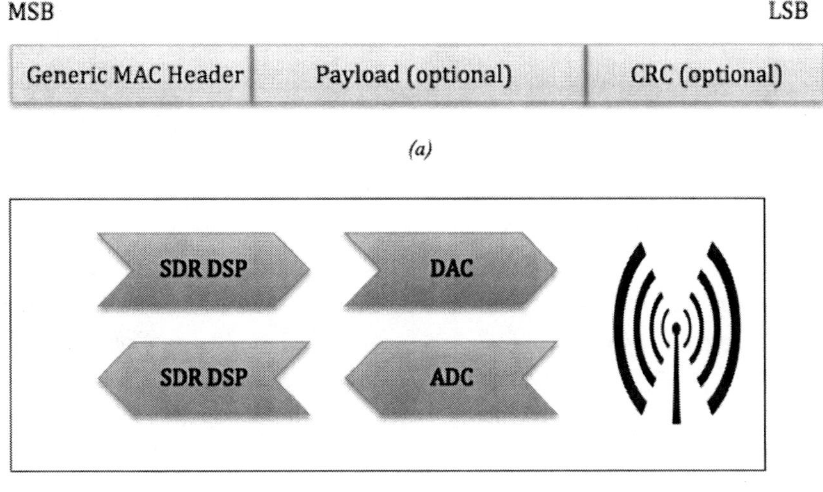

by Mitola (1992), who published the first paper on the topic in 1992. From the First International Workshop on Software Radio (1998), though the concept was first proposed in 1991, SDRs have their origins in the defense sector since late 1970s in both the US and Europe. Reed (2002) has defined the SDR as a radio that is substantially defined in software and whose physical layer behavior can be significantly altered through changes to its software. In other words, SDR denotes a completely configurable radio that can be programmed in software to reconfigure the physical hardware.

Worldwide interest and investment in the SDR technologies is growing significantly, with key standardization and development efforts now taking place throughout Europe, US, Japan, Korea, and China (Trinadh, 2009). The radio frequency range available is very broad from 0.1 kHz up to 4 GHz (Wikipedia, 2012). The SDR platforms are very compact and completely self-contained. Their mobility also makes them easy to test in the field. The wide-reaching benefits of SDR technology realized by radio service providers and product developers through to end users, which are as below (Wireless Innovation, 2012):

- New features and capabilities to be added to existing infrastructure without requiring major new capital expenditures allows service providers to quasi-future proof their networks.
- The logistical support and operating expenditure can be reduced significantly with the use of a common radio platform for multiple markets.
- Remote software downloads through which capacity can be increased, capability upgrades can be activated and new revenue generating features can be inserted.
- A family of radio products to be implemented using a common platform architecture allows new products to be faster introduced into the market.
- Software to be reused across radio products reduces development costs dramatically.
- Remote reprogramming allows 'bug fixes' to occur while a radio is in service, whereas the time and costs associated with the operation and maintenance can be reduced.
- Reduce costs in providing end users with access to ubiquitous wireless communications.

By referring to Source Lyrtech (2012), there are three different modules implemented in SDR platform: 1) Digital processing module – uses Texas Instruments System-on-Chip (SoC) for baseband processing and Xilinx Field Programmable Gate Array (FPGA). This offers developers the performance necessary to implement custom intellectual property and acceleration functions with varying requirements from one protocol to another supported on the same hardware. 2) Data conversion module – equipped with very fast dual-channel Analog to Digital Converter (ADC) and Digital to Analog Converter (DAC) for signal conversion. 3) Radio Frequency (RF) module – supports many tunable from low band frequency up to WiMAX frequency, making it capable of supporting a wide array of applications.

The ideal SDR concept is illustrated in Figure 1(b) based on Trinadh (2009). In RF communication receiver, the ideal scheme would be to attach an Analog to Digital Converter (ADC) to an antenna. The digital data would be read by a Digital Signal Processor (DSP). Then, its software would transform the stream of data from the converter to any other form of application requires. Similar with the transmitter, where the DSP would generate a stream of data defining the transmission waveform. This data would be sent to a Digital to Analog Converter (DAC) connected to a radio antenna.

With the capabilities of SDR, the verification of WiMAX applications on SDR platform should be no problem. Since this chapter also focused on battery power consumption for cryptography algorithms, the SDR platform is the best choice because it offers the power measurement capabilities that allow developers to precisely monitor the power consumption of individual components such as the Digital Signal Processing (DSP) or FPGA cores. It gives the developers an incredible insight into their systems and allows them to develop less power-efficient applications (Source Lyrtech, 2012).

MAIN FOCUS OF THE CHAPTER

Cryptography Algorithms

Cryptography algorithms are divided into three types: symmetric, asymmetric and hash functions. Symmetric algorithms are preferred because they only use one key to encrypt and decrypt the data. Hence, the memory used could be reduced and this is suitable for WiMAX-based small mobile devices. Meanwhile, asymmetric algorithms are highly computationally intensive compared to symmetric algorithms because they use different keys for encryption and decryption processes. That is why these asymmetrics are unpopular for low power mobile communication. The hash functions also called message digests and one-way encryption. This algorithm is typically used to provide a digital fingerprint of a file's contents, also employed by many operating systems to encrypt passwords (Kessler, 2010). The classification of cryptography algorithms can be shown in Figure 2.

Previously, there are a few research works have been done on various types of cryptography algorithm in terms of battery power consumption, memory size, and speed. Earlier, Schneier and Whiting (1997) has done a comparison on Blowfish, Khufu, Rivest Ciphers (RC5), Data Encryption Standard (DES) and 3DES as a block cipher algorithms. The performance analysis of these algorithms was done on Intel Pentium. The results showed that Blowfish is the fastest algorithm among others with 18 clocks per byte of output. On the other hand, the energy consumptions of different symmetric key encryption on hand held devices were shown in Ruangchaijatupon and Krishnamurthy (2001). The authors found that after only 600 encryptions of a 5 MB file using 3DES, the remaining battery power is 45% and subsequent encryptions are not possible as the battery dies rapidly.

Hirani (2003) mentioned in his thesis that increasing research is being done towards developing wireless systems with built in security. He has

Figure 2. Overview of the types of cryptography

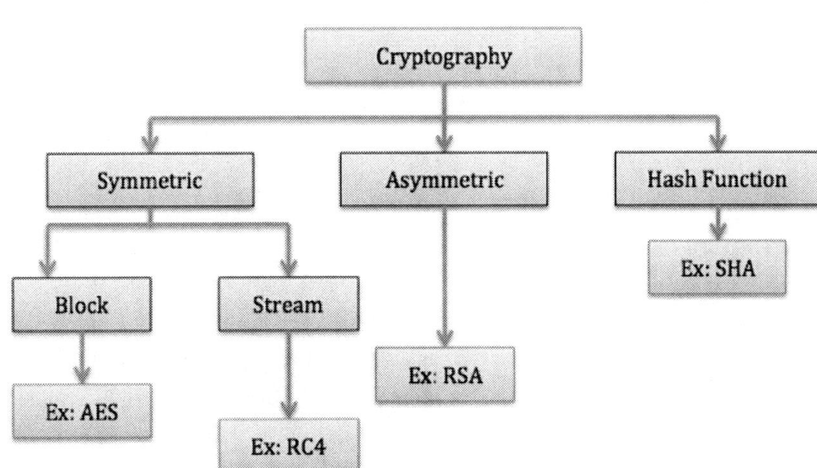

studied the performance and energy consumption of symmetric key schemes likes AES, CAST and International Data Encryption Algorithm (IDEA), and asymmetric schemes RSA, ElGamal and ECIES; and also the effect of changing key size for AES and RSA. It was concluded that AES is faster and more efficient than other symmetric encryption algorithms, and RSA is more efficient than the other asymmetric encryption algorithms with lowest time and power consumptions. The experiment results also showed that higher key sizes (AES-256) will increase power and time consumptions. The author had confirmed that the AES implementation can be done by means of 12 rotate byte operations and 16 XOR operations per round. The processing time for different key size of AES was summarized as follows:

$$\begin{aligned} AES-128 &= 120\ t_{rotByte} + 164\ t_{xor} \\ AES-192 &= 144\ t_{rotByte} + 196\ t_{xor} \\ AES-256 &= 168\ t_{rotByte} + 226\ t_{xor} \end{aligned} \qquad (1)$$

where $t_{rotByte}$ is a time for single rotate byte operation and t_{xor} is a time for XOR operation. However, for normal application of 128 bits key is considered very secure. Hence, targeting for higher key sizes would mean unnecessary wastage of resources for the added security that is actually not required.

The performance evaluation of DES, 3DES and Blowfish algorithms based on CPU execution time have been done in Kofahi et al. (2004) with 10 MB file. Java and Java Cryptography Architecture (JCA) are used in designing the algorithms. The three encryption algorithms were tested on a Sun workstation with a SunOS 5.7, with CPU speed of 400 MHz and 1024 MB of RAM. These algorithms were analyzed in four steps including table initialization, key initialization, data encryption and data decryption. The simulation results show that the Blowfish is the fastest algorithm with the average of CPU executing time in 14 ns, followed by DES and 3DES.

Four secret key encryption algorithms consist of DES, 3DES, AES-Rijndael and Blowfish have been implemented in Nadeem and Javed (2006). Their performances are compared by encrypting input files of varying contents and sizes, on different hardware platforms such as Pentium-II 266 MHz and Pentium-4 2.4 GHz. The comparison was done on Electronic Codebook (ECB) and Cipher Feedback (CFB) modes. Based on the experiments, Blowfish has been found as the best performing algorithm among the other algorithms. The implementation results showed that Blowfish had a very good performance with lowest execution time compared to other algorithms, followed by AES, 3DES, and DES. This is shown in Figure 3(a).

A comprehensive study of the energy consumption characteristic on cryptography algorithms was done in Potlapally et al. (2006). For symmetric algorithms, DES, 3DES, AES, IDEA, SACT, RC2, RC4, RC5, and Blowfish were analyzed. The comparison results were summarized in Table 1. From the table, Blowfish shows a big contrast between the energy consumptions of key setup and encryption/decryption. This means that Blowfish has a simple cryptography operation but a complex key setup operation. This algorithm is suitable for applications that need key exchanged scarcely. Meanwhile, AES shows the lowest energy consumption for key setup and encryption/decryption. The authors had concluded that: 1) The energy consumption of a symmetric algorithm depends on both the data encryption/decryption cost and the key-set-up cost; 2) The level of security provided by a cryptography algorithm can be traded off for energy savings by tuning parameters such as key size and number of rounds.

In Bielecki and Burak (2007), an automatic parallelization method of DES, 3DES, IDEA, AES, RC5, Blowfish, LOKI91, GOST, RSA, and data encryption standard modes of operation: ECB, Output Feedback (OFB), Counter (CTR), Cipher Block Chaining (CBC), CFB are presented. The method is based on the data depen-

Figure 3. Summaries of cryptography algorithms' processing times in the previous works: (a) comparative execution time of encryption algorithms in CFB mode on different platform (Nadeem & Javed, 2006); (b) speed-up measurements of the most time-consuming functions of symmetric encryption algorithms in ECB mode (Bielecki & Burak, 2007); (c) the time measurements of encryption algorithms for different web browsers with 50 characters text length (Idrus & Aljunid, 2008)

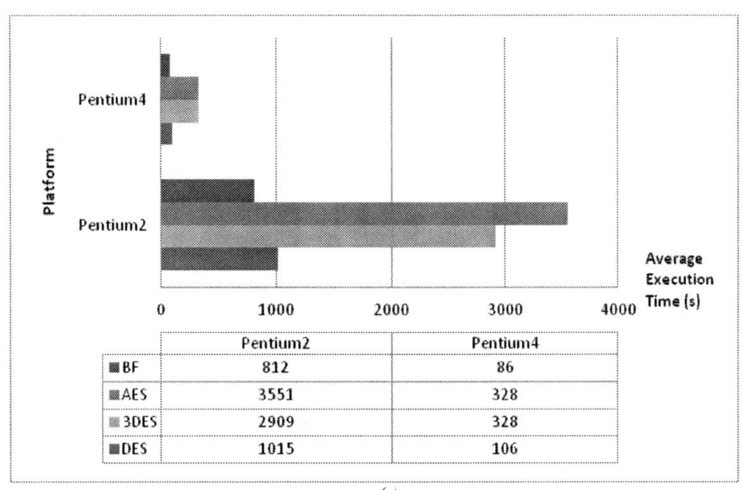

(a)

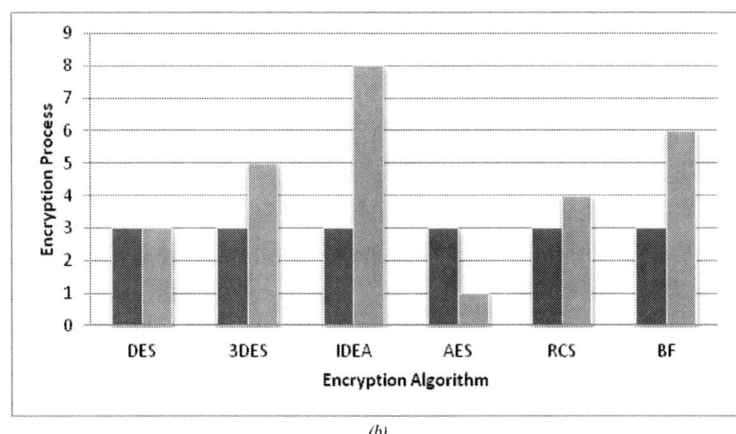

(b)

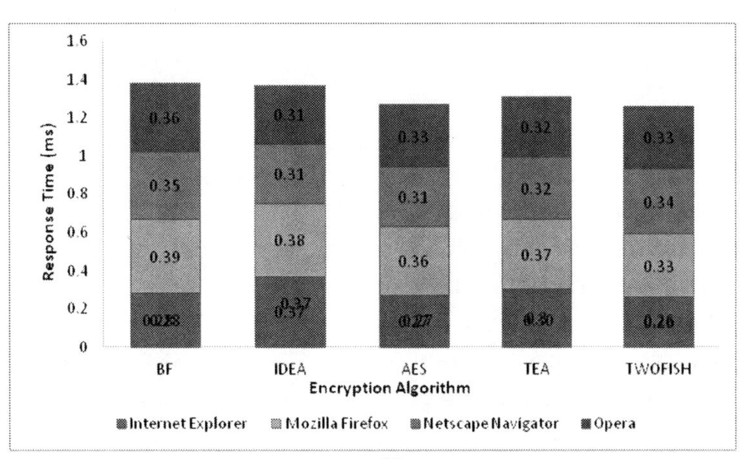

(c)

Table 1. The energy consumption result for various encryption algorithms (Potlapally, et al., 2006)

Algorithms	Energy Consumption (μJ)	
	Key Setup	Encryption/Decryption
DES	27.5	2.1
3DES	87.0	6.0
IDEA	8.0	1.5
CAST	37.6	1.5
AES	7.9	1.2
RC2	32.9	1.7
RC4	96.0	3.9
RC5	66.5	0.8
BLOWFISH	3166.3	0.8

dency analysis of loops and well-known loop parallelization techniques. Sixteen processors of SGI Altix 3700 were used for program execution, along with Intel C++ Compiler that supports the OpenMP 2.5 API. The plaintext of about 10 MB was used in this analysis. The speed-up measurements confirmed that the loops parallelized in accordance with the automatic parallelization method has sufficient efficiencies. Figure 3(b) shows the comparison of speed-up measurement among symmetric encryption algorithms in ECB mode.

A comprehensive study about the implementation of encryption and hashing algorithms in FPGA-based has been done in Gonzalez et al. (2008). The encryption algorithms include DES, 3DES, IDEA, Blowfish and AES, and hashing algorithms like MD5 and SHA-1. The analysis had been done based on two well- known soft core processors: the Xilinx MicroBlaze and open source LEON2. From the implementation results, Blowfish has the smallest execution time with 17.3 s after MD5. The comparison summary of these algorithms in MicroBlaze is shown in Table 2.

A Web programming language has been proposed in Idrus and Aljunid (2008), which was analyzed with four Web browsers in terms of their performances to process the encryption of the programming's script. The algorithms used are Blowfish, IDEA, AES, Tiny Encryption Algorithm (TEA) and Twofish. These algorithms able to support 128-bit key size. The performance analysis showed that Twofish performs better compared to others and sustain lower response time via Internet Explorer and Mozilla Firefox. Figure 3(c) summarizes the time measurements for these algorithms through different Web browsers.

Elminaam et al. (2010) have evaluated the performance of AES-Rijndael, DES, 3DES, RC2, Blowfish, and RC6 encryption algorithms. These algorithms have been conducted at different settings such as different sizes of data blocks, different data types, battery power consumption, different key sizes and encryption/decryption speed. In their experiment, the text data, audio data, and video files were encrypted at different sizes. The experiment results showed the superiority of Blowfish algorithm over other algorithms in terms of processing time and throughput. The higher the value of throughput, then the lower power consumption for that algorithm. The RC2 is found to be worse over all other algorithms in term of time consumption. In case of RC6, the result showed that higher key size leads to clear change in the battery and time consumption. Table 3 shows the analysis results of the encryption algorithms using hexadecimal encoding plaintext.

Table 2. Comparison of implementation results for encryption algorithms in MicroBlaze (Gonzalez, et al., 2008)

Algorithm	No. of Cycles	Execution Time (s)	Slices Used
DES	1,787	18.7	3,920
3DES	19,085	200.1	11,786
IDEA	4,209	44.1	1,744
BLOWFISH	1,648	17.3	2,496
AES	7,076	37.1	4,497
MD5	5,502	7.2	2,412
SHA-1	21,978	28.8	2,134

Table 3. The time consumption and throughput results for various encryption algorithms based on hexadecimal encoding (Elminaam, et al., 2010)

Algorithms	Time Consumption (ms)	Throughput (MB/s)
BLOWFISH	100	25
RC6	800	6
AES	1,400	4
DES	1,700	4
3DES	1,800	3
RC2	1,900	3

Another comparison on AES-Rijndael, DES, 3DES, and Blowfish in terms of power consumption with different data types like text, image, audio and video have been done in Umaparvathi and Varughese (2010). Their simulation results showed that AES has a better performance among the other algorithms used, in terms of encryption time, decryption time and throughput. This makes AES is the best choice for Mobile Ad Hoc Network (MANET) nodes due to limited battery power. Table 4 summarizes the analysis results for these algorithms.

Meanwhile, Nie et al. (2010) have evaluated the performance of two symmetric key encryption algorithms: DES and Blowfish. These algorithms were analyzed in the aspect of security, speed and power consumption. C language was used to programme the algorithms under Windows XP operating system. The test platform is a laptop with 1.8 GHz AMD Athlon and 1 GB memory. The plaintext of 128-byte size block was generated as an input data. Experimental results showed that Blowfish runs faster than DES with 7.4 ~ 7.5 MB/s. The power consumptions for both algorithms are almost similar where the remained battery is 85% for 50 cycles, and 65% for 100 cycles. The author claimed that Blowfish maybe more suitable for wireless network application which exchanges small size packages.

Verma et al. (2011) made a comparison on four encryption algorithms: DES, 3DES, Blowfish, and AES-Rijndael. All the algorithms were coded in C++, and ran on a Pentium-4 with 2.1 GHz processor under Windows XP operating system. The comparison was conducted on different sizes of data blocks to evaluate the encryption and decryption speed. The analysis results concluded that the Blowfish is the best performing algorithm among others with the speed of 64.4 MB/s. The summary of analysis results is shown in Table 5. However, the comparison made was not very accurate since the data sizes used were not similar.

In Ganesan and Selvakumar (2011), the authors have studied DES, 3DES, AES, CAST, UMARAM, RC2, Blowfish, and RC6 encryption algo-

Table 4. The execution time and throughput results for encryption process on different data (Umaparvathi & Varughese, 2010)

Algorithms		Image (2.5 MB)	Audio (4.0 MB)	Video (11.1 MB)	Text (4.8 MB)
DES	Execution time (ms)	274	399	998	455
	Throughput (MB/s)	9.3	10.0	11.1	10.6
3DES-168	Execution time (ms)	674	1,032	2,817	1,244
	Throughput (MB/s)	3.8	3.9	4.0	3.9
AES	Execution time (ms)	125	180	405	205
	Throughput (MB/s)	20.5	22.1	27.4	23.5
BLOWFISH	Execution time (ms)	167	234	568	274
	Throughput (MB/s)	15.3	17.1	19.6	17.6

Table 5. The comparison results using C++ language (Verma, et al., 2011)

Algorithms	Data Size (MB)	Run Time (s)	Throughput (MB/s)
BLOWFISH	256	4.0	64.4
AES-128	256	4.2	61.0
AES-192	256	4.8	53.1
AES-256	256	5.3	48.2
DES	128	6.0	21.3
3DES	128	6.2	20.8

rithms. A comparison has been conducted for these algorithms at different settings such as different sizes of data blocks, different data types, battery power consumption, different key sizes and cryptography speed. A new symmetrical encryption algorithm that avoids the key exchange between users and reduces the time taken for encryption, decryption and authentication processes, has also been proposed and implemented for WiFi applications. This algorithm uses S-box generation, XOR and AND gates to encrypt a plaintext of size 64-bits by a key size of 64-bits. The encryption and decryption processes are applied using Pentium-4 with 2.8 GHz processor, wireless USB adapter and wireless access point. The results showed that it operates at higher data rate (750 B/s) than DES, 3DES, AES, CAST, UMARAM, and RC6 algorithms when applied on text and image files. However, their research works also shows that Blowfish is better than other algorithms in terms of throughput, power consumption, and processing time.

Through the study of symmetric cryptography algorithms from the recent research works, most of the analysis results can be concluded as in Table 6. This table shows the results' comparison in terms of execution time. However, the data block sizes and data types of models are different. Some of the models were analyzed without transmission. From the table, it can be seen that most of the works proved that Blowfish is the fastest algorithms among others. So, it is up to researchers to decide which algorithm is suitable for their applications.

Table 7 summarized the comparison of previous research works in terms of design approach. This table shows that most of previous works only simulated their cryptography algorithms. There are only a few of researchers who had validated their algorithms. The technologies used for their

Table 6. The comparison of various cryptography algorithms in terms of execution time

Reference	Time (ms)					
	AES	DES	3DES	CAST	IDEA	BLOWFISH
(Hirani, 2003)	3.0×10^2	-	-	5.0×10^2	6.0×10^2	-
(Kofahi et al., 2004)	-	2.5×10^4	3.8×10^4	-	-	1.4×10^4
(Nadeem & Javed, 2006)	3.3×10^5	1.1×10^5	3.3×10^5	-	-	8.6×10^4
(Gonzalez et al., 2008)	3.7×10^4	1.9×10^4	2.0×10^5	-	4.4×10^4	1.7×10^4
(Idrus & Aljunid, 2008)	2.7×10^{-1}	-	-	-	3.7×10^{-1}	2.8×10^{-1}
(Elminaam et al., 2010)	1.4×10^3	1.7×10^3	1.8×10^3	-	-	1.0×10^2
(Umaparvathi & Varughese, 2010)	2.1×10^2	4.6×10^2	1.2×10^3	-	-	2.7×10^2
(Nie et al., 2010)	-	1.1×10^6	-	-	-	8.7×10^5
(Verma et al., 2011)	4.2×10^3	6.0×10^3	6.2×10^3	-	-	4.0×10^3
(Ganesan & Selvakumar, 2011)	2.5×10^2	2.4×10^2	2.6×10	-	-	5.5×10

Table 7 The comparison of design approach on cryptography algorithms

Reference	Simulation	Validation Device
(Schneier & Whiting, 1997)	Intel Pentium	-
(Hirani, 2003)	IBM-Athlon Processor	Wireless devices & Cisco AP1200
(Kofahi et al., 2004)	Sun Workstation	-
(Nadeem & Javed, 2006)	Java Platform	-
(Gonzalez et al., 2008)	Xilinx MicroBlaze	FPGA
(Idrus & Aljunid, 2008)	Web programming language	-
(Elminaam et al., 2010)	Intel Pentium	-
(Umaparvathi & Varughese, 2010)	Java Platform	-
(Nie et al., 2010)	AMD Athlon	-
(Verma et al., 2011)	Intel Pentium	-
(Ganesan & Selvakumar, 2011)	Intel Pentium	CPU & 54 M wireless access point

validation processes such as in Hirani (2003) and Ganesan and Selvakumar (2011) were only suitable for WiFi communication. Furthermore, the wireless devices used could not be configured for different cryptography algorithms in one time. This means a further work needs to be done through implementation to validate these algorithms in real time environment for different communication standard, which can be done in a short time through SDR platform. As in Gonzalez et al. (2008), the validation of cryptography algorithms was based on FPGA configuration without going through the real transmission process. This is apparently not enough to prove that these algorithms are really optimized for certain communication standards.

SDR Implementations

Since the SDR is proposed as a validation technique for cryptography algorithm in WiMAX communication, this sub-section will study the SDR implementations of wireless security systems and WiMAX applications based on the recent research works. However, there are very few researchers who had verified their designs either for cryptography algorithms or WiMAX communication. Perhaps from this analysis, researchers will know how far is the SDR capability for WiMAX-based security.

To the best of our knowledge, only two papers presented the implementation of cryptography algorithms on SDR platform. One of them is by Mihaljevic and Kohno (2002), where this paper considers a framework for reconfiguration of cryptographic algorithms employed in SDR. This paper also discussed about cryptographic keys management issue in the context of techniques for delivering the software for SDR reconfiguration. The types of algorithm implemented are hash function (SHA and MD5), asymmetric (RSA and ECC), and symmetric (both stream and block ciphers). However, no data transmission over any communication standard was done in this paper.

Meanwhile in Jenkins et al. (2007), a hardware design and instruction set extensions for AES processing on a multithreaded SDR platform, the Sandbridge SB3010 system was presented. The research work was done based on five constraints: 1) The ability to operate efficiently in a multithreaded microarchitecture. 2) Encryption and decryption have similar performance. 3) No hidden state is added to the programming model. 4) A throughput of at least 50 Mbps at the highest cipher level. 5) Small area compared to the DSP core. The authors claimed that they were the first who proposed and analyzed the instruction set extensions for AES on a multithreaded SDR.

The SDR implementations for WiMAX applications were also done in the previous research works which none of them were related to cryptography. Zlydareva and Sacchi (2007) had used SDR platform for proving the SDR reconfigurable and reprogramming abilities by implementing Universal Mobile Telecommunications System (UMTS) and WiMAX PHY layers without changing the hardware. A verification for UMTS and WiMAX was carried out based on single carrier and multicarrier techniques, respectively.

SDR also had been used in Gao and Farrell (2008) to embed WiMAX channels within Terrestrial Trunked Radio (TETRA) framework. TETRA is a private mobile radio (PMR) standard that has been developed by the European Telecommunications Standard Institute (ETSI) for the needs of the transport, civil and emergency services. The SDR platform had enabled the investigation on various radio architectures that are fixed to TETRA-WiMAX system. The architectures are including a homodyne (direct-to-RF) transmitter and receiver, and a homodyne transmitter with a heterodyne receiver. In conclusion, this paper showed that RF module of the SDR platform works well within the system.

In Kadhim and Ismail (2010), an investigation was done on physical layer performance in terms of bit error rate and signal-to-noise ratio. The physical layer is comprised of three main sections: transmitter, receiver, and channel. The system model was tested for all modulations function with a different value of Additive White Gaussian Noise (AWGN) channel. The authors had implemented successfully the WiMAX IEEE 802.16e baseband transceiver on SDR platform through the DSP module. This module uses Virtex-4 FPGA and DM6446 System on Chip (SoC) for necessary performance in implementing custom IP and acceleration functions with varying requirements from one protocol to another supported on the same hardware.

Then, Kadhim and Ismail (2010) used the SDR platform for implementation of WiMAX Space-Time Block Coding (STBC) Orthogonal Frequency Division Multiplexing (OFDM) baseband transceiver. This paper had analyzed the performance results of Single-Input Single-Output (SISO), Multiple-Input Multiple-Output (MIMO), and Multiple-Input Single-Output (MISO) systems. The proposed design was proved to achieve much lower bit error rates and better performance for multi-path channels.

Suarez-Casal et al. (2010) presented an implementation of a fully reconfigurable downlink for WiMAX transceivers. Their architecture is made up of Commercial-Off-The-Shelf (COTS) modules available in the market and includes a DSP, three different models of FPGAs, DACs, and ADCs. This architecture is proved capable to support all the functionalities of the downlink subframe of the OFDMA WiMAX PHY layer, including Partial Usage of Subcarriers (PUSC) symbol structure and Forward Error Correction (FEC). The design had been implemented on SDR platform without modification or restart, at different levels in terms of bandwidth, size of Fast Fourier Transform (FFT), modulation, and code rate.

These previous works showed that SDR platform could be used for validation of WiMAX communication in real-time. This means, it is also possible to implement the cryptography algorithms for WiMAX applications using SDR technology although nobody has done this before. Hence, Figure 4 is proposed to analyze cryptography algorithms' performance. Meanwhile, Table 8 summarizes the types of design that were successfully implemented through SDR platform before, in the aspects of design type and wireless communication standard.

Figure 4. The proposed implementation of cryptography algorithms on SDR platform

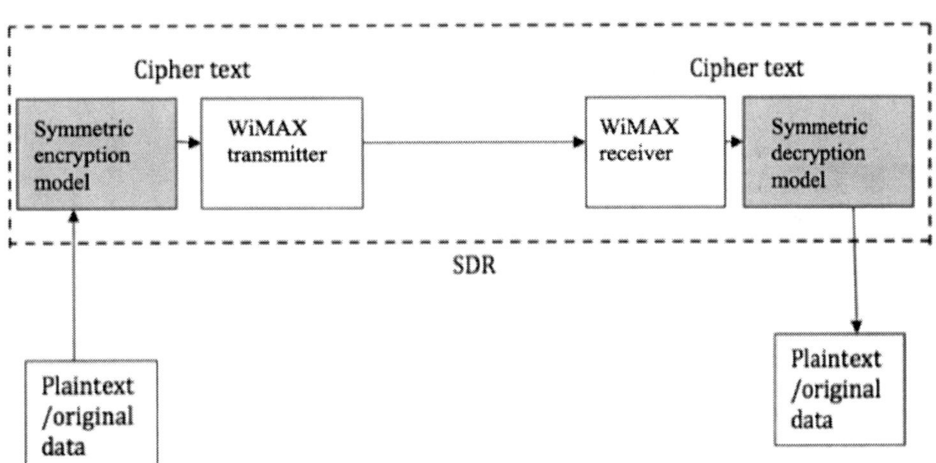

Table 8. Comparison of the related design implementations using SDR platform done by previous researchers

Reference	Design Type	Communication Standard	SDR Module
(Mihaljevic & Kohno, 2002)	Cryptographic components	-	DSP
(Jenkins et al., 2007)	AES algorithm	-	DSP
(Zlydareva & Sacchi, 2007)	PHY layers	UMTS/WiMAX	DSP
(Gao & Farrell, 2008)	TETRA-WiMAX system	WiMAX	RF
(Kadhim & Ismail, 2010)	IEEE 802.16e baseband transceiver	WiMAX	DSP
(Kadhim & Ismail, 2010)	STBC-OFDM baseband transceiver	WiMAX	DSP
(Suarez-Casal et al., 2010)	Downlink of PHY layer	WiMAX	DSP

CONCLUSION

This chapter has studied the high performance symmetric cryptography algorithms as the alternative protection in WiMAX applications. From the previous research works, it can be concluded that it is possible to use another algorithms such as Blowfish instead of AES. Blowfish has lower power consumption, faster processing time, and less memory where these characteristics are suitable for mobile devices. In order to validate this algorithm through WiMAX communication, the SDR platform is proposed as a new technique for implementation. SDR technology has been identified as a key enabler of reconfiguration capabilities within user equipment or network operator nodes enhancing flexibility, configuration granularity and providing a fully seamless experience for the user. SDR also provides solutions in keeping manufacturing costs low and reducing power consumption as more signal processing stages move to digital domain where the benefits of new reduced size manufacturing processes may be expected.

REFERENCES

Ahson, S., & Ilyas, M. (2008). *WiMAX standards and security*. Boca Raton, FL: CRC Press.

Airspan. (2007). Mobile WiMAX security. *Airspan Networks Inc*. Retrieved May 15, 2012, from http://www.airspan.com

Barbeau, M. (2005). WiMAX/802.16 threat analysis. In *Proceedings of Q2SWinet 2005,* (pp. 1-8). Quebec, Canada: Q2SWinet.

Bielecki, W., & Burak, D. (2007). Parallelization method of encryption algorithms. In *Advances in Information Processing and Protection* (pp. 191–204). Berlin, Germany: Springer-Verlag. doi:10.1007/978-0-387-73137-7_17

Elminaam, D. S. A., Kader, H. M. A., & Hadhoud, M. M. (2010). Evaluating the performance of symmetric encryption algorithm. *International Journal of Network Security*, *10*(3), 216–222.

Ganesan, R., & Selvakumar, A. A. L. (2011). A new balanced encryption algorithm with elevated security base on key update. *European Journal of Scientific Research*, *60*(2), 177–194.

Gao, L., & Farrell, R. (2008). *Using SDR to embed WIMAX channels within the TETRA framework*. Retrieved May 15, 2012, from http://eprints.nuim.ie/1413/1/GaoL_final.pdf

Gonzalez, I., Lopez-Buedo, S., & Gomez-Arribas, F. J. (2008). Implementation of secure applications in self-reconfigurable systems. *Microprocessors and Microsystems*, *32*, 23–32. doi:10.1016/j.micpro.2007.04.001

Hirani, S. (2003). *Energy consumption of encryption schemes in wireless devices*. (Unpublished Master of Science Dissertation). University of Pittsburgh. Pittsburgh, PA.

Idrus, S. Z. S., & Aljunid, S. A. (2008). Performance analysis of encryption algorithms text length size on web browsers. *International Journal of Computer Science and Network Thesis, 8*(1), 20-25.

Jenkins, C., Mamidi, S., Schulte, M., & Glossner, J. (2007). Instruction set extensions for the advanced encryption standard on a multithreaded software defined radio platform. *International Journal on Embedded Systems*, *2*(3-4), 203–214.

Jha, R. K., & Dalal, U. D. (2010). A journey on WiMAX and its security issues. *International Journal of Computer and Information Technologies*, *1*(4), 256–263.

Joseph, C. M. T. (2011). Improving security in the IEEE 802.16 standard. In *Proceedings of Eighth International Conference on Information Technology: New Generations,* (pp. 408-412). IEEE.

Kadhim, M. A., & Ismail, W. (2010a). Implementation of WIMAX IEEE 802.16e baseband transceiver on multi-core software defined radio platform. *International Journal of Computer Theory and Engineering*, *2*(5), 1793–8201.

Kadhim, M. A., & Ismail, W. (2010b). Implementation of WIMAX STBC-OFDM (IEEE802.16.d) baseband transceiver on a multi-core software-defined radio platform. *Australian Journal of Basic and Applied Sciences*, *4*(7), 2125–2133.

Kessler, G. C. (2010). *An overview of cryptography*. Retrieved March 20, 2012, from http://people.eecs.ku.edu/~saiedian/teaching/Fa10/710/Readings/An-Overview-Cryptography.pdf

Khan, A. A., & Zaman, N. (2009). Comparative analysis of broadband wireless access from Wi-Fi to WiMAX. In *Proceedings of International Bhurban Conference on Applied Sciences & Technology,* (pp. 8-14). Islamabad, Pakistan: IEEE.

Kofahi, N. A., Al-Somani, T., & Al-Zamil, K. (2004). Performance evaluation of three encryption/decryption algorithms. *Proceedings of the IEEE, 2*, 790–793.

Kumar, A. (2008). *Mobile broadcasting with WiMAX*. New York, NY: ScienceDirect.

Mihaljevic, M. J., & Kohno, R. (2002). On a framework for employment of cryptographic components in software defined radio. *Proceedings of Wireless Personal Communications, 2*, 835–839. IEEE Press.

Mishra, A., & Glore, N. (2008). Privacy & security in WiMAX networks. In *WiMAX Standards and Security* (pp. 205–228). Boca Raton, FL: CRC Press.

Mitola, J. (1992). *The software radio*. Paper presented at the IEEE National Telesystems Conference. New York, NY.

Nadeem, A., & Javed, M. Y. (2006). A performance comparison of data encryption algorithms. In *Proceedings of IEEE Information and Communication Technologies*, (pp. 84-89). IEEE Press.

Nie, T., Song, C., & Zhi, X. (2010). Performance evaluation of DES and blowfish algorithms. In *Proceedings of Biomedical Engineering and Computer Science* (pp. 1–4). IEEE Press. doi:10.1109/ICBECS.2010.5462398

Potlapally, N. R., Ravi, S., Raghunathan, A., & Jha, N. K. (2006). A study of the energy consumption caharacteristics of cryptographic algorithms and security protocols. *IEEE Transactions on Mobile Computing, 5*(2), 128–143. doi:10.1109/TMC.2006.16

Reed, J. H. (2002). *Software radio: A modern approach to radio engineering*. Upper Saddle River, NJ: Prentice Hall.

Ruangchaijatupon, P., & Krishnamurthy, P. (2001). Encryption and power consumption in wireless LANs-N. In *Proceedings of the 3rd IEEE Workshop on Wireless LANs,* (pp. 148-152). Newton, MA: IEEE Press.

Scarfone, K., Tibbs, C., & Sexton, M. (2010). *Guide to securing WiMAX wireless communications*. Gaithersburg, MD: National Institute of Standards and Technology.

Schneier, B., & Whiting, D. (1997). *Fast software encryption: Designing encryption algorithms for optimal software speed on the Intel Pentium processor*. Berlin, Germany: Springer. doi:10.1007/BFb0052351

Source Lyrtech. (2012). Lyrtech launches new small form factor SDR development platforms at SDR forum technical conference. *Lyrtech, Inc*. Retrieved May 15, 2012, from http://www.prnewswire.com/news-releases/lyrtech-launches-new-small-form-factor-sdr-development-platforms-at-sdr-forum-technical-conference-56330167.html

Suarez-Casal, P., Carro-Lagoa, A., Garcia-Naya, J. A., & Castedo, L. (2010). *A multicore SDR architecture for reconfigurable WiMAX downlink*. Retrieved May 15, 2012, from http://gtec.des.udc.es/web/images/pdfConferences/2010/dsd_suarez_2010.pdf

Trimintzios, P., & Georgiou, G. (2010). Review article: WiFi and WiMAX secure deployments. *Journal of Computer Systems, Networks, and Communications. Hindawi Publishing Corporation, 2010*, 1–28.

Trinadh, P. (2009). Emergence of software defined radio, SDR. *Home Tecnology Magazine*. Retrieved May 15, 2012, from http://hometoys.com/emagazine.php?url=/ezine/09.04/trinadh/index.htm

Umaparvathi, M., & Varughese, D. K. (2010). Evaluation of symmetric encryption algorithms for MANETs. In *Proceedings of Computational Intelligence and Computer Research* (pp. 1–3). IEEE Press. doi:10.1109/ICCIC.2010.5705754

Verma, O. P., Agarwal, R., Dafouti, D., & Tyagi, S. (2011). Performance analysis of data encryption algorithms. In *Proceedings of Electronics and Computer Technology* (pp. 399–403). IEEE Press.

Wikipedia. (2012). *List of software-defined radios*. Retrieved May 15, 2012, from http://en.wikipedia.org/wiki/List_of_software-defined_radios

Wireless Innovation. (2012). *Benefits of SDR*. Retrieved May 15, 2012, from http://www.wirelessinnovation.org/Benefits_of_SDR

Xu, X., Bosisio, R. G., & Wu, K. (2006). Analysis and implementation of six-port software defined radio receiver platform. *IEEE Transactions on Microwave Theory and Techniques*, *54*(7), 2937–2943. doi:10.1109/TMTT.2006.877449

Yuksel, E. (2007). *Analysis of the PKMv2 protocol in IEEE 802.16e-2005 using static analysis informatics and mathematical modeling*. (Masters Thesis). Technical University of Denmark. Copenhagen, Denmark.

Zlydareva, O., & Sacchi, C. (2007). SDR application for implementing an integrated UMTS/WiMAX PHY-layer architecture. In *Proceeding of the 3rd International Conference on Mobile Multimedia Communications (MobiMedia 2007)*, (pp. 1-7). IEEE.

KEY TERMS AND DEFINITIONS

Advanced Encryption Standard (AES): A specification for the encryption of electronic data. It is a symmetric key algorithm, which means the same key is used for both encrypting and decrypting the data.

Cryptography Algorithm: Sequences of processes, or rules, used to encipher and decipher messages in a cryptographic system. In simple terms, they are processes that protect data by making sure that unwanted people cannot access it. Most cryptography algorithms used encryption, which allows two parties to communicate while preventing unauthorized third parties from understanding those communications. Encryption transforms human readable plaintext into something unreadable known as ciphertext.

High Performance: A high data processing in an algorithm that involves very low power consumption, short processing time, and small design memory.

Reprogrammable: Can be reprogrammed, whose program can be replaced with another without exchanging its hardware.

Software Defined Radio (SDR): A radio that is substantially defined in software and whose physical layer behavior can be significantly altered through changes to its software. SDR attempts to place much or most of the complex signal handling involved in communications receivers and transmitters into the DSP style.

Wireless Attack: Attack on wireless network such as jamming, man-in-the-middle, eavesdropping, flooding, and DoS.

Worldwide Interoperability for Microwave Access (WiMAX): A wireless digital communications system, also known as IEEE 802.16, that is intended for WMAN. This standard designed to provide 30 to 40 Mbps data rates, with the 2011 update providing up to 1 Gbps for fixed stations. It is a part of 4G wireless communication technology.

Chapter 12
Genetic Algorithms for Decision-Making in Cognitive Radio Networks

Tommy Hult
Lund University, Sweden

Abbas Mohammed
Blekinge Institute of Technology, Sweden

ABSTRACT

Efficient use of the available licensed radio spectrum is becoming increasingly difficult as the demand and usage of the radio spectrum increases. This usage of the spectrum is not uniform within the licensed band but concentrated in certain frequencies of the spectrum while other parts of the spectrum are inefficiently utilized. In cognitive radio environments, the primary users are allocated licensed frequency bands while secondary cognitive users dynamically allocate the empty frequencies within the licensed frequency band according to their requested QoS (Quality of Service) specifications. This dynamic decision-making is a multi-criteria optimization problem, which the authors propose to solve using a genetic algorithm. Genetic algorithms traverse the optimization search space using a multitude of parallel solutions and choosing the solution that has the best overall fit to the criteria. Due to this parallelism, the genetic algorithm is less likely than traditional algorithms to get caught at a local optimal point.

INTRODUCTION

Background and History of Cognitive Radios

Cognitive Radio can be defined as a radio that can adapt its transmitter parameters based on interactions with the environment in which it is operating. Cognitive Radio is an emerging technology aimed for the efficient use of the limited radio frequency spectrum.

The term "cognitive radio" was first used and defined by Joseph Mitola III in an article published in 1999 (Mitola, et al., 1999). He described the way a cognitive radio could enhance the flexibility of personal wireless services through a new language called Radio Knowledge Representation Language (RKRL). The idea of RKRL was further expanded

DOI: 10.4018/978-1-4666-2812-0.ch012

in Mitola's doctoral dissertation, presented at the Royal Institute of Technology, Sweden, in May 2000 (Mitola, 2000). This dissertation presented a conceptual overview of cognitive radio as an exciting multidisciplinary subject. In 2002, FCC in the United States aimed at the changes in technology and the profound impact that those changes would have on spectrum policy. This report set the stage for a workshop on cognitive radio, held in Washington DC, in May 2003 (FCC, 2003). This workshop was immediately followed by a Conference on Cognitive Radios held in Las Vegas, NV, in March 2004.

Cognitive Decision Making

In cognitive radio, decisions have to be made, e.g., whether or not to stay on a specified carrier frequency or if the criteria is fulfilled move to another carrier frequency. This decision has to be made to facilitate a minimal interference policy and at the same time giving a high degree of spectrum efficiency. This approach could be extended to include decision on which type of modulation to use and also deciding on the configuration of a smart antenna system (e.g., SISO, SIMO, or MIMO) depending on what type of environment the transmission is propagating through.

Spectrum Sensing

The radio is searching for empty frequency bands in the licensed spectrum. These empty frequency bands constitute unlicensed slots inside the licensed frequency bands. The handling of these unlicensed bands is a coordinated effort among the equipments using these bands. All the transmitting units reports its own spectral usage and the distribution of these empty bands are decided upon based on some decision parameters. The secondary users can then acquire the usage of the empty bands until a primary user requires it.

Spectrum Reallocation

If a primary user is detected in a frequency band that is used by a secondary user, the secondary user will be allocated another frequency band to avoid interfering with the primary user. The hand over will be handled without affecting the communication.

Spectrum Management Policy

When the empty frequency bands have been detected the QoS (Quality-of-Service) of the secondary application is considered. This means that the secondary user should use these frequency bands if they meet the QoS requirement.

Spectrum Sharing

A scheme for scheduling will be required between the primary and secondary users of a frequency band. This is necessary to minimize the interference between the primary and secondary users. This is done by assigning a holding time to each user.

Overview of Genetic Algorithms

The Genetic Algorithm (GA) is a heuristic method of finding an approximate solution to optimization problems and belongs to a class of algorithms called Evolutionary Algorithms. These algorithms draw on the evolutionary theories of Charles Darwin's survival of the fittest by choosing the best solution from a set of evolved possible solutions. The history of evolutionary computation goes back to 1970s when Rechenberg first described it in his work "Evolution strategies" (Rechenberg, 1971). The GA was first developed by John Holland (1975) in 1975 and in 1992, John Koza (1992) introduced Genetic Programming (GP).

The genetic algorithm works by tagging each candidate solution with an array of bits, were each bit represents a parameter. A fitness function is defined which estimates the quality of the chosen solution. The "fitness" value of a particular solution is the sum of all the parameters of this solution if it valid, otherwise it is zero. The algorithm is initialized by randomly generate a set of initial solutions. During each iteration of the algorithm, a part of the set is chosen based on the quality compared to the fitness function. This subset is then used to generate a new set of solutions by taking two solutions from the new set and swap parts of the tags between them. Another method is to set the probability that each parameter bit will be changed. The algorithm continues until a minimum criterion of the solutions is fulfilled.

The goal in optimization problems is to achieve an optimal solution, but the problem is that the search may get complicated and one may not know where to look for the solution or where to start. There are many methods that can help finding a suitable solution but the solution might not be the best one. Some of these methods are gradient descent, hill climbing, tabular search, simulated annealing and genetic algorithms. The solutions found using these methods are often considered as good solutions as it is not always possible to define the optimal solution. Game Theory (Neel, 2002) that is also at a nascent stage and is used for interactive decision situations provides analytical tools to predict the outcome of complex interactions among the rational entities based on perceived result. It works on predictions of probabilities but demands a precise knowledge of the total number of nodes, but in the dynamic nature of real networks, it is not even possible to have knowledge of what nodes enter or leave the network in real-time. In addition, the definition of a steady state and an undesirable drift with increasing number of nodes can make the implementation of Game Theory concept in the cognitive radios very complex. Another approach is fuzzy logic, but it has severe limitations in the dynamic nature of the wireless environment. It permits the approximate solutions to be found when there are uncertainties in the inputs but its logic for approximation does not have an evolutionary ability to allow it to change over time with the environment encountered at real-time.

GENETIC ALGORITHMS FOR COGNITIVE RADIO

Using the genetic algorithms in software radios require an optimization. Representing the radio in the above mentioned genetic terms allow it to adapt to a constantly changing electromagnetic environment in real-time. The genetic algorithm is used as the optimizing part of the decision-making module of the cognitive radio since they are especially well suited to handle multi-objective functions due to their convergence behavior towards the optimal solution. These types of algorithms also allow for multiple parameter optimization and a flexible problem analysis as long as the objective function and the "chromosomes" are properly defined. Genetic algorithms can have a long convergence time in reaching the optimal solutions, but are usually fast in reaching a very good approximate solution.

Using genetic algorithms for software radio (Rondeau, 2004) can be separated into the following actions:

- "Chromosome" definition
- Definition of the objective function
- Objective evaluation
- Multi-criteria optimization
- Decision-making

"Chromosome" Definition

The first step in the design of a genetic algorithm is the definition of the "chromosome" structure. The chromosome definition must represent the software radio behavior in the decision-making process to achieve the required optimization. There are many possible parameters that can be considered in this respect but only some of the basic parameters for software radio are considered here. The main parameters are the; momentary usage of a frequency band, spectral efficiency, power consumption and data rate. The "chromosomes" in the genetic algorithm are represented as vectors of data structures with different data types defining their "genes." In this chapter, the "genes" and "chromosomes" are defined in terms of bit arrays. In this first study, a minimum number of bits are used for the sake of simplicity. A software radio "chromosome" can have several different types of "genes" to represent its structure, but to keep it simple only a few basic parameters are set up as "genes"; carrier frequency, transmit power, bit error rate and the modulation scheme.

Definition of the Objective Function

The definition of the objective function is utilized for the "fitness" analysis of a population of "chromosomes." The objective function is defined based on the performance parameters of the momentary radio channel. The "genes" representing the "chromosome" will be analyzed for its "fitness" by the defined objective function according to its associated weights. These associated weights represent the relative importance that the user has assigned to each parameter definition.

The Pareto front (Kung, Luccio, & Preparata, 1975; Godfrey, Shipley, & Gryz, 2006) will move so that the optimal solution provides the most efficient performance for the user QoS requirements under combinatorial constraints. This multi-objective decision-making procedure according to the defined objective function provides the best performance with minimal use of the spectrum resources. This implies that the solution is obtained on the user preferences (i.e., the application that the user has to run on the identified virtually unlicensed frequency band). This will prohibit the overconsumption of spectrum resources, as it considers the QoS and QoS requirements specified by the secondary user or the application. For example, allocating a 100Mbps link with a 30 dB CNR (Carrier-to-Noise Ratio) for a user to send only an e-mail would be an overconsumption of the allocated resources. So, this makes the definition of the objective function critical for the best optimization in the decision-making process. The defined objective evaluation functions should reflect the current quality, both at the PHY and the MAC layer of the radio protocol and would therefore include; the average transmitting power, data rate, BER (Bit-Error-Rate), PER (Packet-Error-Rate), spectral efficiency, bandwidth, interference avoidance, packet latency and packet jitter and many more to provide for the QoS requirements, but for simplicity only a few of them are utilized here.

Another attribute of the genetic algorithm is the dynamic "fitness" definition and evaluation, where the weights assigned to the "fitness" of each individual "gene" can be adjusted dynamically. A database of all the objective functions is maintained and the "fitness" for each individual solution can dynamically be weighed and included in the "fitness" evaluation for a particular QoS requested by the user. This involves some higher-layer application to evaluate the radio and the network performance.

Fitness Evaluation

The objective functions are defined individually considering the current user QoS specifications. These fitness functions are applied on a randomly selected population of chromosomes in a multi-objective decision-making process with the use of stochastic processes. This implies the existence of a trade-off among the parameters for a particular channel.

This is analyzed by the corresponding weights assigned by the user to each of them. For example, for loss sensitive applications the BER would have a higher weight than the data rate in its fitness function, due to its importance indicated by the user application in its QoS specifications. Similarly, a higher weight assigned to the data rate than the BER, by the user will be beneficial in case of time sensitive applications. So, the weight associated with each of the objective functions enables a relative measurement among different parameters and the one with higher weight in a particular channel would pass the fitness test. So, an evaluation of the fitness of the randomly generated population of chromosomes is carried out in multiple dimensions and the ones that survive these fitness tests are passed to the next generation as a reproduction of the current one. This serves as a filtering process, carried out at each generation that results in an optimal solution at the end of the evaluation by the genetic algorithm. This is described as the Pareto front (Kung, Luccio, & Preparata, 1975; Godfrey, Shipley, & Gryz, 2006).

The genetic algorithms may result in long and tedious calculations for the optimal solution in some cases but the diversity of the solutions in the selected population allows multiple (some of them turn out to be very good) solutions to be tired at each generation. This prevents the genetic algorithm from getting stuck, at any stage during the decision-making process. This implies that even though the genetic algorithms may not come up with an exact solution in some cases, they do provide the closest possibility among the set of available solutions. This will be discussed in more detail in the coming sections.

Multi-Criteria Optimization

Multi-criteria optimization is a process of analyzing multiple conflicting objectives subject to certain constraints. Multi-criteria optimization problems are found in various fields where maximization or minimization of some functions is required and a trade-off exists among the objectives. Generally, there is no single solution to a multi-criteria optimization problem, as maximizing one of the objectives often affects the other objectives in the same problem. So, there is usually more than one solution as the optimal result. Then it is possible to pick the best among the solutions that conform to our particular case. This gives the best optimization for the problem at hand. A generalized form of the multi-criteria optimization problem can be represented as follows (Steuer, 1986; Sawaragi, Nakayama, & Tanino, 1985):

$$\begin{aligned} &\min/\max(y), \left[f_1(x), f_2(x), \cdots, f_n(x)\right] \\ &\textit{subject to}: \\ &x = \left[x_1, x_2, x_3, \cdots, x_m\right] \in X \\ &y = \left[y_1, y_2, y_3, \cdots, y_n\right] \in Y \end{aligned} \quad (1)$$

where there are n dimensions to consider in the search space and $f_n(x)$ defines the mathematical function to evaluate dimension n. Both x, the set of input parameters, and y, the set of dimensions, may be constrained to some space; X and Y, where X and Y represent the functions for the spaces X and Y, respectively. The optimal solutions to the above problem are a set of Pareto points, where improvement in one of the objectives may impair the others. Accordingly, there exists a trade-off among the individual genes. The Pareto points move in such a way that the optimal solution provides the most efficient performance for the requested QoS in the radio domain and according to the regulatory constraints. Further, there should not be any excessive use of the radio resources while decision-making. A higher QoS than requested by the user would be a waste of resources and should not be allocated.

Decision-Making

The unused or inefficiently utilized spectrum bands that are spread over a wide frequency range in the RF environment can include both unlicensed and licensed frequencies. The empty frequencies detected in a frequency band can have diverse characteristics due to the time-varying radio environment and due to some parameters like carrier frequency and bandwidth. The radio should decide on the best spectrum band to use, that meets the QoS requirements with an assumption that the user specifies the QoS requirements or that the sensing equipment of the radio provides this information.

The receiver in the cognitive radio senses the RF environment and then involves itself in the decision-making process to accommodate a new user requesting spectrum allocation in that environment. It considers the QoS requirements of the user (e.g., the modulation scheme, channel coding, power consumption, and data rate). It also has information about the channel conditions, the network interactions and the radio environment such as the empty frequency bands, transmission and reception power for a specific application and the amount of interference that would be produced with the induction of the new user in that environment with the primary users already using the spectrum. The sensor gives the spectrum status as an input to the spectrum allocator and a spectrum test is performed. If the user is satisfied with the frequency band it takes no further action and terminates the process, otherwise the appropriate frequency bands are selected and certain spectrum decisions are taken. This spectrum test must comply with the regulations of the spectrum allocations and tend to minimize the interference with the other users. These regulations are considered in the decision-making process to meet the QoS requirements of the secondary (new) user. One of the most important factors that should be considered during the decision-making is that the primary user should always be preferred if it needs that particular frequency band.

SETUP OF THE ALGORITHM

The genetic algorithm approach begins with the definition of the chromosome structure. The chromosome size is kept as small as possible to avoid making the selection process for the best solution complex and too slow.

The Chromosome

To define the chromosome structure it is necessary to have information and understanding of the genes of the chromosome that will constitute its structure. The "genes" in this particular case would be the individual parameters that will be considered for the decision-making process. These genes are basically a part of the solution. Four parameters (genes) are considered here, and these are:

1. Carrier frequency
2. Transmit power
3. Bit Error Rate (BER)
4. Modulation techniques

The Frequency Band Gene

Any equipment that is able to transmit and receive information via electromagnetic waves requires a particular frequency band (here, a frequency band is defined as a pre-defined bandwidth of a transmitted signal on a pre-defined carrier frequency) to communicate.

These frequency bands are represented in terms of bits, which will help carrying out certain operations during the decision-making process, like the mutation operation on certain defined chromosomes. The number of bits required to represent all frequency bands is denoted as N_b^f, since this can represent up to $2^{N_b^f} = N_f$ frequency bands to fulfill our requirement. This implies that each frequency band in the range is represented in terms of an integer. The first frequency band would be given the integer value of 0. The second band with the integer value of 1 and so on. So, the bands ranging 0-N_f with a step size BW_{req}, gives integer values ranging according to:

$$\left[g_n^f\right]_b = \left[\left\lfloor \frac{\left\lfloor \frac{f_c}{BW_{req}} \right\rfloor}{2^{b-1}} \right\rfloor \bmod 2\right]_b, \qquad b = \left\{1, 2, 3, 4, 5, \cdots, N_b^f\right\} \tag{2}$$

where $\left[g_n^f\right]_b$ is the frequency gene for chromosome n in binary format and b denotes the position of each binary digit. N_b^f is the number of bits required to represent the number of frequency bands available.

Transmit Power Gene

Another important parameter that is considered for the chromosome structure definition is the transmit power. This is the power required by the application to provide for a good transmission, a power less than that may result in the transmission of a weak signal. The range of power values should be specified in such a way that allows the users to communicate without too many errors, and as increasing the power for a transmission of a signal increases the signal strength and hence increases the chances of a successful communication process. In addition, increasing the power of a signal decreases the number of errors. The power range given in dBm varies from P_{MIN} to P_{MAX} with a step size of ΔP in between them. This gives a total of N_P values to be represented with an integer value. The representation of these values will require a total of N_b^P bits to enable the user to know whether the given input is within the correct range or not. The power gene would be the second one in the order of our chromosome structure:

$$\left[g_n^P\right]_b = \left[\left\lfloor \frac{\left\lfloor \frac{P_{dBm} + \left|P_{MIN}\right|}{\Delta P_{dBm}} \right\rfloor}{2^{b-1}} \right\rfloor \bmod 2\right]_b, \qquad b = \left\{1, 2, 3, \cdots, N_b^P\right\} \tag{3}$$

where $\left[g_n^P\right]_b$ is the power gene for chromosome n in binary format and b denotes the position of each binary digit. This mapping of the actual values in the integer form makes the decision-making process much simpler to manipulate.

Bit Error Rate (BER)

The Bit Error Rate (BER) represents the average number of received erroneous bits for a transmission and it depends on the application that requires the spectrum allocation. There are some applications (error sensitive applications) that require a low bit error rate for the transmission and these applications can sacrifice a high bandwidth for that, e.g., VoIP (Voice over Internet Protocol) applications. On the other hand there are applications that can compromise with a higher bit error rate but demand a high bandwidth (known as bandwidth sensitive applications), for example video streaming applications. Since there are more chances of error in the wireless communications than the wired, these applications need to have some procedure to reduce the occurrence of errors. The bit error rate can either be reduced by the use of certain coding schemes at the receiver and/or the transmitter. Another way to reduce the bit error rates is to increase the transmission power of the device. Each of these values is represented by an integer value, in the same manner as with the frequency and power parameters, for the chromosome definition.

$$\left[g_n^{BER}\right]_b = \left[\frac{\left\|\log_{10}\left(BER\right)\right| - \left|\log_{10}\left(BER_{MAX}\right)\right\|}{2^{b-1}} \bmod 2\right]_b, \quad b = \left\{1, 2, \cdots, N_b^{BER}\right\} \tag{4}$$

where $\left[g_n^{BER}\right]_b$ is the BER gene for chromosome n in binary format and b denotes the position of each binary digit. N_b^{BER} is the number of bits required to represent the total number of BER steps available.

Modulation Scheme

The last of the four genes to be considered in the chromosome structure is the modulation gene. There are four modulation schemes considered here. They are BPSK, QPSK, GMSK, and 16QAM. Therefore, to represent these four schemes we need just 4 integers (see Table 1).

$$\left[g_n^{M}\right]_b = \left[\frac{K_n}{2^{b-1}} \bmod 2\right]_b, \quad b = \left\{1, \cdots, N_b^{M}\right\} \tag{5}$$

where $\left[g_n^{M}\right]_b$ is the Modulation gene for chromosome n in binary format, b denotes the position of each binary digit and K_n is the integer number assigned to each modulation scheme. N_b^{M} is the number of bits required to represent the total number of modulation schemes that are available. The application provides its requirements for the modulation scheme that will be used. The cognitive radio, by its decision-making process, can then provide the best possible match from the solutions available. It maintains a trade-off among all the genes in the chromosome structure according to the corresponding weight assigned to each of them by the user.

Table 1. The four modulation schemes used in this investigation

Modulation Scheme	Integer Value
BPSK	0
QPSK	1
GMSK	2
16 QAM	3

The Chromosome Structure

The above-mentioned four genes all together define our chromosome's structure. The whole structure of the chromosome and the order of each of the genes are given by:

$$C_n = \left[g_n^{f}\right]_{1-N_b^{f}} \left[g_n^{P}\right]_{1-N_b^{P}} \left[g_n^{BER}\right]_{1-N_b^{BER}} \left[g_n^{M}\right]_{1-N_b^{M}} \tag{6}$$

The frequency band is the first "gene" in the chromosome structure and can have any value between 0-999. The power gene follows in the order and can have any value between 0-60. The third gene in the chromosome structure is BER and can have any value between of 0-7. The last gene in the chromosome structure is the modulation that can have a value between 0 and 3. The user or the application specifies its requirements to the cognitive radio as a chromosome and asks for an optimized solution that fulfills these requirements. The optimized solution also has the same structure as that of the chromosome. The representation of these four parameters in a "chromosome" requires 21 bits in total, i.e. 10+ 6+ 3+ 2=21 bits.

The First Population Chromosomes Generation

The chromosome structure is now defined and we can generate the first population of chromosomes. This is done randomly and there are no particular rules about the size limitations in the first population of the chromosomes. To begin with, the generation of an initial population of N_c chromosomes is used as a starting set. This population size can be increased if the results are not satisfactory. This randomly generated first population of chromosomes then undergoes a set of operations; selection, crossover and mutation, which results in the next generation of chromosomes.

Therefore, this stochastically generated first generation population has diversity among the solutions, and some of these solutions may be the good and others may be bad. In order to increase the possibility of good solutions in the next generation we apply the objective function on the initial population and perform a set of operations; selection, crossover and mutation. The initial population of the chromosomes are illustrated in Box 1.

Before performing operations like "crossover" and "mutation," it is necessary to define an objective function, which is responsible for generating the next generation of chromosome population. The objective function assess the fitness of the chromosomes in the initial population of N_c chromosomes and selects the fittest of them to the next generation population.

The Fitness Function

When the structure of the chromosome has been defined and when it has generated the first generation of the chromosome population, the fitness of each chromosome in the population will be evaluated. This first population of chromosomes consists of a diverse set of solutions that meet the specified Qos requirements of the user. Some of these solutions may satisfy the Qos requirement to a high degree while others only get arbitrarily close to those specifications. So, we can choose among this set on the basis of the trade-off conditions mentioned earlier. Consequently, there is a need for the definition of a test for these individual chromosomes such that the probability of chromosomes leading to bad solutions in the next generation can be minimized. It is therefore necessary to fully identify and define an objective function according to the genes of the "chromosomes.

To start with, we can define the simplest possible objective function, i.e., a fitness function that is equally dependent on all four parameters, defined above as its genes. This might not be a practical case in communication systems, as a change in any of these four parameters affects the other parameters (genes), but it provides for a simple starting point. The four parameters will have an equal weight i.e., 25% each and the given

Box 1. Initial population of "chromosomes"

$$C_{init} = \begin{bmatrix} \left[g_1^f\right]_{1-N_b^f} & \left[g_1^P\right]_{1-N_b^P} & \left[g_1^{BER}\right]_{1-N_b^{BER}} & \left[g_1^M\right]_{1-N_b^M} \\ \left[g_n^f\right]_{1-N_b^f} & \left[g_n^P\right]_{1-N_b^P} & \left[g_n^{BER}\right]_{1-N_b^{BER}} & \left[g_n^M\right]_{1-N_b^M} \\ \vdots & \vdots & \vdots & \vdots \\ \vdots & \vdots & \vdots & \vdots \\ \vdots & \vdots & \vdots & \vdots \\ \left[g_{N_c-1}^f\right]_{1-N_b^f} & \left[g_{N_c-1}^P\right]_{1-N_b^P} & \left[g_{N_c-1}^{BER}\right]_{1-N_b^{BER}} & \left[g_{N_c-1}^M\right]_{1-N_b^M} \\ \left[g_{N_c}^f\right]_{1-N_b^f} & \left[g_{N_c}^P\right]_{1-N_b^P} & \left[g_{N_c}^{BER}\right]_{1-N_b^{BER}} & \left[g_{N_c}^M\right]_{1-N_b^M} \end{bmatrix} \quad (7)$$

user QoS requirements are compared against the population of "chromosomes." Their fitness is calculated as the cumulative sum of the individual fitness of each gene (parameter) according to the procedure described in the following sub-sections.

Fitness of the Gene

The fitness of the parameter gene is calculated by taking the distance between the parameter gene pattern of the chromosome and the parameter gene pattern representing the value requested by the user:

$$d_i(g_n^i) = \left| g_n^i - g_{req}^i \right| \tag{8}$$

where i denotes the parameter of the gene and n is the number of a particular gene in the set.

If the distance calculated above is less than a specified threshold g_{th}^i, then the gene is considered acceptable and is assigned a fitness weight, according to:

$$L^g(i) = \begin{cases} \dfrac{w_i \cdot d_i(g_n^i)}{g_{th}^i}, & d_i(g_n^i) < g_{th}^i \\ w_i, & d_i(g_n^i) \geq g_{th}^i \end{cases} \tag{9}$$

where w_i is the associated weight assigned to the parameter gene of the chromosome. For example, w_i having a value of 25 implies that the parameter gene counts for 25% of the total fitness of the respective chromosome. So, we can vary the assignment of weight by changing this value of w_i according to its participation in the total fitness calculation for the chromosome. The higher, the fitness of the parameter gene is, the less suitable the solution is.

Total Fitness

The chromosome fitness L^c is calculated by taking the average of the M gene fitness values L^g in a single chromosome. To make the values of total fitness of the chromosomes more intuitive they are converted into percentages of maximum fitness:

$$L^c(n) = 1 - \frac{1}{M}\sum_{m=1}^{M} L^g(m) \tag{10}$$

The sum of all the weighting factors of the genes should be equal to unity.

DECISION-MAKING

The unused or inefficiently utilized frequency bands that are spread over a wide frequency range in the RF environment may include both unlicensed and licensed bands. The empty frequencies detected in the frequency band may show diverse characteristics due to the time-varying radio environment and some parameters like carrier frequency and bandwidth. The radio should decide the best frequency band to meet the QoS requirements.

The focus of this book chapter is in the spectrum management of cognitive radio; with an assumption that the user specifies the QoS requirements or that the radio sensing equipment provides this information. The radio receiver senses the RF environment and proceeds with the decision-making process to accommodate a new user requesting spectrum allocation in that environment. It considers the user QoS requirements, such as the modulation scheme, BER, power consumption and data rate. It also has the information about the channel conditions, the network interactions and the radio environment such as the empty frequency bands, transmission and reception power for a specific application and the amount of interference that would be produced with the induction of a new user in that environment, with the primary users already using the spectrum. The sensor gives the spectrum status as an input to the spectrum allocator and a spectrum test is then performed. If the user is satisfied with the frequency band it takes no further action and terminates the process, otherwise appropriate frequency bands are selected and certain spectrum decisions are made. This spectrum test must comply with the regulations of the spectrum allocations and to minimize the interference with other users. These regulations are considered in the decision-making process to meet the QoS requirements of the secondary user. One of the most important factors that should be considered during the decision-making is that the primary user should always be preferred if it needs that particular frequency band.

Two approaches are used here for the decision-making that lead to the optimization of the best solution. These two approaches only differ slightly from each other in their implementation part, but both aim at the optimization for the best solution.

Method 1

The procedure followed in this method involves the selection of the fittest of the available chromosomes in the set of solutions. Certain operations; crossover and mutation are performed on these selected chromosomes. This results in an evolution of a new generation of chromosome populations and we select the best of the chromosomes on the basis of our defined objective function. The process is explained in detail in the following sub-sections.

Selection

Selection involves the selection of the best "offspring" within the set of available chromosomes. This selection is done in accordance with our own defined objective function, described in the previous section. The "Roulette wheel selection" method, also known as "fitness proportionate selection" method, is used for the selection operation. It is a genetic operator generally used in the genetic algorithms for the selection of potentially useful solutions for recombination in the next generation. It associates a probability of selection with each individual chromosome, using the following equation.

$$p(j) = \frac{Fitness(j)}{\sum_{n=1}^{N} Fitness(n)} \tag{11}$$

where $p(j)$ is the associated probability of the individual chromosome in the set, $Fitness(j)$ is the "fitness" of an individual j in the set and N is the total number of "chromosomes" in the population. So, the chromosomes with the highest probability value would be transferred to the next generation. This enables a mechanism that does not to allow the transfer of the lower fitness value chromosomes to the new population of chromosomes. After each generation, the fittest chromosomes are transferred to the new population until the new population reaches the maximum defined limit. Once the fittest chromosomes in the set have been selected, we can perform operations like crossover and mutation as we move towards the optimization.

Crossover

The next step after the selection of the fittest chromosomes is to perform the crossover operation. The crossover operation is performed on a pair of chromosomes, selected randomly. The operation is done at the crossover points that define the junction of the genes in the chromosome structure. Equation 12 in Box 2 illustrates explicitly the range that the crossover points may exist.

In a single iteration, the crossover is performed randomly at one of these crossover points. The part of both the chromosomes after the crossover

Box 2. Crossover points

Crossover points

$$C_n = \left[\left[g_n^f \right]_{1-N_b^f} \left[g_n^P \right]_{1-N_b^P} \left[g_n^{BER} \right]_{1-N_b^{BER}} \left[g_n^M \right]_{1-N_b^M} \right] \quad (12)$$

points is swapped with each other. This results in totally new combination with new values for each of the genes.

The crossover operation, thereby gives rise to new solutions in the solution space of the problem. A crossover probability of 100% means that all chromosomes from the previous generation are crossed with each other and 0% means that all chromosomes are an exact copy of the chromosomes from the previous generation. Usually, the probability for the crossover operation to occur between two chromosomes is set to about 90%. This means that out of every N_c individual chromosomes, the crossover operation will occur in $0.9{\cdot}N_c$ "individuals" and in $0.1{\cdot}N_c$ there will be no crossover. The crossover rate is a user controlled parameter. N_c is the number of chromosomes in a population.

Mutation

The next step that follows the crossover is the mutation operation. The crossover operation did not need any conversion, i.e. the operation was performed on a randomly selected pair of chromosomes on the same decimal values of the genes. This is not the case with the mutation operation, which involves the conversion to corresponding binary values with consideration of only one chromosome at a time. In mutation, a randomly selected bit belonging to any of the four genes is inverted. The random selection of the bit indicates that it can be any of the bits from the total of 21 bits representing the "chromosome." The equation in Box 3 illustrates the binary structure of an example chromosome.

The mutation operation is performed by picking one bit of the chromosome code in Equation 1.11, randomly with a uniform distribution, and inverting it (change from 0 to 1 in this case).

The probability for the mutation operation to occur, are not as high as that for the crossover operation. Usually it is between 0.1% and 5%. For example if the mutation probability is 3% it means that mutation will occur in three out of every 100 individual chromosome and if the mutation is not performed then the chromosome remains the same. The purpose of the mutation is to avoid any local optimal points along the search path of the algorithm.

The New Population Generation

The crossover and mutation operations resulted in the production of new chromosomes in the solution space of the problem. The same fitness test, as explained above, is used to ensure the selection of the fittest chromosomes among the population, which are then transferred to a new set. This is now the new generation of the popu-

Box 3. Binary structure of example "chromosome"

Change to 1

$$C_n = [10000100\boxed{0}11101111011] \quad (13)$$

lation set. The size of this set is equal to the new generation population. The values in the genes for certain chromosomes differ from those in the initial population, due to the operations performed during the process.

The Optimal Solution

The above-mentioned steps are repeated and a new population of chromosomes is generated at every generation and the fitness tests are performed on them. This process continues unless the maximum number of specified populations has been reached. After the application of the fitness test on the last population, the chromosome with the maximum fitness value among these is chosen. This is our desired optimal result. It may or may not exactly match the requested solution but it will definitely be the closest of all the available solutions in the solution set. This completes the decision-making process, for this method.

Method 2

Most of the operations like "crossover" and "mutation" are the same as in method 1, but they differ with respect to their implementation. The main differences are that the "Roulette wheel selection" is not, used; instead the concept of "elitism" is used for the "selection" in method 2.

Elitism

The word "Elitism" comes from the word "Elite," that means the ones with the most distinguished attributes. "Elitism" used here in our decision-making process selects the best among the population of chromosomes and transfers them to the next generation, using the defined objective function. It copies a few of the best chromosomes to the new population before the other operations are performed on that population and can increase the performance of the algorithm as it prevents the loss of the best possible solution. The number of chromosomes to be transferred to the next generation after the "Elitism" operation is controlled by the user and can be varied. Only two of the fittest chromosomes from a generation of population is considered here and transferred to the next generation. This does not require the crossover or mutation operations. So, the two fittest chromosomes among the population are selected, using Elitism.

Crossover

The crossover operation is performed in the same manner as described for method 1. There is no difference in the operation, except in the meaning of the crossover rate. The crossover rate here in this method, stands for the number of chromosomes that will undergo the crossover operation, rather than the probability that the operation will occur. Therefore, a crossover rate of 90% in this method would mean that $0.9 \cdot N_c$ chromosomes among the population will undergo the crossover operation and would be transferred to the next generation.

Mutation

Mutation operation is also performed in the same manner as in method 1, but the difference here is the way it is implemented in this method. In addition, as with crossover, only the meaning of mutation rate is different. The mutation rate here in method 2, stands for the number of chromosomes that will undergo the mutation operation, rather than the probability that the operation will occur. Therefore, a mutation rate of 3% in this method would mean that $3 \cdot N_c$ chromosomes among the population will undergo the mutation operation and will be transferred to the next generation. The bit position for the mutation operation is chosen randomly with an even distribution. The rest of the procedure after the mutation operation is the same as described for method 1.

SIMULATIONS

The QoS specifications are specified by the user and include the four parameter "genes": required frequency band, transmit power, bit error rate, and modulation scheme. The frequency bands considered in this simulation ranges from zero to 10 MHz. The bandwidth of each band considered here is preset to 10 kHz, which gives a total of 1000 available frequency bands. The power range varies from -30 dBm to 30 dBm with a step size of 1 dBm and gives a total of 61 values of the available power. The bit error rates considered here are within a range of 10^{-1} and 10^{-8}. This range fulfills the requirements of the most mobile applications. The step size considered for the bit error is 10^{-1} bps.

The following QoS specifications in Table 2 were given as an input and calculations were performed using the same input for both method 1 and method 2. The optimal solution found by the genetic algorithm might not match exactly the QoS specifications requested by the user but is the closest possible in the solution set, for each gene.

The corresponding outputs of the "total fitness" error produced by method 1 and method 2 can be seen in Figure 1. The number of "generations" (i.e., number of iterations in the algorithm) is set to 2500 and the "population" size of each generation is set at 15 since Figure 2 show that a large population size do not give any benefits (trials revealed that a "population" size of 10 – 30 is adequate for this simulation). The crossover and mutation probabilities have been empirically found to be 95% and 0.4%, respectively, for method 1, 90% and 3%, respectively, for method 2.

Table 2. Input specification for both method 1 and 2

Frequency	Power	BER	Modulation
547	53	6	2

In this simulation, all four parameter are assumed equally important, so the weight factor is set to 0.25 for each parameter "gene."

The progress of each parameter gene of the chromosome is shown for method 1, in Figure 3. Each gene parameter is getting closer to the optimal value as the error of the total fitness in Figure 1 decreases. The behavior of the parameters in method 2 is similar. The final parameter values that the algorithm has converged to are shown in Table 3 for both method 1 and 2. When comparing these values to the requested parameters it is evident that we have a very close approximation for all four values.

Note that, the output or the optimal solution found by the genetic algorithm is also a chromosome having the same structure as that of the input QoS specification. In addition, both methods work using random values for each gene and can produce very different values for the chromosome structure each time they are run. So, for comparison the two methods are initiated with the same initial chromosome population.

CONCLUSION AND FUTURE RESEARCH DIRECTIONS

The aim of this chapter is to show a solution for the decision-making process in cognitive radios by using genetic algorithms. The simulations of the simple cognitive radio system show that a genetic algorithm for the decision making process is a viable solution and should be investigated further. The results show that the "fitness" of the individual "genes" (i.e., the suitably selected frequency, power, bit-error rate and modulation) increase with the increase of the number of generations; however, the fine-tuning of the algorithm is still in an 'ad-hoc' state and needs more investigation. In addition, it is shown that the population size of the "chromosomes" do not have to be very large (e.g., 10 to 30 "chromosomes") for a reasonably good convergence of the algorithm.

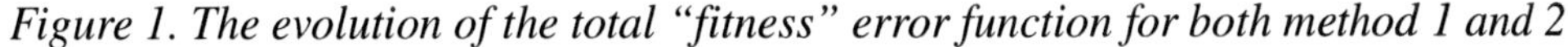

Figure 1. The evolution of the total "fitness" error function for both method 1 and 2

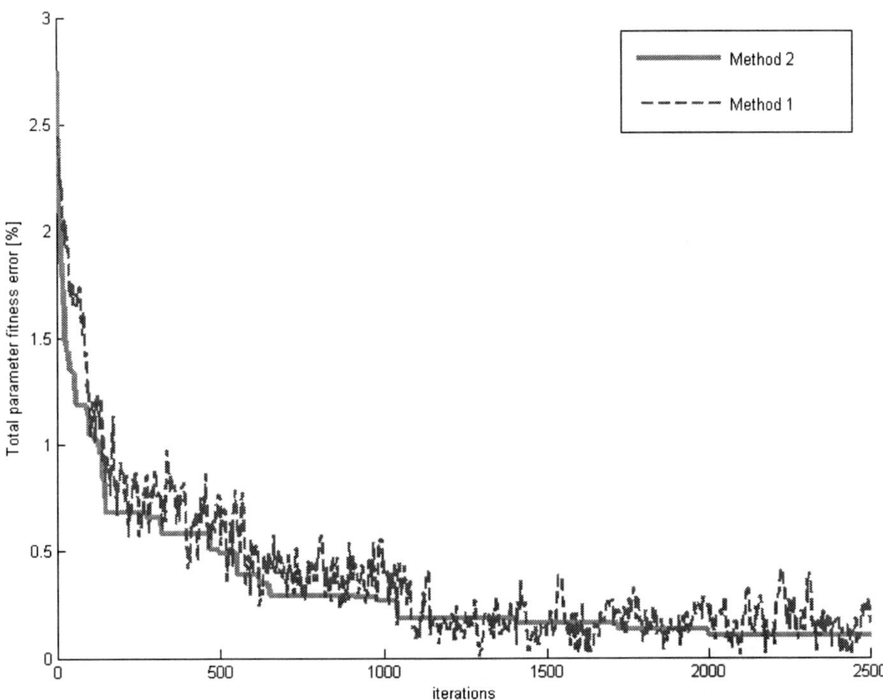

Figure 2. The total "fitness" minimum mean error function with respect to the population size for both methods

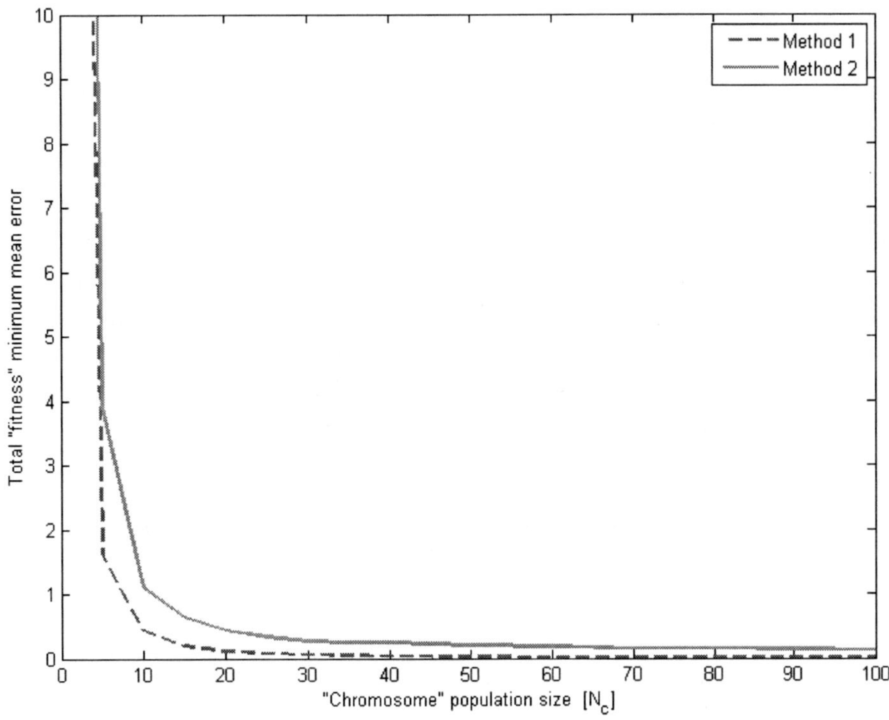

Figure 3. These four subfigures show the progress of each of the four parameters using method 1 optimization process

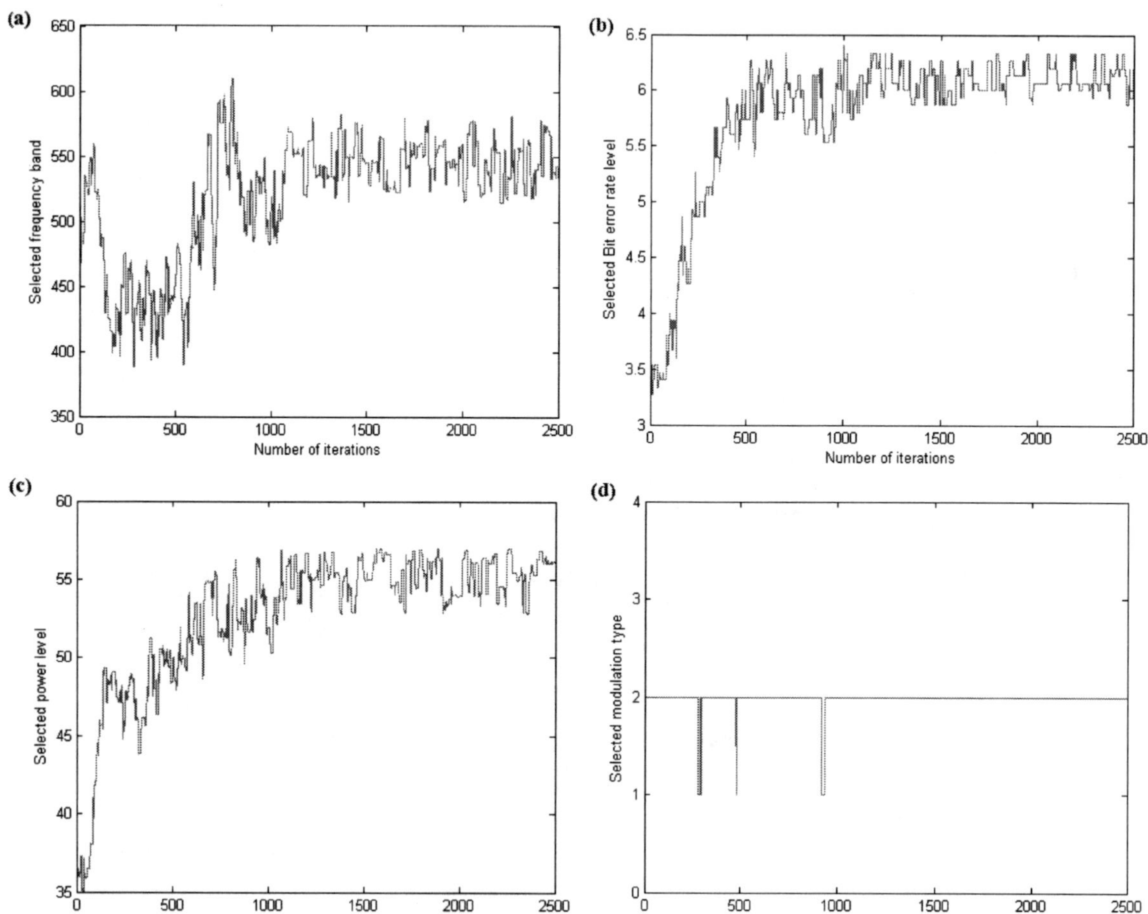

The allocation of the parameters for multiple users at the same time is the next task to incorporate into the process. Besides, there was no mechanism defined here to vacate and reallocate another band in the spectrum for the secondary user if a primary user is detected in the same band. Hence, future research will be extended for this purpose as well.

Table 3. Output results from the genetic algorithm

	Frequency	Power	BER	Modulation
Requested	547	53	6	2
Method 1	543	56	6.1	2
Method 2	544	55	6.3	2

This could involve the definition of holding times for the bands allocated to the secondary users so that the primary users requesting the band should always be proffered, even after the allocation. The parameters considered for the objective function were independent of each other and another possible future extension of the research is the consideration of interdependency among the parameters that exist in the objective function. In addition, only four parameters were considered for our objective function (i.e. frequency band, power, bit-error-rate, and modulation scheme). In the future, other parameters like data rate, spectral efficiency, and signal to noise ratio could also be added to the "fitness" function definition.

REFERENCES

FCC. (2003). *Cognitive radio workshop*. Retrieved from http://www.fcc.gov/searchtools.html

Godfrey, P., Shipley, R., & Gryz, J. (2006). Algorithms and analyses for maximal vector computation. *The VLDB Journal, 16*, 5–28. doi:10.1007/s00778-006-0029-7

Holland, J. H. (1975). *Adaptation in natural and artificial systems*. Englewood Cliffs, NJ: Prentice Hall.

Koza, J. R. (1990). *Genetic programming: A paradigm for genetically breeding populations of computer programs to solve problems*. Technical Report STAN-CS-90-1314. Palo Alto, CA: Stanford University.

Kung, H. T., Luccio, F., & Preparata, F. P. (1975). On finding the maxima of a set of vectors. *International Journal of the ACM, 22*(4), 469–476. doi:10.1145/321906.321910

Mitola, J. (2000). *Cognitive radio: An integrated agent architecture for software defined radio* (unpublished doctoral dissertation). Royal Institute of Technology (KTH). Stockholm, Sweden.

Mitola, J. (1999). Cognitive radio: Making software radios more personal. *IEEE Personal Communications, 6*(4), 13–18. doi:10.1109/98.788210

Neel, J., et al. (2002). *Game theoretic analysis of a network of cognitive radios*. Midwest Symposium on Circuits and Systems 2002. Columbus, OH.

Rondeau, T. W. (2004). Cognitive radios with genetic algorithms: Intelligent control of software defined radios. *Proceedings of SDR, 2004*, C3–C8. Phoenix, AZ: SDR.

Sawaragi, Y., Nakayama, H., & Tanino, T. (1985). *Theory of multiobjective optimization*. Orlando, FL: Academic Press Inc.

Steuer, R. E. (1986). *Multiple criteria optimization: Theory, computations, and application*. New York, NY: John Wiley & Sons, Inc.

Chapter 13
A Survey of Radio Resource Management in Cognitive Radio Networks

Chengshi Zhao
China Mobile Group Shanxi Co., China

Jing Li
China Mobile Group Shanxi Co., China

Wenping Li
China Mobile Group Shanxi Co., China

Zheng Zhou
Beijing University of Posts and Telecommunications, China

Kyungsup Kwak
Inha University, Korea

ABSTRACT

The framework of "green communications" has been proposed as a promising approach to address the issue of improving resource-efficiency and the energy-efficiency during the utilization of the radio spectrum. Cognitive Radio (CR), which performs radio resource sensing and adaptation, is an emerging technology that is up to the requests of green communications. However, CR networks impose serious challenges due to the fluctuating nature of the available radio resources corresponding to the diverse quality-of-service requirements of various applications. This chapter provides an overview of radio resource management in CR networks from several aspects, namely dynamic spectrum access, adaptive power control, time slot, and code scheduling. More specifically, the discussion focuses on the deployment of CR networks that do not require modification to existing networks. A brief overview of the radio resources in CR networks is provided. Then, three challenges to radio resource management are discussed.

DOI: 10.4018/978-1-4666-2812-0.ch013

INTRODUCTION

Some related researches, termed "multi-dimension resource allocation," are focused on the joint allocation of radio resources in CR networks. Whereas we will classify the resources and discuss the techniques used for each single resource in this chapter to exhibit the technical details. This can be termed "one-dimension resource allocation."

Diverse methods are generally adopted during resource management/allocation in CR networks to use these radio resources efficiently. These methods have been extensively studied in specified scenarios to achieve efficient and fair solutions for different network architectures (centralized/distributed) or user behaviors (cooperative/non-cooperative). This chapter presents an overview on the methods of dynamic resource management in CR networks. The three main resources are presented in this chapter: Dynamic Spectrum Access (DSA), Adaptive Power Control (APC), Time Slot and Code Scheduling (TSS).

BACKGROUND

The data rates in wired and wireless networks are driven by Moore's Law and are thus rising approximately tenfold, every five years (Zhang, 2011). Conversely, the requirements for multimedia high rate transmissions in Information and Communications Technology (ICT) drive the greatly increasing power consumption. According to a recent Ericsson research report (Ericsson Press Release, 2008), energy costs account for as much as half of a mobile operators operating expenses. The price paid for the enormous energy growth is a doubling of the power consumption in cellular networks infrastructure (base stations and core network) every 4–5 years. This was is about 60 TWh (billion kWh) in 2007. The radio access network accounts for 80% of this energy consumption (GreenComm, 2009). Most importantly, currently 3% of energy world-wide is consumed by ICT infrastructure. This causes about 2% of CO2 emissions world-wide, comparable to CO2 emissions by airplanes or one quarter of the world-wide CO2 emissions by cars (W-GREEN, 2008).

The steadily rising energy cost and the need to reduce global CO2 emission to protect our environment are economic and ecological drivers for the consideration of energy consumption in all fields of our manufacturing and daily lives. Therefore, radio networking solutions that improve energy efficiency, as well as resource efficiency (green communications), are of benefit to the global environment. It also makes commercial sense for telecommunication operators supporting sustainable and profitable businesses. Within the framework of "green communications," a number of paradigm-shifting technical approaches can be expected. These include, but are not limited to, energy-efficient network architecture and protocols, energy-efficient wireless transmission techniques (e.g., reduced transmission power and reduced radiation), cross-layer optimization methods, and opportunistic spectrum sharing without causing harmful interference pollution (i.e., green spectrum) (Zhang, 2011).

Cognitive radio (Mitola, 1999; Haykin, 2005) is one of the best candidates to achieve the targets of green communications. CR is a promising radio access method to increase resource utilization by exploiting unused or low utilization spectrum which has been already authorized to primary systems. Traditional wireless technologies allocate the wireless resources statically, whilst CR technology works in a sensing, learning, reasoning and acting manner, and hence allocates the resource dynamically. CR system is expected to dynamically allocate the power to the users who can use it most efficiently in dynamically changing environments. CR users adapt to the dynamic radio environments, then make decisions on their operational parameters, such as working spectrum, transmission power, time slot and even modulation and coding type. In CR frameworks, future wireless devices will not operate on statically assigned resources, but dynamically acquire spectrum according to the spectral environment and allocate the power/timeslot/code in a most efficient way, by which the resource-efficiency and energy-efficiency of the CR system will be eventually increased.

Secondary Users (SUs) of CR systems opportunistically lease spare spectrum from Primary Users (PUs), which have the exclusive right to use the spectrum, without disturbing the PUs' operations. In this manner, cognitive networks are envisioned to maximize the spectrum utilization and use the radio resources efficiently via heterogeneous wireless architectures. To make users utilize wireless resources efficiently, one of the main challenges in cognitive networks is realizing the dynamic resource management.

As shown in Figure 1, the key components of the resource management in CR systems includes Dynamic Spectrum Access (DSA), which is responsible for providing efficient and fair spectrum allocation solutions among primary and secondary users, Adaptive Power Control (APC), which enables SUs to obtain maximum capacities and simultaneously ensure PUs bear the minimum interference, and Time Slot Scheduling (TSS), which is to evade the collisions between a PU and a SU when they are supposed to be accessing to the same spectral band. The last, but not least, aspect is code scheduling, such as Multi-Carrier Code Division Multiple Access (MCCDMA) (Kondo, 1996; Attar, 2008) that avoid a SU's interference to others by allocating the appropriate spreading codes to the SU.

DSA solves a limited resource distribution problem. It is designed to dynamically assign available spectrum holes (Weiss, 2004) to CR users in a fair and efficient manner, where a spectrum hole is a spectral band assigned to a primary user but the band is not being utilized by that user at a particular time and a specific geographic location. The term DSA has broad connotations that encompass various approaches to spectrum reform. The diverse ideas presented by prior research suggest the extent of this term. As illustrated in Figure 1, DSA strategies can be broadly categorized under three models: sale-based sharing, underlay sharing, and opportunistic sharing.

The power allocation issue for SUs in cognitive networks includes two aspects: (1) the multiple-SU case; (2) the single-SU case. The multiple-SU case focuses on the joint power optimization of multiple SUs, where there exists at least one constraint forcing the SUs to cooperatively achieve the global optimum. Thus, this case is suitable for scenarios where there exits a network coordinator, e.g., downlink power allocation or central-controlled uplink power allocation (closedloop power control). The single-SU case considers power allocation for a single SU without considering the positive/negative effects induced by this SU to other SUs. Thus, this case is suitable for scenarios of non-central-controlled

Figure 1. The hierarchy for radio resource management in CR systems

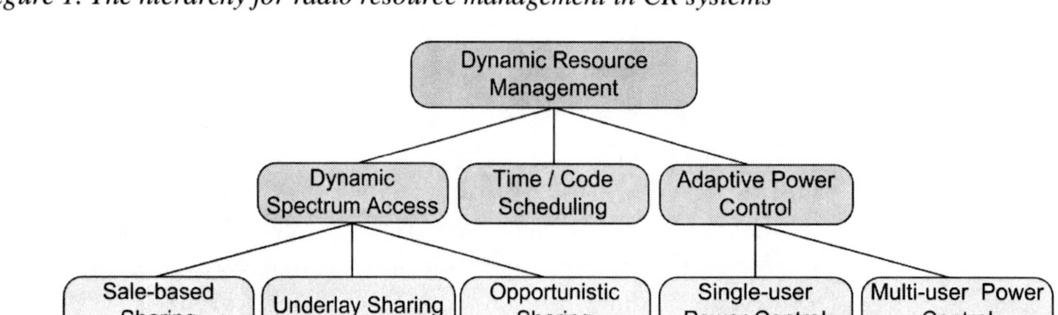

power allocation (open-loop power control) or the power allocation in a point-to-point link between a secondary transmitter and its receiver.

When SUs observe that some spectrum holes are available, time slot scheduling is an important consideration. The existing approaches for TSS can be classified into (1) random access protocols, (2) time slotted protocols, and (3) hybrid protocols (Cormio, 2009).

As for the code scheduling, different spreading codes, i.e., signature sequences, can be assigned to each user's sub-band due to the variable sub-band bandwidth in the direct sequence MCCDMA based networks as depicted in Attar (2008). The variable bandwidth alters the spreading code length and makes it impossible to use the same spreading code in all sub-bands for a given user. Thus allocating the codes to different CR users to make them communicate without interference with any other PU or SU is a major challenge in this kind of CR network.

STANDARDIZATION EFFORTS OF CR

The activities of standardization organizations aim at facilitating the development of research ideas into standards to expedite the use of research results for public use (Prasad, 2008).

Growing interest in CR was demonstrated by start in 2005 of the IEEE communications society technical committee on cognitive networks (IEEE Technical, 2009). Moreover, due to the importance of CR, IEEE initiated in 2004 a set of standardization projects related to CR called IEEE P1900, which evolved in 2006 into IEEE Standards Coordinating Committee 41 (IEEE SCC41) "dynamic spectrum access networks" (IEEE Standard, 2009). CR standardization as well as research has dramatically evolved since recent discussions on the topic (Sherman, 2008; Ma, 2005). For example, the utility of CR has been recognized by ETSI and 3GPP as an important enabler for future wireless services, and CR has been also recognized by many other non-IEEE CR standardization organizations, such as ITU-R, OMG, SDR Forum.

IEEE SCC41

IEEE SCC41 is currently structured into six active WGs. Each WG is responsible for drafting a standard for a specific topic described below:

IEEE 1900.1: Standard dictionary of terms. Define the concepts for spectrum management, policy defined radio, cognitive radio and SDR definitions and concepts for dynamic spectrum access.

IEEE 1900.2: Recommended practice for interference and coexistence between radio systems.

IEEE 1900.3: Recommended practice for conformance evaluation of SDR software modules using formal concepts and methods analysis.

IEEE 1900.4: Architectural building blocks enabling network-device distributed decision making for optimized radio resource usage in heterogeneous wireless access network.

1900.4a: Architectural building blocks enabling network-device distributed decision making for optimized radio resource usage in heterogeneous wireless access networks – Amendment.

1900.4.1: Interfaces and protocols enabling distributed dcision making for optimized radio resource usage in heterogeneous wireless networks.

IEEE 1900.5: Policy language and policy architectures for managing cognitive radio for dynamic spectrum access applications.

IEEE 1900.6: Spectrum sensing interfaces and data structures for dynamic spectrum access and other advanced radio communication systems discussed next.

IEEE 802.22

IEEE 802.22 is thought of as an alternative technology to IEEE 802.11, operating in a spectrum range that allows better and further propagation (54-863 MHz). Physical and MAC layer of IEEE 802.22 is similar to IEEE 802.16 with the amendments related to the identification of the PUs and defining the power levels so as to not to interfere with the adjacent bands. The two important entities defined in the standard are the Base Station (BS) and Customer Premises Equipment (CPE). BS controls all the CPEs decisions as to when to send data and the channels to use. CPEs sense the spectrum in its vicinity, helping in distributed detection of PU activity.

IEEE 802.16

IEEE 802.16 group has its own set of standards that support CR-like functionalities. IEEE 802.16.2-2004, which superseded IEEE 802.16.2-2001 in March 2003, describes engineering practices to mitigate interference in fixed Broadband Wireless Access (BWA) systems. This document explained methods of efficient coexistence of multiple BWA systems. IEEE 802.16a-2003, as an amendment to IEEE 802.16-2001 and completed in March 2003, described the operation of Wireless MAN interface of BWA networks in license-exempt bands. It also discussed interference analysis and coexistence issues for BWA networks in these bands. Finally, improved coexistence of IEEE 802.16-based networks working in the unlicensed bands is covered by IEEE 802.16h WG.

IEEE 802.11

Coexistence mechanisms are also included in IEEE 802.11 standards: IEEE 802.11-2007 and IEEE 802.11y-2008. IEEE 802.11-2007, published in July 2007, describes dynamic frequency selection and transmit power control for coexistence with satellite and radar systems operating in 5 GHz band, first described in IEEE 802.11h. IEEE 802.11y-2008, approved in September 2008, is an extension of IEEE 802.11 to 3650-3700MHz frequency range, which is shared with satellite earth stations.

IEEE 802.19

This standard defines general coexistence metrics for all IEEE 802 networks working in the unlicensed bands. Although focusing on IEEE 802 networks, the guidelines of the standard can be applicable to other unlicensed wireless systems. Currently, IEEE 802.19 technical advisory group is evaluating coexistence between IEEE 802.11y and IEEE 802.16h.

ITU-R

In the ITU-R Resolution 805 (Agenda for 2011 World Radio Communication Conference) in Agenda Item 1.19 ITU-R decided "to consider regulatory measures and their relevance, in order to enable the introduction of software-defined radio and cognitive radio systems, based on the results of ITU-R studies, in accordance with Resolution 956 (Regulatory measures and their relevance to enable the introduction of software defined radio and cognitive radio systems)." In a similar Resolution 951 (Enhancing the international spectrum regulatory framework), ITU-R concluded "that evolving and emerging radio communication technologies may enable sharing possibilities and may lead to more frequency-agile and interference tolerant equipment and consequently to more flexible use of spectrum." To address WRC-11 Agenda Item 1.19 Study Group 1 (Spectrum Management) has been assigned to be the lead organizational entity within ITU-R.

SDR Forum

SDR Forum is also involved in several activities related to CRs, CNs and DSA. The group "Test Guidelines and Requirements for Secondary Spectrum Access of Unused TV Spectrum" will aim at use cases and test requirements for the use of CR techniques to allow unlicensed secondary spectrum access for unused TV bands. Many reports were published by the WGs in SDR forim to identify the unique test challenges created by systems with SDR/CR features and propose solutions in such a framework, CR market, certification of CR technologies, CR architecture recommendations, design processes and tools and hardware abstraction layer for CR.

ETSI

In February 2008, European Telecommunications Standards Institute (ETSI) started a new technical committee on Reconfigurable Radio Systems (RRS), with liaison with IEEE SCC41 and SDR Forum. RRS has four groups: WG1 (System Aspects), WG2 (Radio Equipment Architecture), WG3 (Functional Architecture for Cognitive Pilot Channel), and WG4 (Public Safety).

Object Management Group (OMG)

The Object Management Group (OMG) is also involved in activities related to next generation radio systems, with the Software Radio Special Interest Group and the Software-Based Communication Domain Task Force. The mission of Software-Based Communication (SBC) Domain Task Force (DTF) is the development of specifications supporting the development, deployment, operation and maintenance of software technology targeted to software defined communication devices. The SBC DTF targets mainly issues related to the use of UML and model drives development technology for SDR, interoperability and exchangeability of software-defined components, and in general attempts to broaden the scope to new related technologies, e.g. CR, Digital Intermediate Frequency, Spectrum Management, etc.

3GPP

The Third Generation Partnership Project (3GPP) is also interested in standardizing CR-like features in its future releases. In particular, 3GPP plans to enhance Long Term Evolution (LTE) standard (radio interface of the UMTS) in Release 10 with CR functionalities. For example the idea of Cognitive Reference Signal is proposed through which each RAN can broadcast interference level, frequency bands, radio access technologies of other networks and other information which can help newly joined user equipment to choose the best RAN.

Different aspects of standardization for CR are summarized in Table 1.

ROUTING CHALLENGES AND SOLUTIONS FOR CR

Most of the research on CRNs was focused on single-hop scenarios, tackling physical layer and/or MAC layer issues, including the definition of effective spectrum sensing, spectrum decision and spectrum sharing techniques. The research community realized the potentials of multi-hop CRNs

Table 1. Summary of CR standardizations and activities

Organizations	Functionalities
IEEE SCC41, ETSI, ITU-R	Definitions
IEEE 802.19, IEEE SCC41	Coexistence
IEEE SCC41, SDR Forum, ITU-R, OMG	SDR
IEEE 802.22, 3GPP	Radio Interfaces
ETSI, IEEE SCC41	Heterogeneous Access
IEEE 802.22, IEEE SCC41	Spectrum Sensing

which can open up new and unexplored service possibilities enabling a wide range of pervasive communication applications. Indeed, the cognitive paradigm can be applied to different scenarios of multi-hop wireless networks including Cognitive Wireless Mesh Networks featuring a semi-static network infrastructure, and cognitive radio *ad hoc* networks characterized by a completely self-configuring architecture, composed of CR users which communicate with each other in a peer to peer fashion through ad hoc connections. To fully unleash the potentials of such networking paradigms, new challenges must be addressed and solved. In particular, effective routing solutions must be integrated into the work already carried out on the lower layers (PHY/MAC), while accounting for the unique properties of the cognitive environment. The main challenges for routing information throughout multi-hop CRNs include (Cesana, 2011):

1. **Spectrum-Awareness:** Designing efficient routing solutions for multi-hop CRNs requires a tight coupling between the routing module(s) and the spectrum management functionalities such that the routing module(s) can be continuously aware of the surrounding physical environment to take more accurate decisions.
2. **Setup of "Quality" Routes in Dynamic Variable Environments:** The very same concept of "route quality" is to be re-defined under CRN scenario. Indeed, the actual topology of multi-hop CRNs is highly influenced by PUs' behavior, and classical ways of measuring/assessing the quality of end-to-end routes (nominal bandwidth, throughput, delay, energy efficiency and fairness) should be coupled with novel measures on path stability, spectrum availability/PU presence.
3. **Route Maintenance/Reparation:** The sudden appearance of a PU in a given location may render a given channel unusable in a given area, thus resulting in unpredictable route failures, which may require frequent path rerouting either in terms of nodes or used channels. In this scenario, effective signaling procedures are required to restore "broken" paths with minimal effect on the perceived quality.

There are two main categories for routing solutions: (1) proposals focused on static network topologies, with fully available topological information on neighboring SUs and spectrum occupancy; (2) proposals based on local radio resource management decisions on partial information about the network state (approaches based on local spectrum knowledge). In the first case, the problem of designing/modeling CRNs scales down to the classical problem of designing static (wireless) networks, where tools of graph theory and mathematical programming can be leveraged extensively. Even if the implementation of these approaches may result complex and may be scarcely scalable, their importance can be seen in the application to all that scenarios where the SUs have access to data bases storing the spectrum maps, as envisaged by the FCC (FCC, 2008). On the other hand there exist several approaches based on local information on spectrum occupancy gathered by each SU through local and distributed sensing mechanisms. In some cases the protocols are able to set up the whole path while in other cases the proposed approaches are based on the selection hop by hop of the next forwarding node. However, a distinguishing characteristic of all routing approaches is that they combine to the routing the selection of the spectrum on each link of the path. This can be done by using different metrics for capturing the characteristics of the available spectrum holes. The most appropriate spectrum bands can be then selected according to both radio environment (interference, power) as well as QoS parameters like throughput, delay, etc. Also the behaviors of the PUs are a key parameter to be considered for routing data in a multi-hop CRNs. In fact, routes must explicitly provide a measure of protection to the ongoing communication of the PUs while at the SUs side must guarantee stability when the PU behavior varies (Cesana, 2011).

This is taken into account in a set of routing solutions where the PUs' statistical behavior and the consequent spectrum fluctuations are considered via suitable models in the routing metrics. Besides this, also the ability to reconfigure the routing paths when a PU becomes active can be a distinguish feature of the routing. The field of modeling/designing CRNs routing still needs major contributions explicitly endorsing network dynamics and variability, which are distinctive features of the multi-hop CRNs.

Dynamic Spectrum Access (DSA)

Dynamic Spectrum Access (DSA) will enable efficient spectrum usage to network users via dynamic spectrum access techniques and heterogeneous network architectures. SUs are able to access spectrum resources from PUs through opportunistic or negotiation-based methods while not causing harmful interference or channel collision. The main features that define DSA and those of interest for dynamic spectrum sharing includes (Ji & Liu, 2007):

1. The network users are equipped with cognitive devices, which enable them to perform various dynamic spectrum access techniques including spectrum sensing, spectrum management, seamless handoff, and spectrum sharing.
2. Different from traditional static or centralized spectrum assignment among different base stations or systems in cellular networks, DSA enable multiple systems to be deployed with overlapping spectrum or coverage. Therefore, flexible spectrum access is possible by allowing secondary users to gain access to multiple primary operators or having multiple secondary users compete for available spectrum.
3. The characteristics of spectrum resources may vary over frequency, time, and space due to user mobility, channel variations, or wireless traffic fluctuations. A management point may exist to handle the billing information for spectrum leasing activities. Control channels are assumed for exchanging spectrum sharing information.
4. Considering different applications of DSA, such as military, emergency, or civilian applications, different types of users coexist in wireless networks, including cooperative, selfish, and malicious users. Cooperative users unconditionally cooperate with each other to serve a common goal; selfish users aim to maximize their own interests; malicious users intend to cause as much damage as they can to the network.

Sale-Based Sharing

In this approach, PUs sell their spectrum to SUs in a limited time whenever it is not used by any PU, and SUs compete to buy these spectrum holes. Thus, there is no need to perform DSA in cognitive networks in this method, because it is performed by the primary network.

Implementation of sale-based spectrum sharing is quite simple. This method is currently used by cell phone networks and is termed roaming. In roaming, a cell phone uses a wireless network other than its own service provider's network and pays some extra fees to use the spectrum. However, for several reasons, the sale-based method for spectrum sharing is not suitable for CR and cannot always be used by CR networks (Khozeimeh, 2009):

- This method requires a primary network willing to sell its unoccupied spectrum to CR users. This makes communications in CR networks unreliable because not all of the primary networks are pleased to sell out their radio resources.

- There is no need for radio scene analysis in this method. Thus, SUs lack awareness of their surrounding environments, which makes it hard for SUs to analyze and utilize the radio resources efficiently.
- Using the sale-based method, SUs are limited to operate in a centralized manner and are controlled by a primary network.

Underlay Sharing

The spectrum underlay technique (Menon, 2005; Zhao, 2007) is proposed for spectrum sharing systems that intend to coexist with static radio systems. It is an interference averaging or spreading-based technique, where the term static radio refers to existing legacy users in the spectrum. Interference averaging refers to the transmission technique where radios spread their signals across the entire bandwidth available to the spectrum sharing system, such as Code Division Multiple Access (CDMA) (Gilhousen, 1991) and Ultra Wide Band (UWB) (Win, 1998) techniques. By spreading the transmitted signals over a wide frequency band, SUs can potentially achieve a short-range high data rate with extremely low transmission power. Based on a worst-case assumption that PUs transmit all the time, this approach does not rely on detection and exploitation of the spectrum white space.

SUs and PUs transmit simultaneously on a common frequency band in the spectrum underlay technique. A SU's transmission is regarded as noise by PUs, thus more than one source influences the interference caused to the static radio systems in this scheme. The underlay sharing approach imposes severe constraints on the transmission power of SUs, so that they operate below the PUs' tolerable noise floors.

A new metric termed interference temperature is recommended by the Federal Communications Commission's (FCC) spectrum policy task force (FCC, 2002; Xing, 2007), which is intended to quantify and manage the interference sources in radio environments. The interference temperature is defined as the RF's Power Spectrum Density (PSD) measured at a receiving antenna. Any transmission is considered to be harmful if it increases the noise floor above the interference temperature threshold. Given a particular frequency band in which the interference temperature is not exceeded, that band could be available to SUs. Let the interference temperature limit be T_{max}, which serves as an upper bound or "cap" on the potential transmission power of a RF, the upper limit on the SUs' permissible PSD becomes to be $T_{max}*k$, where k is the Boltzmann's constant. Therefore, a SU might attempt to coexist with PUs by spectrum underlay sharing, such that the presence of the SU goes unnoticed by any PU.

Opportunistic Sharing

Spectrum opportunistic sharing is a method suggested for CRs in the literatures (Brik, 2005; Hoang, 2006). In the opportunistic sharing scheme, CR users utilize spectrum holes whenever they are available in an opportunistic manner without any cooperation or communication with the primary network. The opportunity sharing scheme is responsible for accurately identifying and intelligently tracking idle spectrum holes that are dynamic in both time and space. It then decides whether and how a transmission should take place. During opportunistic sharing, the best spectrum holes and the most appropriate operating parameters are selected by SUs, thus the spectrum holes are exploited and efficiently shared by SUs without interfering with the PUs. If some spectrum holes are further used by a PU, the SUs move to another spectrum hole or stay in the same frequency band, altering their transmission power level or coding/modulation scheme to avoid interference with the PU.

As shown in Figure 2, we assume that two SUs, named s1 and s2, need to opportunistically access to the idle spectrum holes. In time slot t_1, there

Figure 2. Opportunistic sharing concept of the spectrum

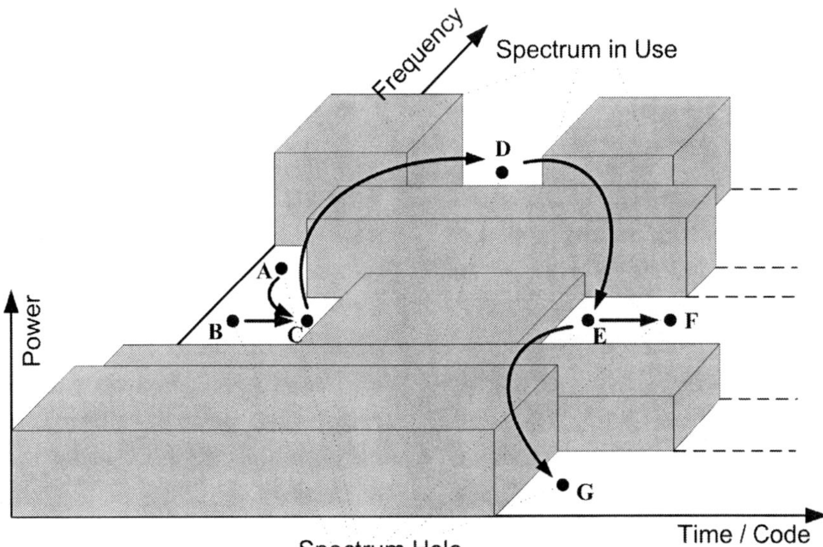

are two spectrum holes available, i.e., *A* and *B*; in time slot t_2, t_3, and t_4, the spectrum hole is *C, D,* and *E*, respectively; in time slot t_5, the spectrum holes are *F* and *G*.

As indicated in the figure, *A* and *B* are available for s_1 and s_2 in t_1. The rule to decide who access to A and who access to another is to make the secondary network achieve the maximum utility (e.g., throughput, fairness, power efficiency, etc.). Only one spectrum hole in time slot t_2, namely C, is available for two SUs. Thus, two SUs will adopt some competition mechanisms to decide which SU should access to the spectrum, while the other is denied to access. In *t3*, the available spectrum hole is changed to *D* for two SUs. Therefore, a new decision procedure should be executed in two SUs, and an appropriate SU is expected to access *D*. A similar procedure is performed in time slot t_4 when the available spectrum hole is changed to *E*. In time slot t_5, since there are two spectrum holes available, two SUs return to a new competition mechanism to make them respectively access to a proper spectrum hole, while enabling the secondary network achieve maximum utility.

During opportunistic sharing, a SU is expected to be interference-free to all PUs at any time; the secondary network should always achieve maximum utility. Therefore, except for the spectrum allocation, the efficient management schemes for other resources are important for the system, e.g., power control, code scheduling, and adaptive modulation.

Therefore, in the opportunistic sharing scheme, SUs do not pay any accessing fee to the spectrum owners, however they have to continuously perform spectrum sensing and radio scene analysis to avoid collision with PUs. The collision probability cannot be zero due to the uncertainty of the radio environment, while it can be decreased by adopting some advanced spectrum sensing schemes. Detection probability can be tremendously increased simultaneously. Meanwhile, the detection probability and the average duration of collisions can be ameliorated by advanced spectrum sensing schemes.

In this section, we provide an overview on challenges and recent developments in opportunistic sharing schemes. We explain this in several aspects.

Research Classification. Existing work on opportunistic spectrum sharing (or access control) can be classified based on their architecture (centralized or decentralized), spectrum allocation behavior (cooperative or non-cooperative) (Akyildiz, 2006). Brik (2005), Zheng (2005), Peng (2006), and Wang (2007) proposed centralized protocols for spectrum sharing. Sharing protocols that allow every user to individually decide to access the spectrum are proposed in Zhao (2007) and Zheng (2005) for decentralized CR systems. Cooperative spectrum sharing is discussed in Brik (2005), Cao (2005), and Zhao (2005, 2007) and non-cooperative spectrum allocation is discussed in Sankaranarayanan (2005), Zhao (2005), Zheng (2005), and Alyfantis (2007), respectively.

Centralized Spectrum Sharing. In centralized sharing, the information of all CR units' spectrum scenes is available in a central entity (base station) that controls the spectrum allocation and access procedures. Generally, a distributed sensing procedure can be used, such that measurements of the spectrum are forwarded to the central entity. An optimization problem is globally solved at the central entity that considers the spectrum set-up data for all CR units in the entire network. Furthermore, the result solving this optimization problem is an optimum spectrum assignment for the entire network. Once the optimum has been obtained, it is sent back to all CR units. However, the complexity and the information overhead of centralized optimization exponentially grow with the size of the network; the centralized system is not scalable. The spectral resources are wasted during transmission of the control information. Therefore, the distributed spectrum sharing is canonized in practical systems.

Distributed Spectrum Sharing. In the distributed/decentralized approach, the optimization problem is solved locally using the local data available to each CR unit. Accordingly, each unit is responsible for the spectrum allocation and the decisions on the access are made based on the local policies performed by each node distributively. In a distributed sharing scheme, based only on the local information available to it, each CR unit chooses the optimum channel from the available spectrum holes in the environment.

Cooperative Spectrum Sharing. During cooperative spectrum sharing, each CR unit estimates its interference to other units. A common technique used in this scheme is to form clusters and share the interference information locally; this is termed a centralized solution. While all the centralized solutions can be regarded as cooperative, there also exist distributed cooperative solutions. In cooperative spectrum sharing, the spectrum allocation algorithms consider theinterference to other users.

Non-Cooperative Spectrum Sharing. Contrary to cooperative sharing, only a single node is considered in non-cooperative solutions. These solutions are also referred to as selfish. Since interference to other CR units is not considered, non-cooperative solutions may result in reduced spectrum utilization. However, non-cooperative sharing does not require frequent message exchanges among neighbor units; this introduces a tradeoff for practical utilization.

Research Methodology. There are extensive researches on spectrum allocation. Many technical methods are adopted in the implementation of the spectrum allocation in CR systems, e.g., optimization theory (Tao, 2009; Urgaonkar, 2009), graph theory (Zheng, 2005; Peng, 2006; Cao, 2005; Zhao, 2008), game theory (Ji & Liu, 2007; Huang, 2006), and queueing theory (Wang, 2007; Zhao, 2007).

Optimization Theory-Based Method. As indicated in Khozeimeh (2009), the spectrum sharing problem can be formulated as an optimization problem, where the objective function is to find the optimum channel assignment subject to the collision constraints with the PUs. Defining the objective function is an important issue, because being based on different criteria such as bandwidth, network coverage or fairness, different spectral sharing problems can be defined. For example, max-sumbandwidth, max-min-bandwidth, max-proportional-fair schemes are proposed in Khozeimeh (2009), while the throughput of the secondary networks are considered as the optimization goal in Tao (2009) and Urgaonkar (2009).

Although the optimization-theory based method is straightforward and flexible during the analysis of network performance, the multi-objective optimization problem needs to be solved in this method. This becomes extremely complex. Therefore, other methods are extensively studied by researchers.

Graph Theory-Based Method. In Zheng (2005), by mapping each frequency band into a color, the spectrum allocation problem is reduced to a heuristics Graph Multi-Coloring (GMC) problem in a given small fixed network topology. The model obtains conflict free spectrum assignments that closely approximate the global optimum in centralized systems where a control channel exists. It is extended to distributed cooperative networks in Peng (2006), Cao (2005), and Zhao (2007). Cao (2008) introduced a distributed spectrum management architecture, in which nodes fairly share spectrum resource by making independent actions following several spectrum rules.

In Zhao (2008), the dynamic spectrum allocation problem is modeled as a weighted bipartite graph matching problem. It is assumed that nearby users self-organize into coordination groups and use the spectrum cooperatively with neighbors by exchanging control messages through a local common channel in each group, where a group is built up according to Cao (2005) and Zhao (2007).

In reality, spectrum allocation occurs between two users, i.e., a link connecting two users. That is the edge of the connecting graph of the network. Edge-vertex transform is introduced before the spectrum allocation to facilitate the consideration. Figure 3 depicts five users in the network, numbered from I to V; neighboring users connected by solid edges denote that they have a link for their communications. Through the edge-vertex transformation, solid lines in the original graph are mapped into vertices in the dashed-line graph, numbered from 1 to 5. Two neighboring vertices connected by a dashed line denote that these two links cannot use the same band simultaneously, because these two links connecting to the same user in the network and a user must use two different bands to communicate with other two users.

As shown in Figure 3, links are transformed into vertices in the transformed-graph. These vertices choose the band from the intersection of the neighboring users' available band lists, e.g., {band list of 1}= band list of I} ∩ {band list of II}.

In the transformed-graph, the CR units that interfere with each other are connected by dashed lines. Using GMC method (Zheng, 2005), we may color the five vertices using minimum number of colors. When the transformed-graph is colored, the corresponding spectrum band to the color of each vertex is assigned the corresponding link of that vertex.

On the other hand, if the utility achievable by matching/allocating a spectrum band to a CR user, spectrum availabilities and the constraints reflecting the interference among users are all known before performing the allocation, spectrum allocation can be considered as a weighted bipartite graph matching problem by matching the available spectrum with the vertices of the transformed-graph (Zhao, 2008).

Game Theory-Based Method. Multiple CR users competing for the spectrum resources are selfish and may compete against each other to

Figure 3. An example of edge-vertex transform

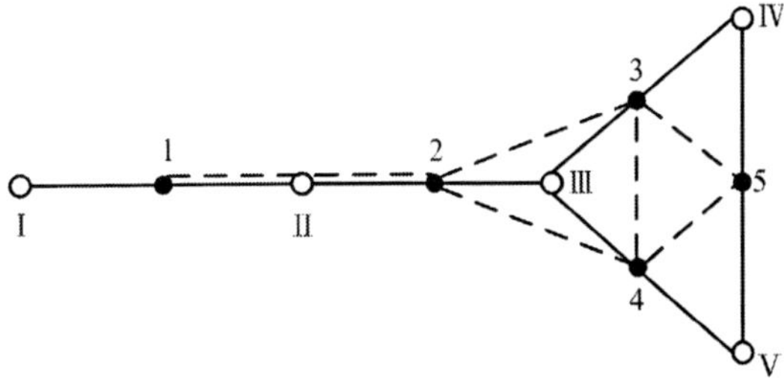

maximize their own utility, e.g., maximizing their own service quality or revenue. Thus, game theory has been advocated as an appropriate framework to study competitive spectrum access in decentralized networks. A game theoretical overview of dynamic spectrum sharing is provided in Ji and Liu (2007) from several aspects: analysis of network users' behaviors, efficient dynamic distributed design and optimality analysis. The dynamic spectrum sharing problem can be formulated as various game models, e.g., one-stage cooperative game (Han, 2005), one stage non-cooperative game (Etkin, 2005; Niyato, 2008), repeated game (Wu, 2008), potential game (Nie, 2005), hierarchical game model (Zhang, 2009; Stanojev, 2008), multi-stage dynamic game and auction game (Ji, 2008; Huang, 2006). By modeling spectrum sharing problem in CR networks (including PUs and SUs) as diverse games, the network users' behaviors and actions can be analyzed in a formalized game structure, by which multi-objective optimization problems are resolved in reduced complexities.

A game model in the strategic form has three elements using incentive structures: the set of players, the strategy space for each player, and the payoff function which measures the outcome for each player. The players in spectrum sharing games are all the network users, including both PUs and SUs. The strategy space for SUs indicates the licensed channels that they will use, the transmission power and the time slot to access, and the accessing cost they would like to pay, etc. For PUs, the strategy space includes the spectrum holes and the price for each spectrum hole. The payoff functions describe the players' communication goals representing their interests.

The basic concept of game theoretical spectrum sharing is described below:

Assume a network scenario, where SUs can opportunistically access to the PUs' unoccupied spectrum holes. With a cost of b, a SU can achieve a basic utility of G_0 without access to any spectrum hole; and by leasing one channel to a SU, the PUs endure a cost of c. The phases of the game is: (1) the PUs makes a price for each channel, as ϕ, which means that PUs can obtain the gain of ϕ by leasing one channel to a SU, while SUs must pay the cost of ϕ to access to a channel; (2) the secondary users observes (and accepts) ϕ and then decide how many channels they should lease from APs, as n; (3) payoffs of PUs and SUs are calculated, as $U_P(\phi, n)$ and $U_S(\phi, n)$, respectively. The process of the game is illustrated as follows:

Step 1: A SU's utility function can be calculated as:

$$U_S(\phi, n)=G_0+G(n)-b-n\times\phi \quad (1)$$

where $G(n)$ is the utility achieved by the SU when it accesses to n channels of PUs and it is a monotonically increasing function of n (according to usual economic laws), and $G'(\infty)=0$. Therefore, when $G'(n)=\phi$, the secondary users obtain the maximum payoff as shown in Figure 4.

On the other hand, since the PUs can anticipate that the SUs' reaction to the price of ϕ is to access $n^*(\phi)$ channels. Thus, the PU's problem amounts to:

$$\mathrm{Max}\{U_P(\phi, n^*(\phi))\} \quad (2)$$

Suppose that the utilities' indifference curves for the PU are as shown in Figure 5. Holding n fixed, the PU does better when ϕ is higher. Higher indifference curves represent higher utility levels for the PUs. in terms to Figure 5, PUs would like to choose the price ϕ and yields the outcome of $(\phi,n^*(\phi))$ which is on the highest possible indifference curve. Assume the solution to the PU's optimal price is ϕ^*, the price of the channel such that the PU's indifference curve. Assume the solution to the PU's optimal price is ϕ^*, the price of a channel such that the PU's indif-

Figure 4. Utility function of a secondary user

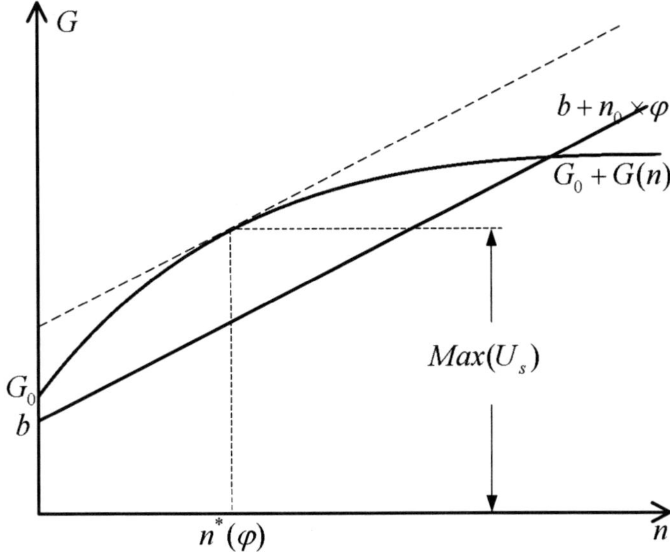

Figure 5. PU's indifference curves

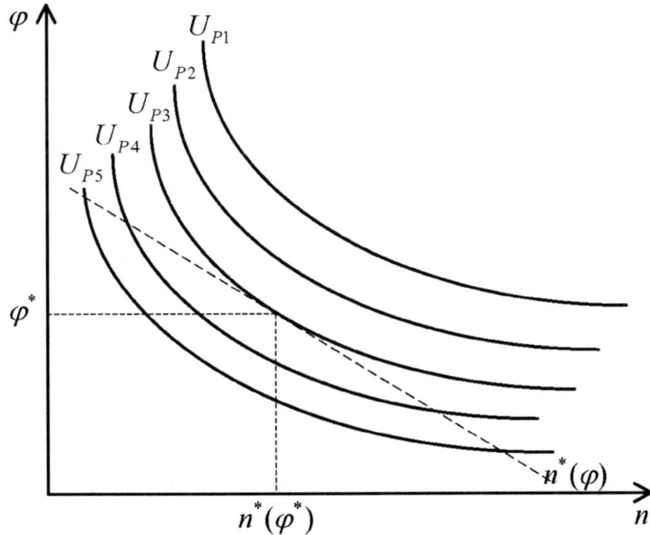

ference curve through the point $(\phi^*, n^*(\phi))$ is tangent to $(\phi^*, n^*(\phi^*))$ at that point, as shown in Figure 5.

Step 2: The PU knows that the SU's choice of n at the price of ϕ is $n^*(\phi)$, so when the PU decides the price of each channel, it will select the price of ϕ to maximum its payoff.

$$U_P(\phi, n) = (\phi^* - c) \times n^*(\phi^*) \tag{3}$$

Therefore, $(\phi^*, n^*(\phi^*))$ is the backwards-induction outcome of this game. This is the Nash Equilibrium (NE) of this game.

Specifically, NE is an important concept in game models to measure the outcome. NE is a set of strategies, one for each player, such that no selfish player has incentive to selfishly change its action.

Queueing Theory-Based Method. In Wang (2007), the interactions between PUs and SUs are modeled using continuous time Markov chains. By studying the SUs' optimal spectrum access probabilities, the spectrum resources are efficiently and fairly shared by the SUs in an opportunistic way without interrupting the PUs' spectrum usage. In Zhao (2007), authors develop an analytical framework for opportunistic spectrum access based on the theory of Partially Observable Markov Decision Process (POMDP). This decision-theoretic approach integrates the design of spectrum access protocols with spectrum sensing at the physical layer. Under this POMDP framework, the proposed protocols optimize the performance of SUs, whilst limiting the interference perceived by PUs.

Adaptive Power Control (APC)

Traditional wireless technologies statically allocate the spectrum, whereas the CR adapts to the dynamic radio environments and dynamically allocates the spectrum. The CR cognizes its environment and then makes decisions on its operation parameters. The CR usually works by sensing, learning, reasoning, and acting. In CR frameworks, future wireless devices will not operate on statically assigned spectrum but dynamically acquire spectra according to their requirements. The Secondary User (SU) opportunistically leases spare spectrum from the Primary User (PU), which has an exclusive right to use the spectrum, without disturbing the PU's operations. In this manner, cognitive networks are envisioned to provide wider bandwidth and maximize spectrum utilization via heterogeneous wireless architectures. Except for the spectrum allocation, one of the main challenges in cognitive networks is how to make the SU obtain maximum capacity and simultaneously ensure that the PU bears the minimum interference during power allocation (Haykin, 2005).

The power allocation issue for SUs in cognitive networks includes two aspects: (1) the multiple-SU case; (2) the single-SU case. The multiple-SU case focuses on the joint power optimization of multiple SUs, where there exists at least one constraint forcing the SUs to cooperatively achieve the global optimum. Thus, this case is suitable for scenarios where there exits a network coordinator, e.g., downlink power allocation or central-controlled uplink power allocation (closedloop power control). The single-SU case considers power allocation for a single SU without considering the positive/negative effects induced by this SU to other SUs. Thus, this case is suitable for scenarios of non-central-controlled power allocation (open-loop power control) or the power allocation in a point-to-point link between a secondary transmitter and its receiver.

Orthogonal Frequency-Division Multiplexing (OFDM) is widely considered one of the best candidates for cognitive networks due to its great flexibility in dynamically allocating unused spectrum for the SUs and the ease of analysis of the PUs' spectral activities (Bansal, 2008). Power allocation in OFDM-based cognitive networks is normally formulated as a constrained optimization proble to distribute the available power over subcarriers in an optimal way that maximizes the SU's capacity under various constraints (Zhao, 2010; Wang, 2007). The target, generally, is to maximize the achievable data rate under a given power budget or minimize the required power under a given target data rate. Specifically, the optimization problems have virous constraints, such as power budget, peak transmission power, average transmission power, SU's interference, PU's capacity loss, and even the access fairness among PUs and SUs.

In previous works (Zhao, 2010; Bansal, 2008, 2007), the power allocation problem is considered in OFDM-based cognitive networks, where the interference induced by the SU to the PU is considered as a constraint. Several suboptimal algorithms are given to decrease the computational complexity in Bansal (2008). However, the SU's instantaneous interference introduced to the PUs is directly set as the constraint of the optimization problem. This makes the acquisition of the optimum resolution of the problem difficult. The authors in Attar (2008) regard the interference threshold level for a set of users (including PUs and SUs) as the fairness in cognitive networks. They develop a constrained optimization model to maximize system throughput (the sum rate of PUs and SUs), where the received interference, considered as the constraint for each, is further carefully resolved.

A temperature model, which is mostly specified to CR systems, is introduced into the loading problem in Cheng (2008). The authors develop a novel effective iterative water filling algorithm for power allocation in cognitive networks, where the power budget and the peak transmission power of the SU are considered as constraints in Wang (2007). The paper introduces a distance model to convert the constraints on the interference that were induced by a SU to the PUs, into those on the SU's transmission power. A novel model is presented in fading channels to maximize the SU's ergodic capacity, where the PU's average capacity loss and the SU's average transmission power are treated as constraints in Wang (2007). The maximized ergodic capacity constrained by the transmission power and interference power is investigated over fading channels, where the interference from the PU to the SU is not taken into consideration in Kang (2008).

In Zhao (2007), the authors point out that a SU and a PU may access to the same frequency band if the distance between them satisfies certain conditions, where only the Co-Channel Interference (CCI) is taken into consideration. In another case, SUs and PUs may exist in side-by-side bands, where the Adjacent Channel Interference (ACI) is the limiting factor for the performance of the network.

A model is suggested to protect the PU from interference in the proposed IEEE 802.22 implementation (Caldwell, 2010). As shown in Figure 6, the sensing region of SU s is defined as D_s. There is a protection contour for each PU. Without loss of the generality, all PUs in the network have the same requirement on the protection contour whose radius is R. The interference to a specified PU, say p, induced by the SUs can be ignored if the interference power at the margin of the PU's protection contour is lower than a specified value η. This can be formulated as:

$$I_{s,n}^{p,n} = E_{s,n} \cdot G_{s,n}^{p,n} \leq \eta \qquad (4)$$

where $I_{s,n}^{p,n}$ is the CCI induced by channel n of SU s to the same channel of its neighboring PU p; $E_{s,n}$ is the transmission power in channel n of SU

Figure 6. Considered CR network model

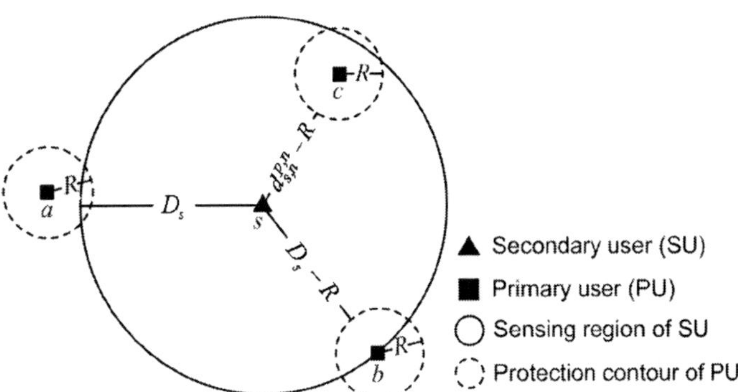

s; $G_{s,n}^{p,n}$ is the channel gain between SU s and the point closest to SU s on the protection contour of PU p using channel n.

Based on this model, authors in Zhao (2010) proposed a power allocation scheme considering the mutual interference among PUs and SUs. If a PU using channel n is sensed inside the sensing region D_s of SU s (e.g., PU c in Figure 6), to ensure that there is no CCI to this PU, radiation power of SU s on channel n should be lower than η at the point $d_{s,n}^{p,n}$ -R away from SU s, where $d_{s,n}^{p,n}$ is the spatial distance between SU s and PU p that are both using channel n. In another case, if no PU is detected using channel n in the sensing range D_s (e.g., PU a and b in Figure 6), the radiation power on channel n of SU s should be lower than η at the point D_s-R away from SU s. Ignoring the small scale fading, SU s follows the transmission power limitation induced by the CCI restriction:

$$E_{s,n}^{\max,CCI} = \begin{cases} \eta \cdot PL(D_s - R) & \min\left\{d_{s,n}^{p,n}\right\} > D_s \\ \eta \cdot PL(\min\left\{d_{s,n}^{p,n}\right\} - R) & \min\left\{d_{s,n}^{p,n}\right\} \le D_s \end{cases} \quad (5)$$

where $E_{s,n}^{\max,CCI}$ is the maximal allowable transmission power on channel n of SU s restrained by the CCI; $PL(x- y)$ represents the difference between two large scale fadings that x meters away and y meters away from a specified position; $\min\left\{d_{s,n}^{p,n}\right\}$ is the spatial distance between SU s and its nearest PU using channel n. Neighboring to SU s, the nearest one is chosen from all of the PUs that are using channel n to restrict the SU's transmission power. This makes the constraint extremely strict.

In particular, if the channel occupation states of the PUs in Figure 6 are as illustrated in Figure 7, the power allocation of SU s should correspondingly follow the rules depicted in Figure 7.

From the above discussion, we can accept that the CCI between the SUs and the PUs is insignificant if Equation 5 is satisfied. However, side-lobe-leakage interference still exists between them. Similarly, the interference introduced to a PU's channel by ACI can be formulated as:

$$I_{s,n}^{p^*,n\pm j^*} = E_{s,n} \cdot G_{s,n}^{p^*,n\pm j^*} \cdot L(d_0) \le \eta \quad (6)$$

where $\{p^*,j^*\} = \arg\max_{\forall p,j} \left\{I_{s,n}^{p,n\pm j}\right\}$ and $I_{s,n}^{p,n\pm j}$ indicates the ACI from the channel n of SU s to channel n+j or n-j of PU p; $G_{s,n}^{p,n\pm j}$ indicates the channel gain between SU s using channel n and PU p using channel n+j or n-j; $L(d_0)$ indicates the power decline caused by the spectral distance between the SU's channel and the PU's channel.

At different spectral distance, i.e., different j, there may be a different PU that endures the ACI from this SU. We choose the PU suffering the

Figure 7. The influence of CCI on SU's power allocation

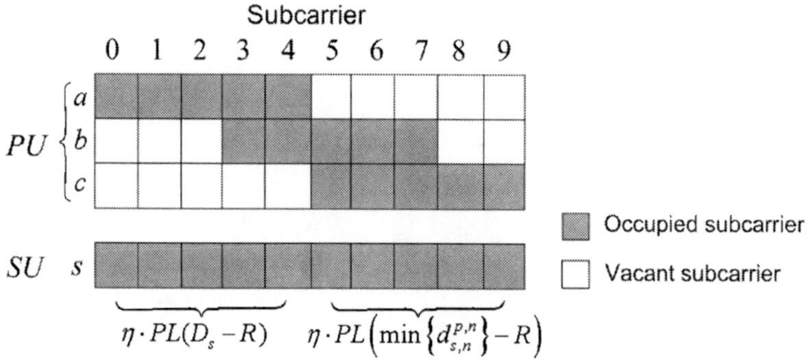

largest ACI induced by the SU's channel n to restrict the transmission power of this channel. Therefore, the largest ACI, namely $I_{s,n}^{p^*,n\pm j^*}$ is considered in the formulation.

Consequently, similar to Equation 5, we obtain the restriction on the SU's transmission power induced by the ACI between SU and PU as:

$$E_{s,n}^{\max,ACI} = \begin{cases} \eta \cdot PL(D_s + d_0 - R) & d_n^{p^*,n\pm j^*} > D_s \\ \eta \cdot PL(d_n^{p^*,n\pm j^*} + d_0 - R) & d_n^{p^*,n\pm j^*} \le D_s \end{cases} \quad (7)$$

By Equation 5 and Equation 7, to achieve the maximum capacity of the system, the problem can be mathematically expressed as:

$$\max\left\{\sum_{n=1}^{N} \log_2(1 + E_{s,n} \cdot SNR_{s,n})\right\}$$
$$s.t. \begin{cases} \sum_{n=1}^{N} E_{s,n} \le E_{s,budget} \\ 0 \le E_{s,n} \le E_{s,n}^{\max}, \forall n \in \{1,...,N\} \end{cases} \quad (8)$$

where $E_{s,budget}$ is the power budget of SU s, $E_{s,n}^{\max}$ $=\min\left\{E_{s,n}^{\max,CCI}, E_{s,n}^{\max,ACI}\right\}$ is the transmission power threshold of the SU's channel n, and $SNR_{s,n}$ is the signal-to-noise ratio on channel n of SU s.

This is a standard constrained optimization problem that can be modeled as a "cap-limited" water-filling scheme (Papandreou, 2008). Various power allocation schemes can be adopted to resolve this problem, whereas the allocation result is optimal without mutual interference between the Pus and SUs.

Specifically, the optimization objective here is set to maximize the total transmission rate of the SU, whilst keeping the instantaneous interference introduced to the PUs below a certain threshold, where the restriction on the SU's interference introduced to the PUs is straightforwardly converted to that on the SU's transmission power. In some other cases, the optimization target can also be set to minimize the required power under a given target data rate.

Time Slot and Code Scheduling (TSS)

Due to the central role of the scheduler in determining the overall system performance, there have been many published studies on LTE scheduling (Kwan, 2010). Simply stated, the scheduling problem in LTE is to determine the allocation of scheduling blocks to a subset of UEs in order to maximize some objective function, for example, overall system throughput or other fairness-sensitive metrics. The identities of the assigned scheduling blocks and the modulation and coding schemes are then conveyed to the UEs via a downlink control channel.

The scheduling in CR is similar to the it in LTE, and the main difference between them is that: The packet schedulers in LTE can be designed using possibly different metrics depending on the characteristics desired of the scheduler, for example, high throughput, fairness, low packet drop rate, and so forth; however, the spectrum sensing, the interference-free access, the ad-hoc-like acces as well as the utility-maxmization must be forth considered in CR networks.

Normal Slot Scheduling

When SUs observe that some spectrum holes are available, time slot scheduling is an important consideration. The existing approaches for TSS can be classified into (1) random access protocols, (2) time slotted protocols, and (3) hybrid protocols (Cormio, 2009).

1. **Random Access Protocols (Zhao, 2005; Ma, 2005; Ma, 2007):** The protocols in this class do not need time synchronization. Although collision occurs, because two or more CR users access to a channel simultaneously, it can be reduced based on

the Collision Sense Multiple Access with Collision Avoidance (CSMA/CA) principle. Each CR user monitors spectrum holes to detect when there is no transmission from the other CR users and transmits after a back-off duration to prevent collisions.

2. **Time Slotted Protocols (Cordeiro, 2006, 2007):** Network-wide synchronization among users is required for these protocols, where time is divided into slots for both the control channel and the data transmission. It is widely used due to its simple implementation, although resource is wasted.
3. **Hybrid Protocols (Chen, 2007; Hamdaoui, 2008):** The superframe of hybrid protocols uses a random access period to control message exchange and a slotted period for data transmission. However, the following data transmission may have random channel access schemes without time synchronization. In a different approach, the durations for control and data transfer may have predefined durations constituting a superframe that is common to all users in the network. Within each control or data duration, the access to the channel may be completely random (Cormio, 2009).

Sensing Slot Scheduling

Spectrum sensing may cause negative effects on the performance of the CR network, since all CR communications has to be postponed during channel sensing. This makes the scheduling of spectrum sensing periods an important issue. There are mainly three kinds of schemes used for sensing slot scheduling.

- **Fixed Scheduling:** The first slot of each frame is always scheduled for spectrum sensing.
- **Greedy Scheduling:** Schedule spectrum sensing only when the users' channel states are below some threshold state, or when it is already at the last slot of a frame.
- **Optimal Scheduling (Hoang, 2007, 2010):** When the instantaneous channel state information is available, sensing periods are adaptively scheduled so that the throughput of the CR network is maximized. This is based on the observation that transmitting data when the channels are good and carrying out spectrum sensing when the channels are in poorer conditions would increase the spectrum efficiency of CR operation. When users experience both stochastic data arrival and time-varying channels, data packets of CR nodes have deadline constraints and the objective is to

Figure 8. An example of sensing slot scheduling in a CR network

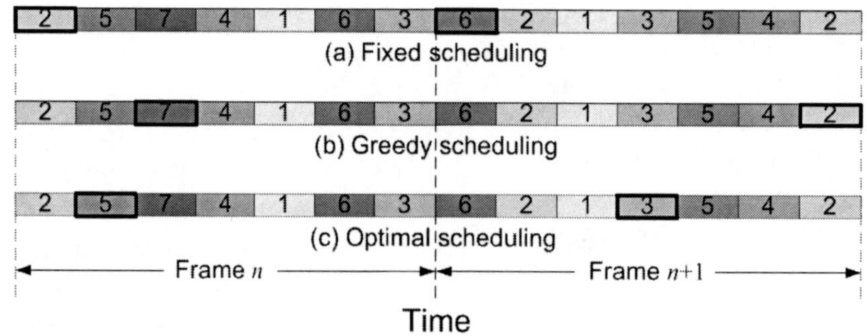

jointly scheduling data transmission and spectrum sensing activities of the CR network so that the packet loss due to deadline violation is minimized. Here, joint data-transmission/spectrumsensing decisions are based on both channel and buffer conditions of the CR nodes. In this case, carrying out spectrum sensing on a particular channel is equivalent to transmitting a virtual sensing packet on that channel.

Figure 8 describes an example of sensing slot scheduling in a CR network. Each frame is divided into 7 slots. During each frame, one sensing period must be scheduled for a channel. The channel states are numbered from 1 to 7, where the larger number indicates the worse channel state (it is also colored, and the darker color indicates the worse channel state). As for the greedy scheduling, the threshold channel state is set to be 6, and the sensing slot is scheduled when the channel state is worse than it. The outlined slots indicate the scheduled sensing slot. It is obvious that diverse schemes obtain different sensing slot in a frame, and different network performances will be correspondingly achieved. Generally, the optimal scheduling always outperforms the other two schemes while the greedy scheduling performs very closed to the optimal scheduling and the fixed scheduling always performs worst (Hoang, 2007, 2010).

Code Scheduling

As for the code scheduling, different spreading codes, i.e., signature sequences, can be assigned to each user's sub-band due to the variable sub-band bandwidth in the direct sequence MCCDMA based networks as depicted in Attar (2008). The variable bandwidth alters the spreading code length and makes it impossible to use the same spreading code in all sub-bands for a given user. Thus allocating the codes to different CR users to make them communicate without interference with any other PU or SU is a major challenge in this kind of CR network.

CONCLUSION AND FUTURE RESEARCH DIRECTIONS

By exploiting the existing radio resources opportunistically, CR networks are being developed to solve current wireless network problems, such as inefficient spectrum-usage and low power efficiency. CR networks, equipped with intrinsic intelligent capabilities, are expected to use flexible radio resources more efficiently and fairly. We provided an overview of major technical issues of radio resource management in CR networks in this contribution. We posit that the appearance of CR in wireless communications will foster a more flexible management scenario and a more permissive allocation policy for radio resources.

Many organizations and researchers are currently engaged in developing the technologies and protocols required for CR networks. However, radio resource management in CR networks is still in its infancy. More research in technical, economic, and regulatory aspects must be conducted to extend the points introduced in this survey to assess and realize the potentials of CR networks.

ACKNOWLEDGMENT

This work was supported by the National Research Foundation of Korea (NRF) grant funded by the Korea government (MEST) (No.2010-0018116).

REFERENCES

Akyildiz, I., Lee, W., Vuran, M., & Mohanty, S. (2006). Next generation/dynamic spectrum access/cognitive radio wireless networks: A survey. *Computer Networking*, *50*(13), 2127–2159. doi:10.1016/j.comnet.2006.05.001

Alyfantis, G., Marias, G., Hadjiefthymiades, S., & Merakos, L. (2007). Non-cooperative dynamic spectrum access for CDMA networks. In *Proceedings of IEEE Global Telecommunications Conference (GLOBECOM),* (pp. 3574–3578). IEEE Press.

Attar, A., Holland, O., Nakhai, M., & Aghvami, A. (2008). Interference-limited resource allocation for cognitive radio in orthogonal frequency division multiplexing networks. *IET Communication*, *2*(6), 806–814. doi:10.1049/iet-com:20070355

Attar, A., Nakhai, M., & Aghvami, A. (2008). Cognitive radio transmission based on direct sequence MC-CDMA. *IEEE Transactions on Wireless Communications*, *7*(4), 1157–1162. doi:10.1109/TWC.2007.060907

Bansal, G., Hossain, M., & Bhargava, V. (2007). Adaptive power loading for OFDM-based cognitive radio systems. *Proceedings of the IEEE, ICC*, 5137–5142. IEEE Press.

Bansal, G., Hossain, M., & Bhargava, V. (2008). Optimal and suboptimal power allocation schemes for OFDM-based cognitive radio systems. *IEEE Transactions on Wireless Communication, 7*(11-2), 4710–4718.

Brik, V., Rozner, E., Banarjee, S., & Bahl, P. (2005). DSAP: A protocol for coordinated spectrum access. In *Proceedings of IEEE DySPAN,* (pp. 611–614). IEEE Press.

Caldwell, W., & Chouinard, G. (2010). *IEEE 802.22 wireless RANs draft recommended practice*. Retrieved December 13, 2011, from https://mentor.ieee.org/802.22/dcn/06/22-06-0242-27-0002-draft-recommended-practice.doc

Cao, L., & Zheng, H. (2005). Distributed spectrum allocation via local bargaining. In *Proceedings of the IEEE Conference on Sensor and Ad Hoc Communications and Networking (SECON),* (pp. 475–486). IEEE Press.

Cao, L., & Zheng, H. (2008). Distributed rule-regulated spectrum sharing. *IEEE Journal on Selected Areas in Communications*, *26*(1), 130–145. doi:10.1109/JSAC.2008.080112

Cesana, M., Cuomo, F., & Ekici, E. (2011). Routing in cognitive radio networks: Challenges and solutions. *Journal Ad Hoc Networks*, *9*(3), 228–248. doi:10.1016/j.adhoc.2010.06.009

Chen, T., Zhang, H., Maggio, G., & Chlamtac, M. (2007) CogMesh: A cluster-based cognitive radio network. In *Proceedings of IEEE DySPAN*, (pp. 168–178). IEEE Press.

Cheng, P., Zhang, Z., Chen, H., & Qiu, P. (2008). Optimal distributed joint frequency, rate and power allocation in cognitive OFDMA systems. *IET Communications*, *2*(6), 815–826. doi:10.1049/iet-com:20070358

Cordeiro, C., & Challapali, K. (2007). C-MAC: A cognitive MAC protocol for multichannel wireless networks. In *Proceedings of IEEE DySPAN,* (pp. 147–157). IEEE Press.

Cordeiro, C., Challapali, K., Birru, D., & Shankar, S. (2006). IEEE 802.22: An introduction into the first wireless standard based on cognitive radio. *The Journal of Communication*, *1*(1), 38–47.

Cormio, C., & Chowdhury, K. (2009). A survey on MAC protocols for cognitive radio networks. *Ad Hoc Networks*, *7*(7), 1315–1329. doi:10.1016/j.adhoc.2009.01.002

CWC. (2008). *The first international workshop on green wireless 2008 (W-GREEN)*. Retrieved from http://www.cwc.oulu.fi/workshops/W-Green2008.pdf

Ericsson Press Release. (2008). *Website*. Retrieved from http://www.ericsson.com/ericsson/press/factsfigures/doc/energy efficiency.pdf

Etkin, R., Parekh, A., & Tse, D. (2005). Spectrum sharing for unlicensed bands. In *Proceedings of IEEE DySPAN,* (pp. 251–258). IEEE Press.

FCC. (2002). *Spectrum policy task force: Report of the interference protection working group*. Washington, DC: FCC.

FCC. (2008). *Unlicensed operation in the TV broadcast bands*. Washington, DC: FCC.

Gilhousen, K. (1991). On the capacity of a cellular CDMA system. *IEEE Transactions on Vehicular Technology*, *40*(2), 303–312. doi:10.1109/25.289411

Green Communications. (2009). *The first international workshop on green communications (GreenComm)*. Retrieved from http://www.green-communications.net/icc09/home.html

Hamdaoui, B., & Shin, K. (2008). OS-MAC: An efficient MAC protocol for spectrum-agile wireless networks. *IEEE Transactions on Mobile Computing*, *7*(8), 915–930. doi:10.1109/TMC.2007.70758

Han, Z., Ji, Z., & Liu, K. (2005). Fair multiuser channel allocation for OFDMA networks using Nash bargaining solutions and coalitions. *IEEE Transactions on Communications*, *53*(8), 1366–1376. doi:10.1109/TCOMM.2005.852826

Haykin, S. (2005). Cognitive radio: Brain-empowered wireless communications. *IEEE Journal on Selected Areas in Communications*, *23*(2), 201–220. doi:10.1109/JSAC.2004.839380

Hoang, A., & Liang, Y. (2007). Adaptive scheduling of spectrum sensing periods in cognitive radio networks. In *Proceedings of IEEE GLOBECOM,* (pp. 3128–3132). IEEE Press.

Hoang, A., Liang, Y., & Zeng, Y. (2010). Adaptive joint scheduling of spectrum sensing and data transmission in cognitive radio networks. *IEEE Transactions on Communications*, *58*(1), 235–246. doi:10.1109/TCOMM.2010.01.070270

Huang, J., Berry, R., & Honig, M. (2006). Auction-based spectrum sharing. *Mobile Networking and Applications*, *11*(3), 405–418. doi:10.1007/s11036-006-5192-y

IEEE Standard Coordinating Committee 41. (2009). *Home page*. Retrieved from http://www.scc41.org

IEEE Technical Committee on Cognitive Networks. (2009). *Home page*. Retrieved from http://www.eecs.ucf.edu/tccn

Ji, Z., & Liu, K. (2007). Cognitive radios for dynamic spectrum access - Dynamic spectrum sharing: A game theoretical overview. *IEEE Communications Magazine*, *45*(5), 88–94. doi:10.1109/MCOM.2007.358854

Ji, Z., & Liu, K. (2008). Multi-stage pricing game for collusion-resistant dynamic spectrum allocation. *IEEE Journal on Selected Areas in Communications*, *26*(1), 182–191. doi:10.1109/JSAC.2008.080116

Kang, X., Liang, Y., & Nallanathan, A. (2008). Optimal power allocation for fading channels in cognitive radio networks under transmit and interference power constraints. *Proceedings of the IEEE, ICC*, 3568–3572. IEEE Press.

Khozeimeh, F., & Haykin, S. (2009). Dynamic spectrum management for cognitive radio: An overview. *Wireless Communications and Mobile Computing*, *9*(11), 1147–1159. doi:10.1002/wcm.732

Kondo, S., & Milstein, B. (1996). Performance of multicarrier DS CDMA systems. *IEEE Transactions on Communications*, *44*(2), 238–246. doi:10.1109/26.486616

Kwan, R., & Leung, C. (2010). A survey of scheduling and interference mitigation in LTE. *Journal of Electrical and Computer Engineering*, *2010*, 1–10. doi:10.1155/2010/273486

Ma, L., Han, X., & Shen, C. (2005a). Dynamic open spectrum sharing MAC Protocol for wireless ad hoc network. In *Proceedings of the IEEE DySPAN,* (pp. 203–213). IEEE Press.

Ma, L., Han, X., & Shen, C. (2005b). Dynamic open spectrum sharing MAC protocol for wireless ad hoc networks. In *Proceedings of IEEE DySPAN,* (pp. 203–213). IEEE Press.

Ma, L., Shen, C., & Ryu, B. (2007). Single-radio adaptive channel algorithm for spectrum agile wireless ad hoc networks. In *Proceedings of IEEE DySPAN,* (pp. 547–558). IEEE Press.

Menon, R., Buehrer, R., & Reed, J. (2005). Outage probability based comparison of underlay and overlay spectrum sharing techniques. In *Proceedings of the IEEE International Symposium on New Frontiers in Dynamic Spectrum Access Networks (DySPAN),* (pp. 101–109). IEEE Press.

Mitola, J., & Maguire, G. (1999). Cognitive radios: Making software radios more personal. *IEEE Personal Communication, 6*(4), 13–18.

Nie, N., & Comaniciu, C. (2005). Adaptive channel allocation spectrum etiquette for cognitive radio networks. In *Proceedings of IEEE DySPAN*, (pp. 269–278). IEEE Press.

Niyato, D., & Hossain, E. (2008). Competitive spectrum sharing in cognitive radio networks: A dynamic game approach. *IEEE Transactions on Wireless Communications*, *7*(7), 2651–2660. doi:10.1109/TWC.2008.070073

Papandreou, N., & Antonakopoulos, T. (2008). Bit and power allocation in constrained multicarrier systems: The single-user case. *EURASIP Journal on Advances in Signal Processing*, *1*, 1–14.

Peng, C., Zheng, H., & Zhao, B. (2006). Utilization and fairness in spectrum assignment for opportunistic spectrum access. *ACM Mobile Networking and Applications*, *11*(4), 555–576. doi:10.1007/s11036-006-7322-y

Prasad, R. V., Pawełczak, P., Hoffmeyer, J., & Berger, H. S. (2008). Cognitive functionality in next generation wireless networks: Standardization efforts. *IEEE Communications Magazine*, *46*(4), 72–78. doi:10.1109/MCOM.2008.4481343

Sankaranarayanan, S., Papadimitratos, P., Mishra, A., & Hershey, S. (2005). A bandwidth sharing approach to improve licensed spectrum utilization. In *Proceedings of IEEE DySPAN,* (pp. 279–288). IEEE Press.

Sherman, M., Mody, A. N., Martinez, R., Rodriguez, C., & Reddy, R. (2008). IEEE standards supporting cognitive radio and networks, dynamic spectrum access, and coexistence. *IEEE Communications Magazine*, *46*(9), 72–79. doi:10.1109/MCOM.2008.4557045

Stanojev, I., Simeone, O., Bar-Ness, Y., & Yu, T. (2008). Spectrum leasing via distributed cooperation in cognitive radio. *Proceedings of the IEEE, ICC*, 3427–3431. IEEE Press.

Tao, S., & Krunz, M. (2009). Coordinated channel access in cognitive radio networks: A multi-level spectrum opportunity perspective. In *Proceedings of the IEEE Conference on Computer Communications (INFOCOM),* (pp. 2976–2980). IEEE Press.

Urgaonkar, R., & Neely, M. (2009). Opportunistic scheduling with reliability guarantees in cognitive radio networks. *IEEE Transactions on Mobile Computing*, *8*(6), 766–777. doi:10.1109/TMC.2009.38

Wang, B., Ji, Z., & Liu, K. (2007). Primary-prioritized Markov approach for dynamic spectrum access. In *Proceedings of IEEE DySPAN,* (pp. 507–515). IEEE Press.

Wang, P., et al. (2007). Power allocation in OFDM-based cognitive radio systems. In *Proceedings of IEEE GLOBECOM,* (pp. 4061–4065). IEEE Press.

Weiss, T., & Jondral, F. (2004). Spectrum pooling: An innovative strategy for the enhancement of spectrum efficiency. *IEEE Communications Magazine, 42*(3), S8–S14. doi:10.1109/MCOM.2004.1273768

Win, M., & Scholtz, R. A. (1998). Impulse radio: How it works. *IEEE Communications Letters, 2*(2), 36–38. doi:10.1109/4234.660796

Wu, Y., Wang, B., & Liu, K. (2008). Repeated spectrum sharing game with self-enforcing truth-telling mechanism. *Proceedings of the IEEE, ICC,* 3583–3587. IEEE Press.

Xing, Y. (2007). Dynamic spectrum access with QoS and interference temperature constraints. *IEEE Transactions on Mobile Computing, 6*(4), 423–433. doi:10.1109/TMC.2007.50

Zhang, H. (2011). *Green Communications, Green Networking, and Green Spectrum.* Retrieved May 13, 2012, from http://mypage.zju.edu.cn/honggangzhang/572794.html

Zhang, J., & Zhang, Q. (2009). Stackelberg game for utility-based cooperative cognitive radio networks. In *Proceedings of the ACM International Symposium on Mobile Ad Hoc Networking and Computing (MobiHoc),* (pp. 23–32). ACM Press.

Zhang, R. (2008). Optimal power control over fading cognitive radio channels by exploiting primary user CSI. In *Proceedings of IEEE GLOBECOM,* (pp. 931–935). IEEE Press.

Zhao, C., & Kwak, K. (2010). Power/bit-loading in OFDM-based cognitive networks with comprehensive interference considerations: The single SU case. *IEEE Transactions on Vehicular Technology, 59*(4).

Zhao, C., Zou, M., Shen, B., Kim, B., & Kwak, K. (2008). Cooperative spectrum allocation in centralized cognitive networks using bipartite matching. In *Proceedings of IEEE GLOBECOM,* (pp. 1-5). IEEE Press.

Zhao, J., Zheng, H., & Yang, G. (2005). Distributed coordination in dynamic spectrum allocation networks. In *Proceedings of IEEE DySPAN*, (pp. 259–268). IEEE Press.

Zhao, J., Zheng, H., & Yang, G. (2007). Spectrum sharing through distributed coordination in dynamic spectrum access networks. *Wireless Communications and Mobile Computing, 7*(9), 1061–1075. doi:10.1002/wcm.481

Zhao, Q., & Sadler, B. (2007). A survey of dynamic spectrum access: Signal processing, networking, and regulatory policy. *IEEE Signal Processing Magazine, 24*(3), 79–89.

Zhao, Q., Tong, L., & Swami, A. (2005). Decentralized cognitive MAC for dynamic spectrum access. In *Proceedings of IEEE DySPAN,* (pp. 224–232). IEEE Press.

Zhao, Q., Tong, L., Swami, A., & Chen, Y. (2007). Decentralized cognitive MAC for opportunistic spectrum access in ad hoc networks: A POMDP framework. *IEEE Journal on Selected Areas in Communications, 25*(3), 589–600. doi:10.1109/JSAC.2007.070409

Zheng, H., & Cao, L. (2005). Device-centric spectrum management. In *Proceedings of IEEE DySPAN,* (pp. 56–65). IEEE Press.

Zheng, H., & Peng, C. (2005). Collaboration and fairness in opportunistic spectrum access. In *Proceedings of the IEEE International Conference on Communications, (ICC),* (pp. 3132–3136). IEEE Press.

ADDITIONAL READING

Bruce, A., & Bruce, F. (2006). *Cognitive radio technology*. New York, NY: Newnes.

Bruce, A. F. (2009). *Cognitive radio technology* (2nd ed.). New York, NY: Academic Press.

Bruce, F. (2009). *Cognitive radio technology* (2nd ed.). New York, NY: Academic Press.

Chen, K., & Ramjee, P. (2009). *Cognitive radio networks*. New York, NY: Wiley-Interscience. doi:10.1002/9780470742020

Chouinard, G., Lei, Z., Hu, W., Shellhammer, S., & Caldwell, W. (2009-01). IEEE 802.22: The first cognitive radio wireless regional area networks (WRANs) standard. *IEEE Communications Magazine*, *47*(1), 130–138.

Cordeiro, C., Challapali, K., Birru, D., & Sai, S. N. (2006). IEEE 802.22: An introduction to the first wireless standard based on cognitive radios. *The Journal of Communication*, *1*(1).

Devroye, N., Mitran, P., & Tarokh, V. (2006). Limits on communication in a cognitive radio channel. *IEEE Communications Magazine*, *44*(6), 44–49. doi:10.1109/MCOM.2006.1668418

Ekram, H., Dusit, N., & Zhu, H. (2009). *Dynamic spectrum access and management in cognitive radio networks*. Cambridge, UK: Cambridge University Press.

Ekram, H., & Vijay, K. B. (2007). *Cognitive wireless communication networks*. Berlin, Germany: Springer.

Frank, H. P. F., & Marcos, D. K. (2007). *Cognitive wireless networks: Concepts, methodologies and visions inspiring the age of enlightenment of wireless communications*. Berlin, Germany: Springer.

Frank, H. P. F., & Marcos, K. (2006). *Cooperation in wireless networks: Principles and applications*. Berlin, Germany: Springer.

Hector, C. (2006). *Focus on cognitive radio technology*. New York, NY: Nova Science Publishers.

Huseyin, A. (2007). *Cognitive radio, software defined radio, and adaptive wireless systems*. Unpublished.

Ivan, C., Friedrich, K. J., Milind, M. B., & Ryuji, K. (2008). *Cognitive radio and dynamic spectrum sharing systems*. New York, NY: Hindawi Publishing Corp.

Joseph, M. (2000). *Software radio architecture: Object-oriented approaches to wireless systems engineering*. New York, NY: John Wiley & Sons, Inc.

Joseph, M. (2006). *Cognitive radio architecture: The engineering foundations of radio XML*. New York, NY: Wiley-Interscience.

Kiani, S., & Gesbert, D. (2006). Maximizing the capacity of large wireless networks: Optimal and distributed solutions. In *Proceedings of the IEEE International Symposium on Information Theory (ISIT),* (pp. 2501–2505). IEEE Press.

Kim, D., Le, L., & Hossain, E. (2008). Joint rate and power allocation for cognitive radios in dynamic spectrum access environment. *IEEE Transactions on Wireless Communications*, *7*(12), 5517–5527. doi:10.1109/T-WC.2008.071465

Lars, B., & Stefan, M. (2009). *Cognitive radio and dynamic spectrum access*. New York, NY: Wiley-Interscience.

Le, L., & Hossain, E. (2008). Resource allocation for spectrum underlay in cognitive radio networks. *IEEE Transactions on Wireless Communications*, *7*(12), 5306–5315. doi:10.1109/TWC.2008.070890

Quan, Z., Cui, S., Sayed, A., & Poor, H. (2009). Optimal multiband joint detection for spectrum sensing in cognitive radio networks. *IEEE Transactions on Signal Processing*, *57*(3), 1128–1140. doi:10.1109/TSP.2008.2008540

Qusay, M. (2007). *Cognitive networks: Towards self-aware networks*. New York, NY: Wiley-Interscience.

Savo, G. G. (2007a). *Advanced wireless communications: 4G cognitive and cooperative* (2nd ed.). New York, NY: Wiley-Interscience.

Savo, G. G. (2007b). *Advanced wireless communications: 4G cognitive and cooperative broadband technologies* (2nd ed.). New York, NY: Wiley-Interscience.

Sendonaris, A., Erkip, E., & Aazhang, B. (2003). User cooperation diversity-part I: System description. *IEEE Transactions on Communications*, *51*(11), 1927–1938. doi:10.1109/TCOMM.2003.818096

Weiss, T., Hillenbrand, J., Krohn, A., & Jondral, F. (2004). Mutual interference in OFDM-based spectrum pooling systems. In *Proceedings of the 59th IEEE Vehicular Technology Conference (VTC),* (pp. 1873–1877). IEEE Press.

Xiao, Y., & Hu, F. (2008). *Cognitive radio networks*. Reading, MA: Auerbach Publications. doi:10.1201/9781420064216

KEY TERMS AND DEFINITIONS

Cognitive Radio (CR): A kind of radio that automatically changes its transmission or reception parameters, in a way where the entire wireless communication network—of which it is a node—communicates efficiently, while avoiding interference with licensed or licensed exempt users.

Dynamic Spectrum Access (DSA): To minimize unused spectral bands by spectral agile radios, it is the efficient utilization of allocated frequency spectrum through opportunistic transmission.

Power Allocation: To allocate the limited power to the user to achieve the maxmum utilization.

Radio Resource Management (RRM): Involves strategies and algorithms for controlling parameters such as transmit power, channel allocation, data rates, handover criteria, modulation scheme, error coding scheme, *etc*. The objective is to utilize the limited radio spectrum resources and radio network infrastructure as efficiently as possible.

Time Slot Scheduling (TSS): To properly schedule the slot to minimize the collision and maximize the network utilization.

Chapter 14
Secondary Use of Radio Spectrum by High Altitude Platforms

Zhe Yang
Blekinge Institute of Technology, Sweden

Abbas Mohammed
Blekinge Institute of Technology, Sweden

ABSTRACT

Traditional spectrum licensing enables guaranteed quality of service but could lead to inefficient use of the spectrum. The quest to achieve higher usage efficiency for the spectrum has been the hottest research topic worldwide recently. More efficient transmission technologies are being developed, but they alone cannot solve problems of spatially and temporally underused spectrum and radio resources. In this chapter, the authors review major challenges in traditional spectrum sharing and mechanisms to optimize the efficiency of spectrum usage. They investigate and assess incentives of a primary terrestrial system and secondary system based on a High-Altitude Platform (HAP) to share spectrum towards common benefits. The primary terrestrial system is defined to have exclusive rights to access the spectrum, which is shared by the secondary HAP system upon request. The Markov chain is presented to model two spectrum-sharing scenarios and evaluate the performance of spectrum sharing between primary terrestrial and secondary HAP systems. Simulation results show that to reserve an amount of spectrum from a primary system could encourage spectrum sharing with a secondary system, which has a frequent demand on requesting spectrum resources.

DOI: 10.4018/978-1-4666-2812-0.ch014

INTRODUCTION

Given the success of mobile communications and broadband Internet, network operators and equipment vendors are changing to focus on various data applications and wireless access to the Internet. With the introduction of packet data switching services in cellular systems, connection can be provided for broadband services. Due to the influence and preferences by fixed broadband services, it is believed that data rates will also be increased in mobile communication systems.

To support high data rates with wide area coverage with an accepted cost would require substantial technological advances to increase efficiently usage of radio network resources, e.g. core radio network, base station, radio spectrum. Different approaches have been introduced and applied to increase the efficiency:

- Advanced antenna systems and radio spectrum in lower frequency bands are the two main ways to improve link budgets in macro cellular systems, which are the dominant architecture for second and third generation of mobile networks.
- Relaying techniques could also be used to increase coverage area for high data rates.
- Dynamic Radio Resource Management (RRM) can be improved to efficiently exploit spectrum usage via different strategies, e.g. channel assignment techniques, cognitive radio spectrum sharing.

Regardless of types of wireless services and technologies, a critical component common to all wireless deployments is the access to radio spectrum, which is generally agreed to be a limited and scarce resource. However, there are several measurement campaigns on the spectrum occupancy over time, space and frequency showing that spectrum is sporadically used, which creates motivations to develop new mechanisms to better utilize the spectrum (Peha & Panichpapiboon, 2004). The 2002 report of the Federal Communications Commission (FCC)'s spectrum Policy Task Force (SPTF) represents a seminal document, which shows FCC is moving away from a traditional "command-and-control" spectrum management to a more market-oriented, dynamic approach enabled by rapid innovations in radio communications technology. Consequently, these activities show that it is possible for a spectrum sharing between different operators (or radio access techniques). Approaches to improve the efficiency of spectrum usage have been motivated by its space-time varying spectrum usage. The improvement can be achieved with collaborative mechanisms between different actors using the spectrum. These actors are typically end user terminals, base stations (or the operators owning base stations and users). Therefore, a dynamic spectrum allocation can be regarded as a multi-actor system where actors can share spectrum. Then it is possible to apply knowledge of economics (e.g., game theory) aided mechanism to manage spectrum sharing.

Secondary Access to Limited Spectrum Based on Cognitive Radio

Secondary access means a secondary device is allowed to transmit if and only if it does not interfere with the primary license-holder. In this scheme, spectrum is licensed. A secondary device gain the right to transmit by explicitly requesting permission from the license-holder as needed. A license-holder can grant the permission to access to the spectrum only if a secondary device does not interfere with the primary license-holder, which is serving other calls simultaneously. The primary license-holder will grant the permission if the requirement of Quality of Service (QoS) which is already underway can be met. The license-holder can charge a fee for the secondary access, which provides an incentive to share the spectrum.

Wireless network is characterized by a fixed spectrum assignment policy. In U.S., more than half of the radio spectrum is controlled by the federal government. A large portion of the assigned spectrum is utilized sporadically and geographically with a high variance both in time and space. With the introduction and adaption to the next generation of multimedia applications in wireless systems, the demand on capacity increases significantly. It is predicated that traffic will be some 20 times 2010 level by 2015. Therefore, the bandwidth needed to operate new techniques and applications is at the risk of overload. In 2010, U.S. national broadband plan called for additional 500 MHz of spectrum to be reallocated for broadband use. Although regulators e.g. FCC in US have held spectrum auctions for operators to buy new spectrums in the last twenty years, most telecom providers demand more spectrum and claim that service will suffer if they don't get enough.

A good example of a primary license-holder is the television broadcaster, which serves the television in a traditional broadcasting. In the case where the television broadcaster does not use the spectrum at all, there is no mechanism approved by the regulator that a secondary device can gain the right to access the spectrum owned by the broadcaster. A permitted secondary access to the spectrum could enable a continuous transmission occupying the spectrum, therefore it improves the efficiency.

Instead of the traditional spectrum reallocation to release more spectrum bands, Cognitive Radio (CR) has emerged to be the most promising solution for the limited spectrum. Currently unlicensed spectrum promotes efficiency through sharing, but quality of service cannot be guaranteed, which creates problems for some applications. CR transmitter and receiver are expected to communicate as a secondary system, which produce acceptable interference to the primary system by sharing the spectrum in a cognitive way. Solutions based on CR usually require low power consumption, low transmission power to reduce interference and high data rates required by applications, therefore they may not be reliable under circumstances e.g. complex fading environment and frequent handoff between secondary base stations due to a shorter transmission range of these devices. A CR system could be more expensive and complex to be deployed in the area where a primary system is dominant. New solutions and developments in the future are expected to achieve a reliable communication with a larger network size and considerable cost of deployment and services.

Market-Oriented Mechanism to Increase Dynamic Spectrum Sharing and Efficiency

FCC in its SPTF report released in 2002 has supported the idea of using the market mechanism to use spectrum dynamically and efficiently. One of the market-oriented approaches to mange spectrum is the creation of secondary spectrum market, which allows a transfer of the rights of using license-holder's spectrum to secondary users who could put it into a better use. A transparent and accountable spectrum trading could impose a clear, market-based opportunity cost upon license-holder to provide with spectrum-sharing in a correct incentive. In this case, a secondary user gets guaranteed quality of service through explicit coordination between license-holder a secondary users.

Spectrum leasing as an option to share spectrum has been authorized by FCC to spectrum licensees to exercise in many wireless services. This policy is aimed at encouraging the development of secondary spectrum markets by facilitating the spectrum leasing arrangements among spectrum users.

In 2010, FCC approved new rules for the use of unlicensed white space spectrum in a move that could pave the way for more unused wireless spectrum to be released in the future. White spaces in telecommunications refer to frequencies allocated to broadcasting services but not used locally. These frequencies, e.g. a guard band between two adjacent analog TV channels in most cases can be permitted for secondary use, which consequently creates technical challenges to avoid interference. In addition, large areas in spectrum between about 50 MHz and 700 MHz will be becoming free as a result of the switchover to digital television in many parts of the world. National and international bodies assign different frequencies for specific uses, and in most cases license the rights to broadcast over these frequencies.

The chapter is organized as follows: firstly, spectrum sharing incentives and issues of HAP and terrestrial systems to share spectrum are investigated. Then two scenarios of spectrum sharing modeled by Markov chain are described. Afterwards performance of spectrum sharing scenarios are evaluated in terms of blocking probability of the primary users and secondary access failure probability of secondary user by the primary system. Finally, conclusions and future research direction are given.

HIGH ALTITUDE PLATFORM AS A SECONDARY SYSTEM

Introduction to High Altitude Platform

A typical wireless communication service is realized by using conventional terrestrial and satellite systems. High Altitude Platforms (HAPs), recently proposed novel aerial platforms to operate at an altitude of 17 km to 22 km to provide communication services, have been suggested by the International Telecommunication Union (ITU) to serve footprints larger than 150 km radius. HAPs have been recently proposed as a novel approach for the delivery of wireless broadband services to fixed and mobile users (Mohammed & Yang, 2010). They can act as base-stations or relay nodes, which may be effectively regarded as a very tall antenna mast or a very Low-Earth-Orbit (LEO) satellite (Karapantazis & Pavlidou, 2005). Communications from the platform possess many useful characteristics, which include high receiver elevation angle, Line Of Sight (LOS) transmission, large coverage area and mobile deployment. These characteristics make HAPs competitive compared to conventional terrestrial and satellite systems, and furthermore they can contribute to a better overall system performance, greater capacity and cost-effective deployment (Yang & Mohammed, 2008). As an emerging technique, HAP preserves many of the advantages of both satellite and terrestrial systems and attracted considerable attention in Europe through the European Union (EU) CAPANINA Project and the recently completed EU COST 297 Action, which was the largest gathering of research community worldwide with interest in HAPs and related technologies.

Secondary access from HAP to a primary has unique advantages. To implement a secondary access from a single aerial platform would replace a large number of secondary terrestrial base stations, therefore the solution could effectively reduce the system complexity and achieve a high system efficiency. In addition, it is a cost-efficient secondary soulution because it can reduce the installation and operational costs and handoffs in the network (Mohammed, et al., 2011, 2008; Yang & Mohammed, 2011). The unique height of the platform can further provide a high elevation angle and a low attention loss, therefore secondary mobile devices can transmit at a lower power, achieve higher energy efficiency and reduce interference to the primary system.

Spectrum Sharing between HAP and Terrestrial System

Enhancing spectrum efficiency and usage is a significant task of regulatory authorities worldwide. There is an increasing demand for a secondary usage of lowly utilized radio spectrum, which allows unused parts of spectrum from a primary user to become temporarily available for commercial purposes with acceptable interference to a primary user. Considering the above advantages and depending on the applications, HAPs is an ideal complement or alternative solution to current terrestrial systems. Therefore it is vital to consider spectrum sharing techniques with terrestrial systems, which are assumed to be the primary system and mostly applicable in the reality. Spectrum licensing enables quality of service guarantees for the primary terrestrial system, but often leads to inefficient use of spectrum. An operator based on HAP system asks for a period of time to access spectrum through a market-oriented spectrum mechanism.

A terrestrial system in our investigation is defined to be a primary system having the exclusively rights to the spectrum. A HAP system is defined to be a secondary user of radio spectrum and needs to share spectrum of terrestrial system. The terrestrial system is assumed to have exclusive spectrum license and adequate incentives to share its excess or unused spectrum if HAP operator is acquiring the spectrum.

Incentives of HAP and Terrestrial Systems to Share Spectrum

Different spectrum incumbents have different incentives to share and trade their spectrum. For instance, government agencies have available spectrum but few incentives since they usually get the spectrum for free or have some specific usage purposes, e.g., security and military communications.

Because the primary operator in the scenario is assumed to have exclusive spectrum license and adequate incentives to share spectrum with the secondary HAP system, a flexible spectrum trading mechanism and profit of leasing spectrum are two basic incentives for terrestrial system. A primary terrestrial system needs to gain profit by sharing the spectrum. Therefore, profit of a terrestrial system comes from the difference between the revenue and cost, which depend on the following factors:

- *Revenue* depends on the following factors:
  - The amount of spectrum leased to a secondary HAP operator.
  - The price of spectrum leased to a secondary HAP operator.
- *Cost* depends on the following factors:
  - Blocking terrestrial user due to leasing available spectrum to secondary system.
  - The interference from secondary users to primary users.
  - Reducing the QoS in order to provide excess spectrum for secondary operators*.

*In this case, a primary operator prefers to lease its spectrum for profit even if there is a traffic demand from a primary system. An example is that the primary system can compare the potential revenue of serving its own users with revenue of leasing the spectrum to a secondary system, e.g. HAP system by increasing the blocking probability to its users in the primary network.

Conditions and Problems of Secondary HAP System to Share Spectrum

A HAP system is considered to be the secondary system, which has a strategy to share spectrum with primary terrestrial system in a cost-effective mechanism to provide services. Here we assume that the secondary system intends to rent the spectrum from primary terrestrial system when the traffic demand is out of serving capability. Since the primary user has exclusive spectrum license, QoS of HAP system cannot usually be guaranteed when renting spectrum from terrestrial operator. Furthermore, it pays a fee to access to the spectrum owned by terrestrial system. Consequently, it provides a solid incentive for secondary system not to waste spectrum in order to increase spectrum efficiency in general. Profit of a secondary system comes from the difference between the revenue and cost, which depend on the following factors:

- *Revenue* for secondary system depends on the following factors:
    - The revenue of service provided from a secondary HAP system to users.
- *Cost* for a secondary system depends on the following factors:
    - Cost of renting spectrum from terrestrial system.
    - Interference from primary system due to spectrum sharing.

SECONDARY SPECTRUM SHARING SCHEMES AND PERFORMANCE

Secondary Spectrum Access Mechanism

The secondary spectrum access is controlled by the mechanism based on the occupancy of spectrum in the terrestrial system. A description of the mechanism is shown Figure 1. The process denotes the request handling process of the secondary spectrum request from the HAP system. A feedback process to update the available spectrum of the primary system is included in the process.

Figure 1. Secondary spectrum request and access

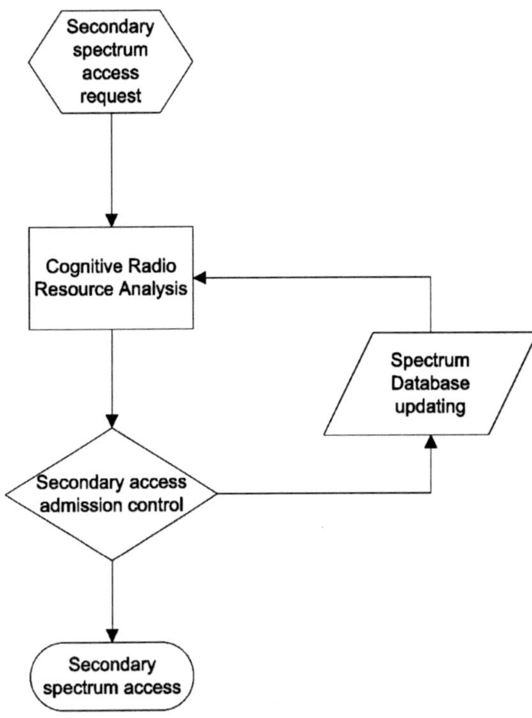

Secondary Access Admission Control

Using a Poisson process to model the spectrum demand in the primary terrestrial system and secondary HAP system, a Markov chain can be used to represent the system transition behavior regarding the spectrum request admission functionality.

We assume the new traffic demand at λ_p in the primary terrestrial system and secondary traffic demand at λ_s from the secondary HAP system following a Poisson process. It is supposed that the holding time t_p is exponentially distributed with density function $f_p(t_p)=\mu e^{-\mu t}$ with mean $1/\mu$. The t_p is the amount of time that the radio channel would continue to be occupied by the primary user without forced termination by the primary system due to spectrum sharing with the secondary system. If a secondary traffic demand is given by the primary system for a radio channel for a period of t_s, which is called the secondary access residence time, before it releases the radio channel. We assume t_s to be exponentially distributed with probability density function $f_s(t_s)= \eta e^{-\eta t}$ and mean $1/\eta$.

If a traffic demand is admitted with a radio channel, the channel would be released either by a completed traffic demand by the primary system or by a completed secondary traffic demand in the secondary system. In this way, the traffic occupancy time t_{co} is the smaller of the holding time t_p and residence time t_s with a probability density function $f_{co}(t)= (\mu+\eta)e^{-(\mu+\eta)t}$. Figure 2 shows all

Figure 2. Primary and secondary resource request model

arrival requests and released resources in both primary and secondary systems.

The blocking probability p_p is defined as the probability that a primary user finds all channels occupied in the primary system; the secondary access failure probability p is defined as the probability that a secondary spectrum access request is rejected by a primary system.

Nonreserved Spectrum Sharing Scheme (NSSS)

A Nonreserved Spectrum Sharing Scheme (NSSS) can be modeled by a Markov process with $s+1$ states, where s is the number of available channels in the primary system. In this case, requests from a secondary system can use all s channels as long as they are available. So that for $0 \leq j < s$, the transition rate from state P_{j-1} to P_j is given by $\lambda_p+\lambda_s$ and a transition rate from state P_{j+1} to P_j is given with rate $(j+1)(\mu+\eta)$ because the radio channel occupancy is exponentially distributed. Suppose that all s channels are occupied and then a new request is demanded by primary system, this request will be rejected. In addition, if a request is demanded by a secondary system at this moment, a secondary access failure is produced. The state diagram is shown in Figure 3.

In this scheme, requests from primary system and secondary system are served with equal priority. Therefore, the blocking probability p_p in the primary system will be equal to the secondary access failure probability p_s. If we assume the secondary system waits some random time interval after its first request is rejected by the primary system, a traffic formula applied to infinite sources and lost call cleared can be adopted:

Figure 3. The state diagram for the nonreserved spectrum sharing scheme

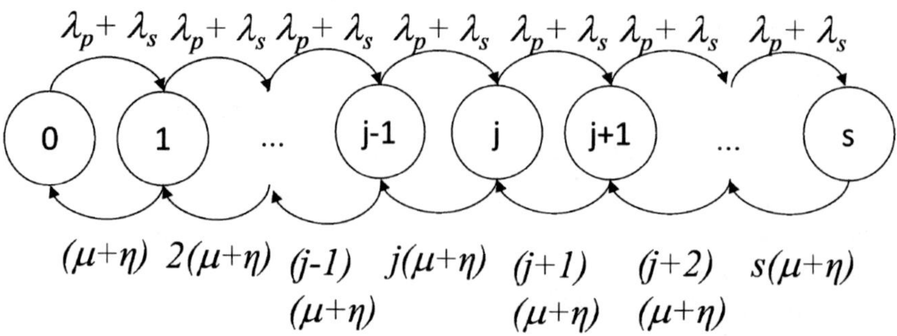

$$p = \frac{\frac{A^N}{N!}}{\sum_{x=0}^{N} \frac{A^x}{x!}} \tag{1}$$

where:

- A = Offered traffic load in Erlangs
- N = Number of channels
- P = Probability of blocking

Considering the requests from primary and secondary systems, A and N can be replaced by corresponding variables below in NSSS:

$$A = (\lambda_p + \lambda_s)\frac{1}{\mu + \eta} \qquad N = s \tag{2}$$

Reserved Spectrum Sharing Scheme (RSSS)

In Reserved Spectrum Sharing Scheme (RSSS), some of the s channels are reserved for secondary access only. Suppose that R channels are reserved of the s total available channels, the state diagram that described the system performance is shown in Figure 4. If there is a new request from the primary system, it will be accepted if the number of idle channels are less than n, where $n = s - R$, otherwise, this request will be rejected. Request from a secondary spectrum access will be rejected if all channels are occupied in the primary system, therefore producing a secondary access failure.

The state diagram of this scheme is shown in Figure 4. Compared with the NSSS, a differentiation is made between the system behavior when the number of occupied channels is less than n and when the number of occupied channels is more than n. For $0 \le j < n$, a transition rate from state P_j to P_{j+1} and from P_{j+1} to P_j is the same as in NSSS. For $n \le j < s$, a transition rate from state P_j to P_{j+1} is given with rate λ_s and a transition from state P_{j+1} to P_j is given with rate $(j+1)(\mu+\eta)$ because any request from primary system is rejected to occupy a channel when there are n or more occupied channels in the primary system.

Assume the steady-state probability is P_j, the blocking probability p_p in the primary system is the sum of probabilities of all steady-state P_j, where $n \le j \le s$. The secondary access failure probability p_s in RSSS is the steady-state probability of P_s, where all s channels are occupied; that is

$$p_p = \sum_{j=n}^{s} P_j \qquad p_s = P_s \tag{3}$$

Figure 4. The state diagram for the reserved spectrum sharing scheme

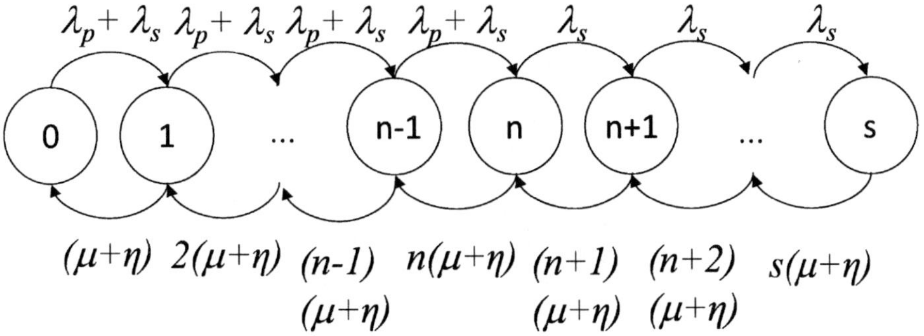

Performance Assessment

In order to simulate the spectrum sharing schemes of NSSS and RSSS for a primary terrestrial system and secondary HAP system, we consider there are 10 channels in each cell with a reuse factor at 7. A uniform traffic model with an offered traffic load in a cell ranging from 1 to 20 Erlangs is considered. We consider the mean channel holding time of the primary user is 3 min. The secondary access from HAP system to occupy an available channel granted by the primary system is initially taken as 3 min and then varied from 0.5 min to 6 min in order to evaluate the influence of the decreasing channel occupancy time of a secondary access.

The results are simulated for NSSS and RSSS, respectively. RSSS has the number of reserved channels for a secondary system *R=1* (RSSS-1) and *R=2* (RSSS-2). Figure 5 shows the blocking probability of a primary user decreases when using NSSS instead of RSSS, which implies a given priority to secondary access from a primary system will increase the blocking probability to primary users. It also shows that the secondary access failure probability decreases as a result of more reserved channels for secondary access.

Figure 6 shows the frequency influence of a secondary access to the primary system. For a constant offered traffic load, the average secondary access residence time varies from 0.5μ to 6μ. With the behavior of more frequent secondary system requests, the blocking probability of primary user decreases. With more frequent requests from secondary system, its secondary access failure probability decreases for NSSS case, but has a slight reduction for RSSS cases on relative small secondary access residence time.

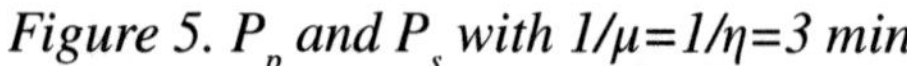

Figure 5. P_p and P_s with $1/\mu=1/\eta=3$ min

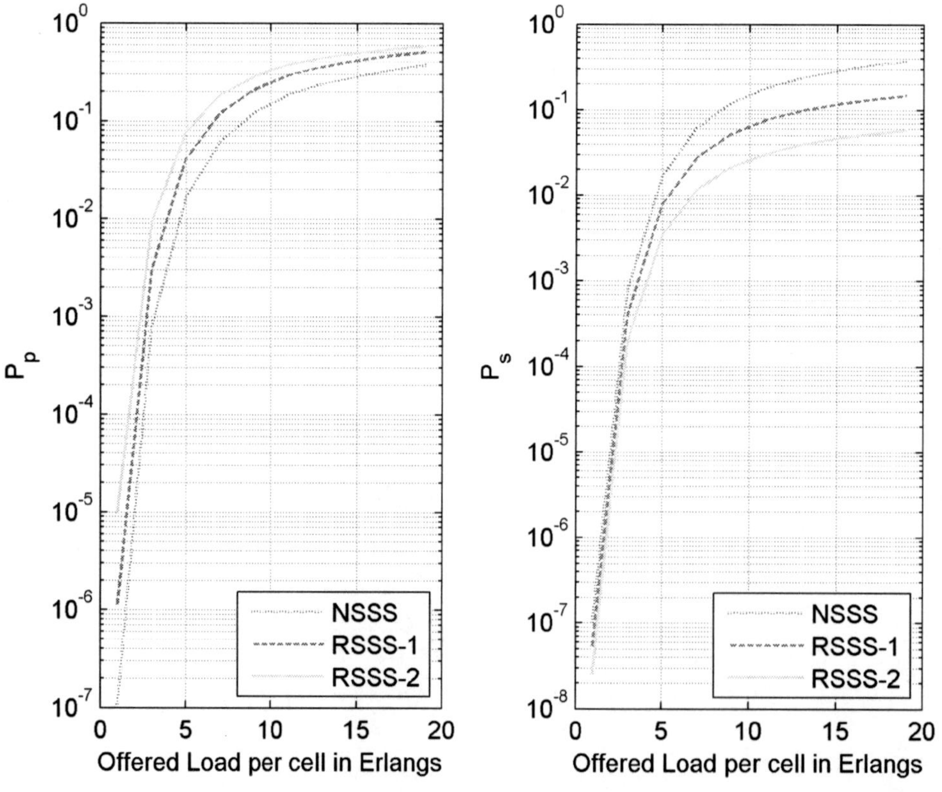

Figure 6. P_p and P_s with varying secondary access residence time

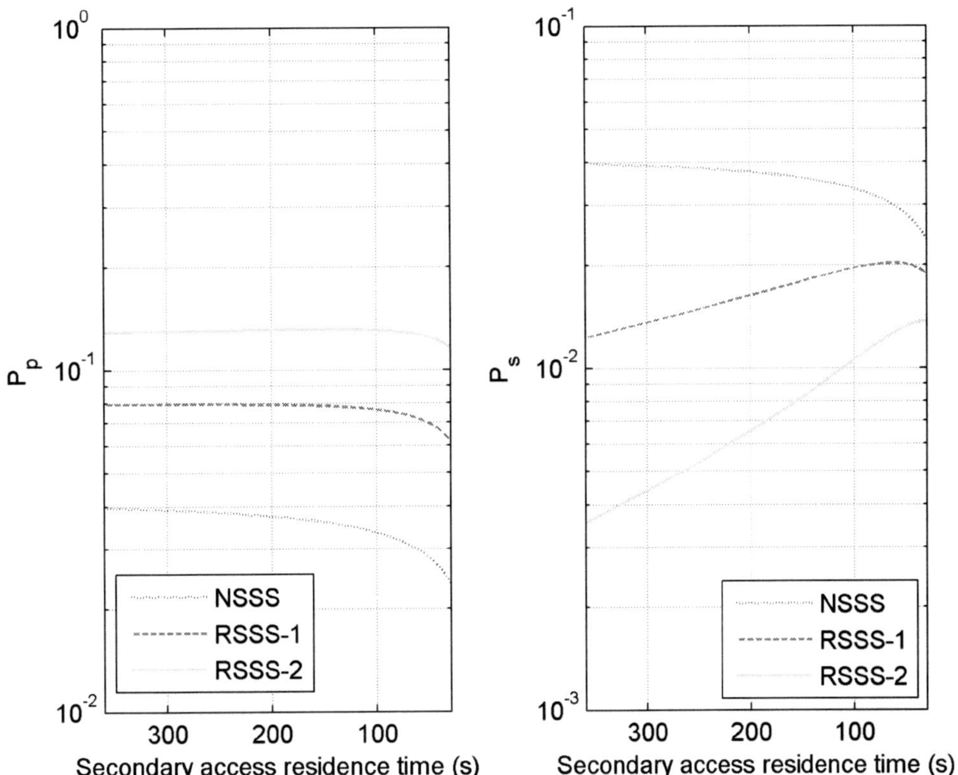

CONCLUSION AND FUTURE RESEARCH DIRECTIONS

In this chapter, we investigated two spectrum sharing schemes and performance of a secondary HAP system (or operator) to access spectrum granted by a primary terrestrial license-holder. A reduced capital expenditure of secondary network operators by gaining revenue from inefficiently utilized spectrum for the primary operator could be achieved. For the secondary operator, e.g. HAPs, it is a cost-effective way to share spectrum via spectrum sharing schemes of renting instead of owning a full spectrum, which was usually inefficiently utilized. Licensing spectrum grants the terrestrial system (or operator) some exclusivity with guaranteed quality of service, but leads to inefficient use of spectrum in terms of idle channels. Spectrum sharing among a primary and secondary system (or operators) could alleviate spectrum scarcity and increase the probability of rejecting spectrum requests from the primary system to an acceptable extent. If requests from a secondary system arrive more frequently, reserving spectrum resources in a primary system could encourage the spectrum sharing incentives of a primary system due to a decreased blocking probability of primary users.

Pricing and revenue through secondary access is a way of encouraging efficient usage of spectrum in a decentralized fashion. Price is ideally set at a level where supply and demand are matched. It further encourages secondary users to acquire sufficient amount of spectrum. In the future, a dynamic version of the pricing mechanism based on the congestion level can potentially adapt to a long-term traffic load change. In addition, it is interesting to develop a dynamic pricing scheme with incremental deployment plan of a multiple HAP system. Considering the impact of a dynamic pricing mechanism, a state transition rate needs to consider an acceptance ratio of secondary access spectrum subject to the price and profit. The pricing and profit based on market-oriented mechanism needs to be updated after granted request from a secondary system. The information needs to be ready before the arrival of next request from a secondary system.

REFERENCES

Karapantazis, S., & Pavlidou, F. (2005). Broadband communications via high-altitude platforms: A survey. *IEEE Communications Surveys*, *7*(1), 2–31. doi:10.1109/COMST.2005.1423332

Mohammed, A., Arnon, S., Grace, D., Mondin, M., & Miura, R. (2008). Advanced communications techniques and applications for high-altitude platforms. *EURASIP Journal on Wireless Communications and Networking*. Retrieved from http://jwcn.eurasipjournals.com/content/pdf/1687-1499-2008-934837.pdf

Mohammed, A., Mehmood, A., Paviudo, N., & Mohorcic, M. (2011). The role of high-altitude platforms (HAPs) in the wireless global connectivity. *Proceedings of the IEEE*, *99*(11), 1939–1953. doi:10.1109/JPROC.2011.2159690

Mohammed, A., & Yang, Z. (2010). Next generation broadband services from high altitude platforms. In *Fourth-Generation Wireless Networks: Applications and Innovations* (pp. 249–267). Hershey, PA: IGI Global. doi:10.4018/978-1-61520-674-2.ch012

Peha, J., & Panichpapiboon, S. (2004). Real-time secondary markets for spectrum. *Telecommunications Policy*, *28*(7-8), 603–616. doi:10.1016/j.telpol.2004.05.003

Yang, Z., & Mohammed, A. (2011). Deployment and capacity of mobile WiMAX from high altitude platform. In *Proceedings of the 2011 IEEE 74th Vehicular Technology Conference*. IEEE Press.

Yang, Z., Mohammed, A., & Hult, T. (2008). Performance evaluation for WiMAX Broadband from high altitude platform cellular system and terrestrial coexisting capability. *EURASIP International Journal on Wireless Communications and Networking*. Retrieved from http://www.researchgate.net/publication/26571586_Performance_Evaluation_of_WiMAX_Broadband_from_High_Altitude_Platform_Cellular_System_and_Terrestrial_Coexistence_Capability

Compilation of References

3 GPP TS 32.500. (2008). *Telecommunication management: Self-organizing networks (SON): Concepts and requirements (rel. 8)*. Retrieved from http://www.3gpp.org

3 Tech, G. P. P. Report. (2010). *Further advancements for e-utra physical layer aspects*. Retrieved from http://www.3gpp.org

Adamis, K. M. (2007). A new MAC protocol with control channel auto-discovery for self-deployed cognitive radio networks. In *Proceedings of the Program for European Wireless 2007, EW 2007*. EW.

Afridi, A. (2011). *Macro and femto network aspects for realistic LTE usage scenarios*. (Masters Dissertation). Royal Institute of Technology (KTH). Stockholm, Sweden.

Ahson, S., & Ilyas, M. (2008). *WiMAX standards and security*. Boca Raton, FL: CRC Press.

Airspan. (2007). Mobile WiMAX security. *Airspan Networks Inc*. Retrieved May 15, 2012, from http://www.airspan.com

Akin, S., & Gursoy, M. (2010). Effective capacity analysis of cognitive radio channels for quality of service provisioning. *IEEE Transactions on Wireless Communications*, *9*(11), 3354–3364. doi:10.1109/TWC.2010.092410.090751

Akyildiz, I. F., & Lee, W. Y. (2006) NeXt generation/dynamic spectrum access/cognitive radio wireless networks: A survey. *Computer Networks Journal*. Retrived March 21, 2012, from http://www.sciencedirect.com

Akyildiz, I. F., Brandon, F., & Balakrishnan, R. (2011). Cooperative spectrum sensing in cognitive radio networks: A survey. *Journal of Physical Communication*. Retrieved January 16, 2012, from http://www.elsevier.com/locate/phycom

Akyildiz, I. F., Lee, W. Y., & Chowdury, K. R. (2009). CRAHNs: Cognitive radio ad hoc networks. *Journal of Ad Hoc Networks*, *7*(5), 810–836. doi:10.1016/j.adhoc.2009.01.001

Akyildiz, I. F., Lee, W. Y., Vuran, M. C., & Mohanty, S. (2006). NeXt generation dynamic spectrum access in cognitive radio wireless networks: A survey. *Computer Networks Journal*, *50*(13), 2127–2159. doi:10.1016/j.comnet.2006.05.001

Akyildiz, I. F., Lee, W. Y., Vuran, M. C., & Mohanty, S. (2008). A survey on spectrum management in cognitive radio networks. *IEEE Communications Magazine*, *46*(4), 40–48. doi:10.1109/MCOM.2008.4481339

Al-Dulaimi, A., & Al-Saeed, L. (2010). An intelligent scheme for first run cognitive radios. In *Proceedings of the IEEE International Conference and Exhibition on Next Generation Mobile Applications, Services, and Technologies (NGMAST 2010)*. Amman, Jordan: IEEE Press.

Al-Dulaimi, A., Al-Rubaye, S., & Cosmas, J. (2011). Adaptive congestion control for mobility in cognitive radio networks. *Proceedings of the Wireless Advanced (WiAD), 2011*, 273–277. doi:10.1109/WiAd.2011.5983268

Ali, S., & Yu, F. (2009). Cross-layer qos provisioning for multimedia transmissions in cognitive radio networks. In *Proceedings of the Wireless Communications and Networking Conference, 2009*. IEEE Press.

Alouini, M. S., & Goldsmith, A. (1999). Area spectral efficiency of cellular mobile radio systems. *IEEE Transactions on Vehicular Technology*, *48*(4), 1047–1066. doi:10.1109/25.775355

Al-Rubaye, S., Al-Dulaimi, A., Al-Saeed, L., Al-Raweshidy, H. S., Kadhum, E., & Ismail, W. (2010). Development of heterogeneous cognitive radio and wireless access network. In *Proceedings of the 24th Wireless World Research Forum (WWRF)*. Penang Island. Malaysia: WWRF.

Al-Rubaye, S., Al-Dulaimi, A., & Cosmas, J. (2011). Cognitive femtocell. *IEEE Magazine of Vehicular Technology*, *6*(1), 44–51. doi:10.1109/MVT.2010.939902

Altman, E., Boulogne, T., Azouzi, R., & Jimenez, T. (2006). A survey on networking games in telecommunications. *Computers & Operations Research*, *33*(2). doi:10.1016/j.cor.2004.06.005

Alyfantis, G., Marias, G., Hadjiefthymiades, S., & Merakos, L. (2007). Non-cooperative dynamic spectrum access for CDMA networks. In *Proceedings of IEEE Global Telecommunications Conference (GLOBECOM)*, (pp. 3574–3578). IEEE Press.

Andrews, J. G., Claussen, H., Dohler, M., Rangan, S., & Reed, M. C. (2012). Femtocells: Past, present, and future. *IEEE Journal on Selected Areas in Communications*, *30*(3), 497–508. doi:10.1109/JSAC.2012.120401

Ansari, J., Zhang, X., Achtzehn, A., Petrova, M., & Mahonen, P. (2011). A flexible MAC development framework for cognitive radio systems. *Proceedings of the, WCNC-2011*, 156–161. IEEE Press.

Arshad, K., Imran, M. A., & Moessner, K. (2010). Collaborative spectrum sensing optimisation algorithms for cognitive radio networks. *Eurasip International Journal of Digital Multimedia Broadcasting*. Retrieved from http://www.hindawi.com/journals/ijdmb/2010/424036/

Ashraf, I., Boccardi, F., & Ho, L. (2011). SLEEP mode technique for small cell deployments. *IEEE Communications Magazine*, *49*(8), 72–79. doi:10.1109/MCOM.2011.5978418

Attar, A., Holland, O., Nakhai, M., & Aghvami, A. (2008). Interference-limited resource allocation for cognitive radio in orthogonal frequency division multiplexing networks. *IET Communication*, *2*(6), 806–814. doi:10.1049/iet-com:20070355

Attar, A., Nakhai, M., & Aghvami, A. (2008). Cognitive radio transmission based on direct sequence MC-CDMA. *IEEE Transactions on Wireless Communications*, *7*(4), 1157–1162. doi:10.1109/TWC.2007.060907

Badic, B., O'Farrell, T., Loskot, P., & He, J. (2009). Energy efficient radio access architectures for green radio: Large versus small cell size deployment. In *Proceedings of the IEEE 70th Vehicular Technology Conference (VTC 2009 Fall)*. Anchorage, AK: IEEE.

Bae, Y. H., Alfa, A., & Choi, B. D. (2010). Performance analysis of modified IEEE 802.11-based cognitive radio networks. *IEEE Communications Letters*, *14*(10), 975–977. doi:10.1109/LCOMM.2010.082310.100322

Baldini, G., & Pons, E. C. (2011). Design of a robust cognitive control channel for cognitive radio networks based on ultra wideband pulse shaped signal. In *Proceedings of the 4th International Conference on Multiple Access Communications*. (pp. 13-23). Berlin, Germany: Springer.

Bansal, G., Hossain, M., & Bhargava, V. (2008). Optimal and suboptimal power allocation schemes for OFDM-based cognitive radio systems. *IEEE Transactions on Wireless Communication*, *7*(11-2), 4710–4718.

Bansal, G., Hossain, M., & Bhargava, V. (2007). Adaptive power loading for OFDM-based cognitive radio systems. *Proceedings of the IEEE, ICC*, 5137–5142. IEEE Press.

Ban, T. W., Choi, W., Jung, B. C., & Sung, D. K. (2009). Multi-user diversity in a spectrum sharing system. *IEEE Transactions on Wireless Communications*, *8*(1), 102–106. doi:10.1109/T-WC.2009.080326

Barbeau, M. (2005). WiMAX/802.16 threat analysis. In *Proceedings of Q2SWinet 2005*, (pp. 1-8). Quebec, Canada: Q2SWinet.

Bejerano, Y., Han, S.-J., & Li, L. E. (2004). Fairness and load balancing in wireless lans using association control. In *Proceedings of ACM MobiCom*. ACM Press.

Bejerano, Y., & Han, S.-J. (2009). Cell breathing techniques for load balancing in wireless lans. *IEEE Transactions on Mobile Computing*, *8*(6). doi:10.1109/TMC.2009.50

Benslimane, A., Ali, A., Kobbane, A., & Taleb, T. (2009). A new opportunistic MAC layer protocol for cognitive IEEE 802.11-based wireless networks. In *Proceedings of 2009 IEEE 20th International Symposium on Personal, Indoor and Mobile Radio Communications,* (pp. 2181 - 2185). Tokyo, Japan: IEEE Press.

Berlemann, L., Dimitrakopoulos, G., & Moessner, K. (2005). Cognitive radio and management of spectrum and radio resources in reconfigurable network. *Wireless World*. Retrieved from http://www.wireless-world-research.org/fileadmin/sites/default/files/about_the_forum/WG/WG6/White%20Paper/WG6_WP7.pdf

Bianchi, P., Najim, J., Maida, M., & Debbah, M. (2009). Performance analysis of some eigen-based hypothesis tests for collaborative sensing. In *Proceedings of the IEEE Workshop Datatistical Signal Processing*, (pp. 5–8). Cardiff, UK: IEEE Press.

Bielecki, W., & Burak, D. (2007). Parallelization method of encryption algorithms. In *Advances in Information Processing and Protection* (pp. 191–204). Berlin, Germany: Springer-Verlag. doi:10.1007/978-0-387-73137-7_17

Biswas, S., & Morris, R. (2004). Opportunistic routing in multi-hop wireless networks. *SIGCOMM Computer Communication Review*, *34*(1), 69–74. doi:10.1145/972374.972387

Bobarshad, H., van der Schaar, M., & Shikh-Bahaei, M. (2010). A low-complexity analytical modeling for cross-layer adaptive error protection in video over wlan. *IEEE Transactions on Multimedia*, *12*(5), 427–438. doi:10.1109/TMM.2010.2050734

Boyd, S., & Vandenberghe, L. (2004). *Convex optimization*. Cambridge, UK: Cambridge University Press.

Brik, V., Rozner, E., Banarjee, S., & Bahl, P. (2005). DSAP: A protocol for coordinated spectrum access. In *Proceedings of IEEE DySPAN,* (pp. 611–614). IEEE Press.

Brooks, R., Ramanathan, P., & Sayeed, A. (2003). Distributed target classification and tracking in sensor networks. *Proceedings of the IEEE*, *91*(8), 1163–1171. doi:10.1109/JPROC.2003.814923

Brown, T. (2005). An analysis of unlicensed device operation in licensed broadcast service bands. In *Proceedings of DySPAN,* (pp. 11–29). IEEE Press.

Buljore, S., Muck, M., Martigne, P., Houze, P., Harada, H., Ishizu, K., … Stametalos, M. (2008). Introduction to IEEE p1900.4 activities. *IEICE Transactions on Communications, E91-B*(1).

Cabric, D., Mishra, S., & Brodersen, R. (2004). Implementation issues in spectrum sensing for cognitive radios. In Proceedings of the Asilomar Conference on Signals, Systems and Computers, (vol. 1, pp. 772–776). Asilomar.

Cabric, D., Tkachenko, A., & Brodersen, R. (2006). Spectrum sensing measurements of pilot, energy, and collaborative detection. In *Proceedings of the IEEE Military Communications Conference*, (pp. 1-7). Washington, DC: IEEE Press.

Cabric, Mishra, & Brodersen. (2004). Implementation issues in spectrum sensing for cognitive radios. In *Proceedings of the Thirty-Eighth Asilomar Conference on Signal, System, and Computer,* (Vol. 1, pp. 272-276). Berkeley, CA: Berkeley Wireless Research Center.

Caldwell, W., & Chouinard, G. (2010). *IEEE 802.22 wireless RANs draft recommended practice*. Retrieved December 13, 2011, from https://mentor.ieee.org/802.22/dcn/06/22-06-0242-27-0002-draft-recommended-practice.doc

Cao, L., & Zheng, H. (2005). Distributed spectrum allocation via local bargaining. In *Proceedings of the IEEE Conference on Sensor and Ad Hoc Communications and Networking (SECON),* (pp. 475–486). IEEE Press.

Cao, L., & Zheng, H. (2008). Distributed rule-regulated spectrum sharing. *IEEE Journal on Selected Areas in Communications*, *26*(1), 130–145. doi:10.1109/JSAC.2008.080112

Cesana, C. E. (2011). Routing in cognitive radio networks: Challenges and solutions. *Ad Hoc Networks*, *9*(3), 228–248. doi:10.1016/j.adhoc.2010.06.009

Cesana, M., Cuomo, F., & Ekici, E. (2011). Routing in cognitive radio networks: Challenges and solutions. *Journal Ad Hoc Networks*, *9*(3), 228–248. doi:10.1016/j.adhoc.2010.06.009

Chachulski, S., Jennings, M., Katti, S., & Katabi, D. (2007). Trading structure for randomness in wireless opportunistic routing. In *Proceedings of the 2007 Conference on Applications, Technologies, Architectures, and Protocols for Computer Communications,* (pp. 169–180). New York, NY: ACM Press.

Chandrasekhar, V. (2009). *Coexistence in femtocell-aided cellular architectures*. (PhD Dissertation). University of Texas at Austin. Austin, TX.

Chandrasekhar, V., & Andrews, J. G. (2008). Spectrum allocation in two-tier femtocell networks. In *Proceedings of the IEEE Conference Signals, Systems and Computers*. Pacific Grove, CA: IEEE Press.

Chandrasekhar, V., Andrews, J., & Gatherer, A. (2008). Femtocell networks: A survey. *IEEE Communications Magazine*, *46*(9), 59–67. doi:10.1109/MCOM.2008.4623708

Chen, H. (2010). Relay selection for cooperative spectrum sensing in cognitive radio networks. In *Proceedings of the IEEE 2010 International Conference on Communications and Mobile Computing*. Shenzhen, China: IEEE.

Chen, K., Peng, Y., Prasad, N., Liang, Y., & Sun, S. (2008). Cognitive radio network architecture: Part I - General structure. In *Proceedings of the IEEE 2nd International Conference on Ubiquitous Information Management and Communication*. Suwon, Korea: IEEE Press.

Chen, T., Zhang, H., Maggio, G. M., & Chlamtac, I. (2007). CogMesh: A cluster-based cognitive radio network. In *Proceedings of the 2007 IEEE Symposium on New Frontiers in Dynamic Spectrum Access Networks (IEEE DySPAN 2007)*. Dublin, Ireland: IEEE Press.

Cheng, S. M., Ao, W. C., & Chen, K. C. (2010). Downlink capacity of two-tier cognitive femto networks. In *Proceedings of the IEEE 21st International Symposium on Personal Indoor and Mobile Radio Communications, PIMRC 2010*, (pp. 1303–1308). Istanbul, Turkey: IEEE.

Cheng, W., Zhang, X., & Zhang, H. (2011). Full-duplex spectrum sensing in non-time-slotted cognitive radio networks. In *Proceedings of the IEEE 2011 Military Communications Conference (MILCOM 2011)*. Baltimore, MD: IEEE Press.

Cheng, P., Zhang, Z., Chen, H., & Qiu, P. (2008). Optimal distributed joint frequency, rate and power allocation in cognitive OFDMA systems. *IET Communications*, *2*(6), 815–826. doi:10.1049/iet-com:20070358

Cheng, S. M., Lien, S. Y., Chu, F. S., & Chen, K. C. (2011). On exploiting cognitive radio to mitigate interference in macro/femto heterogeneous networks. *IEEE Magazine of Wireless Communications*, *18*(3), 40–47. doi:10.1109/MWC.2011.5876499

Cheng, S., & Yang, Z. (2008). Adaptive power control algorithm based on SIR in cognitive radios. *Journal of Electronics & Information Technology*, *1*, 15.

Chen, Y. (2010). Improved energy detector for random signals in Gaussian noise. *IEEE Transactions on Wireless Communications*, *9*(2), 558–563. doi:10.1109/TWC.2010.5403535

Chen, Y.-S., Cho, C.-H., You, I., & Chao, H.-C. (2011). A cross-layer protocol of spectrum mobility and handover in cognitive LTE networks. *Simulation Modelling Practice and Theory*, *19*(8), 1723–1744. doi:10.1016/j.simpat.2010.09.007

Chen, Y., Zhao, Q., & Swami, A. (2008). Joint design and separation principle for opportunistic spectrum access in the presence of sensing errors. *IEEE Transactions on Information Theory*, *54*(5), 2053–2071. doi:10.1109/TIT.2008.920248

Chouhan, A. T. (2011). Cognitive radio netwoks: Application and implemetation issues. In *Proceedings of the 26th M.P. Young Scientist Congress,* (p. 65). Jabalpur, India: MPCOST.

Chouhan, A. T. (2011). Cognitive radio networks: Implementation and application issues in India. In *Proceedings of the Seminar on Next Generation Networks – Implementation and Implications*. New Delhi, India: Telecom Regulatory Authority of India (TRAI). Retrieved from http://www.trai.gov.in/WriteReadData/trai/upload/misc/174/Lokesh_Chauhan.pdf

Chouhan, L. (2012). *Design and analysis of mac protocols for cognitive radio networks.* (Unpublished Dissertation). ABV-Indian Institute of Information Technology and Management. Gwalior, India.

Chowdbury, K. R., & Akyildiz, I. F. (2008). Cognitive wireless mesh networks with dynamic spectrum access. *IEEE Journal on Selected Areas in Communications*, *26*(1), 168–181. doi:10.1109/JSAC.2008.080115

Chu, M., Haussecker, H., & Zhao, F. (2002). Scalable information-driven sensor querying and routing for ad hoc heterogenous sensor networks. *International Journal of High Performance Computing Applications*, *16*(3), 293–313. doi:10.1177/10943420020160030901

Cordeiro, C., & Challapali, K. (2007). C-MAC: A cognitive MAC protocol for multi-channel wireless networks. In *Proceedings of 2nd IEEE International Symposium on New Frontiers in Dynamic Spectrum Access Networks, 2007,* (pp. 147-157). IEEE Press.

Cordeiro, C., Challapali, K., Birru, D., & Sai Shankar, N. (2005). IEEE 802.22: The first worldwide wireless standard based on cognitive radios. In *Proceedings of the First IEEE International Symposium on New Frontiers in Dynamic Spectrum Access Networks, 2005,* (pp. 328 - 337). Baltimore, MD: IEEE Press.

Cordeiro, C., Challapali, K., Birru, D., & Shankar, S. (2006). IEEE 802.22: An introduction into the first wireless standard based on cognitive radio. *The Journal of Communication*, *1*(1), 38–47.

Cormio, C., & Chowdhury, K. (2009). A survey on MAC protocols for cognitive radio networks. *Ad Hoc Networks*, *7*(7), 1315–1329. doi:10.1016/j.adhoc.2009.01.002

Cormio, C., & Chowdhury, K. R. (2010). Common control channel design for cognitive radio wireless ad hoc networks using adaptive frequency hopping. *Ad Hoc Networks*, *8*, 430–438. doi:10.1016/j.adhoc.2009.10.004

Cost Europe. (2012). *Cost actions*. Retrieved May 5, 2012, from http://www.cost.eu/domains_actions/ict/Actions

Costa, G., Cattoni, A., Roig, V. A., & Mogensen, P. E. (2010). Interference mitigation in cognitive femtocells. In *Proceedings of the 2010 IEEE GLOBECOM Workshops (GC Wkshps)*. Miami, FL: IEEE.

Coucheney, P., Touati, C., & Gaujal, B. (2009). Fair and efficient user-network association algorithm for multi-technology wireless networks. In *Proceedings of the 28th Conference on Computer Communications Miniconference (INFOCOM)*. IEEE Press.

Cox, D. (1967). *Renewal theory*. London, UK: Butler & Tanner Ltd.

CWC. (2008). *The first international workshop on green wireless 2008 (W-GREEN)*. Retrieved from http://www.cwc.oulu.fi/workshops/W-Green2008.pdf

Damljanovi, Z. (2010). Mobility management strategies in heterogeneous cognitive radio networks. *Journal of Network and Systems Management*, *18*, 4–22. doi:10.1007/s10922-009-9146-0

Das, D., & Ramaswamy, V. (2009). Co-channel femtocell-macrocell deployments-access control. In *Proceedings of the IEEE 70th Vehicular Technology Conference (VTC-2009 Fall)*. Chelmsford, MA: IEEE Press.

Das, S., Chandhar, P., Mitra, S., & Ghosh, P. (2011). *Issues in femtocell deployment in broadband OFDMA networks: 3GPP–LTE a case study.* Paper presented at IEEE Vehicular Technology Conference (VTC-2011 Fall). San Francisco, CA.

De Couto, D., Aguayo, D., Bicket, J., & Morris, R. (2005). A high-throughput path metric for multi-hop wireless routing. *Wireless Networking*, *11*(4), 419–434. doi:10.1007/s11276-005-1766-z

De Domenico, A., Strinati, E. C., & Di Benedetto, M. (2012). A survey on MAC strategies for cognitive radio networks. *IEEE Communications Surveys & Tutorials*, *14*(1), 21–44. doi:10.1109/SURV.2011.111510.00108

De Domenico, E. C.-G. (2010). A survey on MAC strategies for cognitive radio networks. *IEEE Communications Surveys & Tutorials*, *14*(1), 1–24.

De la Roche, G., Valcarce, A., Lopez-Perez, D., & Zhang, J. (2010). Access control mechanisms for femtocells. *IEEE Communications Magazine*, *48*(1), 33–38. doi:10.1109/MCOM.2010.5394027

Deng, Z., Shen, L., Bao, N., Su, B., Lin, J., & Wang, D. (2011). Autocorrelation based detection of DSSS signal for cognitive radio system. In *Proceedings of the International Conference on Wireless Communications and Signal Processing (WCSP), 2011*. WCSP.

Digham, F., et al. (2007). On the energy detection of unknown signals over fading channels. *IEEE Transactions on Communications*, *55*, 21–24. doi:10.1109/TCOMM.2006.887483doi:10.1109/TCOMM.2006.887483

Donvito, M., & Kassam, S. (1979). Characterization of the random array peak sidelobe. *IEEE Transactions on Antennas and Propagation*, *27*(3), 379–385. doi:10.1109/TAP.1979.1142097

Dubey, S. S. (2012). Security for cognitive radio networks. In Lin, M.-L. K.-C. (Ed.), *Cognitive Radio and Interference Management: Technology and Strategy*. Hershey, PA: IGI Global. doi:10.4018/978-1-4666-2005-6.ch013

Durrani, S., & Bialkowski, M. (2004). Effect of mutual coupling on the interference rejection capabilities of linear and circular arrays in CDMA systems. *IEEE Transactions on Antennas and Propagation*, *52*(4), 1130–1134. doi:10.1109/TAP.2004.825640

Elayoubi, S. E., Altman, E., Haddad, M., & Altman, Z. (2010). A hybrid decision approach for the association problem in heterogeneous networks. In *Proceedings of IEEE Infocom*. San Diego, CA: IEEE Press.

Elminaam, D. S. A., Kader, H. M. A., & Hadhoud, M. M. (2010). Evaluating the performance of symmetric encryption algorithm. *International Journal of Network Security*, *10*(3), 216–222.

Enserink, S., & Cochran, D. (1994). A cyclostationary feature detector. In *Proceedings of the 28th Asilomar Conerence on Signals, Systems, and Computers*, (vol. 2., pp. 806–810). Pacific Grove, CA: Asilomar.

Ericsson Press Release. (2008). *Website*. Retrieved from http://www.ericsson.com/ericsson/press/factsfigures/doc/energy efficiency.pdf

Etkin, R., Parekh, A., & Tse, D. (2005). Spectrum sharing for unlicensed bands. In *Proceedings of IEEE DySPAN*, (pp. 251–258). IEEE Press.

European Commission. (2012). *Frame work 7, ICT*. Retrieved May 5, 2012, from http://cordis.europa.eu/fp7/ict/home en.html

FCC. (2002). *Spectrum policy task force: Report of the interference protection working group*. Washington, DC: FCC.

FCC. (2003). *Cognitive radio workshop*. Retrieved from http://www.fcc.gov/searchtools.html

FCC. (2003). *Notice of proposed rulemaking and order, ET Docket No 03-222*. Washington, DC: The Federal Communications Commission (FCC).

FCC. (2005). *Rules for wireless broadband services in the 3650-3700 MHz band (FCC 05-56)*. Washington, DC: Federal Communications Commission (FCC).

FCC. (2008). *Unlicensed operation in the TV broadcast bands*. Washington, DC: FCC.

Federal Communications Commission. (2002). *Spectrum policy task force report*. Washington, DC: Federal Communications Commission (FCC).

Federal Communications Commission. (2003). *Notice of proposed rule making and order*. Washington, DC: Federal Communications Commission (FCC).

Federal Communications Commission. (2003). *Establishment of an interference temperature metric. FCC 03-289*. Washington, DC: FCC.

Federal Communications Commission. (2003). *ET docket no 03-222 notice of proposed rule making and order*. Washington, DC: FCC.

Felegyhazi, M., & Hubaux, J. (2006). *Game theory in wireless networks: A tutorial. EPFL Technical Report, LCA-REPORT-2006-002*. New York, NY: EPFL.

Femto Forum. (2012). *An overview of the femtocell concept*. Retrieved May 5, 2012, from http://www.femtoforum.org

Fette, B. (2006). *Cognitive radio technology*. New York, NY: Newnes.

Fettweis, G. P., & Zimmermann, E. (2008). ICT energy consumption – Trends and challenges. In *Proceedings of the 11th International Symposium on Wireless Personal Multimedia Communications*. Lapland, Finland: IEEE.

Foh, E. W. (2009). Analysis of cognitive radio spectrum access with finite user population. *IEEE Communications Letters, 13*(5), 294–296. doi:10.1109/LCOMM.2009.082113

Fortuna, C., & Mohorcic, M. (2009). Trends in the development of communication networks: Cognitive networks. *Computer Networks, 53*(9), 1355–1376. doi:10.1016/j.comnet.2009.01.002

Foukalas, F., Gazis, V., & Alonistioti, N. (2008). Cross-layer design proposals for wireless mobile networks: A survey and taxonomy. *IEEE Communications Surveys & Tutorials, 10*(1), 70–85. doi:10.1109/COMST.2008.4483671

FP7-ICT-248891. (2010). *Femtocell-based network enhancement by interference management and coordination of information for seamless connectivity*. STP FREEDOM.

Fudenberg, D., & Tirole, J. (1991). *Game theory*. Cambridge, MA: MIT Press.

Gandetto, A. F. (2005). A distributed approach to mode identification and spectrum monitoring for cognitive radios. In *Proceedings of the SDR Forum Technical Conference*. Orange County, CA: SDR.

Gandetto, A. F. (2007). Spectrum sensing: A distributed approach for cognitive terminals. *IEEE Journal on Selected Areas in Communications, 25*(3), 546–557. doi:10.1109/JSAC.2007.070405

Ganesan, G., & Li, Y. (2007). Cooperative spectrum sensing in cognitive radio, part I: Two user networks. *IEEE Transactions on Wireless Communications, 6*(6), 2204–2213. doi:10.1109/TWC.2007.05775

Ganesan, R., & Selvakumar, A. A. L. (2011). A new balanced encryption algorithm with elevated security base on key update. *European Journal of Scientific Research, 60*(2), 177–194.

Gao, L., & Farrell, R. (2008). *Using SDR to embed WIMAX channels within the TETRA framework*. Retrieved May 15, 2012, from http://eprints.nuim.ie/1413/1/GaoL_final.pdf

Gao, P., Wang, J., & Shaoqian, L. (2010). Non-contiguous CI/OFDM: A new data transmission scheme for cognitive radio context. In *Proceedings of the IEEE International Conference on Wireless Information Technology and Systems (ICWITS)*, (pp. 1-4). IEEE Press.

Gastpar, M. (2004). On capacity under received-signal constraints. In *Proceedings of the 42nd Annual Allerton Conference on Communication, Control and Computing*. Allerton.

Gel'fand, I., Gindikin, S., & Graev, M. (2003). *Selected topics in integral geometry*. New York, NY: American Mathematical Society.

Ghasemi, A., & Sousa, E. (2006). Capacity of fading channels under spectrum-sharing constraints. In *Proceedings of the IEEE International Conference on Communications*, (vol. 10, pp. 4373-4378). IEEE Press.

Ghasemi, A., & Sousa, E. (2008). Spectrum sensing in cognitive radio networks: Requirements, challenges and design trade-offs. *IEEE Communications Magazine, 46*(4), 32–39. doi:10.1109/MCOM.2008.4481338doi:10.1109/MCOM.2008.4481338

Gilhousen, K. (1991). On the capacity of a cellular CDMA system. *IEEE Transactions on Vehicular Technology, 40*(2), 303–312. doi:10.1109/25.289411

Godfrey, P., Shipley, R., & Gryz, J. (2006). Algorithms and analyses for maximal vector computation. *The VLDB Journal, 16*, 5–28. doi:10.1007/s00778-006-0029-7

Gonzalez, I., Lopez-Buedo, S., & Gomez-Arribas, F. J. (2008). Implementation of secure applications in self-reconfigurable systems. *Microprocessors and Microsystems, 32*, 23–32. doi:10.1016/j.micpro.2007.04.001

Govil, J., & Govil, J. (2007). *4G mobile communication systems: Turns, trends and transition.* Paper presented at IEEE International Conference on Convergence Information Technology (ICCIT). New York, NY.

Green Communications. (2009). *The first international workshop on green communications (GreenComm)*. Retrieved from http://www.green-communications.net/icc09/home.html

Guandr, G., Bayhan, S., & Alagoandz, F. (2010). Cognitive femtocell networks: An overlay architecture for localized dynamic spectrum access. *IEEE Magazine of Wireless Communications, 17*(4), 62–70. doi:10.1109/MWC.2010.5547923

Guo, D., & Wang, X. (2004). Dynamic sensor collaboration via sequential Monte Carlo. *IEEE Journal on Selected Areas in Communications*, *22*(6), 1037–1047. doi:10.1109/JSAC.2004.830897

Gupta, P., & Kumar, P. R. (2003). Towards an information theory of large net-works: An achievable rate region. *IEEE Transactions on Information Theory*, *49*(8), 1877–1894. doi:10.1109/TIT.2003.814480

Haddad, M., Hayar, A., & Debbah, M. (2008). Spectral efficiency for spectrum pooling systems. *IET*, *2*(6), 733–741. doi:10.1049/iet-com:20070469

Hai Ngoc, P., et al. (2010). Energy minimization approach for optimal cooperative spectrum sensing in sensor-aided cognitive radio networks. In *Proceedings of the 5th International Conference on Wireless Internet (WICON 2010)*. Piscataway, NJ: IEEE.

Haines, R. J. (2010). Cognitive pilot channels for femtocell deployment. In *Proceedings of the 7th International Symposium on Wireless Communication Systems (ISWCS 2010)*. York, UK: IEEE Press.

Hamdaoui, B., & Shin, K. (2008). OS-MAC: An efficient MAC protocol for spectrum-agile wireless networks. *IEEE Transactions on Mobile Computing*, *7*(8), 915–930. doi:10.1109/TMC.2007.70758

Hamid, M., & Mohammed, A. (2012). MAC layer spectrum sensing in cognitive radio networks. In *Self-Organization and Green Applications in Cognitive Radio Networks*. Hershey, PA: IGI Global. doi:10.1109/CMC.2010.342doi:10.1109/CMC.2010.342

Han, Z., Ji, Z., & Liu, K. (2005). Fair multiuser channel allocation for OFDMA networks using Nash bargaining solutions and coalitions. *IEEE Transactions on Communications*, *53*(8), 1366–1376. doi:10.1109/TCOMM.2005.852826

Hardin, G. (1968). The tragedy of the commons. *Science*, *162*(3859), 1243–1248. doi:10.1126/science.162.3859.1243

Harley, P. (1989). Short distance attenuation measurements at 900 MHz and 1.8 GHz using low antenna heights for microcells. *IEEE Journal on Selected Areas in Communications*, *7*, 5–11. doi:10.1109/49.16838

Haykin, S. (2005). Cognitive radio: Brain-empowered wireless communications. *IEEE Journal on Selected Areas in Communications*, *23*(2), 201–220. doi:10.1109/JSAC.2004.839380doi:10.1109/JSAC.2004.839380

Haykin, S., Thomson, D. J., & Reed, J. H. (2009). Spectrum sensing for cognitive radio. *Proceedings of the IEEE*, *97*(5), 849–877. doi:10.1109/JPROC.2009.2015711

Hirani, S. (2003). *Energy consumption of encryption schemes in wireless devices*. (Unpublished Master of Science Dissertation). University of Pittsburgh. Pittsburgh, PA.

Ho, L. T. W., & Claussen, H. (2007). Effects of user-deployed, co-channel femtocells on the call drop probability in a residential scenario. In *Proceedings of the IEEE 18th International Symposium on Personal, Indoor and Mobile Radio Communications (PIMRC 2007)*. Athens, Greece: IEEE Press.

Hoang, A., & Liang, Y. (2007). Adaptive scheduling of spectrum sensing periods in cognitive radio networks. In *Proceedings of IEEE GLOBECOM*, (pp. 3128–3132). IEEE Press.

Hoang, A., Liang, Y., & Zeng, Y. (2010). Adaptive joint scheduling of spectrum sensing and data transmission in cognitive radio networks. *IEEE Transactions on Communications*, *58*(1), 235–246. doi:10.1109/TCOMM.2010.01.070270

Holland, J. H. (1975). *Adaptation in natural and artificial systems*. Englewood Cliffs, NJ: Prentice Hall.

Horrich, S., Elayoubi, S.-E., & Jemaa, S. B. (2008). On the impact of mobility and joint rrm policies on a cooperative wimax/hsdpa network. In *Proceedings of IEEE WCNC*. Las Vegas, NV: IEEE Press.

Hossain, E., Niyato, D., & Han, Z. (2009). *Dynamic spectrum access and management in cognitive radio networks*. Cambridge, UK: Cambridge University Press. doi:10.1017/CBO9780511609909

Hossain, L. L. (2008). OSA-MAC: A MAC protocol for opportunistic spectrum access in cognitive radio networks. *Proceedings of IEEE WCNC*, *2008*, 1426–1430. IEEE Press.

How, K.-C., & Ma, M. (2011). Routing and qos provisioning in cognitive radio networks. *Computer Networks*, *55*(1), 330–342. doi:10.1016/j.comnet.2010.09.008

Hoydis, J., & Kobayashi, M. (2011). Green small-cell networks. *IEEE Vehicular Technology Magazine*, *6*(1), 37–43. doi:10.1109/MVT.2010.939904

HP. (2004). *MAC level (link layer)*. Retrieved February 28, 2012, from http://www.hpl.hp.com/personal/Jean_Tourrilhes/Linux/Linux.Wireless.mac.html

Huang, S., Ding, Z., & Liu, X. (2007). Non-intrusive cognitive radio networks based on smart antenna technology. In *Proceedings of the IEEE Conference on Global Communications, GLOBECOM 2007*, (pp. 4862–4867). Washington, DC: IEEE Press.

Huang, J., Berry, R., & Honig, M. (2006). Auction-based spectrum sharing. *Mobile Networking and Applications*, *11*(3), 405–418. doi:10.1007/s11036-006-5192-y

Huang, R. B. (2005). Spectrum sharing with distributed interference compensation. *Proceedings of IEEE DySPAN, 2005*, 88–93. IEEE Press.

Hu, D., Mao, S., Hou, Y., & Reed, J. (2010). Scalable video multicast in cognitive radio networks. *IEEE Journal on Selected Areas in Communications*, *28*(3), 334–344. doi:10.1109/JSAC.2010.100414

Hu, Y. Y.-D., & Yang, Z. (2012). Cognitive medium access control protocols for secondary users sharing a common channel with time division multiple access primary users. *Wireless Communications and Mobile Computing*, *12*(4), 20–36.

Hyoil, K., & Shin, K. (2008). Efficient discovery of spectrum opportunities with MAC-layer sensing in cognitive radio networks. *IEEE Transactions on Mobile Computing*, *7*, 533–545. doi:10.1109/TMC.2007.70751doi:10.1109/TMC.2007.70751

Idrus, S. Z. S., & Aljunid, S. A. (2008). Performance analysis of encryption algorithms text length size on web browsers. *International Journal of Computer Science and Network Thesis*, *8*(1), 20-25.

IEEE 802.11. (2012). *Wikipedia*. Retrieved March 1, 2012, from http://en.wikipedia.org/wiki/IEEE_802.11

IEEE Standard Coordinating Committee 41. (2009). *Home page*. Retrieved from http://www.scc41.org

IEEE Standards Coordinating Committee 41. (2012). *Dynamic spectrum access networks*. Retrieved from http://grouper.ieee.org/groups/scc41

IEEE Technical Committee on Cognitive Networks. (2009). *Home page*. Retrieved from http://www.eecs.ucf.edu/tccn

IEEE. (2011). *IEEE 802.22 working group on wireless regional area networks*. Retrieved March 10, 2012, from http://www.ieee802.org/22/

Intanagonwiwat, C., Govindan, R., Estrin, D., Heidemann, J., & Silva, F. (2002). Directed diffusion for wireless sensor networking. *IEEE/ACM Transactions on Networking*, *11*(1), 2–16. doi:10.1109/TNET.2002.808417

Iyer, G., & Lim, Y. C. (2011). Efficient multi-channel MAC protocol and channel allocation schemes for TDMA based cognitive radio networks. In *Proceedings of the International Conference on Communications and Signal Processing (ICCSP), 2011*, (pp. 394-398). Kerla, India: IEEE Press.

Jenkins, C., Mamidi, S., Schulte, M., & Glossner, J. (2007). Instruction set extensions for the advanced encryption standard on a multithreaded software defined radio platform. *International Journal on Embedded Systems*, *2*(3-4), 203–214.

Jha, R. K., & Dalal, U. D. (2010). A journey on WiMAX and its security issues. *International Journal of Computer and Information Technologies*, *1*(4), 256–263.

Jha, S., Rashid, M., Bhargava, V., & Despins, C. (2011). Medium access control in distributed cognitive radio networks. *IEEE Wireless Communications*, *18*(4), 41–51. doi:10.1109/MWC.2011.5999763

Jha, U. P. (2011). Design of omc-mac: An opportunistic multi-channel mac with qos provisioning for distributed cognitive radio networks. *IEEE Transactions on Wireless Communications*, *10*(10), 3414–3425. doi:10.1109/TWC.2011.072511.102196

Jia, J., Zhang, Q., & Shen, X. (2008). HC-MAC: A hardware-constrained cognitive MAC for efficient spectrum management. *IEEE Journal on Selected Areas in Communications*, *26*(1), 106–117. doi:10.1109/JSAC.2008.080110

Jiang, L., Parekh, S., & Walrand, J. C. (2008). Base station association game in multi-cell wireless networks. In *Proceedings of IEEE WCNC*. IEEE Press.

Jiang, T., Grace, D., & Mitchell, P. D. (2009). Improvement of pre-partitioning on reinforcement learning based spectrum sharing. In *Proceedings of IET International Communication Conference on Wireless Mobile and Computing (CCWMC 2009),* (pp. 299-302). IET.

Ji, Z., & Liu, K. (2007). Cognitive radios for dynamic spectrum access - Dynamic spectrum sharing: A game theoretical overview. *IEEE Communications Magazine*, *45*(5), 88–94. doi:10.1109/MCOM.2007.358854

Ji, Z., & Liu, K. (2008). Multi-stage pricing game for collusion-resistant dynamic spectrum allocation. *IEEE Journal on Selected Areas in Communications*, *26*(1), 182–191. doi:10.1109/JSAC.2008.080116

Joseph, C. M. T. (2011). Improving security in the IEEE 802.16 standard. In *Proceedings of Eighth International Conference on Information Technology: New Generations,* (pp. 408-412). IEEE.

Kadhim, M. A., & Ismail, W. (2010). Implementation of WIMAX IEEE 802.16e baseband transceiver on multi-core software defined radio platform. *International Journal of Computer Theory and Engineering*, *2*(5), 1793–8201.

Kadhim, M. A., & Ismail, W. (2010). Implementation of WIMAX STBC-OFDM (IEEE802.16.d) baseband transceiver on a multi-core software-defined radio platform. *Australian Journal of Basic and Applied Sciences*, *4*(7), 2125–2133.

Kaimaletu, S., Krishnan, R., Kalyani, S., Akhtar, N., & Ramamurthi, B. (2011). Cognitive interference management in heterogeneous femto-macro cell networks. In *Proceedings of the IEEE International Conference on Communications (ICC 2011)*. Kyoto, Japan: IEEE Press.

Kamal, K., & Mohammed, A. (2012). Game theory for cognitive radio. In *Cognitive Radio and Interference Management: Technology and Strategy*. Hershey, PA: IGI Global. doi:10.4018/978-1-4666-2005-6.ch007

Kamruzzaman, S. (2010). An energy efficient multichannel MAC protocol for cognitive radio ad hoc networks. *International Journal of Communication Networks and Information Security*, *2*(2), 112–119.

Kandeepan, S., Sierra, A., Campos, J., & Chlamtac, I. (2010). Periodic sensing in cognitive radios for detecting UMTS/HSDPA based on experimental spectral occupancy statistics. In *Proceedings of the Wireless Communications and Networking Conference (WCNC),* (pp. 1-6). Sydney, Australia: IEEE Press.

Kang, X., Liang, Y., & Nallanathan, A. (2008). Optimal power allocation for fading channels in cognitive radio networks under transmit and interference power constraints. *Proceedings of the IEEE, ICC*, 3568–3572. IEEE Press.

Kapoor, S., Rao, S., & Singh, G. (2011). Opportunistic spectrum sensing by employing matched filter in cognitive radio network. In *Proceedings of the International Conference on Communication Systems and Network Technologies (CSNT)*. CSNT.

Karapantazis, S., & Pavlidou, F. (2005). Broadband communications via high-altitude platforms: A survey. *IEEE Communications Surveys*, *7*(1), 2–31. doi:10.1109/COMST.2005.1423332

Kauffmann, B., Baccelli, F., Chaintreau, F., Mhatre, V., Papagiannaki, K., & Diot, C. (2007). Measurement-based self organization of interfering 802.11 wireless access networks. In *Proceedings of IEEE INFOCOM*. IEEE Press.

Kay, S. M. (1998). *Fundamentals of statistical signal processing: Detection theory*. Englewood Cliffs, NJ: Prentice-Hall.

Kessler, G. C. (2010). *An overview of cryptography*. Retrieved March 20, 2012, from http://people.eecs.ku.edu/~saiedian/teaching/Fa10/710/Readings/An-Overview-Cryptography.pdf

Khan, A. A., & Zaman, N. (2009). Comparative analysis of broadband wireless access from Wi-Fi to WiMAX. In *Proceedings of International Bhurban Conference on Applied Sciences & Technology,* (pp. 8-14). Islamabad, Pakistan: IEEE.

Khan, S., Peng, Y., Steinbach, E., Sgroi, M., & Kellerer, W. (2006). Application-driven cross-layer optimization for video streaming over wireless networks. *IEEE Communications Magazine*, *44*(1), 122–130. doi:10.1109/MCOM.2006.1580942

Khozeimeh, F., & Haykin, S. (2009). Dynamic spectrum management for cognitive radio: An overview. *Wireless Communications and Mobile Computing*, *9*(11), 1147–1159. doi:10.1002/wcm.732

Kim, H., & Shin, K. G. (2008). Fast discovery of spectrum opportunities in cognitive radio networks. In *Proceedings of the 3rd IEEE Symposia on New Frontiers in Dynamic Spectrum Access Networks*, (pp. 1-12). Chicago, IL: IEEE Press.

Kim, K., Xin, Y., & Rangarajan, S. (2010). Energy detection based spectrum sensing for cognitive radio: An experimental study. In *Proceedings of the Global Telecommunications Conference (GLOBECOM*), 1-5. IEEE.

Kim, H., & Shin, K. (2008). Efficient discovery of spectrum opportunities with MAC-layer sensing in cognitive radio networks. *IEEE Transactions on Mobile Computing*, *7*(5), 33–45.

Kim, R., Kwak, J., & Etemad, K. (2009). WiMAX femtocell: Requirements, challenges, and solutions. *IEEE Communications Magazine*, *47*(9), 84–91. doi:10.1109/MCOM.2009.5277460

Knisely, D. (2010). *Femtocell standardization.* White Paper. Chelmsford, MA: Airvana.

Knisely, D., Yoshizawa, T., & Favichia, F. (2009). Standardization of Femtocells in 3GPP2. *IEEE Communications Magazine*, *47*(9), 76–82. doi:10.1109/MCOM.2009.5277459

Kofahi, N. A., Al-Somani, T., & Al-Zamil, K. (2004). Performance evaluation of three encryption/decryption algorithms. *Proceedings of the IEEE*, *2*, 790–793.

Kondareddy, Y., & Agrawal, P. (2008). Synchronized MAC protocol for multi-hop cognitive radio networks. In *Proceedings of the International Conference on Communications, 2008,* (pp. 3198-3202). Beijing, China: IEEE Press.

Kondo, S., & Milstein, B. (1996). Performance of multicarrier DS CDMA systems. *IEEE Transactions on Communications*, *44*(2), 238–246. doi:10.1109/26.486616

Koutsopoulos, I., & Tassiulas, L. (2008). The impact of space division multiplexing on resource allocation: A unified treatment of TDMA, OFDMA and CDMA. *IEEE Transactions on Communications*, *56*(2), 1–10. doi:10.1109/TCOMM.2008.050102

Koza, J. R. (1990). *Genetic programming: A paradigm for genetically breeding populations of computer programs to solve problems.* Technical Report STAN-CS-90-1314. Palo Alto, CA: Stanford University.

Krenik, W., & Batra, A. (2005). Cognitive radio techniques for wide area networks. In *Proceedings of Design Automation Conference 42nd,* (pp. 409-412). IEEE.

Kumar, D., Altman, E., & Kelif, J.-M. (2007). Globally optimal user-network association in an 802.11 wlan and 3G umts hybrid cell. In *Proceedings of 20th International Teletraffic Congress (ITC 20).* Ottawa, Canada: ITC.

Kumar, S., & Sengupta, J. (2010). AODV and OLSR routing protocols for wireless ad-hoc and mesh networks. In *Proceedings of the International Conference on Computer & Communication Technology, ICCCT,* (pp. 402-407). ICCCT.

Kumar, A. (2008). *Mobile broadcasting with WiMAX.* New York, NY: ScienceDirect.

Kung, H. T., Luccio, F., & Preparata, F. P. (1975). On finding the maxima of a set of vectors. *International Journal of the ACM*, *22*(4), 469–476. doi:10.1145/321906.321910

Kwan, R., & Leung, C. (2010). A survey of scheduling and interference mitigation in LTE. *Journal of Electrical and Computer Engineering*, *2010*, 1–10. doi:10.1155/2010/273486

Kyungtae, K., et al. (2010). Energy detection based spectrum sensing for cognitive radio: An experimental study. In *Proceedings of the 2010 IEEE Global Communications Conference (GLOBECOM 2010)*. Piscataway, NJ: IEEE Press.

Landstrom, A., Furuskar, A., Johansson, K., Falconetti, L., & Kronestedt, F. (2011). Heterogeneous networks increasing cellular capacity. *Journal of the Ericson Review*, *89*, 4–9.

Le, L., & Hossain, E. (2008). A MAC protocol for opportunistic spectrum access in cognitive radio networks. In *Proceedings of the Wireless Communications and Networking Conference, 2008,* (pp. 1426-1430). IEEE Press.

Lee, J., Lee, E., Park, S., Park, H., & Kim, S.-H. (2010). Destination-initiated geographic multicasting protocol in wireless ad hoc sensor networks. In *Proceedings of the IEEE 71st Vehicular Technology Conference (VTC 2010-Spring),* (pp. 1 - 5). IEEE Press.

Lee, W.-Y. Kausbik, Cbowdbury, & Mebmet. (2008). Spectrum sensing algorithms for cognitive radio networks. In Y. Xiao & F. Hu (Eds.), *Cognitive Radio Networks*. New York, NY: Auerbach Publications.

Lee, B.-H., & Wong, C.-M. (2011). Coordinated non-sensing MAC protocol in dynamic spectrum access networks. *Wireless Personal Communications*, *58*, 867–887. doi:10.1007/s11277-010-9999-2

Lee, W., & Akyildiz, I. F. (2008). Optimal spectrum sensing framework for cognitive radio networks. *IEEE Transactions on Wireless Communications*, *7*(10), 3845–3857. doi:10.1109/T-WC.2008.070391

Lee, W.-Y., & Akyildiz, I. F. (2008). Optimal spectrum sensing framework for cognitive radio networks. *IEEE Transactions on Wireless Communications*, *7*(10), 2845–3857.

Le, L., & Hossain, E. (2008). Resource allocation for spectrum underlay in cognitive radio networks. *IEEE Transactions on Wireless Communications*, *7*(12), 5306–5315. doi:10.1109/T-WC.2008.070890

Li, D., Dai, X., & Zhang, H. (2008). Cross-layer scheduling and power control in cognitive radio networks. In *Proceedings of the 4th International Conference on Wireless Communications, Networking and Mobile Computing, 2008,* (pp. 1 - 3). IEEE Press.

Li, L., Khan, F. A., Pesavento, M., & Ratnarajah, T. (2011). Power allocation and beamforming in overlay cognitive radio systems. In *Proceedings of the IEEE 73rd Vehicular Technology Conference, VTC-Spring 2011*. Budapest, Hungary: IEEE Press.

Li, X., & Cadeau, W. (2011). Anti-jamming performance of cognitive radio networks. In *Proceedings of the IEEE 45th Annual Conference on Information Sciences and Systems (CISS)*. Baltimore, MD: IEEE Press.

Li, Y., Feng, Z., Zhang, Q., Tan, L., & Tian, F. (2010). Cognitive optimization scheme of coverage for femtocell using multi-element antenna. In *Proceedings of the IEEE 72nd Vehicular Technology Conference, VTC-Fall 2010*, (pp. 1–5). Ottawa, Canada: IEEE Press.

Li, Z. H. (2009). Queuing analysis of dynamic spectrum access subject to interruptions from primary users. In *Proceedings of the 5th International ICST Conference on Cognitive Radio Oriented Wireless Networks and Communications (CROWNCOM),* (pp. 1-5). IEEE Press.

Liang, E. P.-C. (2007). Optimization for cooperative sensing in cognitive radio networks. In *Proceedings of the Wireless Communications and Networking Conference,* (pp. 27-32). Hong Kong, China: IEEE Press.

Liang, Y. C., Chen, K. C., Li, G. Y., & Mahonen, P. (2011). Cognitive radio networking and communications: An overview. *IEEE Transactions on Vehicular Technology*, *60*(7), 3386–3407. doi:10.1109/TVT.2011.2158673

Liang, Y.-C., & Zeng, Y. (2008). Sensing-throughput tradeoff for cognitive radio networks. *IEEE Transactions on Wireless Communications*, *7*(4), 1326–1337. doi:10.1109/TWC.2008.060869

Liang, Y., Chen, K., Li, G., & Mahonen, P. (2011). Cognitive radio networking and communications: An overview. *IEEE Transactions on Vehicular Technology*, *60*(7), 3386–3407. doi:10.1109/TVT.2011.2158673

Lin, Y., & Chen, K. C. (2010). Distributed spectrum sharing in cognitive radio networks - Game theoretical view. In *Proceedings of the 7th IEEE Consumer Communications and Networking Conference (CCNC 2010)*. Las Vegas, NV: IEEE Press.

Lin, P., Zhang, J., Chen, Y., & Zhang, Q. (2011). Macro-femto heterogeneous network deployment and management: From business models to technical solutions. *IEEE Magazine of Wireless Communications*, *18*(3), 64–70. doi:10.1109/MWC.2011.5876502

Liu, K. (2011). Cognitive radio and game theory. *IEEE Spectrum*. Retrieved from http://spectrum.ieee.org/telecom/wireless/cognitive-radio-and-game-theory/0

Liu, X., & Zhai, X. (2008). Feature detection based on multiple cyclic frequencies in cognitive radios. In *Proceedings of the 2008 China-Japan Joint Microwave Conference (CJMW 2008)*, (pp. 290-393). Piscataway, NJ: CJMW.

Liu, Q., Zhou, S., & Giannakis, G. (2004). Cross-layer combining of adaptive modulation and coding with truncated arq over wireless links. *IEEE Transactions on Wireless Communications*, *3*(5), 1746–1755. doi:10.1109/TWC.2004.833474

Lopez-Perez, D., Guvenc, I., Roche, G., Kountouris, M., Quek, T., & Zhang, J. (2011). Enhanced intercell interference coordination challenges in heterogeneous networks. *IEEE Wireless Communications*, *18*(3), 22–30. doi:10.1109/MWC.2011.5876497

Lopez-Perez, D., Valcarce, A., de la Roche, G., & Zhang, J. (2009). OFDMA femtocells: A roadmap on interference avoidance. *IEEE Communications Magazine*, *47*(9), 41–48. doi:10.1109/MCOM.2009.5277454

Lo, Y. (1964). A mathematical theory of antenna arrays with randomly spaced elements. *IEEE Transactions on Antennas and Propagation*, *12*(3), 257–268. doi:10.1109/TAP.1964.1138220

Luo, H., Ci, S., Wu, D., & Tang, H. (2010). Cross-layer design for real-time video transmission in cognitive wireless networks. In *Proceedings of INFOCOM IEEE Conference on Computer Communications Workshops*. IEEE Press.

Luo, C., Yu, F., Ji, H., & Leung, V. (2010). Cross-layer design for tcp performance improvement in cognitive radio networks. *IEEE Transactions on Vehicular Technology*, *59*(5), 2485–2495. doi:10.1109/TVT.2010.2041802

Luo, H., Argyriou, A., Wu, D., & Ci, S. (2009). Joint source coding and network supported distributed error control for video streaming in wireless multihop networks. *IEEE Transactions on Multimedia*, *11*(7), 1362–1372. doi:10.1109/TMM.2009.2030639

Ma, L., Han, X., & Shen, C. (2005). Dynamic open spectrum sharing MAC Protocol for wireless ad hoc network. In *Proceedings of the IEEE DySPAN*, (pp. 203–213). IEEE Press.

Ma, L., Han, X., & Shen, C. (2005). Dynamic open spectrum sharing MAC protocol for wireless ad hoc networks. In *Proceedings of IEEE DySPAN*, (pp. 203–213). IEEE Press.

Ma, L., Shen, C., & Ryu, B. (2007). Single-radio adaptive channel algorithm for spectrum agile wireless ad hoc networks. In *Proceedings of IEEE DySPAN*, (pp. 547–558). IEEE Press.

Ma, R.-T., Hsu, Y.-P., & Feng, K.-T. (2009). A POMDP-based spectrum handoff protocol for partially observable cognitive radio networks. In *Proceedings of the Wireless Communications and Networking Conference, 2009*, (pp. 1-6). IEEE Press.

Ma, C.-C. S. (2007). Single-radio adaptive channel algorithm for spectrum agile wireless ad hoc networks. *Proceedings of the DySPAN*, *2007*, 547–558. IEEE Press.

Mahmoud, H., & G¨uven, I. (2010). *A comparative study of different deployment modes for femtocell networks*. Paper presented at IEEE 20th Symposium on Personal, Indoor and Mobile Radio Communications. Istanbul, Turkey.

Ma, J., Li, G., & Juang, B. H. (2009). Signal processing in cognitive radio. *Proceedings of the IEEE*, *97*(5), 805–823. doi:10.1109/JPROC.2009.2015707

Maldonado, D., Le, B., Hugine, A., Rondeau, T. W., & Bostian, C. W. (2005). Cognitive radio applications to dynamic spectrum allocation: A discussion and an illustrative example. In *Proceedings of the First IEEE International Symposium on New Frontiers in Dynamic Spectrum Access Networks*, (pp. 597–600). Baltimore, MD: IEEE Press.

Masrub, A., Al-Raweshidy, H., & Abbod, M. (2011). Cognitive radio based MAC protocols for wireless ad hoc networks. In *Proceedings of the 4th International Conference on Developments in eSystems Engineering*, (pp. 465-469). Dubai, UAE: IEEE.

Ma, X. H.-C. (2005). Dynamic open spectrum sharing MAC protocol for wireless ad hoc networks. *Proceedings of the IEEE DySPAN, 2005*, 203–213. IEEE Press.

Mehanna, O., Sultan, A., & El Gamal, H. (2009). Blind cognitive MAC protocols. In *Proceedings of the IEEE International Conference on Communications,* (pp. 1-5). Dresden, Germany: IEEE Press.

Meko, S. F., & Chaporkar, P. (2009). Channel partitioning and relay placement in multi-hop cellular networks. In *Proceedings of the 6th International Symposium on Wireless Communication Systems, ISWCS 2009,* (pp. 66-70). Piscataway, NJ: IEEE Press.

Menon, R., Buehrer, R., & Reed, J. (2005). Outage probability based comparison of underlay and overlay spectrum sharing techniques. In *Proceedings of the IEEE International Symposium on New Frontiers in Dynamic Spectrum Access Networks (DySPAN),* (pp. 101–109). IEEE Press.

Mihaljevic, M. J., & Kohno, R. (2002). On a framework for employment of cryptographic components in software defined radio. *Proceedings of Wireless Personal Communications, 2*, 835–839. IEEE Press.

Mishra, S., Ten, S., Mahadevappa, R., & Brodersen, R. (2007). Cognitive technology for ultra-wideband/wimax coexistence. *Proceedings of IEEE DySPAN, 2007*. IEEE Press, 179–186.

Mishra, A., & Glore, N. (2008). Privacy & security in WiMAX networks. In *WiMAX Standards and Security* (pp. 205–228). Boca Raton, FL: CRC Press.

Mitola, J. (1992). *The software radio*. Paper presented at the IEEE National Telesystems Conference. New York, NY.

Mitola, J. (1995). The software radio architecture. *IEEE Communications Magazine*, *33*(5), 26–38. doi:10.1109/35.393001doi:10.1109/35.393001

Mitola, J. (2000). *Cognitive radio: An integrated agent architecture for software defined radio* (unpublished doctoral dissertation). Royal Institute of Technology (KTH). Stockholm, Sweden.

Mitola, J., & Maguire, G. (1999). Cognitive radios: Making software radios more personal. *IEEE Personal Communication, 6*(4), 13–18.

Mitola, J. (1999). Cognitive radio: Making software radios more personal. *IEEE Personal Communications*, *6*(4), 13–18. doi:10.1109/98.788210

Mitola, J. III. (1999). Software radio architecture: A mathematical perspective. *IEEE Journal on Selected Areas in Communications*, *17*(4), 514–538. doi:10.1109/49.761033

Mitola, J., & Maguire, G. (1999). Cognitive radio: Making software radios more personal. *IEEE Personal Communications*, *6*(4), 13–18. doi:10.1109/98.788210

Mo, R., Quek, T. Q. S., & Heath, R. W. (2011). Robust beamforming and power control for two-tier femtocell networks. In *Proceedings of the IEEE 73rd Vehicular Technology Conference, VTC-Spring 2011*. Budapest, Hungary: IEEE Press.

Mobile Virtual Network Operator. (2012). *Wikipedia*. Retrieved March 12, 2012, from http://en.wikipedia.org/wiki/MVNO

Mohammed, A., Arnon, S., Grace, D., Mondin, M., & Miura, R. (2008). Advanced communications techniques and applications for high-altitude platforms. *EURASIP Journal on Wireless Communications and Networking*. Retrieved from http://jwcn.eurasipjournals.com/content/pdf/1687-1499-2008-934837.pdf

Mohammed, A., Mehmood, A., Paviudo, N., & Mohorcic, M. (2011). The role of high-altitude platforms (HAPs) in the wireless global connectivity. *Proceedings of the IEEE*, *99*(11), 1939–1953. doi:10.1109/JPROC.2011.2159690

Mohammed, A., & Yang, Z. (2010). Next generation broadband services from high altitude platforms. In *Fourth-Generation Wireless Networks: Applications and Innovations* (pp. 249–267). Hershey, PA: IGI Global. doi:10.4018/978-1-61520-674-2.ch012

Moore, J., Keiser, T., Brooks, R., Phoha, S., Friedlander, D., Koch, J., et al. (2003). Tracking targets with self-organizing distributed ground sensors. In *Proceedings of the IEEE Areospace Conference*. IEEE Press.

Murthy, C. B. (2004). *Ad hoc wireless networks: Architectures and protocols*. Upper Saddle River, NJ: Prentice Hall PTR.

Musavian, L., & Aissa, S. (2009). Adaptive modulation in spectrum-sharing systems with delay constraints. In *Proceedings of the Communications, 2009*. IEEE Press.

Musavian, L., & Aissa, S. (2009). Capacity and power allocation for spectrum-sharing communications in fading channels. *IEEE Transactions on Wireless Communications, 8*(1), 148–156. doi:10.1109/T-WC.2009.070265

Nadeem, A., & Javed, M. Y. (2006). A performance comparison of data encryption algorithms. In *Proceedings of IEEE Information and Communication Technologies*, (pp. 84-89). IEEE Press.

Nasipuri, A., Zhuang, J., & Das, S. R. (1999). A multichannel CSMA MAC protocol for multihop wireless networks. In *Proceedings of the IEEE Wireless Communications and Networking Conference,* (vol. 3, pp. 1402-1406). New Orleans, LA: IEEE Press.

Neel, J., et al. (2002). *Game theoretic analysis of a network of cognitive radios*. Midwest Symposium on Circuits and Systems 2002. Columbus, OH.

Ni, W., & Collings, I. (2009). Centralized inter-network spectrum sharing with opportunistic frequency reuse. In *Proceedings of the IEEE Global Telecommunications Conference (GLOBECOM 2009)*. Hawaii, HI: IEEE Press.

Nie, N., & Comaniciu, C. (2005). Adaptive channel allocation spectrum etiquette for cognitive radio networks. In *Proceedings of IEEE DySPAN*, (pp. 269–278). IEEE Press.

Nie, T., Song, C., & Zhi, X. (2010). Performance evaluation of DES and blowfish algorithms. In *Proceedings of Biomedical Engineering and Computer Science* (pp. 1–4). IEEE Press. doi:10.1109/ICBECS.2010.5462398

Niyato, D., & Hossain, E. (2008). Competitive spectrum sharing in cognitive radio networks: A dynamic game approach. *IEEE Transactions on Wireless Communications, 7*(7), 2651–2660. doi:10.1109/TWC.2008.070073

Office of Communications. (2005). *Technology Research Programme*. Washington, DC: Office of Communications.

Oh, D.-C., Lee, H.-C., & Lee, Y.-H. (2010). Cognitive radio based femtocell resource allocation. In *Proceedings of the 2010 International Conference on Information and Communication Technology Convergence (ICTC 2010),* (pp. 274-279). Jeju Island, Korea: ICTC.

OMNeT++. (2012). *Website*. Retrieved from http://www.omnetpp.org

Ozgur, E. (2008). Association games in IEEE 802.11 wireless local area networks. *IEEE Transactions on Wireless Communications, 7*(12), 5136–5143. doi:10.1109/T-WC.2008.071418

Papandreou, N., & Antonakopoulos, T. (2008). Bit and power allocation in constrained multicarrier systems: The single-user case. *EURASIP Journal on Advances in Signal Processing, 1*, 1–14.

Park, J., Park, S., Kim, D., Cho, P., & Cho, K. (2003). Experiments on radio interference between wireless LAN and other radio devices on a 2.4 GHz ISM band. In *Proceedings of the 57th IEEE Semiannual Vehicular Technology Conference,* (vol. 3, pp. 1798-1801). IEEE Press.

Pawełczak, G. J. (2006). Performance measures of dynamic spectrum access networks. In *Proceedings of the Global Telecommunications Conference (Globecom)*. San Francisco, CA: IEEE Press.

Peha, J., & Panichpapiboon, S. (2004). Real-time secondary markets for spectrum. *Telecommunications Policy, 28*(7-8), 603–616. doi:10.1016/j.telpol.2004.05.003

Peng, C., Zheng, H., & Zhao, B. (2006). Utilization and fairness in spectrum assignment for opportunistic spectrum access. *ACM Mobile Networking and Applications, 11*(4), 555–576. doi:10.1007/s11036-006-7322-y

Pennanen, H., Tolli, A., & Latva-aho, M. (2011). Decentralized coordinated downlink beamforming for cognitive radio networks. In *Proceedings of the IEEE 22nd International Symposium on Personal Indoor and Mobile Radio Communications, PIMRC 2011*, (pp. 566–571). Toronto, Canada: IEEE.

Perez, D., Valcarce, A., Ladanyi, A., Roche, G., & Zhang, J. (2010). Intracell handover for interference and handover mitigation in OFDMA two-tier macrocell-femtocell networks. *EURASIP Journal on Wireless Communications and Networking*, *1*, 1–16.

Poor, H. V. (1994). *An introduction to signal detection and estimation.* New York, NY: Springer-Verlag.

Potlapally, N. R., Ravi, S., Raghunathan, A., & Jha, N. K. (2006). A study of the energy consumption caharacteristics of cryptographic algorithms and security protocols. *IEEE Transactions on Mobile Computing*, *5*(2), 128–143. doi:10.1109/TMC.2006.16

Prasad, R. V., Pawełczak, P., Hoffmeyer, J., & Berger, H. S. (2008). Cognitive functionality in next generation wireless networks: Standardization efforts. *IEEE Communications Magazine*, *46*(4), 72–78. doi:10.1109/MCOM.2008.4481343

Qaraqe, K., Ekin, S., Agarwal, T., & Serpedin, E. (2011). Performance analysis of cognitive radio multiple-access channels over dynamic fading environments. *Wireless Personal Communications.* Retrieved from http://www.qscience.com/doi/abs/10.5339/qfarf.2012.CSPS2

Qing Zhao, S. (2007). A decision-theoretic framework for opportunistic spectrum access. *IEEE Transactions on Wireless Communications*, *14*(4), 14–20. doi:10.1109/MWC.2007.4300978

Quan, Z., Cui, S., Poor, H. V., & Sayed, A. H. (2008). Collaborative wideband sensing for cognitive radios. *IEEE Signal Processing Magazine*, *25*(6), 60–73. doi:10.1109/MSP.2008.929296

Quan, Z., Cui, S., & Sayed, A. H. (2008). Optimal linear cooperation for spectrum sensing in cognitive radio networks. *IEEE Journal on Selected Topics in Signal Processing*, *2*(1), 28–40. doi:10.1109/JSTSP.2007.914882

Quan, Z., Cui, S., Sayed, A. H., & Poor, H. V. (2009). Optimal multiband joint detection for spectrum sensing in cognitive radio networks. *IEEE Journal Transaction on Signal Processing*, *57*(3), 1128–1140. doi:10.1109/TSP.2008.2008540

Quan, Z., Zhang, W., Shellhammer, S. J., & Sayed, A. H. (2011). Optimal spectral feature detection for spectrum sensing at very low SNR. *IEEE Journal Transaction on Communications*, *59*(1), 201–202. doi:10.1109/TCOMM.2010.112310.090306

Qu, Q., Milstein, L., & Vaman, D. (2008). Cognitive radio based multi-user resource allocation in mobile ad hoc networks using multi-carrier CDMA modulation. *IEEE Journal on Selected Areas in Communications*, *26*(1), 70–82. doi:10.1109/JSAC.2008.0801007

Radios, O. P. M. (2012). *Website.* Retrieved from http://www.omeshnet.com

Raj, R., & Gagneja, A. (2012). 4G wireless technology. *International Journal of Computer Science and its Applications*, 263-270.

Ramanath, S., Kavitha, V., & Altman, E. (2010). Impact of mobility on call block, call drops and optimal cell size in small cell networks. In *Proceedings of the IEEE 21st International Symposium on Personal, Indoor and Mobile Radio Communications Workshops (PIMRC Workshops).* Istanbul, Turkey: IEEE Press.

Rangan, S. (2010). Femto-macro cellular interference control with subband scheduling and interference cancelation. In *Proceedings of the 2010 IEEE GLOBECOM Workshops (GC Wkshps).* Miami, FL: IEEE Press.

Rashid, M. (2009). Opportunistic spectrum scheduling for multiuser cognitive radio: A queueing analysis. *IEEE Transactions on Wireless Communications*, *8*(10), 5259–5269. doi:10.1109/TWC.2009.081536

Reed, J. H. (2002). *Software radio: A modern approach to radio engineering.* Upper Saddle River, NJ: Prentice Hall.

Rondeau, T. W. (2004). Cognitive radios with genetic algorithms: Intelligent control of software defined radios. *Proceedings of SDR, 2004*, C3–C8. Phoenix, AZ: SDR.

Ruangchaijatupon, P., & Krishnamurthy, P. (2001). Encryption and power consumption in wireless LANs-N. In *Proceedings of the 3rd IEEE Workshop on Wireless LANs,* (pp. 148-152). Newton, MA: IEEE Press.

Runsheng, G., Zhongyu, H., & Tao, S. (2008). Adaptive CRN spectrum sensing scheme with excellence in topology and scan scheduling. In *Proceedings of the 3rd International Conference on Sensing Technology,* (pp. 384-391). IEEE.

Sachs, J., Maric, I., & Goldsmith, A. (2010). Cognitive cellular systems within the TV spectrum. In *Proceedings of the 2010 IEEE Symposium on New Frontiers in Dynamic Spectrum*. Singapore, Singapore: IEEE Press.

Sahai, N. H. (2004). Some fundamental limits on cognitive radio. In *Proceedings of the Allerton Conference on Communication, Control and Computing*, (pp. 1662–1671). IEEE Press.

Salami, G., Durowoju, O., Attar, A., Holland, O., Tafazolli, R., & Aghvami, H. (2011). A comparison between the centralized and distributed approaches for spectrum management. *IEEE Communications Surveys Tutorials, 13*(2), 274–290. doi:10.1109/SURV.2011.041110.00018

Salous, S. (2010). Chirp sounder measurements for broadband wireless networks and cognitive radio. In *Proceedings of the 7th International Symposium on Communication Systems Networks and Digital Signal Processing (CSNDSP), 2010,* (pp. 846-851). IEEE Press.

Sankaranarayanan, S., Papadimitratos, P., Mishra, A., & Hershey, S. (2005). A bandwidth sharing approach to improve licensed spectrum utilization. In *Proceedings of IEEE DySPAN,* (pp. 279–288). IEEE Press.

Saunders, S., Carlaw, S., Giustina, A., Bhat, R., Rao, V., & Siegberg, V. (2009). *Femtocells book: Femtocells opportunities and challenges for business and technology*. New York, NY: John Wiley & Sons Ltd.

Sawaragi, Y., Nakayama, H., & Tanino, T. (1985). *Theory of multiobjective optimization*. Orlando, FL: Academic Press Inc.

Scarfone, K., Tibbs, C., & Sexton, M. (2010). *Guide to securing WiMAX wireless communications*. Gaithersburg, MD: National Institute of Standards and Technology.

Schedule. (2012). *Wikipedia*. Retrieved from http://en.wikipedia.org/wiki/Schedule_%28resource%29

Schneier, B., & Whiting, D. (1997). *Fast software encryption: Designing encryption algorithms for optimal software speed on the Intel Pentium processor*. Berlin, Germany: Springer. doi:10.1007/BFb0052351

Shakir, M. Z., & Alouini, M. S. (2012). On the area spectral efficiency improvement of heterogeneous network by exploiting the integration of macro-femto cellular networks. In *Proceedings of the IEEE International Conference on Communications, ICC 2012*, (pp. 1–6). Ottawa, Canada: IEEE Press.

Shakir, M. Z., Rao, A., & Alouini, M. S. (2011). Collaborative spectrum sensing based on the ratio between largest eigenvalue and Geometric mean of eigenvalues. In *Proceedings of the International Conference on Global Communications, GLOBECOM 2011*. Houston, TX: IEEE Press.

Shakkottai, S., Altman, E., & Kumar, A. (2006). The case for non-cooperative multihoming of users to access points in IEEE 802.11 wlans. In *Proceedings of IEEE INFOCOM*. IEEE Press.

Shakkottai, S., Altman, E., & Kumar, A. (2007). Multihoming of users to access points in wlans: A population game perspective. *IEEE Journal on Selected Areas in Communications, 25*(6). doi:10.1109/JSAC.2007.070814

Shao, H., & Beaulieu, N. (2011). Direct sequence and time-hopping sequence designs for narrowband interference mitigation in impulse radio UWB systems. *IEEE Transactions on Communications, 59*(7), 1957–1965. doi:10.1109/TCOMM.2011.060911.100581

Shen, H., Zhang, W., & Kwak, K. S. (2008). Cognitive implementation of chirp waveform in UWB system. *IEICE Transactions on Communications, E91.B*(1), 147-150.

Sherman, M., Mody, A. N., Martinez, R., Rodriguez, C., & Reddy, R. (2008). IEEE standards supporting cognitive radio and networks, dynamic spectrum access and coexistence. *IEEE Communications Magazine, 46*(7), 72–79. doi:10.1109/MCOM.2008.4557045

Shetty, N., Parekh, S., & Walrand, J. (2009). Economics of femtocells. In *Proceedings of the IEEE Conference on Global Communications, GLOBECOM 2009*. Honolulu, HI: IEEE Press.

Shi, Y., Hou, Y., Zhou, H., & Midkiff, S. (2010). Distributed cross-layer optimization for cognitive radio networks. *IEEE Transactions on Vehicular Technology*, *59*(8), 4058–4069. doi:10.1109/TVT.2010.2058875

Singh, M. B. (2011). Cooperative spectrum sensing with an improved energy detector in cognitive radio network. In *Proceedings of the National Conference on Communications (NCC)*, (pp. 1-5). IEEE Press.

Song, L. (2008). Cognitive networks: Standardizing the large scale wireless systems. In *Proceedings of IEEE Consumer Communications and Networking Conference*, (pp. 988 – 992). Las Vegas NV: IEEE Press.

Song, L. (2008). Mesh infrastructure supporting broadband Internet with multimedia services. In *Proceedings of the IEEE International Conference on Circuits and Systems for Communications*. Shanghai, China: IEEE Press.

Song, L., & Hatzinakos, D. (2008). Real-time communications in large scale wireless networks. *International Journal of Multimedia Broadcasting*. Retrieved from http://www.hindawi.com/journals/ijdmb/2008/586067/

Song, L., & Hatzinakos, D. (2007). A cross-layer architecture of wireless sensor networks for target tracking. *IEEE/ACM Transactions on Networking*, *15*(1), 145–158. doi:10.1109/TNET.2006.890084

Song, L., & Hatzinakos, D. (2009). Cognitive networking of large scale wireless systems. *International Journal of Communication Networks and Distributed Systems*, *2*(4), 452–475. doi:10.1504/IJCNDS.2009.026558

Song, L., & Hatzinakos, D. (2011). Wireless sensor networks: From application specific to modular design. In Foerster, A. (Ed.), *Emerging Communications for Wireless Sensor Networks*. New York, NY: IN-TECH.

Source Lyrtech. (2012). Lyrtech launches new small form factor SDR development platforms at SDR forum technical conference. *Lyrtech, Inc.* Retrieved May 15, 2012, from http://www.prnewswire.com/news-releases/lyrtech-launches-new-small-form-factor-sdr-development-platforms-at-sdr-forum-technical-conference-56330167.html

Sousa, A. G. (2005). Collaborative spectrum sensing for opportunistic access in fading environments. In *Proceedings of the Symposium on Dynamic Spectrum Access Networks (DySPAN 2005)*, (pp. 131–136). Baltimore, MD: IEEE Press.

Sousa, A. G. (2007). Asymptotic performance of collaborative spectrum sensing under correlated log-normal shadowing. *IEEE Communications Letters*, *11*(1), 34–36.

Spachos, P., Bui, F., Song, L., Lostanlen, Y., & Hatzinakos, D. (2011). Performance evaluation of wireless multihop communications for an indoor environment. In *Proceedings of IEEE International Symposium on Personal, Indoor and Mobile Radio Communications*. Toronto, Canada: IEEE Press.

Spachos, P., Song, L., & Hatzinakos, D. (2011). Performance comparison of opportunistic routing schemes in wireless sensor networks. In *Proceedings of the Ninth Annual Communication Networks and Services Research Conference*, (pp. 271-277). IEEE.

Spachos, P., Song, L., & Hatzinakos, D. (2012). Opportunistic multihop wireless communications with calibrated channel model. In *Proceedings of IEEE International Conference on Communications (ICC)*. Ottawa, Canada: IEEE Press.

Spread Spectrum. (2012). *Wikipedia*. Retrieved from http://en.wikipedia.org/wiki/Spread_spectrum

Srinivasa, S., Jafar, S., & Jindal, N. (2006). On the capacity of the cognitive tracking channel. In *Proceedings of the IEEE International Symposium on Information Theory*, (pp. 2077-2080). Seattle, WA: IEEE Press.

Srinivasa, S. (2008). How much spectrum sharing is optimal in cognitive radio networks? *IEEE Transactions on Wireless Communications*, *7*(10), 4010–4018. doi:10.1109/T-WC.2008.070647

Stanojev, I., Simeone, O., Bar-Ness, Y., & Yu, T. (2008). Spectrum leasing via distributed cooperation in cognitive radio. *Proceedings of the IEEE, ICC*, 3427–3431. IEEE Press.

Steuer, R. E. (1986). *Multiple criteria optimization: Theory, computations, and application*. New York, NY: John Wiley & Sons, Inc.

Stevens-Navarro, E., Lin, Y., & Wong, V. W. S. (2008). An MDP-based vertical handoff decision algorithm for heterogeneous wireless networks. *IEEE Transactions on Vehicular Technology*, *57*(2), 1243–1254. doi:10.1109/TVT.2007.907072

Stotas, S., & Nallanathan, A. (2011). Enhancing the capacity of spectrum sharing cognitive radio networks. *IEEE Transactions on Vehicular Technology*, *60*(8), 3768–3779. doi:10.1109/TVT.2011.2165306

Suarez-Casal, P., Carro-Lagoa, A., Garcia-Naya, J. A., & Castedo, L. (2010). *A multicore SDR architecture for reconfigurable WiMAX downlink*. Retrieved May 15, 2012, from http://gtec.des.udc.es/web/images/pdfConferences/2010/dsd_suarez_2010.pdf

Su, H., & Zhang, X. (2008). Cross-layer based opportunistic MAC protocols for QoS provisionings over cognitive radio wireless networks. *IEEE Journal on Selected Areas in Communications*, *26*(1), 118–129. doi:10.1109/JSAC.2008.080111

Sung, K., Haas, H., & McLaughlin, S. (2010). A semi-analytical PDF of downlink SINR for femtocell networks. *EURASIP Journal on Wireless Communications and Networking*, 9.

Sutton, K. N. (2008). Cyclostationary signatures in practical cognitive radio applications. *IEEE Journal on Selected Areas in Communications*, *26*(1), 13–24. doi:10.1109/JSAC.2008.080103

Tandra, R., & Sahai, A. (2005). Fundamental limits on detection in low SNR under noise uncertainty. In *Proceedings of the IEEE International Conference on Wireless Networks, Communication, and Mobile Computing*, (vol. 1, pp. 464-469). Maui, HI: IEEE Press.

Tang, H. (2005). Some physical layer issues of wide-band cognitive radio systems. In *Proceedings of the 2005 1st IEEE International Symposium on New Frontiers in Dynamic Spectrum Access Networks*, (pp. 151-159). Piscataway, NJ: IEEE Press.

Tao, S., & Krunz, M. (2009). Coordinated channel access in cognitive radio networks: A multi-level spectrum opportunity perspective. In *Proceedings of the IEEE Conference on Computer Communications (INFOCOM)*, (pp. 2976–2980). IEEE Press.

Technology Watch Report, I. T. U.-T. 10. (2009). *The future internet*. Retrieved from http://www.itu.int

Telecom Regulatory Authority of India. (2012). *Telecom regulatory authority of India (TRAI) online*. Retrieved from http://www.trai.gov.in

Thanabalasingham, T. (2006). *Resource allocation in OFDM cellular networks*. (PhD Dissertation). University of Melbourne. Melbourne, Australia.

Thomas, R. W., DaSilva, L. A., & MacKenzie, A. B. (2005). Cognitive networks. In *Proceedings of the First IEEE International Symposium on New Frontiers in Dynamic Spectrum Access Networks*. Baltimore, MD: IEEE Press.

Tombaz, S., Vastberg, A., & Zander, J. (2011). Energy- and cost-efficient ultra-high-capacity wireless access. *IEEE Wireless Communications*, *18*(5), 18–24. doi:10.1109/MWC.2011.6056688

TR 36.942. (2009). *Radio frequency (RF) system scenarios (release 8)*. 3GPP Technical Report, V. 8.2.0. Retrieved from http://www.3gpp.org

TR. (2008). Interference management in UMTS femtocells. *Femto Forum- Technical Report*. Retrieved from http://www.femtoforum.org

Trimintzios, P., & Georgiou, G. (2010). Review article: WiFi and WiMAX secure deployments. *Journal of Computer Systems, Networks, and Communications. Hindawi Publishing Corporation*, *2010*, 1–28.

Trinadh, P. (2009). Emergence of software defined radio, SDR. *Home Tecnology Magazine*. Retrieved May 15, 2012, from http://hometoys.com/emagazine.php?url=/ezine/09.04/trinadh/index.htm

Tung, A. H. (2008). Dynamic spectrum access with prioritization in open spectrum wireless networks. In *Proceedings of the 11th IEEE Singapore International Communication Systems, 2008*, (pp. 1026 –1030). IEEE Press.

Tu, S.-Y., Chen, K.-C., & Prasad, R. (2009). Spectrum sensing of OFDMA systems for cognitive radio networks. *IEEE Transactions on Vehicular Technology*, *58*(7), 3410–3425. doi:10.1109/TVT.2009.2014775

Umaparvathi, M., & Varughese, D. K. (2010). Evaluation of symmetric encryption algorithms for MANETs. In *Proceedings of Computational Intelligence and Computer Research* (pp. 1–3). IEEE Press. doi:10.1109/ICCIC.2010.5705754

Urgaonkar, R., & Neely, M. (2009). Opportunistic scheduling with reliability guarantees in cognitive radio networks. *IEEE Transactions on Mobile Computing*, *8*(6), 766–777. doi:10.1109/TMC.2009.38

Urkowitz, H. (1967). Energy detection of unknown deterministic signals. *Proceedings of the IEEE*, *55*(4), 523–531. doi:10.1109/PROC.1967.5573doi:10.1109/PROC.1967.5573

Uyanik, G., & Oktug, S. (2011). A priority based cooperative spectrum utilization considering noise in common control channel in cognitive radio networks. In *Proceedings of the 7th International Wireless Communications and Mobile Computing Conference (IWCMC), 2011,* (pp. 477-482). IEEE Press.

Vamsi Krishna, A. D. (2009). A survey on MAC protocols in OSA networks. *Computer Networks*, *53*(9), 1377–1394. doi:10.1016/j.comnet.2009.01.003

Verma, O. P., Agarwal, R., Dafouti, D., & Tyagi, S. (2011). Performance analysis of data encryption algorithms. In *Proceedings of Electronics and Computer Technology* (pp. 399–403). IEEE Press.

Vigneron, B. C. (2006). Multiband frequency hopping for high data-rate communications with adaptive use of spectrum. In *Proceedings of the IEEE 63rd Vehicular Technology Conference, 2006,* (pp. 251-255). IEEE Press.

Vivier, G., Kamoun, M., Becvar, Z., de Marinis, E., Lostanlen, Y., & Widiawan, A. (2010). Femtocells for next-G wireless systems: The FREEDOM approach. In *Proceedings of Future Network and Mobile Summit*. Paris, France: La Défense.

Volcano Lab. (2012). *Website.* Retrieved from http://www.siradel.com

Vu, M., Devroye, N., Sharif, M., & Tarokh, N. (2007). Scaling laws of cognitive networks. In *Proceedings of Cognitive Radio Oriented Wireless Networks and Communications, 2007.* CrownCom. doi:10.1109/CROWNCOM.2007.4549764

Wang, B., Ji, Z., & Liu, K. (2007). Primary-prioritized Markov approach for dynamic spectrum access. In *Proceedings of IEEE DySPAN,* (pp. 507–515). IEEE Press.

Wang, J., Huang, Y., & Jiang, H. (2009). Improved algorithm of spectrum allocation based on graph coloring model in cognitive radio. In *Proceedings of the International Conference on Communication and Mobile Computing*, (vol. 3, pp. 353-357). Yunnan, China: IEEE.

Wang, L., & Wang, C. (2008). Spectrum handoff for cognitive radio networks: Reactive-sensing or proactive-sensins? In *Proceedings of the IEEE International Performance, Computing and Communications Conference 2008 (IPCCC 2008)*, (pp. 343-348). IEEE Press.

Wang, L.-C., & Chen, A. (2007). A cognitive MAC protocol for QoS provisioning in overlaying ad hoc networks. In *Proceedings of the 4th IEEE Consumer Communications and Networking Conference, CCNC 2007,* (pp. 1139-1143). IEEE Press.

Wang, P., et al. (2007). Power allocation in OFDM-based cognitive radio systems. In *Proceedings of IEEE GLOBECOM*, (pp. 4061–4065). IEEE Press.

Wang, S.-Y., Huang, Y.-M., Lau, L.-C., & Lin, C.-C. (2011). Enhanced MAC protocol for cognitive radios over IEEE 802.11 networks. In *Proceedings of the Wireless Communications and Networking Conference (WCNC)*, (pp. 37 - 42). Cancun, Mexico: IEEE Press.

Wang, B., & Liu, K. J. (2011). Advances in cognitive radio networks: A survey. *IEEE Journal on Selected Topics in Signal Processing*, *5*(1), 5–23. doi:10.1109/JSTSP.2010.2093210

Wang, C. W.-T. (2011). A queueing-theoretical framework for qos-enhanced spectrum management in cognitive radio networks. *IEEE Wireless Communications*, *18*(6), 18–26. doi:10.1109/MWC.2011.6108330

Wang, J., Ghosh, M., & Challapali, K. (2011). Emerging cognitive radio applications: A survey. *IEEE Communications Magazine*, *49*(3), 74–81. doi:10.1109/MCOM.2011.5723803

Wang, K. R. (2011). *Cognitive radio networking and security: A game theoreitic view*. Cambridge, UK: Cambridge University Press.

Wang, L. N. (2011). Timeslot allocation scheme for cognitive satellite networks. *Advanced Materials Research, 230-232*, 40–43. doi:10.4028/www.scientific.net/AMR.230-232.40

Weiss, T., Hillenbrand, J., & Jondral, F. (2003). A diversity approach for the detection of idle spectral resources in spectrum pooling systems. In *Proceedings of the 48th International Scientific Colloquium*. Ilmenau, Germany: IEEE.

Weiss, T., & Jondral, F. (2004). Spectrum pooling: An innovative strategy for the enhancement of spectrum efficiency. *IEEE Communications Magazine*, *42*(3), S8–S14. doi:10.1109/MCOM.2004.1273768

Wellens, M., Wu, J., & Mahonen, P. (2007). Evaluation of spectrum occupancy in indoor and outdoor scenario in the context of cognitive radio. In *Proceedings of the 2nd International ICST Conference on Cognitive Radio Oriented Wireless Networks and Communications (CROWNCOM 2007)*, (pp. 420-427). Orlando, FL: CROWNCOM.

Westphal, C. (2006). Opportunistic routing in dynamic ad hoc networks: The oprah protocol. In *Proceedings of the IEEE International Conference on Mobile Adhoc and Sensor Systems*, (pp. 570 –573). IEEE Press.

Wikipedia. (2012). *List of software-defined radios*. Retrieved May 15, 2012, from http://en.wikipedia.org/wiki/List_of_software-defined_radios

Wild, B., & Ramchandran, K. (2005). Detecting primary receivers for cognitive radio applications. In *Proceedings of DySPAN First IEEE International Symposium*, (pp. 124-130). IEEE Press.

Win, M., & Scholtz, R. A. (1998). Impulse radio: How it works. *IEEE Communications Letters*, 2(2), 36–38. doi:10.1109/4234.660796

Wireless Innovation. (2012). *Benefits of SDR*. Retrieved May 15, 2012, from http://www.wirelessinnovation.org/Benefits_of_SDR

Wong, C.-M., Chen, J.-D., & Hsu, W.-P. (2009). Coordinated non-sensing MAC protocol for dynamic spectrum access networks. In *Proceedings of the IEEE 20th International Symposium on Personal, Indoor and Mobile Radio Communications, 2009*, (pp. 471-475). IEEE Press.

Won-Yeol, L., & Akyildiz, I. F. (2008). Optimal spectrum sensing framework for cognitive radio networks. *IEEE Transactions on Wireless Communications*, *7*, 3845–3857. doi:10.1109/T-WC.2008.070391doi:10.1109/T-WC.2008.070391

Wu, A., Yang, C., & Huang, D. (2010). Cooperative sensing of wideband cognitive radio: A multiple-hypothesis-testing approach. *IEEE Transactions on Vehicular Technology*, *59*(4), 1835–1846. doi:10.1109/TVT.2010.2043967

Wu, D., & Negi, R. (2003). Effective capacity: A wireless link model for support of quality of service. *IEEE Transactions on Wireless Communications*, *2*(4), 630–643.

Wu, Y., & Tsang, D. (2009). Dynamic rate allocation, routing and spectrum sharing for multi-hop cognitive radio networks. In *Proceedings of Communications Workshops, 2009*. IEEE Press. doi:10.1109/ICCW.2009.5208054

Wu, Y., Wang, B., & Liu, K. (2008). Repeated spectrum sharing game with self-enforcing truth-telling mechanism. *Proceedings of the IEEE, ICC*, 3583–3587. IEEE Press.

Wyglinski, M. N. (Ed.). (2010). *Cognitive radio communications and networks: Principels and practice*. New York, NY: Elsevier Academic Press.

Xia, P., Chandrasekhar, V., & Andrews, J. (2010). Open vs. closed access femtocells in the uplink. *IEEE Transactions on Wireless Communications*, *9*(12), 798–809. doi:10.1109/TWC.2010.101310.100231

Xie, J. X., & Zhou, K. (2012). QoS multicast routing in cognitive radio ad hoc networks. *International Journal of Communication Systems*, *25*, 30–46. doi:10.1002/dac.1285

Xin, C., Xie, B., & Shen, C.-C. (2005). A novel layered graph model for topology formation and routing in dynamic spectrum access networks. *Proceedings of New Frontiers in Dynamic Spectrum Access Networks, 2005*, 308–317. IEEE Press.

Xing, Y. (2007). Dynamic spectrum access with QoS and interference temperature constraints. *IEEE Transactions on Mobile Computing*, *6*(4), 423–433. doi:10.1109/TMC.2007.50

Xu, X., Bosisio, R. G., & Wu, K. (2006). Analysis and implementation of six-port software defined radio receiver platform. *IEEE Transactions on Microwave Theory and Techniques*, *54*(7), 2937–2943. doi:10.1109/TMTT.2006.877449

Yan Zhang, J. Z.-H. (Ed.). (2010). *Cognitive radio networks*. Boca Raton, FL: CRC Press. doi:10.1201/EBK1420077759

Yan, Y., Zhang, B., Mouftah, H., & Ma, J. (2008). Practical coding aware mechanism for opportunistic routing in wireless mesh networks. In *Proceedings of the IEEE International Conference on Communications,* (pp. 2871 –2876). IEEE Press.

Yang, Z., & Mohammed, A. (2011). Deployment and capacity of mobile WiMAX from high altitude platform. In *Proceedings of the 2011 IEEE 74th Vehicular Technology Conference*. IEEE Press.

Yang, Z., Mohammed, A., & Hult, T. (2008). Performance evaluation for WiMAX Broadband from high altitude platform cellular system and terrestrial coexisting capability. *EURASIP International Journal on Wireless Communications and Networking*. Retrieved from http://www.researchgate.net/publication/26571586_Performance_Evaluation_of_WiMAX_Broadband_from_High_Altitude_Platform_Cellular_System_and_Terrestrial_Coexistence_Capability

Yang, K., & Wang, X. (2008). Cross-layer network planning for multi-radio multi-channel cognitive wireless networks. *IEEE Transactions on Communications*, *56*(10), 1705–1714. doi:10.1109/TCOMM.2008.4641901

Yao, N. Y. (2009). Joint design of frequency and power adaptation in FHSS systems based on cognitive radio. In *Proceedings of the 5th International Conference on Wireless Communications, Networking and Mobile Computing, WiCom 2009*. IEEE Press.

Ying-Jie, M. A. (2011). Cognitive UWB adaptive pulse design for interference suppression. *Journal of Beijing University of Posts and Telecommunications*, *4*, 61–69.

Yonghong, Z., & Ying-Chang, L. (2007). Maximum-minimum eigenvalue detection for cognitive radio. In *Proceedings of the 2007 IEEE 18th International Symposium on Personal, Indoor and Mobile Radio Communications,* (pp. 1165-1169). Piscataway, NJ: IEEE Press.

Yuan, H. F., & Du, S. (2012). Resource allocation for multiuser cognitive OFDM networks with proportional rate constraints. *International Journal of Communication Systems*, *25*, 254–269. doi:10.1002/dac.1272

Yucek, T., & Arslan, H. (2009). A survey of spectrum sensing algorithms for cognitive radio applications. *IEEE Communications Surveys Tutorials*, *11*(1), 116–130. doi:10.1109/SURV.2009.090109doi:10.1109/SURV.2009.090109

Yuksel, E. (2007). *Analysis of the PKMv2 protocol in IEEE 802.16e-2005 using static analysis informatics and mathematical modeling*. (Masters Thesis). Technical University of Denmark. Copenhagen, Denmark.

Zamat, H., & Natarajan, B. (2009). Practical architecture of a broadband sensing receiver for use in cognitive radio. *Physical Communication*, *2*(12), 87–102. doi:10.1016/j.phycom.2009.02.005

Zeng, Y. H., & Liang, Y. C. (2007). Covariance based signal detections for cognitive radio. In *Proceedings of the IEEE Dynamic Spectrum Access Networks,* (pp. 202–207). Dublin, Ireland: IEEE Press.

Zeng, Y., Koh, C. L., & Liang, Y. C. (2008). Maximum eigenvalue detection: Theory and application. In *Proceedings of the IEEE International Conference on Communications,* (pp. 4160–4164). Beijing, China: IEEE Press.

Zeng, Y. H., & Liang, Y. C. (2009). Spectrum sensing algorithms for cognitive radio based on statistical covariance. *IEEE Transactions on Vehicular Technology*, *58*(4), 1804–1815. doi:10.1109/TVT.2008.2005267

Zeng, Y., & Liang, Y. C. (2009). Eigenvalue-based spectrum sensing algorithms for cognitive radio. *IEEE Transactions on Communications*, *57*(6), 1784–1793. doi:10.1109/TCOMM.2009.06.070402

Zhang, H. (2011). *Green Communications, Green Networking, and Green Spectrum*. Retrieved May 13, 2012, from http://mypage.zju.edu.cn/honggangzhang/572794.html

Zhang, J., & Zhang, Q. (2009). Stackelberg game for utility-based cooperative cognitive radio networks. In *Proceedings of the ACM International Symposium on Mobile Ad Hoc Networking and Computing (MobiHoc),* (pp. 23–32). ACM Press.

Zhang, R. (2008). Optimal power control over fading cognitive radio channels by exploiting primary user CSI. In *Proceedings of IEEE GLOBECOM,* (pp. 931–935). IEEE Press.

Zhang, W., Bai, S., Liu, Y., & Tang, J. (2010). Cognitive radio scheduling for overwater communications. In *Proceedings of the 53rd IEEE Global Communications Conference, GLOBECOM 2010,* (pp. 1-5). Miami, FL: IEEE Press.

Zhang, Y. (2008). Dynamic spectrum access in cognitive radio wireless networks. In *Proceedings of the IEEE International Conference on Communications,* (pp. 4927–4932). IEEE Press.

Zhang, J. X. (2010). Medium access control in cognitive radio networks. In Zhang, J. X., & Yan Zhang, J. Z.-H. (Eds.), *Cognitive Radio Netwoks* (pp. 89–120). Boca Raton, FL: CRC Press. doi:10.1201/EBK1420077759

Zhang, Q. (2009). Cooperative relay to improve diversity in cognitive radio networks. *IEEE Communications Magazine, 47*(2), 111–117. doi:10.1109/MCOM.2009.4785388

Zhang, R. (2010). On active learning and supervised transmission of spectrum sharing based cognitive radios by exploiting hidden primary radio feedback. *IEEE Transactions on Communications, 58*(10), 2960–2970. doi:10.1109/TCOMM.2010.082710.090412

Zhang, W., & Cao, G. (2004). Optimizing tree reconfiguration for mobile target tracking in sensor networks. In *Proceedings of IEEE INFOCOM* (pp. 2434–2445). IEEE Press.

Zhang, X., & Su, H. (2011). CREAM-MAC: Cognitive radio-enabled multi-channel MAC protocol over dynamic spectrum access networks. *IEEE Journal of Selected Topics in Signal Processing, 5*(1), 110–123. doi:10.1109/JSTSP.2010.2091941

Zhang, X., & Su, H. (2011). Opportunistic spectrum sharing schemes for CDMA-based uplink MAC in cognitive radio networks. *IEEE Journal on Selected Areas in Communications, 29*(4), 716–730. doi:10.1109/JSAC.2011.110405

Zhao, B., Seshadri, R., & Valenti, M. (2004). Geographic random forwarding with hybrid-arq for ad hoc networks with rapid sleep cycles. In *Proceedings of IEEE Global Telecommunications Conference,* (pp. 3047 – 3052). IEEE Press.

Zhao, C., Hu, J., & Shen, L. (2010). A MAC protocol of cognitive networks based on IEEE 802.11. In *Proceedings of the 2010 12th IEEE International Conference on Communication Technology (ICCT),* (pp. 1133 - 1136). IEEE Press.

Zhao, C., Zou, M., Shen, B., Kim, B., & Kwak, K. (2008). Cooperative spectrum allocation in centralized cognitive networks using bipartite matching. In *Proceedings of IEEE GLOBECOM,* (pp. 1-5). IEEE Press.

Zhao, H., Luo, T., Yue, G., & He, X. (2009). Myopic sensing for opportunistic spectrum access using channel correlation. In *Proceedings of the International Conference on Wireless Communications and Mobile Computing,* (pp. 512-516). Leipzig, Germany: IEEE.

Zhao, J., Zheng, H., & Yang, G. (2005). Distributed coordination in dynamic spectrum allocation networks. In *Proceedings of IEEE DySPAN,* (pp. 259–268). IEEE Press.

Zhao, Q., & Krishnamachari, B. (2007). Structure and optimality of myopic sensing for opportunistic spectrum access. In *Proceedings of the IEEE International Conference on Communications,* (pp. 6476-6481). Glasgow, UK: IEEE Press.

Zhao, Q., Tong, L., & Swami, A. (2005). Decentralized cognitive MAC for dynamic spectrum access. In *Proceedings of IEEE DySPAN,* (pp. 224–232). IEEE Press.

Zhao, C., & Kwak, K. (2010). Power/bit-loading in OFDM-based cognitive networks with comprehensive interference considerations: The single SU case. *IEEE Transactions on Vehicular Technology, 59*(4).

Zhao, F., Shin, J., & Reich, J. (2002). Information-driven dynamic sensor collaboration. *IEEE Signal Processing Magazine, 19*(2), 61–72. doi:10.1109/79.985685

Zhao, H. Z.-H. (2005). Distributed coordination in dynamic spectrum allocation networks. *Proceedings of the IEEE DySPAN, 2005,* 259–268. IEEE Press.

Zhao, J., Zheng, H., & Yang, G. (2007). Spectrum sharing through distributed coordination in dynamic spectrum access networks. *Wireless Communications and Mobile Computing*, *7*(9), 1061–1075. doi:10.1002/wcm.481

Zhao, L. T. (2005). Decentralized cognitive MAC for dynamic spectrum access. *Proceedings of IEEE DySPAN, 2005*, 224–232. IEEE Press.

Zhao, Q., & Sadler, B. (2007). A survey of dynamic spectrum access. *IEEE Signal Processing Magazine*, *24*(3), 79–89. doi:10.1109/MSP.2007.361604

Zhao, Q., Tong, L., Swami, A., & Chen, Y. (2007). Decentralized cognitive MAC for opportunistic spectrum access in ad hoc networks: A POMDP framework. *IEEE Journal on Selected Areas in Communications*, *25*(3), 589–600. doi:10.1109/JSAC.2007.070409

Zheng, H., & Cao, L. (2005). Device-centric spectrum management. In *Proceedings of IEEE DySPAN*, (pp. 56–65). IEEE Press.

Zheng, H., & Peng, C. (2005). Collaboration and fairness in opportunistic spectrum access. In *Proceedings of the IEEE International Conference on Communications, (ICC)*, (pp. 3132–3136). IEEE Press.

Zhong, Z., Wang, J., Nelakuditi, S., & Lu, G. (2006). On selection of candidates for opportunistic anypath forwarding. *SIGMOBILE Mobile Computer Communication Review*, *10*(4), 1–2. doi:10.1145/1215976.1215978

Zhu Han, H. J., & Fan, R. (2009). Replacement of spectrum sensing in cognitive radio. *IEEE Transactions on Wireless Communications*, *8*(6), 2819–2826. doi:10.1109/TWC.2009.080603

Zhu, L. S. (2007). Analysis of cognitive radio spectrum access with optimal channel reservation. *IEEE Communications Letters*, *11*(4), 304–306. doi:10.1109/LCOM.2007.348282

Zlydareva, O., & Sacchi, C. (2007). SDR application for implementing an integrated UMTS/WiMAX PHY-layer architecture. In *Proceeding of the 3rd International Conference on Mobile Multimedia Communications (MobiMedia 2007)*, (pp. 1-7). IEEE.

Zorzi, M., & Rao, R. (2003). Geographic random forwarding (geraf) for ad hoc and sensor networks: Energy and latency performance. *IEEE Transactions on Mobile Computing*, *2*(4), 349–365. doi:10.1109/TMC.2003.1255650

Zorzi, M., & Rao, R. (2003). Geographic random forwarding (geraf) for ad hoc and sensor networks: Multihop performance. *IEEE Transactions on Mobile Computing*, *2*(4), 337–348. doi:10.1109/TMC.2003.1255648

Zyren, J., & McCoy, W. (2007). *Overview of the 3GPP long term evolution physical layer*. White Paper. LTD Free scale Semiconductor. Retrieved from http://www.3gpp.org

About the Contributors

Anwer Al-Dulaimi obtained his PhD degree in Cognitive LTE Radio Systems in 2012 from Brunel University, London, UK, after being awarded BSc and MSc honours degrees in Telecommunication Engineering. Prior to this, he worked as a laboratory assistant and assistant lecturer in many universities, where he taught many courses in communications and electronics engineering. He worked also as a technical manager for the Canadian Chambers of Trade and Technology Federation. His research interests lie in the area of 4G wireless systems with special focus on dynamic spectrum access, adaptive transmission domains, mobility, and alternative routing algorithms that consider the energy savings and information exchange between peer radios. His research has been documented in high quality IEEE publications and has been referenced by USA patents. He is a member in the IEEE 1900.5, 7, and SE43 standardization committees for the future cognitive radio systems. He is a member of the COST Action IC0905 TERRA, European Alliance for Innovation (EAI), and the European Technology Platform (Photonics21). Dr. Anwer is a reviewer for many IEEE journals and conferences. Besides being a member in the IEEE and the IET, he was recognised as a Charted Engineer (CEng) in April 2010 and Associate Practitioner (AHEA) of the British Higher Education Academy in 2012.

John Cosmas received his BEng in Electronic Engineering at Liverpool University, UK, in 1978, and a PhD in Image Processing at Imperial College, University of London, UK, in 1986. He worked for five years in industry first with Tube Investments and then with Fairchild Camera and Instruments. After completing his PhD, he worked for 13 years as a Lecturer in Digital Systems and Telecommunications at Queen Mary College, University of London. Since 1999, he has worked for Brunel University, first as a Reader and then in 2002 as a Professor of Multimedia Systems. He is currently the Director of the Wireless Networks and Communications Centre (WNCC) at Brunel University, London, UK. His current research interests are communications systems and mobile systems. Prof. Cosmas currently serves as Associate Editor for *IEEE Transactions on Broadcasting*. He also contributes to the Digital TV Group's "Mobile Applications" sub-group and to the Digital Video Broadcast's Technical Module: Converged Broadcast and Mobile Services (CBMS).

Abbas Mohammed is a Professor of Telecommunications Theory at the School of Engineering, Blekinge Institute of Technology (BTH), Sweden. He was awarded the PhD degree from Liverpool University, UK, in 1992, and the Swedish Docent degree in Radio Communications and Navigation from BTH in 2001. He is the recipient of the Blekinge Research Foundation Award "Researcher of the Year Award and Prize" for 2006. From 1993 to 1996, he was a Research Fellow with the Radio Navigation Group, University of Wales (Bangor), UK. From 1996 to 1998, he was with the University of Newcastle, UK, working on a European collaborative project within the ACTS (Advance Communications Technologies and Services) FP4 Research Programme that investigated 3G Satellite-UMTS systems. He was also employed by Ericsson AB, where he consulted on power control standardization issues for 3G mobile communication systems. He has been a visiting lecturer to several universities. He is a Fellow of the Institution of Electrical Engineering since 2004. In 2006, he received Fellowship of the UK's Royal Institute of Navigation "in recognition of his significant contribution in navigation and in particular to advanced signal processing techniques that have enhanced the capability of Loran-C." He is a Senior Member of IEEE and a Life Member of the International Loran Association, USA. He is the Editor-in-Chief of *ACEEE International Journal on Communication* (USA) and the *Journal of Selected Areas in Telecommunications of Cyber Journals* (Canada). He is a Board Member of the IEEE Signal Processing Swedish Chapter since its inception in 2000, and the UK's *IET Teknologia Journal* to promote the research from the Middle East/North Africa region. He is an Associate Editor to the *International Journal of Navigation and Observation* and the *Mediterranean Journal of Electronics and Communications*, and a former Editorial Board Member of the *Radio Engineering Journal.* He has also been a Guest Lead Editor for several special issues of international journals. He is the author of over 200 publications, including many invited articles in peer reviewed journal special issues and over 25 book chapters, in the fields of telecommunications, navigation systems, and signal processing. He has also developed techniques for measuring skywave delays in Loran-C receivers and received a Best Paper Award from the International Loran Association, USA, in connection to this work. He is the Swedish representative and member of the management committee to the European Community COST 280, 296, 297, IC0802, IC0902, and IC1102 Projects. From 2002-2010, he was the BTH Director of the Graduate School of Telecommunications (a collaborative consortium in research and education between 4 Swedish universities), chaired by the Royal Institute of Technology (KTH). Professor Mohammed's research interests are in space-time signal processing and MIMO systems, channel modeling, antennas and propagation, satellite and high altitude platform communications, cognitive radio and dynamic spectrum access, and radio navigation systems.

* * *

Rafidah Ahmad received the Master of Science in Electrical and Electronic Engineering from the Universiti Sains Malaysia, Malaysia, in 2005. Since 2006, she works as a Research Officer on Integrated Circuit (IC) and Field Programmable Gate Array (FPGA) designs for Collaborative MicroElectronic Design Excellence Centre (CEDEC) at Universiti Sains Malaysia. Until today, she has been the main author for more than 10 international conference papers and journals. In August 2011, she continues her study as a part time PhD student in Electrical and Electronic Engineering at Universiti Sains Malaysia. Her current research interests are in digital signal processing, wireless communications, and applied cryptography. She is also a member of IEEE.

Mohamed-Slim Alouini was born in Tunis, Tunisia. He received the Ph.D. degree in electrical engineering from the California Institute of Technology (Caltech), Pasadena, CA, USA, in 1998. He was with the department of Electrical and Computer Engineering of the University of Minnesota, Minneapolis, MN, USA, then with the Electrical and Computer Engineering Program at the Texas A&M University at Qatar, Education City, Doha, Qatar. Since June 2009, he has been a Professor of Electrical Engineering in the Division of Physical Sciences and Engineering at KAUST, Saudi Arabia, where his current research interests include the design and performance analysis of wireless communication systems.

Saba Al-Rubaye received her BSc and MSc degrees in Telecommunications from University of Technology. She was a Laboratory Instructor and then Assistant Lecturer in the Department of Telecommunication Engineering, in different universities. She taught different courses in Electrical and Electronic Engineering. She is currently a PhD candidate in the Electronic and Computer Engineering Department at Brunel University, London, UK. Her research interests lie in the area of 4G wireless systems for next generation network with special focus on femtocell, cognitive radio system, LTE, and power consumption. She has published over twenty journal and conference papers in the area of telecommunications.

Eitan Altman received the B.Sc. degree in Electrical Engineering (1984), the B.A. degree in Physics (1984), and the Ph.D. degree in Electrical Engineering (1990), all from the Technion-Israel Institute, Haifa. In 1990, he received his B.Mus. degree in Music Composition in Tel-Aviv University. Since 1990, he has been with INRIA (National Research Institute in Informatics and Control) in Sophia-Antipolis, France. His current research interests include performance evaluation and control of telecommunication networks and in particular congestion control, wireless communications, and networking games. He is on the editorial board of several scientific journals: JEDC, COMNET, DEDS, and WICON. He has been the (co)chairman of the program committee of several international conferences and workshops on game theory, networking games, and mobile networks.

Zwi Altman received the B.Sc. and M.Sc. degrees in Electrical Engineering from the Technion-Israel Institute of Technology, in 1986 and 1989, and the Ph.D. degree in Electronics from the INPT France in 1994. He was a Laureate of the Lavoisier Scholarship of the French Foreign Ministry in 1994, and from 1994 to 1996, he was a Post-Doctoral Research Fellow in the University of Illinois at Urbana Champaign. In 1996, he joined France Telecom R&D, where he has been involved in different projects related to network engineering, including radio resource management, automatic cell planning, and self-organizing networks. In 2004, he co-received the France Telecom Innovation Prize, and in 2005 the IEEE Wheeler Award. From 2005 to 2007, Dr. Altman was the Coordinator of the Eureka Celtic Gandalf Project "Monitoring and Self-Tuning of RRM Parameters in a Multi-System Network" that received the Celtic Excellence Award.

Rachad Atat received his B.E. degree in Computer Engineering with distinction from Lebanese American University (LAU), Byblos, Lebanon, in 2010. He obtained M.Sc. in Electrical Engineering from King Abdullah University of Science and Technology (KAUST), Thuwal, Saudi Arabia in 2012 after being awarded the KAUST Discovery Scholarship Award. He is currently working toward his Ph.D. degree in Energy Efficient Wireless Communications at the Royal Institute of Technology (KTH), Stockholm, Sweden. His research interests are in the general area of wireless communications with emphasis on energy efficient networks, LTE network, cooperative communications, relays, heterogeneous networks and cognitive radios. He is a member of the order of Engineers in Beirut, and a member of IEEE Communications Society.

Niclas Björsell was born in Falun, Sweden, in 1964. He received his B.Sc. in Electrical Engineering and his Lic. Ph. in Automatic Control from Uppsala University, Sweden, in 1994 and 1998, respectively. His Ph. D. in Telecommunication was received at the Royal Institute of Technology, Stockholm, Sweden, in 2007. He has several years of experience from research and development projects that fostered collaborations between industry and the academy. For more than 15 years, he has held positions in the academy as well as in industry. Between 2006 and 2009, he served as the head of Division of Electronics at the Department of Technology and Built Environment, University of Gävle, Sweden. He is currently Associate Professor at the University of Gävle and Guest Professor at the Vrije Universiteit Brussel, Belgium. He has published more than 50 papers in peer-review journals and conferences, and his research interests include radio frequency measurement technology, analog-to-digital conversion, and cognitive radio. Dr. Björsell is Associate Editor of *IEEE Transactions on Instrumentation and Measurement* and voting member of the IEEE, Instrumentation and Measurement, TC-10.

Lokesh Chouhan received his Bachelor of Engineering (B.E.) in Computer Engineering, from SGSITS, Indore, India, in 2005, and Master of Technology degree (with distinction) in Information Technology from School of Information Technology, Bhopal, India, in 2009. Currently, he is the Research Scholar in ABV-Indian Institute of Information Technology and Management, Gwalior, India. His research interests are focused on wireless networking and information security. The major topics include cognitive radio networks, ad hoc networks, and MAC protocols design and security for various wireless systems. He is reviewer for *Journal of Supercomputing*, and various international and national conferences. He has also authored and co-authored in number of international and national papers.

Salah Eddine Elayoubi received the Engineering diploma in Telecommunications and Computer Science from the Lebanese University at Tripoli, Lebanon, in 2000 and the Master's degree in Telecommunications and Networking from the National Polytechnic Institute at Toulouse, France, in 2001. He completed his Ph.D. degree in Computer Science and Telecommunications at the University of Paris VI in 2004. Since then, he has been working in Orange Labs, the research and development division of the France Telecom group. His research interests include radio resource management, performance evaluation, and dimensioning of mobile networks.

Maria Erman, MSc EE, is a Doctoral Student in Signal Processing, and is employed at Blekinge Institute of Technology in Karlskrona, Sweden. In her research, she focuses on soft computing techniques for the improvement of fields, such as cognitive radio.

Majed Haddad received his Electrical Engineering diploma in 2004 from the National Engineering School of Tunis (ENIT), Tunisia. He received the Master degree from the University of Nice Sophia-Antipolis in France in 2005, and a Doctorate in Electrical Engineering from Eurecom Institute in 2008. In 2009, he joined France Telecom R&D as a Post-Doctoral Research Fellow. Dr. Haddad is currently pursuing his postdoctoral research at INRIA Sophia Antipolis in France, where his research interests include advanced design of PHY/MAC layer of wireless communication systems with particular attention on Quality of Experience (QoE) for mobile streaming services.

Mohamed Hamid was born in Al-Dweem, Sudan, in 1983. He received his BSc in Electrical Engineering from University of Khartoum, Khartoum, Sudan, in 2005, and his MSc in Radio Communications from Blekinge Institute of Technology (BTH), Karlskrona, Sweden, in 2009. Currently, he is working towards his PhD in the Royal Institute of Technology, KTH, Stockholm, Sweden, and University of Gävle (HiG), Gävle, Sweden, since April, 2010. From 2005-2007, he worked at University of Khartoum as a Teaching Assistant. In 2006, he joined MTN-Sudan as a Radio Network Optimization (RNO) Engineer, where he worked for one year. Hamid's research interests include dynamic spectrum access and cognitive radio, signal detection techniques, quantitative analysis of spectrum opportunities, and secondary system interference modeling.

Dimitrios Hatzinakos received the Diploma degree from the University of Thessaloniki, Greece, in 1983, the M.A.Sc degree from the University of Ottawa, Canada, in 1986, and the Ph.D. degree from Northeastern University, Boston, MA, in 1990, all in Electrical Engineering. In September 1990, he joined the Department of Electrical and Computer Engineering, University of Toronto, where now he holds the rank of Professor with tenure. Since November 2004, he is the holder of the Bell Canada Chair in Multimedia, at the University of Toronto. In addition, he is the co-founder and since 2009 the Director and the Chair of the Management Committee of the Identity, Privacy, and Security Institute (IPSI) at the University of Toronto. His research interests and expertise are in the areas of multimedia signal processing, multimedia security, multimedia communications, and biometric systems. Among other significant research activities, currently, he is the leading investigator of a major ORF grant in the area of Energy Efficient and self-Sustainable Sensor Networks and a leading investigator in a research initiative on "Smart Data" cosponsored by IPSI, the University of Toronto, and the Privacy Commissioner of Ontario. He is author/co-author of more than 250 papers in technical journals and conference proceedings; he has contributed to 17 books and he has 7 patents in his areas of interest. He is a Fellow of the Engineering Institute of Canada, a senior member of the IEEE, and member of EURASIP, the Professional Engineers of Ontario (PEO), and the Technical Chamber of Greece.

Tommy Hult received his M.Sc. degree in Electrical Engineering with emphasis on Signal Processing from Blekinge Institute of Technology, Sweden, in 2002, and his Ph.D in Telecommunication, 2008. From 2009-2011, he was working as a post doc in the Radio Systems Group at the Department of Electrical and Information Technology, Lund University, Sweden. Currently, he is working as a Researcher at the Swedish Defence Research Institute. He is the author of more than 75 international book chapters, conference and journal papers in the areas of smart antenna systems, wave propagation, channel modeling, and wireless networks. He is also currently the representative of Sweden in two EU COST actions; IC0802 – "Propagation Tools and Data for Integrated Telecommunication, Navigation, and Earth Observation systems" and IC0902 – "Cognitive Radio and Networking for Cooperative Coexistence of Heterogeneous Wireless Networks." He is also a member of the Swedish National Radio Science Association SNRV. His research areas of interest are mainly in space-time processing, electromagnetic wave propagation, smart antenna systems, channel modeling, and channel measurements for a variety of different telecommunication systems.

Widad Ismail graduated from University of Huddersfield, UK, in 1999, and earned First Class Honors in Electronics and Communications Engineering, and she received her PhD in Electronics Engineering from University of Birmingham, UK, in 2004. She is currently an Associate Professor at the School of Electrical and Electronic Engineering, Universiti Sains Malaysia in Nibong Tebal, Penang, Malaysia. She has contributed extensively in research and in the areas of Radio Frequency Identification (RFID), Active Integrated Antennas (AIA), RF Systems, and Wireless Systems Design. She has initiated Auto-ID Laboratory (AIDL), Malaysia in 2008, as a research and commercially oriented centre where the main objective is to become a hub for research and commercialization activities. These research works have produced 8 filed patents, 4 international awards, 3 commercial products, and more than 60 publications, including international journal papers, conference/seminars and other publications. She is also a member of IEEE and Wireless World Research Forum (WWRF).

Kyungsup Kwak received the B.S. degree from Inha University, Incheon, Korea, in 1977, the M.S. degree from the University of Southern California, Los Angeles, in 1981, and the Ph.D. degree from the University of California at San Diego, La Jolla, in 1988, under the Inha University Fellowship and the Korea Electric Association Abroad Scholarship Grants, respectively. From 1988 to 1989, he was with Hughes Network Systems, San Diego. From 1989 to 1990, he was with the IBM Network Analysis Center, Research Triangle Park, NC. Since 1990, he has been a Professor with the School of Information and Communication Engineering, Inha University, where he was the Chairman of the School of Electrical and Computer Engineering from 1999 to 2000, the Dean of the Graduate School of Information Technology and Telecommunications from 2001 to 2002, and is currently the Director of the UWB Wireless Communications Research Center, which is a key national information technology research center in Korea. His research interests include multiple access communication systems, ultra-wideband radio systems, wireless personal area network medium-access control, and sensor networks. Mr. Kwak is a member of the Institute of Electrical, Information, and Communication Engineers, the Korean Institute of Communication Sciences (KICS), the Korea Institute of Intelligent Transport Systems, and the Institute of Electronics Engineers of Korea (IEEK). Since 1994, he has served as a member of the Board of Directors and the Vice President and was elected as the President of KICS in 2006 and as the President of KITS in 2009. He received the Engineering College Achievement Award from Inha University in 1993, a distinguished service medal from IEEK in 1996, and distinguished service medal from the KICS in 1999. He received the LG Paper Award in 1998 and the Motorola Paper Award in 2000. He received awards of research achievements on UWB radio technology from the Minister of Information and Communication and Prime Minister of Korea in 2005 and 2006, respectively. In addition, he received the Haedong Paper Award in 2007 and Haedong Prize of Research Achievement in 2009. In 2008, he was elected for Inha Fellow Professor.

Jing Li, Senior Engineer, received the B.S. degree in Physical Education from the Department of Physics, Shaanxi Normal University, Xi'an, China, in July 1998, and the M.S. degree in Computer Application from the Department of Computer Science, Shanxi University, Taiyuan, China. He is currently the Manager of the IT Support Department at Network Operation and Maintenance Center, China Mobile Group Shanxi Co., Ltd., Taiyuan, China. His research interests include project management in communication engineering, resource management, and the IT support system construction.

Wenping Li, Senior Engineer, received the M.S. degree from Dalian University of Technology, Dalian, China, in July 1998. His research interests include but are not limited to image processing, mobile Internet, Internet security, and cloud computing. He is currently the Deputy General Manager of Network Operation and Maintenance Center, China Mobile Group Shanxi Co., Ltd., Taiyuan, China.

Abdullah Masrub received the B.Eng. degree in Electronic & Computer Engineering from Faculty of Electronic Engineering, Libya, in 1994, and the M.Sc. degree in Computer & Communication Networks from University of Versailles, France, in 2007. He worked as a teaching assistant before 2007, and then as assistant lecturer from May 2007 to Sep. 2008, in the Department of Electrical and Computer

Engineering, Faculty of Engineering, University of Al-Mergib, Libya. He is currently working towards his Ph.D. degree at School of Engineering and Design, Brunel University, UK. His current research interests include Mobile Ad Hoc Networks, Cognitive Radio Networks, and Media Access Control & Resource Allocation in Wireless Networks.

Krishna Nehra completed her MSc and MTech from Banasthali Vidyapith and DA-IICT, India, respectively, in Electronics and Information and Communications Technology. She received her PhD in Wireless Communications at Kings College London, UK, in 2012. During her PhD study, she has served as an organisation committee member and publication chair of the international conference "Wireless Advanced" for 2 years. Since October 2011, she has been working with the company "AlternativeSoft" as a senior quantitative analyst.

Sami Salih received the B.Sc. (First Class Honors, 2003) and M.Sc. (2007) degrees from the Sudan University of Science Technology (SUST), all in Electronics Engineering. He spent several years in industry and academia until he joined NTC (the Sudanese telecom. regulatory) in 2005. His main responsibility is in communication network standards and the assurance of its QoS. During his career, he participated in many ITU activities and helped to carry out projects to bridge the ICT gap in Africa and the Middle East. Now, he is a Visiting Ph.D. Student at Blekinge Institute of Technology (BTH), Sweden. His research interests are on the radio reconfigurable platforms and cognitive radio, especially for the broadband wireless access systems.

Muhammad Zeeshan Shakir received B.E. degree in Electrical Engineering from NED University of Engineering and Technology, Karachi, Pakistan, in 2002. He obtained M.Sc. in Communications, Control, and Digital Signal Processing (CCDSP) with distinction and Ph.D. in Electrical Engineering degrees, in 2005 and 2010, respectively, from University of Strathclyde, Glasgow, Scotland, UK. In 2009, he joined King Abdullah University of Science and Technology (KAUST), Saudi Arabia, where he currently works as a Research Fellow. Dr. Shakir's research interests are in the areas of performance analysis of contemporary and traditional wireless communication systems, which particularly includes heterogeneous networks, cognitive and cooperative networks, and wireless sensor networks. He is a member of IEEE, IEEE-SA, and IEEE ComSoc. At present, he is serving as Secretary to IEEE DySPAN 1900.7 (white space radio working group) since Jan 2012.

Mohammad R. Shikh-Bahaei completed his BSc and MSc in Tehran University and Sharif University of Technology, Iran, respectively, in Electrical Engineering. He received his PhD in Wireless Communications at Kings College London, UK, in 2000. During his PhD study, he worked for Algorex Ltd. and MobilSoft Ltd. on advanced signal processing for wireless communication and cell planning for third generation mobile systems. He then joined National Semiconductor Corp, CA, USA (now part of Texas Instrument), wherein he worked on design of third generation mobile terminals until 2002. In March 2002, he returned to Kings College London, where he is a Senior Lecturer in Centre for Telecommunications Research. His research interest is in the areas of cross-layer optimization, stochastic optimization techniques, multimedia wireless communications, and biomedical applications of signal processing and wireless communications. He is the founder of the high profile IEEE Conference on Wireless Advanced (formerly known as SPWC) and has chaired and organized it for eight years. He is a Senior Member of IEEE.

Liang Song received the Bachelor degree from Shanghai Jiaotong University, China, in 1999, the Master of Science degree from Fudan University, China, in 2002, and the Ph.D. degree from the University of Toronto, Canada, in 2005, all in Electrical Engineering. He was a Senior Systems Engineer with Canamet Inc., Canada, from 2003 to 2006, and a Research Faculty in the Department of Electrical and Computer Engineering, University of Toronto, Canada, from 2006 to 2012. Since 2008, he has been the Founder and Chief Technology Officer of OMESH Networks Inc., Canada, which supplies high-performance radio products with advanced cognitive wireless networking for smart infrastructures. Since 2011, he is also the Founder and President of Huzhou ESOUL Communications Co. Ltd., China. His research areas include communications, networking, and signal processing, where he is best known as the inventor of OPM (Opportunistic Mesh) technology, which realizes the cognitive networking architecture in a wireless mesh network. He is author/co-author of about 40 papers in technical journals and conference proceedings, and he has contributed to 5 books and 9 patents in his areas of interest. He is a recipient of the 2012 University of Toronto Inventor of the Year Award, and a senior member of IEEE.

Petros Spachos received the Diploma degree in Electronics and Computer Engineering from the Technical University of Crete, Greece, in 2008, the M.A.Sc. degree in Electrical and Computer Engineering from the University of Toronto, Canada, in 2010, and is currently working toward the Ph.D. degree in Electrical and Computer Engineering at the University of Toronto. His research interests include communications, networking, and network protocols with the focus on the applications in wireless sensor and cognitive networks. He is a student member of the IEEE.

Aditya Trivedi is a Professor in the ICT Department of ABV-Indian Institute of Information Technology and Management, Gwalior, India. He received his Bachelor degree (with distinction) in Electronics Engg. from the Jiwaji University. He did his M.Tech. (Communication Systems) from Indian Institute of Technology (IIT), Kanpur. He obtained his Doctorate (PhD) from IIT Roorkee in the area of Wireless Communication Engineering. His teaching and research interests include digital communication, CDMA systems, signal processing, and networking. He is a fellow of the Institution of Electronics and Telecommunication Engineers (IETE) and a member of Institution of Electrical and Electronics Engineers (IEEE), USA. He has published more than 60 papers in various prestigious international/national journals and conferences. In 2007, he was given the IETE's K.S. Krishnan Memorial Award for best system oriented paper.

Zhe Yang received the M.Sc. degree with distinction in communications from the Department of Electronics, University of York, UK, in 2006. He is working toward the Ph.D. degree in the Department of Electrical Engineering, Blekinge Institute of Technology, Sweden. In 2007, 2008, 2009, and 2011, he was an invited researcher with Ben-Gurion University in Israel, University of York in UK, Peking University and Beijing Jiaotong University in China. He is a member of IEEE and the author of over fifty publications including book chapters, journal and conference papers. He currently participates in European projects about cognitive radio and developing broadband communications from high-altitude platforms. His research interests include radio communication techniques, communications techniques from high-altitude platforms, cognitive radio, and spectrum management, and economical models of infrastructure deployment and analysis.

Chengshi Zhao received the B.S. degree from Shandong University, Ji'Nan, China, in July 2002, and the M.S. degree in Communication and Information Systems from Chongqing University of Posts and Telecommunications, Chongqing, China, in July 2006. He got the joint Ph.D. degrees with the Wireless Network Lab, School of Information and Communications Engineering, Beijing University of Posts and Telecommunications, Beijing, China, with honor, and the UWB Wireless Communications Research Center (INHA UWB ITRC), Graduate School of Information Technology and Telecommunications, Inha University, Incheon, Korea, with the president award. His research interests include cognitive radio, orthogonal frequency-division multiplexing, multiple-input–multiple-output, cooperative communication, and ultra-wideband techniques. He is currently working at Network Operation and Maintenance Center, China Mobile Group Shanxi Co., Ltd., Taiyuan, China.

Zheng Zhou received the BS degree from Harbin Institute of Technology, Harbin, China, in 1967. He received the MS and PhD degrees from Beijing University of Posts and Telecommunications (BUPT), in 1982 and 1988, respectively. Since then, he has been with the School of Information and Communication Engineering, BUPT. From 1993 to 1995, he was a Visiting Research Fellow with the Chinese University of Hong Kong, supported by the Hong Kong Telecom International Postdoctoral Fellowship. In 2000, he visited Japan Kyocera DDI Research Institute as an Invited Overseas Researcher supported by the Japan Key Technology Center. He is a member of IEEE, a voting member of IEEE 802.15, and a senior member of the China Institution of Communications (CIC), the Radio Application and Management Committee of CIC, the Sensor Network Technical Committee of the China Computer Federation (CCF), and the Chinese National Technical Committee for Standardization on Radio Interference (CTCSRI/H). His research interests include short-range wireless technology, UWB wireless communications, and intelligent signal processing in telecommunications.

Index

A

B

C

D

N

O

P

Q

R

S

T

U

V

W

Y

CPSIA information can be obtained at www.ICGtesting.com
Printed in the USA
BVOW021411180113

310834BV00007B/94/P

9 781466 628120